THE ION CHANNEL *FactsBook* I

Extracellular Ligand-Gated Channels

Other books in the FactsBook Series:

A. Neil Barclay, Albertus D. Beyers, Marian L. Birkeland, Marion H. Brown, Simon J. Davis, Chamorro Somoza and Alan F. Williams
The Leucocyte Antigen FactsBook

Robin Callard and Andy Gearing
The Cytokine FactsBook

Steve Watson and Steve Arkinstall
The G-Protein Linked Receptor FactsBook

Rod Pigott and Christine Power
The Adhesion Molecule FactsBook

Shirley Ayad, Ray Boot-Handford, Martin J. Humphries, Karl E. Kadler and C. Adrian Shuttleworth
The Extracellular Matrix FactsBook

Robin Hesketh
The Oncogene FactsBook

Grahame Hardie and Steven Hanks
The Protein Kinase Factsbook

THE ION CHANNEL *FactsBook* I

Extracellular Ligand-Gated Channels

Edward C. Conley
Molecular Pathology, c/o Ion Channel/Gene Expression
University of Leicester/Medical Research Council
Centre for Mechanisms of Human Toxicity, UK

with contributions from

William J. Brammar
Department of Biochemistry,
University of Leicester, UK

Academic Press
Harcourt Brace & Company, Publishers
LONDON SAN DIEGO NEW YORK BOSTON
SYDNEY TOKYO TORONTO

This book is printed on acid-free paper

ACADEMIC PRESS LIMITED
24–28 Oval Road
LONDON NW1 7DX

United States Edition Published by
ACADEMIC PRESS INC.
San Diego, CA 92101

A catalogue record for this book is available from the British Library

ISBN 0-12-184450-1

Typeset by Alden Multimedia, Oxford and Northampton
Printed and bound in Great Britain by
WBC, Bridgend, Mid Glam.

Contents

VOLUME I

EXTRACELLULAR LIGAND-GATED CHANNELS (ELG)

*Note: A **set of supporting appendices** (Resources) and a **cumulative subject index** for Volumes I to IV appears at the end of Volume IV. An **on-line glossary** of terms marked with the dagger symbol (†) will be accessible from the Cell-Signalling Network 'home page' from mid-1996.*

Cumulative table of contents for Volumes I to IV

VOLUME III INWARD RECTIFIER AND INTERCELLULAR CHANNELS

VOLUME IV VOLTAGE-GATED CHANNELS

ION CHANNEL RESOURCES

Feedback: Comments and suggestions regarding the scope, arrangement and other matters relating to the coverage/contents can be sent to the e-mail feedback file CSN-01@le.ac.uk. *(see field 57 of most entries for further details)*

Acknowledgements

Thanks are due to the following people for their time and help during compilation of the manuscripts: Professors Peter Stanfield, Nick Standen and Gordon Roberts (Leicester), and Ole Petersen (Liverpool) for advice; to Allan Winter, Angela Baxter, Shelly Hundal, Phil Shelton and Sue Robinson for help with photocopying, to Chris Hankins and Richard Mobbs of the Leicester University Computer Centre, and to Dr Tessa Picknett and Chris Gibson of Academic Press for their enthusiasm and patience.

Gratitude is also expressed to all of the anonymous manuscript readers who supplied much constructive feedback, as well as the following who provided advice, information and encouragement: Stephen Ashcroft (Oxford), Eric Barnard (London), Dale Benos (Harvard), William Catterall (Washington), K. George Chandy (UC Irvine), Peter Cobbold (Liverpool), David Clapham (Mayo Foundation), Noel Davies (Leicester), Dario DiFrancesco (Milano), Ian Forsythe (Leicester), Sidney Fleischer (Vanderbilt), George Gutman (UC Irvine), Richard Haugland (Molecular Probes, Inc.), Bertil Hille (Washington), Michael Hollmann (Göttingen), Anthony Hope (Dundee), Benjamin Kaupp (Jülich), Jeremy Lambert (Dundee), Shigetada Nakanishi (Kyoto), Alan North (Glaxo Institute for Molecular Biology), John Peters (Dundee), Olaf Pongs (Hamburg), David Spray (Yeshiva), Kent Springer (Institute for Scientific Information), Steve Watson (Oxford), Paul Van Houlte (I.R.I.S.) and Steven Wertheim (Harvard).

Thanks are also due to the Department of Pathology at the University of Leicester, Harcourt Brace, the Medical Research Council and Zeneca Pharmaceuticals, for providing generous sponsorship, equipment and facilities.

We would like to acknowledge the authors of all those papers and reviews which in the interest of completeness we have quoted, but have not had space to cite directly.

ECC would like to thank Professors Denis Noble in Oxford and Anthony Campbell in Cardiff, Tony Buzan in Winton, Dorset and Richard Gregory in Bristol for help and inspiration, and would like to dedicate his contributions to Paula, Rebecca and Katharine for all their love and support over the past four years.

Left: *Edward Conley*, Right: *William Brammer*

Introduction & layout of entries

Edward C. Conley **Entry 02**

The Ion Channels FactsBook is intended to provide a 'summary of molecular properties' for all known types of ion channel protein in a cross-referenced and 'computer-updatable' format. Today, the subject of ion channel biology is an extraordinarily complex one, linking several disciplines and technologies, each adding its own contribution to the knowledge base. This diversity of approaches has left a need for accessible information sources, especially for those reading outside their own field. By presenting 'facts' within a **systematic framework**, the *FactsBook* aims to provide a 'logical place to look' for specific information when the need arises. For students and researchers entering the field, the weight of the existing literature, and the rate of new discoveries, makes it difficult to gain an overview. For these readers, *The Ion Channels FactsBook* is written as a **directory**, designed to identify similarities and differences between ion channel types, while being able to accommodate new types of data within the framework. The main advantages of a systematic format is that it can speed up identification of **functional links** between any 'facts' already in the database and maybe provide a *raison d'être* for specific experiments where information is not known. Although such 'facts' may not go out-of-date, interpretations based on them may change considerably in the light of additional, more direct evidence. This is particularly true for the explosion of new information that is occurring as a direct consequence of the **molecular cloning of ion channel genes**. It can be anticipated that many more ion channel genes will be cloned in the near future, and it is also likely that their functional diversity will continue to exceed expectations based on pharmacological or physiological criteria alone.

An emphasis on properties emergent from ion channel molecular functions

Understanding how the interplay of currents through many specific ion channel molecules determines complex electrophysiological behaviour of cells remains a significant scientific challenge. The approach of the *FactsBook* is to associate and relate this complex cell phenotypic behaviour (e.g. its physiology and pharmacology) to ion channel **gene expression-control** wherever possible *even where the specific gene has not yet been cloned*. Thus the ion channel *molecule* becomes the **central organizer**, and accordingly arbitrates whether information or topics are included, emphasized, sketched-over or excluded. In keeping with this, ion channel characteristics are described in relation to known structural or genetic features wherever possible (or where they are ultimately **molecular characteristics**). Invariably, this relies on the availability of sequence data for a given channel or group of channels. However, a number of channel types exist which have not yet been sequenced, or display characteristics in the **native form** which are not precisely matched by existing clones expressed in heterologous cells (or are otherwise ambiguously classified). To accommodate these channel types, summaries of characteristics are included in the **standard entry field format**, with inappropriate fieldnames omitted. Thus the present 'working arrangement' of entries and fields is broad enough to include both the 'cloned' and 'uncloned' channel types, but in due course will be gradually supplanted by a comprehensive classification based on gene locus, structure, and relatedness of primary sequences. In all cases, the scope of the *FactsBook* entries is limited to those proteins forming (or predicted to form) membrane-bound, **integral ionic channels** by folding and association of their primary protein sequences. Activation or suppression of the channel current by a specified ligand or voltage step is generally included as part of the channel description or name *(see below)*. Thus an emphasis is made throughout the book on **intrinsic features** of channel molecule itself and not on those of separately encoded, co-expressed proteins. In the present edition, there is a bias towards descriptions of **vertebrate ion channels** as they express the full range of channel types which resemble characteristics found in most eukaryotes.

Anticipated development of the dataset – Integration of functional information around molecular types

Further understanding of complex cellular electrical and pharmacological behaviour will not come from a mere catalogue of protein properties alone. This book therefore begins a process of specific **cross-referencing** of molecular properties within a **functional framework**. This process can be extended to the interrelationships of ion channels and other classes of **cell-signalling molecules** and *their* functional properties. Retaining protein molecules (i.e. gene products) as 'fundamental units of classification' should also provide a framework for understanding complex physiological behaviour resulting from co-expressed *sets* of proteins. Significantly, many **pathophysiological phenotypes** can also be linked to selective molecular 'dysfunction' within this type of framework. Finally, the anticipated growth of raw sequence information from the **human genome project** may reveal hitherto unexpected classes and subtypes of cell-signalling components – in this case the task then will be to integrate these into what is already known *(see also description of Field number 06: Subtype classifications and Field number 05: Gene family)*.

The Cell-Signalling Network (CSN)

From the foregoing discussion, it can be seen that establishment and consolidation of an integrated '**consensus database**' for the many diverse classes of cell signalling molecules (including, for example, receptors, G proteins, ion channels, ion pumps, etc.) remains a worthwhile goal. Such a resource would provide a focus for identifying unresolved issues and may avoid unnecessary duplication of research effort. Work has begun on a prototype **cell-signalling molecule database** co-operatively maintained and supported by contributions from specialist groups world-wide: The ***Cell-Signalling Network (CSN)*** operating from mid-1996 under the **World Wide Web**† of the Internet† has been designed to disseminate **consensus properties** of a wide range of molecules involved in cell signal transduction. While it may take some time (and much good-will) to establish a comprehensive network, the many advantages of such a co-operative structure are already apparent. Immediately, these include an 'open' mechanism for **consolidation** and **verification** of the dataset, so that it holds a 'consensus' or 'validated' set of information about what is known about each molecule and practical considerations such as **nomenclature recommendations** (see, for example, the IUPHAR nomenclature sections under the CSN 'home page').

The CSN also allows **unlimited cross-referencing** by pointing to related information sets, even where these are held in multiple centres around the world. **On-line** support for technical terms (**glossary items**, indicated by **dagger symbols** (†) throughout the text) and reference to explanatory **appendices** (e.g. on associated signalling components such as G protein†-linked receptors†) are already supported for use with this book. Eventually, benefits could include (for instance) direct 'look-up' of graphical resources for protein structure, *in situ* and developmental **gene expression atlases**†, interactive **molecular models** for structure/function analysis, DNA/protein sequences linked to feature tables, **gene mapping** resources and other pictorial data. These developments (not all are presently supported) will use **interactive electronic media** for efficient browsing and maintenance. *For a brief account of the Cell-Signalling Network, see Feedback & CSN access, entry 12. For a full specification, see Resource J – Search criteria & CSN development, entry 65.*

HOW TO USE *THE ION CHANNEL FACTSBOOK*

Common formats within the entries

A proposed organizational hierarchy for information about ion channel molecules

Information on named channel types is grouped in **entries** under common headings which repeat in a fixed order – e.g. for ion channel molecules which have been sequenced, there are broad **sections** entitled NOMENCLATURES, EXPRESSION, SEQUENCE ANALYSES, STRUCTURE & FUNCTIONS, ELECTROPHYSIOLOGY, PHARMACOLOGY, INFORMATION RETRIEVAL and REFERENCES, in that order. Within each section, related **fieldnames** are listed, always in **alphabetical order** and indexed by a **field number** *(see below)*, which makes electronic cross-referencing and 'manual' comparisons easier.

While the sections and fields are *not* rigid categories, an attempt has been made to remain consistent, so that corresponding information for two different channels can be looked up and compared directly. If a field does not appear, either the information was not known or was not found during the compilation period. Pertinent information which has been published but is absent from entries would be gratefully received and will be added to the 'entry updates' sections within the CSN *(see Feedback & CSN access, entry 12)*. Establishment of this 'field' format has been designed so that every available 'fact' should have its logical 'place'. In the future, this arrangement may help to establish 'universally accepted' or '**consensus**' properties of any given ion channel or other cell-signalling molecule. This **validation** process critically depends on user feedback to contributing authors. The CSN *(above)* establishes an efficient electronic mechanism to do this, for continual refinement of entry contents.

Independent presentation of 'facts' and conventions for cross-referencing

The *FactsBook* departs from a traditional review format by presenting its information in related groups, each under a broader heading. *Entries are not designed or intended to be read 'from beginning to end', but each 'fact' is presented independently under the* ***most pertinent fieldname***. Independent citation of 'facts' may sometimes result in some **repetition** (redundancy) of general principles between fields, but if this is the case some effort has been made to 're-phrase' these for clarity (suggested improvements for presentation of any 'fact' are welcome – *see Field number 57: Feedback*).

For readers unfamiliar with the more general aspects of ion channel biology, some introductory information applicable to whole groups of ion channel molecules is needed, and this is incorporated into the '**key facts' sections** preceding the relevant set of entries. These sections, coupled with the '**electronically updated' glossary items** (available on-line, and indicated by the dagger† symbol, *see below*) provide a basic overview of principles associated with detailed information in the main entries of the book.

Extensive **cross-referencing** is a feature of the book. For example, cross-references between fields of the *same* entry are of the format *(see Fieldname, xx-yy)*. Cross-references between fields of *different* channel type entries are generally of the format *see fieldname under SORTCODE, xx-yy;* for example – *see mRNA distribution, under ELG Cl GABA*$_A$*, 10-13*. This **alphabetical 'sortcode'** and **numerical 'entry numbers'** (printed in the **header** to each page) are simply devices to make cross-referencing more compact and to arrange the entries in an approximate **running order** based on **physiological features** such as mode of gating†, ionic selectivity†, and agonist† specificity. A 'sort order' based on physiological features was judged to be more intuitive for a wider readership than one based on gene structure alone, and enables 'cloned' and 'uncloned' ion channel types to be listed together. The use and criteria for sortcode designations are described under the subheading *Derivation of the sortcode (see Field number 02: Category (sortcode))*. Entry 'running order' is mainly of importance in book-form publications. **New entries** (or mergers/subdivisions between existing entries) will use serial entry numbers as 'electronic pointers' to appropriate files.

Cross-references are frequently made to an on-line index of **glossary items** by **dagger symbols**† wherever they might assist someone with technical terms and concepts *when reading outside their own field*. The glossary is designed to be used side-by-side with the *FactsBook* entries and is accessible in updated form over the Internet†/World Wide Web† with suitable browsing† software *(for details, see Feedback & CSN access, entry 12)*.

Contextual markers and styles employed within the entries

Throughout the books, a **six-figure index number** (xx-yy-zz, e.g. 19-44-01:) separates groups of facts about different aspects of the channel molecule, and carries information about **channel type/entry number** (e.g. **19-** ~ $InsP_3$ receptor–channels), **information type/field number** (e.g. **-44-**, *Channel modulation*) and **running paragraph number (datatype)** (e.g. **-01**). This simple 'punctate' style has been adopted for maximum flexibility of **updating** (both error-correction and consolidation with new information), **cross-referencing** and **multi-authoring**. The CSN specification *(see entry 65)* includes longer term plans to structure field-based information into convenient **data-types** which will be indexed by a zz numerical designation.

Italicized subheadings are employed to organize the facts into related topics where a field has a lot of information associated with it. Specific illustrated points or features within a field are referenced to adjacent figures. Usage of abbreviations and common symbols are defined in context and/or within the main **abbreviations index** at the front of each book. Abbreviated **chemical names** and those of proprietary pharmaceutical compounds are listed within the electronically updated *Resource C – Compounds & proteins*, also available via the 'home page' of the *Cell-Signalling Network*.

Generally, highlighting of related **subtopics** emergent from the molecular properties ('facts') associated with the ion channel under description are indicated within a field by **lettering in bold**. All subtopics are cross-referenced by means of a large **cumulative subject index** (entry 66), which can permit retrieval of information by topic *without requiring prior knowledge of ion channel properties*. Throughout the main text, *italics* draw attention to special cases, caveats, hypotheses and exceptions. The '*Note:* ' prefix has been used to indicate **supplemental** or comparative information of significance to the quoted data in context.

Special considerations for integrating properties derived from 'cloned' and 'native' channels

While a certain amount of introductory material is given to set the context, the emphasis on **molecular properties** means the treatment of many important biological processes or phenomena is reduced to a bare outline. References given in the *Related sources and reviews* field and the electronically updated *Resource F – Supplementary ion channel reviews* accessible via the CSN *(see Feedback & CSN access, entry 12)* are intended to address this imbalance.

For summaries of key molecular features, a central channel '**protein domain topography model**' is presented. Individual features that are illustrated on the protein domain topography model are identified within the text by the symbol **[PDTM]**.

Wherever **molecular subtype-specific data** are quoted (such as the particular behaviour of a ion channel gene family† member or isoform†) a convention of using the underlined trivial or systematic name as a prefix has been adopted – e.g. <u>mIRK1</u>: ; <u>RCK1</u>: ; <u>Kv3.1</u>: etc.

GUIDE TO THE PLACEMENT CRITERIA FOR EACH FIELD

Criteria for NOMENCLATURES sections

This section should bring together for comparison present and previous names of ion channels or currents, with brief distinctions between similar terms. Where ***systematic names*** *have already been suggested or adopted by published convention, they should be included and used in parallel to trivial names.*

Field number 01: *Abstract/general description:* This field should provide a summary of the most important functional characteristics associated with the channel type.

Field number 02: *Category (sortcode):* The **alphabetical 'sortcode'** should be used for providing a logical **running order** for the individual entries which make up the book. It is *not* intended to be a rigorous channel classification, which is under discussion, but rather a **practical index** for finding and cross-referencing information, in conjunction with the **six-figure index number** *(see above)*. The *Category (sortcode)* field also lists a designated **electronic retrieval code** (unique embedded identifier or **UEI**) for 'tagging' of new articles of relevance to the contents of the entry. For further details on the use and implementation of UEIs, *see the description for Resource J* (in this entry) and for a full description, *see Resource J – Search criteria & CSN development, entry 65.*

Derivation of the sortcode: Although we do not yet have a complete knowledge of all ion channel primary† structures, knowledge of ion channel gene family† and superfamily† structure allows a *working* **sort order** to be established. To take an example, the extracellular ligand-gated (ELG) receptor–channels share many structural features, which reflects the likely duplication and divergent evolution of an ancestral gene. The present-day forms of such channels reflect the changes that have occurred through adaptive radiation† of the ancestral type, particularly for gating† mechanism and ionic selectivity† determinants. Thus, the **entry running order** (alphabetical, via the sortcode) of the *FactsBook* entries should depend primarily on these two features. The sortcode therefore consists of several groups of letters, each denoting a characteristic of the channel molecule: Entries are sorted first on the principal means for **channel gating**† (first three letters), whether this is by an extracellular ligand† (ELG), small intracellular ligand† (ILG) or transmembrane voltage (VLG). For convenience, the ILG entries also include certain channels which are obligately dependent on *both* ligand binding *and* hydrolysis for their activation – e.g. channels of the ATP-binding cassette (ABC) superfamily. Other channel types may be subject to direct mechanical gating (MEC) or sensitive to changes in osmolarity (OSM) – *see the Cumulative tables of contents* and the first page of each entry for descriptions and scope. Due to their unusual gating characteristics, a separate category (INR) has been created for **inward rectifier-type channels.**

The second sort (the next three letters of the sortcode) should be on the basis of the principal **permeant ions**, and may therefore indicate high selectivity for **single ions** (e.g. Ca, Cl, K, Na) or **multiple ions** of a specified charge (e.g. cations – CAT). Indefinite **sortcode extensions** can be assigned to the sortcode if it is necessary to distinguish similar but separately encoded groups of channels *(e.g. compare ELG Cl GABA*$_A$*, entry 10 and ELG Cl GLY, entry 11).*

Field number 03: *Channel designation:* This field should contain a shorthand designation for the ion channel molecule – mostly of the form X_Y or $X_{(Y)}$ where X denotes the major **ionic permeabilities**† (e.g. K, Ca, cation) and Y denotes the principal **mechanism of gating**† *where this acts directly on the channel molecule itself* (e.g. cGMP, voltage, calcium, etc.). Otherwise, this field contains a shorthand designation for the channel which is used in the entry itself.

Field number 04: *Current designation:* This field should contain a shorthand designation for ionic currents conducted by the channel molecule, which is mostly of the form $I_{X(Y)}$, $I_{X,Y}$ or $I_{X\text{-}Y}$ where X and Y are defined as above.

Field number 05: *Gene family:* This field should indicate the known **molecular relationships** to other ion channels or groups of ion channels at the level of

amino acid primary sequence homology†, within gene families† or gene superfamilies†. Where **multiple channel subunits** are encoded by separate genes, a summary of their principal features should be tabulated for comparison. Where the gene family is particularly large, or cannot be easily described by functional variation, a **gene family tree**† derived by a **primary sequence alignment algorithm**† *(see Resource D – 'Diagnostic' tests, entry 59)* may be included as a figure in this field.

Field number 06: *Subtype classifications:* This field should include supplementary information about any schemes of classification that have been suggested in the literature. Generally, the most robust schemes are those based on complete knowledge of **gene family**† **relationships** *(see above)* and this method can identify similarities that are not easily discernible by pharmacological or electrophysiological criteria alone – see, for example, the entries *JUN (connexins), entry 35,* and *INR K (subunits), entry 33.* Note, however, that some native† channel types are more conveniently 'classified' by functional or cell-type expression parameters which take into account interactions of channels with other co-expressed proteins (see, for example, discussion pertaining to the cyclic nucleotide-gated (CNG-) channel family in the entries *ILG Key facts, entry 14, ILG CAT cAMP, entry 21,* and *ILG CAT cGMP, entry 22.* Debate on the 'best' or 'most appropriate' channel classification schemes is likely to continue for some time, and it is reasonable to suppose that alternative subtype classifications may be applied and used by different workers for different purposes.

Since the 'running order' of the *FactsBook* categories depends on inherent molecular properties of channel cDNAs†, genes† or the expressed proteins, future editions will gradually move to classification on the basis of **separable gene loci**†. Thus multiple channel protein variants resulting from processes of **alternative RNA splicing**† but encoded by a **single gene locus**† will only ever warrant one 'channel-type' entry (e.g. see BK_{Ca} variants under *ILG K Ca, entry 27*). Distinct proteins resulting from transcription† of **separable gene loci**, for example in the case of different gene family members, will (ultimately) warrant separate entries. For the time being, there is insufficient knowledge about the precise phenotypic† roles of many 'separable' gene family members to justify separate entries (as in the case of the *VLG K Kv* series entries).

Classification by **gene locus designation** *(see Field number 18: Chromosomal location)* can encompass all structural and functional variation, while being 'compatible' with efforts directed to identifying phenotypic and pathophysiological† roles of individual gene products (e.g. by gene-knockout†, locus replacement† or disease-linked gene mapping† procedures – *see Resource D – 'Diagnostic' tests, entry 59*). Subtype classifications based on gene locus control can also incorporate the marked developmental changes which pertain to many ion channel genes *(see Field number 11: Developmental regulation)* and can be implemented when the 'logic' underlying **gene expression-control**† for each family member is fully appreciated. A 'genome-based' classification of *FactsBook* entries may also help comprehend and integrate **equivalent information** based for other ('non-channel') cell-signalling molecules *(see Resources G, H and I, entries 62, 63 and 64).*

Field number 07: *Trivial names:* This field should list **commonly used names** for the ion channel (or its conductance†). Often a channel will be (unsystematically) named by its tissue location or unusual pharmacological/physiological properties, and these are also listed in this field. While **unsystematic names** do not indicate molecular relatedness, they are often more useful for comparative/descriptive purposes. For these and historical reasons, trivial names (e.g. clone/isolate names for K^+ channel isoforms) are used side-by-side with **systematic names**, where these exist. A standardized nomenclature for ion channels is under discussion, e.g. see the series of articles by Pongs, Edwards, Weston, Chandy, Gutman, Spedding and Vanhoutte in *Trends Pharmacol Sci* (1993) **14**: 433–6. Future recommendations on **standardized nomenclature** will appear in files accessible under the **IUPHAR** entry of the *Cell-Signalling Network (see Feedback & CSN access, entry 12).*

Criteria for EXPRESSION sections

*This section should bring together information on **expression patterns** of the ion channel gene, indicating functional roles of specific channels in the cell type or organism. The complex and profound roles of ionic currents in vertebrate development (linking plasma membrane signalling and genome activation) are also emphasized within the fields of this section.*

Field number 08: *Cell-type expression index:* Comprehensive systems relating the *expression* of specified molecular components to specified anatomical and developmental loci ('**expression atlases**') are being developed in a number of centres and in due course will form a superior organizational framework for this type information *(see discussion below)*. In the meantime, the range of **cell-type expression** should be indicated in this field in the form of alphabetized listings. Notably, there is a substantial literature concerned with the electrophysiology of ion channels where the tissue or **cell type** forms the main focus of the work. In some cases, this has resulted in detailed **'expression surveys'**, revealing properties of *interacting sets* of ion channels, pumps, transporters and associated receptors. Such review-type information is of importance when discussing the contribution of individual ion channel molecules to a **complex electrophysiological phenotype**† and/or overall function of the cell. For further references to 'cell-type-selective' reviews, *see Resource H – Listings of cell types, entry 63 accessible via the CSN (see Feedback & CSN access, entry 12).*

Problems and opportunities in listing ion channel molecules by cell type: Understanding the roles which individual ionic channels play in the complex electrophysiological phenotypes of native† cells remains a significant challenge. The overwhelming range of studies covering aspects of ion channel expression in vertebrate cells offers unique problems when compiling a representative overview. Certainly the linking of specific **ion channel gene expression** to cell type is a first step towards a more comprehensive indexing, and towards this goal, cell-type-selective studies are useful for a number of reasons. First, they can help visualize the *whole range* of channel expression by providing an **inventory of conductances**† observed. Secondly, these studies generally define the **experimental conditions** required to observe a given conductance. Thirdly, they include much information directly relating **specified ionic conductances** to the functions of the cell type concerned. Collated information such as this should be of increasing utility in showing the relationship of electrophysiological phenotype to mechanistic information on their **gene structure and expression-control** (which largely correlates with cell-type lineage). At this time it is difficult to build a definitive catalogue of ion channel gene expression patterns mapped to cell type, not only because the determinants of gene expression are scarcely explored, but also because there remain many unavoidable **ambiguities in phenotype definition**. Some of these problems are discussed below.

Problems of uneven coverage/omissions: Certain cell preparations have been intensely studied for ion channel expression while others have received very little attention for technical, anatomical or other reasons. Furthermore, a large number of native† ionic currents can be induced or inhibited by agonists† that bind to co-expressed G protein†-coupled receptors†. Thus a difficulty arises in deciding whether channel *currents* can be unambiguously defined in terms of action at a **separately encoded** receptor protein. While it is valid to report that an agonist-sensitive current is expressed in a defined cell type, the factors of crosstalk† and receptor-transducer† subtype specificities in signalling systems are complex and may produce an ambiguous classification. Receptor-coupled agonist-sensitivities *are* an important factor contributing to **cell-pharmacological and -electrical phenotype**†, but the treatment here has been limited to a number of tabular summaries of ion channel regulation through coupling to G protein-linked effector† molecules *(see Resource A – G protein-linked receptors, entry 56)*. As stated earlier, the entries are *not* sorted on agonist specificity except where the underlying ion channel protein sequence would be expected to form an **integral ionic channel** whose gating† mechanism is also part of the assembled protein complex.

Cell preparation methods are variable: A further problem inherent in classifying ion channels by their patterns of expression is that the choice of tissue or cell preparation method may influence phenotype†. The behaviour of channel-mediated ionic currents can be measured in native† cells, e.g. in the **tissue slice**, which has the advantages of extracellular ionic control, mechanical stability, preserved anatomical location, lack of requirement for anaesthetics and largely undisturbed intercellular communication. **Cell-culture techniques** show similar advantages, with the important exceptions that normal developmental context, anatomical organization and synaptic arrangements are lost and (possibly as a consequence) the 'expression profile' of receptor and channel types might change. Cultured cell preparations may also be affected by '**de-differentiation**†' processes and (by definition) cell lines† are uncoupled from normal processes of **cell proliferation**†, **differentiation**† and **apoptosis**†. **Acutely dissociated cells** from native† tissue may provide cell-type-specific expression data without anomalies introduced by intercellular (gap junctional) conductances, but the enzymatic or dispersive treatments used may also affect responses in an unknown way.

Verbal descriptions of cell-type expression divisions are arbitrary and are not rigorous: Definitive mapping of specific ion channel subtype expression patterns has many variables. Localization of specific gene products are most informative when *in situ* localizations are linked to the regulatory factors controlling their expression *(see glossary entry on Gene expression-control†)*. The complexity of this task can extend to processes controlling, for example, developmental regulation, co-expressed protein **subunit stoichiometries** and **subcellular localizations**.

Complete integration of all structural, anatomical, co-expression and modulatory data for ion channels could eventually be accommodated within interactive **graphical databases** which are capable of providing 'overlays' of separately collected *in situ* expression data linked to functional properties of the molecules. By these methods, new data can be mathematically transformed to superimpose on fixed tissue or cell co-ordinates for comparison with existing database information.

Software development efforts focused on the acquisition, analysis and exchange of complex datasets in neuroscience and mouse development have been described, and the next few years should hopefully see their implementation. For further information, see

Baldock, R., Bard, J., Kaufman, M. and Davidson, D. (1992) A real mouse for your computer, *Bioessays* **14:** 501–2
Bloom, F. (1992) *Brain Browser, v 2.0.* Academic Press (Software).
Kaufman, M. (1992) *The Atlas of Mouse Development*, Academic Press
Wertheim, S. and Sidman, R. (1991) Databases for Neuroscience, *Nature* **354:** 88–9

To help rationalize the choices available for selection of these 'prototype' classifications, see *Resource H – Listings of cell types, entry 63.* These listings may also have some practical use for sorting the subject matter of journal articles into functionally related groups. A proposed integration of information resources relating different aspects of **cell-signalling molecule gene expression** is illustrated in Fig. 4 of the section headed *Feedback & CSN access, entry 12.*

Field number 09: ***Channel density:*** This field should contain information about estimated **numbers of channel molecules** per unit area of membrane in a specified preparation. This field lists information derived from local patch-clamp 'sampling' or autoradiographic detection in membranes using anti-channel antibodies. The field should also describe unusually high densities of ion channels ('clustering') in specified membranes where these are of functional interest.

Field number 10: ***Cloning resource:*** This field should refer to cell preparations relatively 'rich' in channel-specific mRNA (although it should be noted that many ion channel mRNAs are of low abundance†). Otherwise, this field defines a **'positive control' preparation** likely to contain messenger† RNA† encoding the channel. Preparations may express only specific subtypes of the channel and therefore related probes (especially PCR† probes) may not work. Alternatively, a genomic† cloning resource may be cited.

Field number 11: ***Developmental regulation:*** This field should contain descriptions of ion channel genes demonstrated (or expected to be) subject to **developmental gene regulation** – e.g. where hormonal, chemical, second messenger† or other environmental stimuli appear to induce (or repress) ion channel mRNA or protein expression in native† tissues (or by other experimental interventions). Protein factors *in trans*† or DNA structural motifs† *in cis*† which influence **transcriptional activation**†, **transcriptional enhancement**† or **transcriptional silencing**† should also be listed under this fieldname. Information about the **timing of onset** for expression should also be included if available, together with evidence for ion channel activity influencing **gene activation**† or **patterning**† during vertebrate development.

Field number 12: *Isolation probe:* This field should include information on probes used to relate distinct gene products by isolation of novel clones following low-stringency **cross-hybridization screens**†. The development of oligonucleotide† sets which have been used to *unambiguously* detect subtype-specific sequences by **PCR**† , **RT-PCR**† or *in situ* hybridization† should be identified with source publication. Both types of sequence *may* be able to serve as unique **gene isolation probes**, dependent upon the library† size, target abundance†, screening stringency† and other factors.

Field number 13: *mRNA distribution:* This field should report either quantitative/ semi-quantitative or presence/absence (±) descriptions of specific channel mRNAs in defined tissues or cell types. This type of information is generally derived from Northern hybridization†, RNAase protection† analysis, RT–PCR† or *in situ* expression assays. *See also notes on expression atlases under Field number 08: Cell-type expression index.*

Field number 14: *Phenotypic expression:* This field should include information on the proposed phenotype† or **biological roles** of specified ion channels where these are discernible from expression studies of native† (wild-type) genes. Phenotypic† consequences of naturally occurring (spontaneous) mutations† in ion channel genes are included where these have been defined, predicted or interpreted *(see also Fields 26–32 of the STRUCTURE & FUNCTIONS section for interpretation of site-directed mutagenesis† procedures as well as Resource D – 'Diagnostic' tests, entry 59).* Associations of ion channels with **pathological states**, or where **molecular 'defects'** could be 'causatory' or contribute to the **progression of disease** should be listed in this field (for links with established cellular and molecular pathology databases, *see Fig. 4 of Feedback & CSN access, entry 12).*

The *Phenotypic expression* field may include references to mutations in other ('non-channel') genes which affect channel function when the proteins are co-expressed. It is also used to link descriptions of specific (cloned) molecular components to native cell-electrophysiological phenotypes. In due course, this field will be used to hold information on phenotypic† effects of transgenic† manipulations of ion channel genes including those based on gene knockout† or gene locus† replacement† protocols.

Field number 15: *Protein distribution:* This field should report results of expression patterns determined with probes such as antibodies raised to channel primary† sequences or radiolabelled affinity ligands†.

Field number 16: *Subcellular locations:* This field should describe any notable arrangements or intracellular locations related to the functional role of the channel molecule, e.g. when the channel is inserted into a **specified subcellular membrane system** or is expressed on one pole of the cell only (e.g. the basolateral† or apical† face).

Field number 17: *Transcript size:* This field should list the main RNA transcript† sizes estimated (in numbers of ribonucleotides) by Northern† hybridization analysis. **Multiple transcript sizes** may indicate (i) alternative processing ('splicing†') of a primary transcript†, (ii) the use of alternative **transcriptional start sites**†, or (iii) the presence of 'pre-spliced' or **'incompletely spliced' transcripts** identified with homologous nucleotide probes† in total cell mRNA† populations. Note that probes can be chosen selectively to identify each of these categories; 'full-length' coding sequence† (**exonic**†) probes are the most likely to identify all variants, while probes based on **intronic**† sequences (where appropriate) will identify 'pre-splice' variants.

Criteria for SEQUENCE ANALYSES sections

This section should bring together data and interpretations derived from the nucleic acid or protein sequence of the channel molecule. The symbol [PDTM] denotes an illustrated feature on the channel monomer ***protein domain topography model****, which is presented as a central figure in some entries for sequenced ion channels. These models are only intended to visualize the relative lengths and positions of features on the whole molecule (see the description for field number 30, Predicted protein topography). The PDTMs as presented are highly diagrammatic – the actual protein structure will depend on patterns of folding, compact packing and multi-subunit associations. In particular, the relative positions of motifs, domain shapes and sizes are subject to re-interpretation in the light of better structural data. Links to information resources for protein and nucleic acid sequence data are described in the* ***Database listings*** *field towards the end of each entry.*

Field number 18: *Chromosomal location:* This field should provide a **chromosomal locus**† designation (chromosome number, arm, position) for channel gene(s) in specified organisms, where this is known. Notes on interactive linking to **gene mapping database resources** appear under an option of the *Cell-Signalling Network* 'home page' *(see Feedback & CSN access, entry 12).*

Field number 19: *Encoding:* This field should report **open reading frame**† lengths as numbers of nucleotides or amino acid residues encoding monomeric channel proteins (i.e. spanning the first A of the ATG translational start codon† to the last base of the translational termination codon†). The field should report and compare any channel protein **length variants** in different tissues or organisms. If considered especially relevant or informative, selected **primary**† **sequence alignments** of different gene family members may appear under this field.

Field number 20: *Gene organization:* This field should describe known **intron**† and **exon**† **junctions** within or outside the protein coding sequence, together with positional information on gene expression-control† elements and polyadenylation† sites where known. *Note: Functional* changes as a result of gene expression-control should be listed under the *Developmental regulation field.*

Field number 21: *Homologous isoforms:* This field should indicate independently isolated and sequenced forms of entire channels which either show virtual **identity** or of such high homology† that they can be considered **equivalent** should also appear in this field (but see note on percentage conservation values under *Field number 28: Domain conservation*). **Isoforms**†* of a channel protein can exist between closely related species or between different tissues of the same species (i.e. the same gene may be expressed in two or more different tissues, sequenced by two groups but named independently). Some tissue-specific variation may also result from alternative splicing†, yielding subtly distinct forms of channel protein. Since small numbers of amino acid changes may exist from individual-to-individual (as a result of normal **sequence polymorphism**† in populations) separate isolates may yield sequence isoforms which can be shown to be 'equivalent' by Southern hybridization† procedures *(see Field number 25: Southerns)*.

Note: In the entries of this book a restrictive definition of molecular identity (or near identity) is used to define an isoform†. In this restricted sense, 'isoforms' would be expected to be the product of the **same gene**† (or **gene variant** produced by, for example, alternative splicing†), and therefore have very similar or identical **molecular constitutions** and **functional roles** in specified cell types of closely related species. Comparative information on different gene family† members or multiple variants affecting particular protein domains† may also be included under the *Gene family* and *Domain conservation* fields respectively.

Field number 22: *Protein molecular weight (purified):* This field should state reported molecular weights estimated from relative protein mobilities using SDS–PAGE† methods (e.g. following affinity† purification from native† or heterologous† cell membranes). Data derived from native† preparations generally includes the weight contribution from oligosaccharide† chains added during post-translational **protein glycosylation**†. In general, extracellular saccharide† components of glycoproteins† may contribute 1–85% by weight, ranging from a few to several hundred oligosaccharide chains per glycoprotein molecule.

Field number 23: *Protein molecular weight (calc.):* This field should list the molecular weight of **monomeric channel proteins** equivalent to the summated (calculated) molecular weights of constituent amino acids in the reported sequence (e.g. derived from open reading frames† of cDNA† sequences). If 'calculated' molecular weights are less than 'purified' molecular weights *(previous field)* this may indicate the existence of post-translational glycosylation† on native† expressed protein subunits *in vivo*.

Field number 24: *Sequence motifs:* This field should report the position of putative **regulatory sites** as deduced from the protein or nucleic acid primary† sequence (with the exception of potential *phosphorylation sites* for protein kinases†, which are listed under *Field number 32: Protein phosphorylation*). Positions of sequence motifs† illustrated on the monomer **protein domain topography model** are denoted by the symbol [PDTM]. Typical consensus† sites include those for enzymes such as glycosyl transferases†, ligand†-binding sites, transcription factor†-binding sites† etc. *N*-glycosylation† motifs are sometimes indicated using the shorthand designation ***N*-gly**:. Signal peptide cleavage sites (sometimes designated by **Sig:**) can be derived by comparing sizes of the signal peptide† and the mature chain†.

Field number 25: *Southerns:* This field should include information which reports the existence of closely related DNA sequences in the genome† or reports the **copy number**† of individual genes via Southern hybridization† procedures. Note that native† diploid somatic† cells will generally maintain two copies of a given ion channel gene locus†, but *stable*† heterologous† expression procedures may result in **multiple locus insertion**†. Multiple locus insertion can be quantitated in Southern† hybridization procedures using two probes of similar length and hybridization affinity†, one specific for a native locus (which will identify two copies) and one for the heterologous gene (which will yield a hybridization signal proportional to the copy number). Note also that the copy number parameter can *not* be equated to the **physiological expression level** of the recombinant† protein unless **locus control regions** are incorporated as part of the channel expression construct *(for details, see the section entitled **Gene copy number** under Resource D – 'Diagnostic' tests, entry 59, and the section describing **heterologous ion channel gene expression** under Resource H – Listings of cell types, entry 63).*

Criteria for STRUCTURE & FUNCTIONS sections

*This section should bring together information based on functional analysis or interpretation of **ion channel structural elements**. This section includes data derived from functional studies following site-directed mutagenesis† of ion channel genes and molecular modelling studies at atomic scale. Future developments linking on-line information resources for protein structure to 'functional datasets' are illustrated in Fig. 5 of Feedback & CSN access, entry 12, and in Resource J – Search criteria & CSN development, entry 65.*

Field number 26: *Amino acid composition:* This field should include information on channel protein hydrophilicity† or hydrophobicity† where this is of structural or functional significance. Similarities to other related proteins should be emphasized.

Field number 27: *Domain arrangement:* This field should describe the predicted number and arrangement of protein domains† when folded in the membrane as determined by hydropathicity analysis† of the primary† sequence. Note that structural predictions of transmembrane domains† on the basis of hydrophobicity† plots may be misleading and prematurely conclusive. For example, high resolution (~9 Å) structural studies of the nicotinic acetylcholine receptor *(nAChR, see ELG CAT nAChR, entry 09)* predict that only *one* membrane-spanning α-helix† (likely to be M2, a pore-lining domain) is present per subunit, with the other hydrophobic regions being present as β-sheets† (see Unwin, *J Mol Biol* (1993) **229**: 1101–24). By contrast, extracellular ligand-gated (ELG) channels such as the nAChR display four *predicted* membrane-spanning regions (M1–M4) on the basis of hydrophobicity plots. *From the foregoing it must be emphasized that all assignments given for the number or arrangement of 'predicted' domains in this field are tentative.*

Field number 28: *Domain conservation:* This field should point out known structural and/or functional motif† sequences which have been conserved as **protein subregions** of ion channel primary† sequences during their evolution (such as those encoding a particular type of protein domain†). Cross-references should be made to functionally related domains conserved in different proteins including 'non ion channel' proteins. Note that '**percentage conservation**' values are not absolute as they depend on which particular subregions of channel sequences are aligned, the numbers and availability of samples, and/or which **sequence alignment algorithms**† are used.

Field number 29: ***Domain functions (predicted):*** This field should indicate predicted functions of channel molecular subregions based on structural or functional data – e.g. regions affecting properties such as voltage-sensitivity, ionic selectivity†, channel gating† or agonist† binding.

Field number 30: ***Predicted protein topography:*** This field should include information on the stoichiometric† assembly† patterns of protein subunits derived from the same or different genes. This field indicates whether channel monomers are likely to form homomultimers†, heteromultimers† or both, and lists estimated physical dimensions of the protein if these have been published. *Note:* **'topography'** is a convenient term borrowed from cartography which when applied to proteins, implies a 'map' at a level of detail or scale *intermediate between* that of an amino acid sequence and a larger-scale representation such as a protein multimeric complex. Topographic maps (or 'models') are therefore particularly useful for displaying selected sets of (inter-related) datatypes within a single 'visual framework'. The **protein domain topography models** (symbolized by [PDTM] throughout the entries) provide prototypes for this form of data representation. The considerable scope for further development of 'shared' topographical models which interactively report and illustrate many different features in the text are described in *Search Criteria & CSN Development (Resource J)*. The terms 'protein topography' and 'protein topology' are often used interchangeably (sic), but the latter should be reserved for those physical or abstract properties of a molecule which are retained when it is subjected to 'deformation'.

Field number 31: ***Protein interactions:*** This field should report well-documented examples of the channel protein working directly *in consort* with separate proteins in its normal cellular role(s). The 'protein interactions' described need not involve physical contact between the proteins (generally referred to as '**protein–protein**' **interactions**), but may involve a **messenger† molecule**. The scope of this field therefore includes notable examples of protein co-localization or functional interaction. For instance, reproduction of native† channel properties in heterologous† cell expression systems may require accessory subunit expression *(e.g. see VLG K Kv-beta, entry 47)*. Common channel–receptor or G protein–channel interactions are described in principle under *Resource A – G protein-linked receptors, entry 56 and Field number 49: Receptor/transducer interactions.*

Field number 32: ***Protein phosphorylation:*** This field should describe examples of experimentally determined '**phosphomodulation**' of ion channel proteins, and if possible list sites and positions of **phosphorylation motifs†** within the channel sequence. Only those consensus sites† explicitly reported in the literature are shown, and these may not be a complete description and may not be based on functional studies. Examples of primary† sequence motifs† for *in vitro* phosphorylation by several kinases† are listed in *Resource C – Compounds & proteins, entry 58 and Resource G – Reported 'Consensus sites' and 'motifs', entry 62* (both updatable via the CSN). Abbreviations used within this field for various enzyme motifs† (e.g. Phos/PKA) are listed in *Abbreviations, entry 03.* Electrophysiological or pharmacological effects of channel protein phosphorylation *in vitro* by use of purified protein kinases† should also be described or cross-referenced in this field.

Criteria for ELECTROPHYSIOLOGY sections

This section should bring together information concerning the ***electrical characteristics*** *of ion channel molecules – how currents are turned on and off, which ions carry them, their sensitivity to applied membrane voltage or agonists, and how individual molecules contribute to total membrane conductance in specified cell types.*

Field number 33: ***Activation:*** This field should contain information on experimental conditions or factors which activate (open) the channel, such as the binding of ligands†, membrane potential changes or mechanical stimulation. Descriptions of characteristic gating† behaviour such as flickering†, bursting†, activation latency† or threshold† of opening are also included. Applicable models of activation and the time course of current flow are briefly described here or referred to *Field number 38: Kinetic model.*

Field number 34: ***Current type:*** Where clarification is required, this field should contain *general* descriptive information on the type, shape, size and direction of ionic current.

Field number 35: ***Current–voltage relation:*** This field should report the behaviour of the channel current passed in response to a series of specified membrane potential shifts from a holding potential† under a specified recording configuration†. For ligand†-gated channels (i.e. those with sortcodes beginning *ELG* and *ILG*) entries should report the current evoked by specific concentrations of agonist† applied at various holding potentials. This field should attempt to illustrate channel behaviour by listing a range of parameters such as slope conductance†, reversal potentials† and steepness† of rectifying† (non-ohmic†) behaviour. The conventions used for labelling the axes of I–V relations for different charge carriers† are outlined in the on-line glossary.

Field number 36: ***Dose–response:*** This field should contain information relating activator 'dose' (e.g. concentration) to channel 'response' parameters (e.g. open time†, open probability†) and whether there are maxima or minima in the response. Agonist† dose–response experiments are used to derive parameters such as the *Hill coefficient*† and *Equilibrium dissociation constant*†.

Field number 37: ***Inactivation:*** This field should describe any inactivation† behaviour of the channel in the *continued presence* of activating stimulus. The field includes information on voltage- and agonist†-dependence, with indications of time course and treatments which extend or remove the inactivation response. Where known, this field will distinguish channel inactivation from receptor desensitization† processes, which are of particular significance for the extracellular ligand†-gated (ELG) channel types *(see ELG Key facts, entry 04).*

Field number 38: ***Kinetic model:*** This field should contain references to major theoretical and functional studies on the kinetic behaviour of selected ion channels. The field contents is limited to a simple description of parameters, terms and fundamental equations.

Field number 39: *Rundown:* This field should collate information on channel 'rundown†' ('washout') phenomena observed during whole-cell† voltage clamp†/ cytoplasm dialysis† or patch-clamp† experiments. Conditions known to accelerate or retard the development of rundown should also be listed.

Field number 40: *Selectivity:* This field should report data on relative **ionic permeabilities**† under stated conditions by means of permeability ratio† and/or selectivity ratio† parameters. The field may also compare measured reversal potentials† in response to ionic equilibrium potentials† with specified charge carriers under physiological conditions. This field also lists estimated physical dimensions of ionic selectivity filters† where derived from ion permeation† or electron micrographic studies.

Field number 41: *Single-channel data:* This field should report *examples* of single-channel current amplitudes and single-channel conductances† measured under stated conditions. In the absence of authentic *single*-channel data, estimates of channel conductances† derived from whole-cell recording† and fluctuation analysis† may be listed.

Field number 42: *Voltage sensitivity:* This field should describe the behaviour of the channel in terms of parameters (e.g. P_{open}†) which are directly dependent upon applied membrane voltage. A distinction should be made between 'voltage sensitivity' resulting from *intrinsic* voltage-gating† phenomena (i.e. applicable to channels possessing integral **voltage sensors**†) and *indirect* effects of applied membrane voltage influencing general physical parameters such as electrochemical driving force†.

Criteria for PHARMACOLOGY sections

This section should bring together information concerning ***pharmacological or endogenous modulators*** *of ion channel molecule activity. Regulatory cascades in cells may simultaneously activate or inhibit many different effector proteins, including ion channels. Analysis of patterns of sensitivity to messengers† and exogenous compounds can help elucidate the* ***molecular signalling pathway*** *in the context of defined cell types.*

Field number 43: *Blockers:* This field should list compounds which reduce or eliminate an ionic current by **physical blockade** of the conductance† pathway. The field should include notes on specificity, sidedness and/or voltage sensitivity of block, together with effective concentrations and *resistance* to classes of blockers where appropriate. Where sites of block have been determined by site-directed mutagenesis†, these should be cross-referenced to *Domain functions, field 29.*

Field number 44: *Channel modulation:* This field should summarize information on effects of important pharmacological or endogenous modulators, including descriptions of extracellular or intracellular processes known to modify channel behaviour. Loci of modulatory sites on the channel protein primary† sequence (as determined by site-directed mutagenesis† procedures) should be cross-referenced to *Domain functions, field 29.*

Field number 45: *Equilibrium dissociation constant:* This field should list published values of K_d for agents whose concentration affects the rate of a specified process. *See also on-line glossary entry for equilibrium dissociation constant*†.

Field number 46: *Hill coefficient:* This field records calculated Hill coefficients† of ligand†-activated processes. The Hill coefficient *(n)* generally estimates the *minimum number* of binding/activating ligands although the *actual* number could be larger. For example, a Hill coefficient reported as $n \geqslant 3$ suggests that complete channel activation requires co-operative binding of at least four ligand molecules *(e.g. see ILG CAT cGMP, entry 22). See also Field number 36: Dose–response.*

Field number 47: *Ligands:* This field should include principal **high-affinity radioligands**† which have been used to investigate receptor–channel function and that are commercially available. Note that numbers of ligand†-binding sites cannot be equated to functional receptors because they only indicate the presence of a ligand-binding entity that may not necessarily be linked to an **effector**† **moiety**†.

Field number 48: *Openers:* This field should list compounds (or other factors) which increase the open probability† (P_{open}) or open time† of the channel in native† tissues.

Field number 49: *Receptor/transducer interactions:* This field should briefly discuss known links to discrete (i.e. separately encoded) receptor and G protein molecules *(see also Resource A – G protein-linked receptors, entry 56, accessible via the CSN)*. Types of **'receptor/transducer/channel' interactions** account for many of the physiological responses of ion channel molecules within complex signalling systems. *Note:* Many pharmacological agents acting at receptor or transducer proteins *(beyond the scope of these entries, but see Watson, S. and Arkinstall, S. (1994) The G-Protein Linked Receptor FactsBook. Academic Press, London)* partially exert their biological effects because these receptor/transducers have ion channel molecules as an ultimate **effector**† protein.

Field number 50: *Receptor agonists (selective):* For the extracellular ligand†-gated (ELG) receptor–channels, this field should list compounds which selectively bind to the **ligand receptor portion** of the molecule and thereby increase the open time†, open probability† or conductance† of the integral channel. Antagonists† should be categorized as competitive†, non-competitive† or uncompetitive† where this has been determined.

Field number 51: *Receptor antagonists (selective):* This field should list agents that selectively bind to the **ligand**† **receptor portion** of integral receptor–channel molecules but do *not* activate a response.

Field number 52: *Receptor inverse agonists (selective):* This field should list compounds which selectively bind (extracellular ligand-gated) receptor–channels but which initiate an *opposite* response to that of an agonist†, i.e. tending to *reduce* the open time†, open probability† or conductance† of the integral channel.

Criteria for INFORMATION RETRIEVAL sections

This section should provide links to other sources of information about the ion channel type, particularly accession to sequence database, gene expression, structure–function and bibliographic resources operating over the Internet[†] *or available on CD-ROM. A full discussion of the potential scope for integration of these resources with molecular-based entries appears in Resource J – Search criteria & CSN development, entry 65. Brief details are given in Feedback & CSN access, entry 12, in each volume.*

Field number 53: *Database listings/primary sequence discussion:* This field should tabulate separately listed items of relevance to the channel type and may include 'retrieval strings' such as **locus names**, **accession numbers**, keyword-containing **identifiers** and other miscellaneous information. Note that terms used by databases are often abbreviated (e.g. K for potassium, Na for sodium etc., therefore only specific identifiers (such as the accession numbers, locus and author names) should be used for retrieval. The actual names and numbers quoted have been sourced from **NCBI-GenBank**® (prefixed **gb:**) or **EMBL** (prefixed **em:**). Since there is now a high concordance between the contents of the EMBL and NCBI-GenBank® nucleic acid databases, the NCBI-GenBank® accession numbers given should retrieve the information from either database. Note that in all of the *Database listings* sections, the lower case prefixes are *not* part of the locus name or accession number, but merely indicate the relevant database.

Sources of pre-translated **protein sequences** are indicated by references to the following databases (given in alphabetical order following the NCBI-GenBank® nucleic acid reference): **SWISSPROT** (prefixed **sp:**), **Protein Identification Resource** (prefixed **pir:**). The journal-scanning component of GenBank uses the NCBI 'Backbone' database (prefixed bbs: for backbone sequence, composed of several individual sequence segments; bbm: for backbone molecule) – these are maintained by the NCBI[†] (National Center of Biotechnology Information).

General notes on sequence retrievals: Updating and error-correction procedures for public domain databases may modify a protein or nucleic acid sequence (retrievable by a given accession number) between releases of a database. Thus, two users performing an analysis on a given database record may come to different conclusions depending upon which release was used. Note also that (i) accession numbers sometimes disappear with no indication of whether a new record has replaced the old one, (ii) multiple databases sometimes each give a different accession number to a single record, and (iii) some databases do not respect the ranges of accession numbers 'reserved' by other databases. Although the 'traditional' format of accession numbers has been a letter followed by five digits (with a maximum space of 2.6 million identifiers), the rapid rate of sequence accumulation will eventually force a different format to be used. Because of these problems, the NCBI now uses **unique integer identifiers (UIDs)** to identify sequence records and encourages their use as the 'real' accession numbers for sequence records. Reference numbers prefixed 'gim' can be read from CD-ROM media, but only refer to a 'GenInfo Import ID' – a *temporary identifier* unique only to a given release of the CD-ROM compilation (such as a numbered release of *Entrez – see below*). Should a sequence supplied by a database change, the record will usually be allocated a new 'gim' number, but the old one will still be available under its UID from the ID database. Because of the transient nature of 'gim' identifiers, they are *not* recommended as search/retrieval parameters and are generally *not* listed in the *Database listings* field (except where an accession number proper has not been found).

In compiling *The Ion Channels FactsBook*, extensive listings of aligned protein or nucleic acids to show **sequence relatedness** have been avoided (as these were judged to be best served by development of on-line data resources specializing in sequence alignments – for a prototype, see Hardison *et al.* (1994) *Genomics* **21**: 344–53). *See also entry 65.* Alternatively or in addition, alignments can be performed according to need by dedicated sequence-manipulation software. Presently available compilations of sequences (e.g. the *Entrez* CD-ROM set or on-line equivalent, for example) can perform powerful **'neighbouring† analyses'** based on *pre-computed* alignments of any sequence against the remainder of the existing database. Establishment of homologous alignments† can proceed by finding a match between the query sequence and any member of the 'neighbouring set'. In practice, comprehensive retrievals can be performed interactively by just one or two rounds of neighbouring analysis. *As indicated at the beginning of each Database listings field, the range of accession numbers provided can be used to initiate relevant searches, but following on from this, neighbouring analysis is strongly recommended to identify newly reported and related sequences.*

Descriptions of features based on primary† sequence data listed within fields of the SEQUENCE ANALYSES or STRUCTURE & FUNCTIONS sections can be more readily interpreted if an **interactive sequence analysis** program is available. Electronic mail servers† at the NCBI can receive specially formatted e-mail† queries, process these queries, and return the search results to the address from which the message was sent out. No specific password or account is needed for these, only the ability to send e-mail to an Internet† site. For local searches, alignment programs such as BLAST can also be retrieved by anonymous file-transfer protocol† or FTP. Detailed information on interactive linking to remote nucleic acid and protein database resources will appear under an option of the Cell-Signalling Network 'home page' *(see Feedback & CSN access, entry 12).* Accession numbers can be *issued* for **newly submitted sequences** (normally within 24 hours) by remote Internet connection or by formatting/submission software (e.g. *Seqwin*, obtainable from the NCBI using an anonymous† FTP†). NCBI-GenBank(R) can also be accessed over the World Wide Web† (http://www.ncbi.nlm.nih.gov).

Sample retrievals in the absence of a CD-ROM resource: For a nucleic acid sequence from the EMBL database, use the e-mail† address below exactly as shown, specifying the appropriate accession number *(nnnnnn)* by the GET NUC command. For example, a database entry can be automatically e-mailed to you by the EMBL server†:

NETSERV@EMBL-HEIDELBERG.DE
GET NUC:*nnnnnn*

An analogous procedure can be used to retrieve protein sequences from the SWISSPROT database, substituting the GET NUC: command with GET PROT:.

Nucleic acid sequences from NCBI-GenBank[R] can be retrieved using the server† at the NCBI. In this case, send an e-mail† message to the service (address below) specifying the name of the database, the command BEGIN and the accession numbers or key words. A sample request is shown below for an accession number *nnnnnn:*

retrieve@ncbi.nlm.nih.gov
DATALIB genbank
BEGIN*nnnnnn*

Protein sequences from the Protein Identification Resource (pir:) can be obtained using an e-mail† request containing the command GET followed by the **database code**. The database code is *distinct* from the accession number but can be obtained by typing the command ACCESSION and then the number. For example, to specify a request for an entry of database code *XXXX* containing the accession number nnnnnn, you would send an e-mail message as follows:

fileserv@gunbrf.bitnet
GET *XXXX*
ACCESSION *nnnnnn*

General information on using these file servers† can be obtained using the above e-mail† addresses followed by the single command HELP. The *Database listings* tables contain short-form references to original research articles which have discussed features of the channel protein and/or nucleic acid primary† sequence(s). Sequences are retrievable with the specified accession number or the author name shown in the short-form reference.

Field number 54: *Gene mapping locus designation:* This field should list references to human gene mapping loci† using terms defined by a human genome mapping workshop (HGMW)† convention where possible. Notes on interactive linking to gene mapping database resources appears under an option of the *Cell-Signalling Network* 'home page' *(see Fig. 4 of Feedback & CSN access, entry 12).* The opportunities for linking to a wide range of genetic information resources are discussed in *Resource J – Search criteria & CSN development, entry 65.*

Field number 55: *Miscellaneous information:* This is a 'catch-all' field used within the entry to reference relevant **peripheral information** or perspectives on the channel molecule or its function. This field also should be used to contain information about ion channels showing *partial* functional relatedness to those in the main entry, but which also possess some features indicating the expression of a distinct gene† (for example, description of *potassium-selective* ligand-gated† channels within an entry describing *non-selective cation* channels gated by the same ligand†, or *vice versa*). Normally, ion channels with distinct properties are covered in 'their own' entry whenever there is sufficient information available to make a clear set of 'defining characteristics'; the *Miscellaneous information* field therefore encompasses those channels which either have been **infrequently reported**, show only **minor variations** with the channel type under description, or are otherwise beyond the scope of the (present) collection of (largely) **vertebrate channel-type entries**.

Field number 56: *Related sources reviews:* For reasons of space, the *FactsBook* cannot provide **citations** for *every* 'fact' within individual entries. Citations within this field should provide a starting point for locating key data through **major reviews** and other primary† sources where these have been quoted extensively within the entry. A full discussion of how future entries could be linked to established on-line bibliographic resources appears in *entries 12 and 65.*

Field number 57: *Feedback:* Information supplementary to the entries but appearing after the publication deadlines will be accessible from the CMHT server† over the Internet† using a World Wide Web† utility from mid-1996 *(see below)*. An aim in compiling this book is that the scope and arrangement of the information should, in time, be **refined** towards containing what is most useful, authoritative and up-to-date: Feedback from individual users is an essential part of this process. The *Feedback* field identifies the appropriate address for e-mail† feedback of significant **corrections, omissions and updates** for the contents of a *specified entry and fieldname*. Comments regarding new or modified field categories (or supplementary reference-type material for incorporation into entries and appendices) would also be most welcome from users *(for details on accessing entry updates via the Cell-Signalling Network, see Feedback & CSN access, entry 12).*

Field number ## (inserted at appropriate points)**: *In-press updates:*** This field has been used occasionally (at the most relevant points in the printed versions of the book) to index publications containing important (direct) evidence which may *significantly* alter several statements or conclusions in the 'finalized' entry as sent to the publishers. It is acknowledged that no 'book-form' information index can ever be *completely* up-to-date, and it is in the nature of scientific progress that 'interpretations' based on reported 'facts' may change considerably in the light of additional or **more direct experimental approaches** to a problem. The scope of the *Cell-Signalling Network* means that users (especially 'non-specialists') can be directed towards citations containing the **'latest' interpretations** (or important **'additional facts'**). The pace of change across all of the fields touched-on by the *FactsBook* means that '**specialists**' in a given area can help '**speed-up**' this indexing process by e-mail† notification **where 're-interpretation' is justified** *(see Feedback & CSN access, entry 12, and Resource J, entry 65)*. According to the original aims and 'philosophy' of the project, the entries will probably never be 'complete' as such. *More appropriately, the framework will continue to evolve towards one which is hopefully more useful, authoritative, and able to* ***comprehensively relate*** *'consensus' knowledge on ion channel molecular signalling.*

Criteria for REFERENCES sections

This section should contain '**short-form**' **references** for **numbered citations** within the entry. *For textbook coverage, refer to the Book references listed under Related sources and Reviews (field 56), Resource E – Ion channel book references, entry 60, Resource F – Supplementary ion channel reviews, entry 61 and Resource H – Listings of cell types, entry 63.* Plans for 'hyperlinking' to full bibliographic databases within the CSN framework are described in *Resource J – Search criteria & CSN development, entry 65.*

Criteria used for compilation of supporting computer-updatable resources

The following reference appendices are referred to within the text and figures of the main entries. Updated versions of these files will be accessible via the 'home page' of the Cell-Signalling Network from mid-1996 – for further details, see Feedback & CSN access, entry 12 and Resource J – Search criteria & CSN development, entry 65.

Resource A – G protein-linked receptors: A large number of ion channels are regulated as part of signalling cascades initiated by activation of **G protein-coupled receptor proteins**. This appendix should describe the basic principles associated with this type of regulation, limiting descriptions to those most relevant to ion channels. Tabulations of known receptor† and G protein† molecules should form a framework of *possible* regulatory mechanisms based on *specific* protein subtypes. The entry may clarify or suggest likely interactions between receptors, transducers† (e.g. G proteins) and ion channel molecules described under the fieldnames *Developmental regulation, field 11, Protein interactions, field 31, Protein phosphorylation, field 32, Channel modulation, field 44, and Receptor/transducer interactions, field 49.*

Resource B – 'Generalized' electrical effects of endogenous receptor agonists: This resource should present a tabulated summary of *general patterns* of agonist†-induced ionic current fluxes that have been reported across a large number of studies, predominantly in the central nervous system. The table may help to indicate whether receptor† agonists tend to act in an **excitatory**† or **inhibitory**† fashion 'or both'.

Resource C – Compounds & proteins: Compounds and proteins mentioned in the entries which are commonly used to investigate **ion channel function and modulation** should be listed, including those used to analyse interactions with other cell-signalling molecules. In general, only frequently reported compounds which are commercially available are described in this appendix.

Resource D – 'Diagnostic' tests: This appendix is intended to be an alphabetical listing of **common experimental manipulations** used to 'implicate or exclude' the contribution of a given signalling component or phenomenon associated with ion channel signalling. For the most part, these approaches use the pharmacological tools listed under Resource C, but may also include sections describing common molecular biological and electrophysiological 'diagnostic' procedures.

Resource E – Ion channel book references: This appendix should list details of **published books** which have addressed themes in ion channel biology or closely related topics. These references complement those of the main entries, which are almost entirely based on citations from scientific journals.

Resource F – Supplementary ion channel reviews: The ion channel literature contains a large number of useful '**minireviews**' which summarize the development of **defined subjects** and which do not necessarily fall into a single channel 'molecular type' category. This appendix should therefore list these 'supplementary' sources, indexed by topic. Updated 're-writes' of subject reviews covering similar areas may replace earlier listings. *Note:* Subject reviews dedicated to aspects of an ion channel type or family can usually be found under the *Related sources & reviews* field of appropriate entries. 'Topic-based' reviews making reference to the basic properties in the 'molecular type' entries are planned for expansion within the CSN framework *(for details, see entry 65)*.

Resource G – Reported 'Consensus sites' and 'motifs': Based on extensive analysis of primary† sequences and determination of substrate specificities for various enzymes, a number of 'consensus' recognition sequences for **post-translational modification**† of proteins (including ion channels) have been determined. While these sites are not absolute, they can be highly conserved across whole families of ion channel proteins and in many cases (e.g. following phosphorylation) can lead to profound changes in ion channel function. However, the presence of '**consensus**' **sites** or **motifs**† (or even demonstrations of substrate specificity *in vitro*) does not necessarily prove that such modifications operate *in vivo*. This appendix should list 'consensus' motifs that are well-characterized, giving examples of 'authentic' sites for comparison. This appendix also contains a subset of consensus† sites from genomic DNA sequences associated with mechanisms of **ion channel gene expression-control**† (e.g. *in trans*† *protein factors which act at DNA structural motifs*† *in cis*†, influencing transcriptional activation†, transcriptional enhancement† or transcriptional silencing† of ion channel genes†).

Resource H – Listings of cell types: Studies of ion channels within the context of **cell-type function** often reflect 'recruitment' of selected genes from the genome† in a cell-developmental lineage†. Because of this, similar 'sets' of ion channel molecules can often be observed in cell types with broadly similar functions. This appendix should describe a framework for describing how *integrated sets* of ion channel molecules (and their associated signalling components) have **co-evolved** for specific functions in terminally differentiated† cell-types. To begin with, a tentative classification of **functional cell types** should be employed, used to cross-reference 'surveys' of ion channel expression wherever possible. This appendix should also contain available information pertaining to efficient and appropriate heterologous expression of ion channel genes in selected cell types, as this is often a limiting factor in biophysical characterization of cloned† ion channel cDNA† or gene products.

Resource I – Framework of cell-signalling molecule types: The flow of information into, within and between cells (signal transduction) generally depends on a *multiplicity* of co-expressed cell-signalling molecules which provide 'measured' responses to stimuli. Communication between different **cellular compartments** (e.g. between the cytoplasm and the nucleus) often requires 'interconversion' or 'transduction' of chemical, electrical (ionic), metabolic and enzymatic signals, with **receptors** and **ion channels** playing key roles in transducing such stimuli. For example, the 'activation' of signal transduction molecules such as kinases† or transcription factors† appear to 'sense' 'activated' conditions which resembles Ca^{2+}-, voltage- or ligand†-gating phenomena commonly observed for ion channels. These modes of **protein activation**† probably have many features in common, and understanding their interrelationship has important consequences for comprehending fundamental links between receptor signalling, cell activation and gene expression.

To facilitate integration of information between these diverse fields of study, this appendix should provide a preliminary listing of **signal transduction molecules**, with some consideration of their inter-dependency in the 'activated' state. By making a rational 'connection' between activation of receptors, ion channels, enzymes and other effector† proteins, it is hoped that some general principles will emerge on the **electrical- and ligand†-control of complex cell phenotypes**† (such as those affecting the **cell cycle**†, **cell proliferation**†, **cell differentiation**† and **apoptosis**†).

The importance of ion channel activation (and activation of receptor/G protein transducers† which modulate ion channel activity) in other **fundamental cell processes** such as **signal transmission/amplification**, **secretion** (multiple forms), **muscular contraction**, **endocytosis**† (and other cellular 'uptake' phenomena) , **sensory transduction** (all types), cell volume control/**osmotic responses**, **mechano-transduction** (various forms), **membrane potential control** (multiple modes) and **developmental compartment formation** are well-documented and multiple examples appear in several fields, notably *Developmental regulation, field 11, Phenotypic expression, field 14, Domain functions, field 29, Protein interactions, field 31, Protein phosphorylation, field 32* and *Channel modulation, field 44.*

Resource J – Search criteria & CSN development. The framework of database entries which form the basis of *The Ion Channel FactsBook* were derived by 'scanning' primary research articles and reviews appearing in a set list of 'principal' journals dealing with ion channel and receptor signalling. A disadvantage of 'journal scanning' by 'keyword' is that search terms used are often ambiguous, and contextual or unconventional grammatical usage of keyword terms within articles often results in failure of specific retrieval. To circumvent this problem, this appendix should suggest new **unique embedded identifiers** (UEIs) which when specified by authors in the **keywords** section of submitted articles should ensure appropriate **electronic retrieval** from the primary literature. The adoption of finalized 'UEI' codes should be open to debate. Their implementation outside the context of the CSN will be difficult unless contributing authors and journal editors acknowledge the benefits. If an alternative system is proposed and accepted by field consensus, then the CSN will move to adopt the system in the interests of simplifying search criteria on specific molecules or topics.

The central principle of unique embedded identifiers is that they can 'automatically' find articles on topics of interest (in for example weekly literature scans). Coupling to an 'expansion' section with further search terms in a conventional order will help enormously in data compilation/consolidation processes on strictly defined subjects within 'validated' databases.

Finally, Resource J should act as a forum for discussing **limitations of data representation** when comparing ion channel properties and suggest improved methods for facilitating **information exchange** (including graphical resources), **diagnostic conventions**, resolution of **'controversial' results**, and identification of areas or highly focused topics requiring **consolidation/extension** of knowledge. The importance of standardized computer software compatible with Internet†-mediated communication should be emphasized *(see also Feedback & CSN access, entry 12)*. Contents organization *within* each 'specialist' field of the *FactsBook* gives further opportunities for comparative data analysis. In due course, the -zz term of the xx-yy-zz index number will be used to indicate such **structured information**.

Criteria used for selection of on-line glossary and index items

Consolidated versions *of the FactsBook support glossary (i.e. extensions, updates and corrected items) are accessible from the Cell-Signalling Network 'home page' (see Feedback & CSN access, entry 12). Entry 65 contains a full specification of the CSN.*

Index of on-line glossary items [†]: To avoid unnecessary duplication of definitions within the text and to provide assistance to readers unfamiliar with a field, the on-line glossary should provide short introductions to technical terms and concepts. Throughout the text, cross-references to the **on-line glossary items** are shown by means of a **dagger symbol**†.

Cumulative subject index for The Ion Channel FactsBook, volumes I to IV.
For the most part, *The Ion Channels FactsBook* should be '**self-indexing**':

1. Locate the channel '**molecular type**' by sortcode, or table of contents
2. Go to the appropriate **section** (NOMENCLATURES, EXPRESSION, SEQUENCE ANALYSES, STRUCTURE & FUNCTIONS, ELECTROPHYSIOLOGY, PHARMACOLOGY, INFORMATION RETRIEVAL or REFERENCES).
3. Look under the most appropriate **fieldname** (as described by the criteria above). Further 'structuring' will arise in due course, when more data are entered *(see previous section)*.

For location of information on ion channel molecules by **miscellaneous related topics**, the **cumulative subject index** should comprehensively list pertinent functional characteristics, concepts, compounds and proteins including those shown in **bold text** under the fieldnames, relating the topic to the **six-figure index number**. The **subject index** should also allow the initial location of entries through alternative names of channels, associated signalling phenomena or commonly reported properties. Electronic cross-relation of topics is intended to be a development focus of the CSN, exploiting the principle of *hyperlinking* between database files stored in 'addressible' loci. For further details on how this might be achieved, *see Resource J – Search criteria & CSN development, entry 65.*

Feedback: Comments and suggestions regarding the scope, arrangement and other matters relating to this introduction can be sent to the e-mail feedback file CSN-02@le.ac.uk. *(see field 57 of most entries for further details)*

Abbreviations

For most abbreviations of **compound names** in use, refer to the *Resource C – Compounds & proteins, entry 58*, as well as the *FactsBook* entries. Abbreviations for ion channel currents are listed under the *Current designation* field of each entry. Terms marked with a dagger symbol appear in the on-line glossary section

$0Ca^{2+}$	Ca^{2+}-free solution
5-HT	5-hydroxytryptamine; serotonin
7TD	7 transmembrane domains
A	ampere†
aa	amino acid†
AHP	afterhyperpolarization†
AP	action potential†
APD	action potential duration†
AV	atrio-ventricular
AVN	atrio-ventricular node† (of heart)
BKCa	large ('big')-conductance calcium-activated K^+ channels
BP	blood pressure
bp	base pairs†
C	coulomb†
C-terminal	carboxyl† terminal† (of protein)
C/A or C-A	cell-attached† (recording configuration)
Ca(mech) or Ca_{mech}	mechanosensitive† Ca^{2+} channel
Ca_V	voltage-gated† Ca^{2+} channels
cds	coding† sequence (used in GenBank†® entries)
CF	cystic fibrosis†
CICR	calcium-induced-calcium-release
Cl(Ca) or Cl_{Ca}	calcium-activated chloride channel
CNG	cyclic-nucleotide-gated (channels)
CNS	central nervous system†
COOH	carboxyl group†
CRC	calcium release channels
cRNA	complementary† RNA
CTK	cytoplasmic tyrosine kinase† (cf. RTK)
Cx or Cxn	connexin
Da	daltons
Dephos/enzyme	putative (consensus†) site for dephosphorylation† by a specified enzyme, e.g. Dephos/PP-1: endogenous protein phosphatase-1; Dephos/PP-2A: protein phosphatase-2A
DHPR	dihydropyridine receptor
DMD	Duchenne muscular dystrophy†
DPSP	depolarizing post-synaptic potential†
E	potential difference†, inside relative to outside
EAA	excitatory amino acid†
E–C	excitation–contraction†

EC_{50}	50% effective concentration
E_K	equilibrium potential† for K^+ ions (analogous nomenclature for other ions)
ELG	extracellular ligand†-gated (as used in *FactsBook* sortcode)
E_m	membrane potential†
EMBL	European Molecular Biology Laboratory†
EMF	electromotive force†
EPP	endplate† potential†
EPSP	excitatory† post-synaptic potential†
ER	endoplasmic reticulum†
E_{rev}	reversal potential†
F	farad†
F	Faraday's constant†
fS	femtosiemens (10^{-15} Siemens†)
G	conductance†
g	conductance (unit – Siemens†, formerly reciprocal ohms† or mho†)
G/G_{max}	peak conductance†
gb:	designation for GenBank® accession number†
g_j, G_j or *G*(j)	gap-junctional conductance†
HGMW	Human Gene Mapping Workshop†
HH	*after* Hodgkin†-Huxley†
h.p.	holding potential
HVA	high-voltage-activated Ca^{2+} channels
I	current†
i	subscript abbreviation for intracellular
I/I_{max}	peak† current†
I/O or I-O	inside-out† (patch†, recording configuration†)
IC_{50}	concentration which gives 50% of maximal inhibition effect in a dose-inhibition response curve†.
ILG	intracellular ligand†-gated (as used in *FactsBook* sortcodes†)
I_{max}	maximal current†
$InsP_{(x)}$ or $InsP_x$	collective abbreviation for inositol polyphosphates† - e.g. $InsP_3$, $InsP_4$
$InsP_3R$ or IP3R	inositol 1,4,5,-trisphosphate-sensitive receptor–channel
IPSC	inhibitory† post-synaptic current†
IPSP	inhibitory† post-synaptic potential†
JCC	junctional channel complex
k	Boltzmann's constant†
K_A or K(A)	A-type† K^+ channels
kb	kilobases† (kbp - kilobase pairs or bp x10^3)
KCa, K_{Ca} or K(Ca)	calcium-activated K^+ channels

K_D	equilibrium dissociation constant†
kDa	kilodaltons† (daltons $\times 10^3$)
K_I	equilibrium dissociation constant† for an inhibitor
K_i ATP	for K_{ATP} channels, the ATP concentration (μM) that produces half-maximal inhibition of channel activity
$K_I(0)$	inhibition constant† at zero voltage†
K_{IR} or K(IR)	inward rectifier†-type K^+ channels
K(mech) or K_{mech}	shorthand designation for mechanosensitive K^+ channels
K_V	voltage-gated K^+ channels (generally delayed rectifiers†)
LTD	long-term depression†
LTP	long-term potentiation†
LVA	low-voltage†-activated Ca^{2+} channels
mAChR	muscarinic acetylcholine receptor
MARCKS	myristoylated†, alanine-rich C-kinase substrate
Mb	megabases† (Mbp - megabase pairs)
MEPC	miniature endplate currents
MDa	megadaltons† (daltons $\times 10^6$)
MH	malignant hyperthermia
M_r	relative molecular mass†
mRNA	messenger RNA†
mV	millivolt (10^{-3} V)
N	number of functional channels *also* - Avogadro's number†
n	Hill coefficient†
nAChR	nicotinic acetylcholine receptor–channel
Na_V	shorthand designation for voltage-gated Na^+ channels
N-gly:	predicted sites for *N*-linked glycosylation† (e.g. *N*-gly: aa122, specifying amino acid number 122 from known glycosylase† substrates)
NH_2	amino† group
NSA	non-selective anion (channel)
NSC	non-selective cation (channel)
NSC(Ca)	non-selective cation channels (calcium-activated)
nt	nucleotides
N-terminal	amino-terminal (of protein)
o	subscript abbreviation for extracellular
O-gly	*O*-linked glycosylation†
OHC	outer hair cells
O/O or O-O	outside-out† (patch†, recording configuration†)
P_{Ca}	permeability† of Ca^{2+} ions (analogous nomenclature for other ions, e.g. P_K, P_{Na}, P_{Rb} etc.)
PCAP	pituitary adenylyl cyclase-activating polypeptide
PDE	phosphodiesterase†

[PDTM]	protein domain topology model; within the text, use of the abbreviation in square brackets denotes a positional feature illustrated on the model
pH_i	intracellular pH†
Phos/enzyme	Putative† (consensus†) site for phosphorylation† by a specified enzyme, e.g. Phos/CaM kinase II - multi-functional (Ca^{2+}/calmodulin)-dependent protein kinase II; Phos/CaseKII: casein kinase II; Phos/GPK: glycogen phosphorylase kinase; Phos/MLCK: myosin light-chain kinase; Phos/PKA: cAMP-dependent protein kinase (PKA); Phos/PKC: protein kinase C (PKC); Phos/PKG: cGMP-dependent protein kinase; Phos/TyrK: tyrosine kinase (TyrK) subtypes
p.i.	post-injection
PIR	Protein Identification Resource† (protein sequence database)
pir:	designation for Protein Identification Resource† accession numbers†
PNS	peripheral nervous system†
poly(A)	polyadenylation† (site)
$poly(A)^+$	polyadenylated† (mRNA) fraction of total cellular RNA
P_{open} or P_o	channel open probability†
pS	picosiemens (10^{-12} Siemens†)
PSC	post-synaptic current†
PSP	post-synaptic potential†
PSS	porcine stress syndrome
Q_{10}	coefficient† for a ten-degree change in temperature
R	receptor
R	resistance† (unit – ohm†), reciprocal of conductance†
r.p.	resting potential†
rRNA	ribosomal RNA†
RTK	receptor tyrosine kinase† (at plasma membrane, cf. CTK)
RyR	ryanodine receptor–channel
S	Siemens† (unit of conductance†; reciprocal ohm† or mho†)
SAN	sino-atrial node† (of heart)
SAPs	signal-activated phospholipases†
s.c.a.	single-channel amplitude†
s.c.c.	single-channel conductance† (symbol, γ)
s.c.p.	single-channel permeability†
SCR	single-channel recording†
SD	standard deviation
SDS–PAGE	sodium dodecyl sulphate–polyacrylamide gel electrophoresis†
SEM	(i) standard error† of the means† or (ii) scanning electron microscopy†

SFA	spike frequency adaptation†
Sig:	indicates the range of amino acids which form the signal peptide of a precursor protein (e.g. Sig: aa1–26); alternatively, the abbreviation indicates the actual cleavage site† forming the signal peptide† and mature chain† from the precursor† protein
SP	substance P
sp:	designation for SWISSPROT protein sequence database accession number†
SR	sarcoplasmic reticulum†
S–S:	disulphide bond†; in sequence database entries, the S–S: symbol is sometimes used to denote positions of a known disulphide bond linkage† or motif† between two residues on a protein molecule, e.g. an experimentally determined link between residues 154 and 182 on the same chain would be written as S–S: 154-bond-182.
TM	transmembrane
$\mathrm{T_m}$	melting temperature†
T_m	upper limit to the amount of material that carrier-mediated transport can move across a membrane
TPeA+	tetrapentylammonium ions
TT	transverse tubule†
V	volt†
V	voltage†
VACaC	voltage-activated calcium channels; analogous nomenclature for other channels, e.g. VAClC, VAKC, VANaC, VDAC
VDAC	voltage-dependent† anion channel
VDCC	voltage-dependent† calcium channel
W/C or W-C	whole-cell† (recording configuration)
WCR	whole-cell† recording
YAC	yeast artificial chromosome†
γ	unitary (single-channel) conductance
γ_j or γ(j)	single-channel junctional conductance†
μA	microamp (10^{-6} Amperes)
Ω	ohm†, unit of electrical resistance†; reciprocal of conductance†
ω-CgTx	omega-conotoxin

Feedback: Comments and suggestions regarding the scope, arrangement and other matters relating to the abbreviations section can be sent to the e-mail feedback file CSN-03@le.ac.uk. *(see field 57 of most entries for further details)*

EXTRACELLULAR LIGAND-GATED CHANNELS (ELG)

Edward C. Conley **Entry 04**

Note: The '***Key facts***' *sections are intended for readers unfamiliar with the more general aspects of ion channel biology, and contain* ***selected*** *introductory information applicable to whole groups of ion channel molecules. These sections, coupled with the* ***glossary items*** *(indicated by dagger symbols) provide a basic overview of principles associated with more detailed information within the main entries of the book. More comprehensive introductory material can be found in references listed under topic names within the* ***Related sources & reviews*** *field appearing at the end of each main entry and in Resource F – Supplementary ion channel reviews, entry 61.*

Extracellular ligand-gated (ELG) receptor-channels rapidly transduce transmitter-binding events into electrical signals

Integrated molecular functions of ELG channels

04-01-01: Fast synaptic neurotransmission, both excitatory and inhibitory, is mediated by extracellular ligand-gated (ELG) receptor channels. These channels combine ion-selective functions with those for agonist binding and signal transduction within a multi-subunit molecular assembly. In general, **excitation** from resting membrane potentials is associated with opening of cation-influx (depolarizing) channels, while **inhibition** of neuronal firing is generally associated with increased chloride ion permeability and hyperpolarization. A number of channel molecular types are responsible for these actions *(Table 1).*

Functional diversity of extracellular ligand-gated receptor channels

04-01-02: ELG receptor–channels which function in fast synaptic transmission include channels directly gated by the **neurotransmitters**, including L-glutamate, acetylcholine, glycine, ATP, serotonin (5-hydroxytryptamine), γ-aminobutyric acid and possibly histamine *(see Table 1).* The excitatory amino acids L-cysteine sulphinate and quinolinate may also act as endogenous neurotransmitters in some brain regions. Furthermore, certain **tastants**† may directly activate (e.g. L-arginine) or block (e.g. H^+ ions) apical non-selective cation channels where the tastant acts as an extracellular ligand *(see also Receptor/transducer interactions under ILG CAT cAMP, 21–49).* In excitable cells, receptor-operated channels may also serve to depolarize the cell to the threshold of action potential generation. In non-excitable cells, receptor-operated channels permit a limited Ca^{2+}-influx during the presence of an agonist *(see also entries describing the intracellular ligand-gated (ILG) channels).*

Time scale of ELG channel signalling

04-01-03: Receptors containing integral ion channels mediate relatively **rapid transduction events**, and are activated on a millisecond time scale, with typical latency† of $\sim<1$ ms. This can be compared to receptors activating G protein-coupled channels which typically operate in the millisecond-to-second range following agonist reception. The rise-time for transmitter

Table 1. *Examples of the extracellular ligand-gated ion channel family (From 04-01-01)*

Extracellular ligand	ELG channel subtype	Principal ionic selectivities[a]	Protein superfamily	Covered under
5-hydroxytryptamine	$5\text{-}HT_3$	Na^+, K^+	Ia	ELG CAT $5\text{-}HT_3$
ATP	P_{2X}	Ca^{2+}, Na^+, Mg^{2+}	2[b]	ELG CAT ATP
Glutamate	non-NMDA	Na^+, K^+, (Ca^{2+})	Ib	ELG CAT GLU AMPA/KAIN
Glutamate	NMDA	Na^+, K^+, Ca^{2+}	Ib	ELG CAT GLU NMDA
Acetylcholine	nAChR (neuronal)	Na^+, K^+, Ca^{2+}	Ia	ELG CAT nAChR
Acetylcholine	nAChR (muscle)	Na^+, K^+, Ca^{2+}	Ia	ELG CAT nAChR
γ-Aminobutyric acid	$GABA_A$	Cl^-, HCO_3^-	Ia	ELG Cl $GABA_A$
Glycine	GlyR	Cl^-, HCO_3^-	Ia	ELG Cl GLY

[a]At resting membrane potential, neuronal excitation is usually associated with influx of sodium ions while inhibition of firing generally results from activation of chloride and potassium conductances. For generalized effects of G protein-coupled receptor agonists on the activation and inhibition of ionic conductances expressed in central neuronal cells, *see Appendix A – Index of G protein-linked receptors, entry 56, and Appendix B – Index of generalized electrical effects of receptor agonism, entry 57.*
[b]Determination of primary sequences for genes encoding P_{2X} purinoceptors[1,2] indicates a distinct structural motif for these receptor–channels consisting of two transmembrane domains per monomer and a pore-forming motif reminiscent of that proposed for potassium channels *(see ELG CAT ATP, entry 06).*

concentrations to levels that activate ELG channels is also relatively brief (a typical diffusion distance across the synaptic cleft† being $\sim\leqslant 2\,\mu m$). Durations of synaptic currents† are generally determined by the **intrinsic molecular properties** of the receptor channels involved (e.g. receptor desensitization† and channel inactivation†).

Distinction of ELG channels from receptor-modulated channels
04-01-04: By definition, ELG channel gating† is independent of any intracellular or membrane-diffusible factor, although phosphorylation is a major mechanism for regulating their function[3] *(see later section)*. In contrast to the ELG channel group, many ion channels coupled to separate receptors (via G proteins and second messengers) can be gated by membrane potential changes in the absence of agonist. The majority of these can be considered as *receptor-modulated channels*, as neurotransmitters activate or block primary voltage-dependent responses. In order to distinguish these parallel signalling responses initiated by single neurotransmitters, receptors which couple to G proteins are often referred to as **metabotropic† receptors** *(see Resource A – G protein-linked receptors, entry 56)*, while receptor protein complexes forming integral ionic channels are known as **ionotropic† receptors**.

ELG channel genes are differentially expressed

Co-ordination of complex overlapping patterns of ELG gene expression
04-01-05: Developmental gene-expression programs co-ordinate the activation and silencing† of ELG ion channel genes, producing complex patterns of expression which underlie functional specialization of individual cell types. Developmental regulation may be effected by multiple signals, including growth factors†, intracellular second messengers†, cell type-specific transcription complexes†, etc.

'Fine-tuning' of ELG channel gene expression
04-01-06: In addition to the specification of appropriate protein subunits, cells employ a variety of mechanisms to modulate ELG channel gene expression, including alternative splicing† of primary transcripts†, RNA editing† and activity-dependent control *(see below)*.

'Darwinian' synaptic interactions in developing brain affect gene expression
04-01-07: Activity-dependent, competitive synaptic interactions, which stabilize some axon branches and dendrites while removing others, centrally involve glutamate receptor–channel expression[4–6] *(for further details see Developmental regulation under ELG CAT GLU AMPA/KAIN, 07-11, or Developmental regulation under ELG CAT GLU NMDA, 08-11)*.

ELG channel expression is silenced in some cell types
04-01-08: Certain cell types can co-express multiple receptor channel types (e.g. neurones), while other cell types do not express any fast ligand-gated channels at all (e.g. epithelial cells).

The ELG receptor–channels form an extended protein sequence 'superfamily'

ELG channel gene evolution

04-01-09: Similar hydropathy plots† and amino acid sequence motifs† are observed at equivalent positions in subunits of different superfamily† members (*see Fig. 1* exemplified for receptor–channels gated by acetylcholine, GABA, glycine and glutamate). Based on closeness of optimal amino acid sequence alignments, the ELG superfamily has been further divided[7] into groups Ia and Ib *(see Table 1)*. Thus ELG channels activated by acetylcholine, γ-aminobutyric acid, glycine and 5-hydroxytryptamine are more similar to each other than those in group Ib (the ionotropic† glutamate receptors). These structural and functional similarities can be explained if all members of the gene superfamily originated by adaptive radiation† from a common gene encoding an ancestral channel type. A striking exception to this pattern of 'divergent†' evolution is shown by the gene/protein domain strtucture of **ATP-gated cation channels**[1,2] which exemplify the 'convergent†' evolution of similar protein functions via markedly different structural characteristics *(see ELG CAT ATP, entry 06)*

The molecular basis of functional similarities and differences in the ELG superfamily

04-01-10: Partial conservation of sequences between gene families can explain how certain features have been retained (e.g. ionic selectivity† and subconductance† levels) while other features have diversified between gene family members (e.g. pharmacological sensitivities). Generally, the molecular heterogeneity of ELG channels shown by **molecular cloning** has been larger than that expected from pharmacological studies of native receptors[8].

ELG channel subunit stoichiometries support a stereotypical model for protein quaternary structure

04-01-11: Investigations of ELG channel quaternary† structure based on electron microscopy, subunit cross-linking, electrophoretic and sedimentation analysis provide direct support for a **pentameric arrangement** of subunits in native receptor–channels. Probably all of the superfamily members possess a 'quasi-symmetrical' arrangement of subunits around the central pore. Subunit stoichiometries of both cation- and anion-selective ELG channels *(e.g. see ELG CAT nAChR, entry 09, and ELG Cl GLY, entry 11)* have been deduced as conforming to a **'3α : 2β' arrangement** of separately encoded α and β subunits. This basic arrangement has therefore been proposed as the **basic quaternary structure** of the ELG superfamily[9] *(see Protein domain topography models within the entries)*. The stoichiometry and native structure of ATP receptor–channels, which display a distinct transmembrane and pore-forming domain structure[1,2], are presently unclear *(see ELG CAT ATP, entry 06)*.

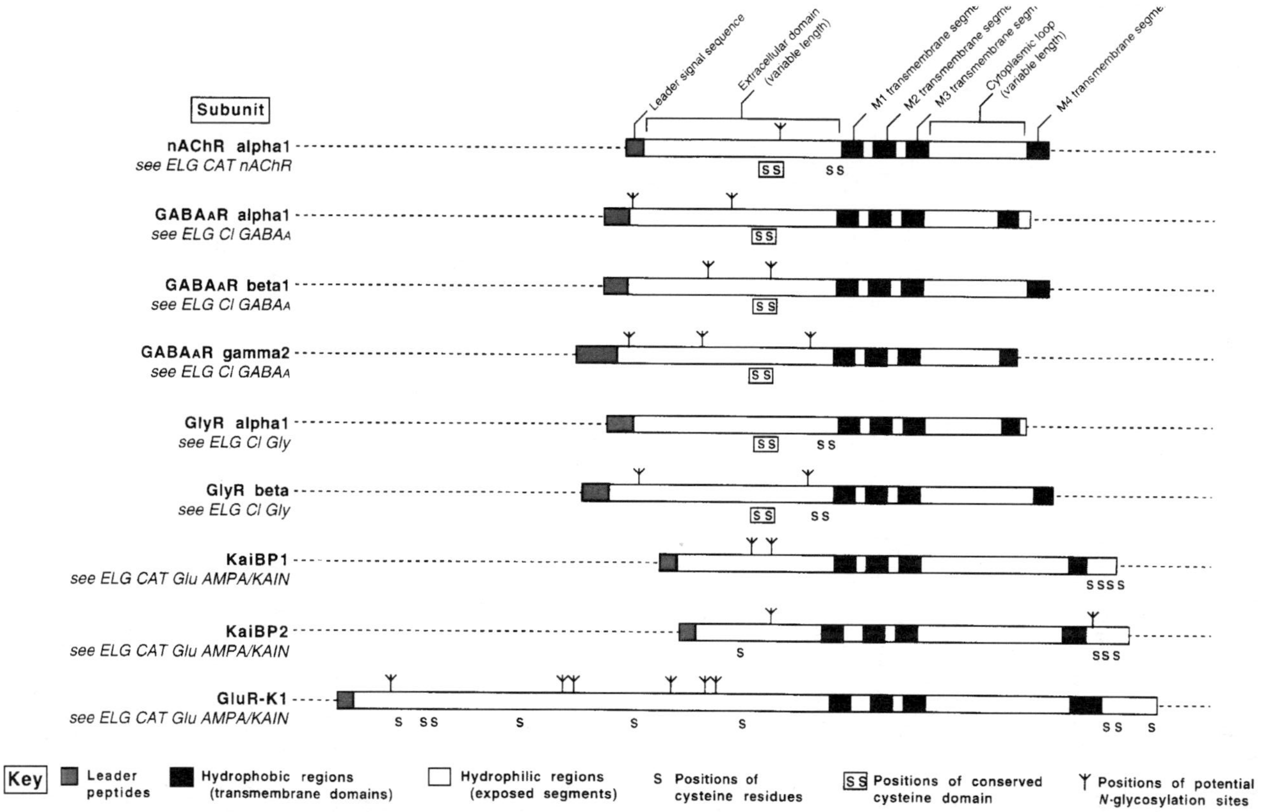

Figure 1. *Structural similarities between primary amino acid sequences for subunits of selected ELG receptor–channel proteins. Hydropathy analyses predict a common transmembrane topology which underlies functional relatedness. (Based on alignments from Betz (1990)* Neuron **5***: 383–92). (From 04-01-09)*

ELG channel properties are largely determined by the specification of co-expressed subunit types and their stoichiometries

'One-gene variant, one-subunit'

04-01-12: Extracellularly activated receptor–channels form relatively large multi-subunit complexes, each subunit generally being derived from the expression of a separate gene or variant of a gene. The same gene product (protein subunit) may be represented more than once in an assembled multi-subunit complex (for clarification, see the [PDTM] protein domain topography models within the entries).

Molecular integration of ligand-binding, signal transduction and ionic conduction

04-01-13: ELG channel genes encode proteins which structurally integrate the distinct functions of ligand-binding, signal transduction and ionic conductance within single macromolecular complexes. The **protein–protein interactions** necessary for monomeric components to co-assemble into these macromolecular complexes are intrinsic to parts of the protein primary† sequence *(see below)*.

Recruitment and assembly of ELG channel subunit proteins

04-01-14: Following 'recruitment' of particular combinations of subunits at the cellular level, specific assemblies of receptor channels with unique functional characteristics may depend on amino acid sequences which mediate inter-subunit contacts and post-translational modifications† of component subunits *(see also section on phosphorylation, below)*.

Transmembrane topography of the ELG receptor–channels is difficult to predict alone from amino acid sequence and mutagenesis studies

Conclusions from high-resolution protein imaging

04-01-15: In general, ELG channels display four *predicted* membrane-spanning regions (M1–M4) on the basis of hydrophobicity† plots. However, structural studies of the nicotinic receptor at high resolution[10] (~9 Å) predict that only *one* membrane-spanning **α-helix**† (presumed to be M2, as a pore-lining domain) is present per subunit, with the other hydrophobic regions being present as **β-sheets**†. Because of these factors, all figures presented as 'protein domain topography models' within the entries should be considered as diagrammatic. In particular, the relative positions of motifs, domain shapes and sizes are subject to re-interpretation in the light of better structural data.

ELG ion channel proteins are invariably regulated by post-translational modifications†

ELG channel proteins are substrates for a variety of protein kinases

04-01-16: ELG channels such as the nicotinic acetylcholine receptor *(see ELG*

CAT nAChR, entry 09) can be phosphorylated by a family of endogenous **protein kinases** including cAMP-dependent protein kinase, protein kinase C, protein tyrosine kinase and calcium/calmodulin-dependent protein kinase[11,12].

An example of functional modulation by phosphorylation – receptor desensitization

04-01-17: Reconstitution into liposomes has permitted functional studies of subunit phosphorylation. Phosphorylation of the nAChR reconstituted into liposomes accelerates **receptor desensitization**, a process by which membrane receptors do not transduce signals even in the continued presence of agonist. Desensitization processes are a result of agonist-induced **conformational changes** that close the ion channel rather than activating (opening) it in the presence of agonist. In addition to neurotransmitters, hormones and other first and second messengers can regulate phosphorylation states of ELG channels, thus acting as physiological regulators. Subunit phosphorylation status may also regulate ELG **channel assembly** processes. *(See also the Protein phosphorylation field in the ELG entries and Resource G – Reported 'Consensus sites' and 'motifs', entry 62.)*

Glutamate is the major excitatory† agonist in the vertebrate brain

Ubiquity of glutamate agonism

04-01-18: Receptors for the neurotransmitter glutamate (GluRs) are expressed on virtually every neuronal cell and some glial cells in the CNS. Ionotropic† GluRs (possessing integral channels) have major roles in **fast synaptic transmission** and are participants in the establishment of synaptic networks, processing of associative/sensory information and co-ordination of motor functions[13].

Subclasses of glutamate receptors with integral ion channels

04-01-19: GluRs have been initially classified into three separate populations, each defined by selective activation with different structural analogues of glutamate[14,15]. Thus, broad categories of channels gated by *N*-methyl-D-aspartate (**NMDA**), α-amino-3-hydroxy-5-methyl-4-isoxazolepropionate (**AMPA**) and kainic acid (**kainate**) have been described extensively (the latter two categories are collectively referred to as the '**non-NMDA receptors**'). The large number of GluR protein subtypes (and the complexity of their potential arrangements in functional channels) indicates that present classifications are not absolute and therefore may be inadequate for describing the range of GluR expression observed *in vivo (for detailed nomenclature within these families, see ELG CAT GLU AMPA/KAIN, entry 07, and ELG CAT GLU NMDA, entry 08).*

Glutamate toxicity and neurodegeneration

04-01-20: In addition to its normal function as an excitatory neurotransmitter, glutamate can kill neurons by prolonged receptor-mediated depolarization, resulting in irreversible disturbances in ionic homeostasis[13,16–18].

Glutamate toxicity has been implicated in the death of neurones following ischaemia, epilepsy, and neurodegenerative disorders such as Alzheimer's, Huntington's and Parkinson's diseases *(for further details, see Phenotypic expression under the ELG CAT GLU entries).*

Glycine and GABA are principal agonists for inhibitory receptor–channels

GABA- and glycine-gated chloride channels

04-01-21: In the spinal cord, most **post-synaptic inhibition** is mediated by glycine *(see ELG Cl GLY, entry 11)* whereas the vast majority of post-synaptic inhibition in the rest of the brain uses GABA (gamma-aminobutyric acid – *see ELG Cl $GABA_A$, entry 10*). Both amino acids activate integral receptor–channels to increase Cl^- conductance. Selective blockade of different populations of inhibitory response can be induced with strychnine (for ionotropic glycine receptors), picrotoxin (for ionotropic GABA receptors) and bicuculline (for G protein-linked $GABA_B$ receptors).

Feedback

Error-corrections, enhancement and extensions

04-57-01: Please notify specific errors, omissions, updates and comments on this entry by contributing to its **e-mail feedback file** (*for details, see Resource J, Search Criteria & CSN Development*). For this entry, send e-mail messages To: **CSN-04@le.ac.uk,** indicating the appropriate paragraph by entering its **six-figure index number** (xx-yy-zz or other identifier) into the **Subject**: field of the message (e.g. Subject: 08-50-07). Please feedback on only **one specified paragraph or figure per message,** normally by sending a **corrected replacement** according to the guidelines in *Feedback & CSN Access* . Enhancements and extensions can also be suggested by this route (*ibid.*). Notified changes will be indexed via 'hotlinks' from the CSN 'Home' page (http://www.le.ac.uk/csn/) from mid-1996.

Entry support groups and e-mail newsletters

04-57-02: Authors who have expertise in one or more fields of this entry (and are willing to provide editorial or other support for developing its contents) can join its support group: In this case, send a message To: **CSN-04@le.ac.uk,** (entering the words "support group" in the Subject: field). In the message, please indicate principal interests (see *fieldname criteria in the Introduction for coverage*) together with any relevant **http://www site links** (established or proposed) and details of any other possible contributions. In due course, support group members will (optionally) receive **e-mail newsletters** intended to **co-ordinate and develop** the present (text-based) entry/fieldname frameworks into a 'library' of interlinked resources covering ion channel signalling. Other (more general) information of interest to entry contributors may also be sent to the above address for group distribution and feedback.

REFERENCES

For major reviews on the individual types of extracellular ligand-gated channels, *see the Related sources & reviews field in the ELG entries*. A classification of membrane receptor classes (including transmitter-gated ion channels) based on structural and functional criteria has appeared in ref.[7]. Review sources on the ionotropic† glutamate receptors which have been quoted above include refs[13, 16, 17, 19].

[1] Valera, *Nature* (1994) **371**: 516–19.
[2] Brake, *Nature* (1994) **371**: 519–23.
[3] Swope, *FASEB J* (1992) **6**: 2514–23.
[4] Lipton, *Trends Neurosci* (1989) **12**: 265–70.
[5] Collingridge, *Trends Pharmacol Sci* (1990) **11**: 290–6.
[6] Bliss, *Nature* (1993) **361**: 31–9.
[7] Barnard, *Trends Biochem Sci* (1992) **17**: 368–74.
[8] Dingledine, *FASEB J* (1990) **4**: 2636–45.
[9] Langosch, *Proc Natl Acad Sci USA* (1988) **85**: 7394–8.
[10] Unwin, *J Mol Biol* (1993) **229**: 1101–24.
[11] Huganir, *Crit Rev Biochem Mol Biol* (1989) **24**: 183–215.
[12] Huganir, *Neuron* (1990) **5**: 555–67.
[13] Gasic, *Annu Rev Physiol* (1992) **54**: 507–36.
[14] Mayer, *Prog Neurobiol* (1987) **28**: 197–276.
[15] MacDermott, *Trends Neurosci* (1987) **10**: 280–4.
[16] Sommer, *Trends Pharmacol Sci* (1992) **13**: 291–6.
[17] Nakanishi, *Science* (1992) **258**: 597–603.
[18] Olney, *Exp Brain Res* (1971) **14**: 61–76.
[19] Wisden, *Curr Opin Neurobiol* (1993) **3**: 291–8.

NOMENCLATURES

Abstract/general description

05-01-01: **5-Hydroxytryptamine** (**serotonin**, 5-HT) is a biogenic amine that functions as a neurotransmitter†, a mitogen† and a hormone. The large number of biological functions for 5-HT is reflected by the existence of an extraordinarily large family of receptor subtypes for 5-HT *(see Subtype classifications, 05-06)*. The great majority of 5-HT receptors couple to separate effector molecules through G proteins *(see Receptor/transducer interactions, 05-49 and Resource A – G protein-linked receptors, entry 56)*. However, the receptor subtype **5-HT$_3$** (for convenience designated as 5-HT$_3$R within this entry) has a rapidly activating, **5-HT ligand-gated, non-selective cation channel** integral to its primary structure.

05-01-02: The genes encoding 5-HT$_3$ receptor–channels form part of the extracellular ligand-gated channel gene superfamily *(see ELG Key facts, entry 04)* and functional 5-HT$_3$ receptors display a number of molecular, pharmacological and physiological similarities to other superfamily members. Notably, 5-HT$_3$ receptor genes *presently* show much less structural diversity than those encoding other types of ELG channel, although RNA splice variants of a single 5-HT$_3$R gene have been characterized. 5-HT$_3$ receptors in different species and preparations often show great variation in electrophysiological and pharmacological properties.

05-01-03: In addition to expression on native central and peripheral neurones, 5-HT$_3$ receptors are also expressed at high density in several neurone-derived clonal cell lines. In native tissues, post-synaptic 5-HT$_3$ receptors display brief, excitatory post-synaptic currents† in response to synaptically released 5-HT. Pre-synaptic 5-HT$_3$ receptors **mediate the release of several neurotransmitters**. 5-HT$_3$ receptors also have important roles in **pain reception, cognition, cranial motor neurone activity, sensory processing** and **modulation of affect**. 5-HT$_3$R-selective antagonists are used clinically as **anti-emetic agents**.

05-01-04: Like other ELG receptor–channels, the 5-HT$_3$Rs display fast desensitization kinetics in the continued presence of agonists, and show co-operative interactions between ligand-binding sites. Under physiological conditions, the 5-HT$_3$R ion channel is **equally permeable to Na$^+$ and K$^+$** ($E_{5\text{-HT}} \approx 0$ mV). 5-HT$_3$Rs in some cell types are permeable to Ca^{2+} ions.

Category (sortcode)

05-02-01: ELG CAT 5-HT$_3$, i.e. extracellular ligand-gated cation channels activated by 5-hydroxytryptamine. The term '5-HT$_3$ receptor' was originally introduced to maintain consistency with the 5-HT$_1$ and 5-HT$_2$ G protein-coupled receptors defined initially by ligand binding. The suggested **electronic retrieval code** (unique embedded identifier or **UEI**) for 'tagging' of new articles of relevance to the contents of this entry is UEI: 5HT3-NAT (for reports or reviews on native† channel properties) and UEI: 5HT3-HET (for reports or reviews on channel properties applicable to heterologously† expressed recombinant† subunits encoded by cDNAs† or genes†). *For a discussion of the advantages of UEIs and guidelines on their implementation, see the section on Resource J under Introduction & layout, entry 02, and for further details, see Resource J – Search criteria &CSN development, entry 65.*

Channel designation

05-03-01: 5-HT_3R; Gaddum's 'M' receptor; 5-HT-M receptors, as originally designated by Gaddum and Picarelli[1] as mediating an excitatory (contractile) response in guinea-pig ileum via enteric neuronal receptors, as inhibited by Morphine.

Designation of cloned splice variants
05-03-02: Properties according to the first cloned subtype[2] are designated within this entry by the prefix 5-HT_3R-A_L. The 'L' subscript indicates a longer form of the two presently known splice variants of the same 5-HT_3R gene *(see Gene organization, 05-20)*. The prefix 5-HT_3R-A_S indicates data collated for the putative **short-splice form**[3], which lacks a short stretch of amino acids (GSDLLP) in the M3–M4 intracellular loop *(see [PDTM], Fig. 1). Note:* Fieldnames without subtype prefixes indicate data applicable to native 5-HT_3R isoforms expressed in specified cell types.

Current designation

05-04-01: Usually of the form $I_{agonist}$, i.e. $I_{5\text{-}HT3}$ or $I_{(5\text{-}HT3)}$.

Gene family

5-HT_3R genes encode proteins sharing basic properties with other ELG receptor–channels.
05-05-01: The prototype gene encoding the 5-HT_3 receptor–channel forms part of the extracellular ligand-gated gene superfamily *(see ELG Key facts, entry 04)*. The 5-HT_3 receptor incorporates a rapidly responding, non-selective cation channel that displays a number of molecular, pharmacological and physiological similarities to the nicotinic acetylcholine receptor. These similarities extend to their predicted **domain arrangements**, approximately **equal permeability to Na^+ and K^+ ions**, blockade by (+)-tubocurarine and **rapid desensitization kinetics**† *(cf. relevant fields of ELG CAT nAChR, entry 09)*.

Subtype classifications

Relationship of 5-HT_3 receptor–channels to other (G protein-linked) 5-HT receptors
05-06-01: Based on pharmacological ('operational')[4] and functional ('transductional')[4] properties, the large number of known **serotonin receptors** have been placed into seven major classes in interim classifications: 5-HT_1, 5-HT_2, 5-HT_3, 5-HT_4, 5-ht_5, 5-ht_6, 5-ht_7 *(see also refs*[5,6]*, Receptor/transducer*

interactions, 05-49 and Resource A – G protein-linked receptors, entry 56). The 5-HT_1 class (five receptor subtypes), 5-HT_2 class (three receptor subtypes) and 5-HT_4 class consist of G protein-linked receptors *(see Resource A, entry 56)* while only 5-HT_3 subtypes mediate rapid excitatory responses in neurones through an integral ion channel[2]. *Note:* For developments in 5-HT receptor nomenclature, refer to the latest IUPHAR Nomenclature Committee recommendations via the CSN *(see Feedback & CSN access, entry 12).*

05-06-02: Variabilities in measured channel conductances, pharmacological selectivities and other properties[7,8] suggest the existence of multiple subtypes of the 5-HT_3 receptor, although relatively few subunit variants have been cloned to date *(see also Single-channel data, 05-41, and Receptor antagonists, 05-51).*

Trivial names

05-07-01: The serotonin-gated receptor–channel; the 5-HT-activated receptor–channel.

EXPRESSION

Cell-type expression index

05-08-01: 5-HT_3 receptors are widely distributed in central and peripheral neurones. For example, virtually all mammalian **autonomic neurones** possess 5-HT_3 receptors, and these may be co-expressed with other ELG channel types. The information in this entry largely refers to well-characterized neuronal types, including:

olfactory bulb
dorsal root ganglia
amygdala
multiple nuclei of trigeminal nerve spinal tract
hypothalamus
brainstem motor neurones
superior cervical ganglion cells
PC12 phaeochromocytoma cells
nodose ganglion neurones
N1E-115 neuroblastoma cells *(see below)*
N18 neuroblastoma cells
NG108-15 hybrid neurones *(see below)*
submucous plexus neurones
coeliac ganglion neurones

Availability of neuronal cell lines expressing 5-HT_3R

05-08-02: In addition to expression on native central and peripheral neurones, 5-HT_3 receptors are also expressed at high density in several neurone-derived clonal cell lines[9], e.g. N1E-115 neuroblastoma cells[10,11], NCB-20 cells[11] *(see Cloning resource, 05-10),* NG108-15 neuronal hybridoma cells[12] and N18 cells[13].

05-08-03: 5-HT_3 receptor-mediated responses have generally shown marked **interspecies differences** *(see Receptor antagonists, 05-51, and Receptor agonists, 05-50).*

Cloning resource

05-10-01: 5-HT_3R-A_L: Isolate 5-HT_3R-A (now designated 5-HT_3R-A_L, *see C1 Channel designation, 05-03*) was **expression-cloned**† in *Xenopus* oocytes from an NCB-20 mouse hybrid (neuroblastoma/Chinese hamster embryonic brain cell) cDNA library[2]. These cells were selected for their ability to express large, robust 5-HT_3 receptor-mediated responses.

05-10-02: 5-HT_3R-A_S: RT-PCR† has been used to isolate a 'short-splice' variant[85] of the 5-HT_3R from mRNA of the rodent neuroblastoma line N1E-115[3] *(see Channel designation, 05-03).* Rat superior cervical ganglion cDNA libraries have been used as resources for isolation of clones encoding 5-HT_3R *(see Database listings, 05-53). Note:* On the basis of many conserved electrophysiological and pharmacological properties, it would be expected that other cell lines expressing 5-HT_3R *(see Cell-type expression index, 05-08)* could also serve as cloning resources.

Developmental regulation

Induction of 5-HT_3R expression by nerve growth factor and cAMP derivatives

05-11-01: Application of **nerve growth factor** (NGF) or 8-bromo-cyclic-adenosine monophosphate (**8-Br-cAMP**) to PC12 cells for long periods (~10 days) expresses 5-HT_3-type receptors, which are of low abundance in untreated cells[14].

05-11-02: The differentiation status of neuronal hybridoma NG108-15 cells alters channel conductance and desensitization properties of the 5-HT_3R[15] *(see Table 2).*

Isolation probe

05-12-01: 5-HT_3R-A_L: As no molecular probe was available, the 5-HT_3R-A_L isoform was isolated[2] using **expression-cloning**† techniques in *Xenopus* oocytes by microinjection of cRNA pools from an NCB-20 cell cDNA expression *(see Cloning resource, 05-10).*

mRNA distribution

mRNA distribution within brain

05-13-01: Using radiolabelled probes specific for the 5-HT_3R-A_L isoform, labelled cells are observed throughout the cortical regions (e.g. piriform, cingulate and entorhinal areas). Distinct labelling is seen in the postventral hippocampus (within the lacunosum molecular layer of CA1, containing inhibitory neurones). These studies[16] also showed 5-HT_3 mRNA to be present in the olfactory bulb, dorsal root ganglia, amygdala and multiple nuclei of trigeminal nerve spinal tract, hypothalamus and brainstem motor neurones.

Implications of brain mRNA distribution patterns
05-13-02: Distributions of 5-HT$_3$ mRNA studied in mouse brain by *in situ* hybridization techniques[16] are consistent with the roles for the 5-HT$_3$ receptor in **cognition**, **cranial motor neurone activity**, **sensory processing** and **modulation of affect**.

Distribution and apparent bias for expression of mRNA splice variants
05-13-03: 5-HT$_3$R-A$_L$/5-HT$_3$R-A$_S$: Cloned isolate 5-HT$_3$R-A$_L$ mRNA is distributed in brain, spinal cord and heart tissue[2]. Approximately 80% of transcripts from mRNA populations derived from NG108 cells and native cortex/brainstem[17] encode the 'short form' of the 5-HT$_3$ receptor.

Phenotypic expression

Distinct roles of pre- and post-synaptic 5-HT$_3$ receptor–channels
05-14-01: Endogenous release of **serotonin** (for example in the lateral amygdala) has been shown to mediate rapid excitatory post-synaptic currents† in response to synaptically released 5-HT. Characteristically, **synaptic potentials are of brief duration** (tens of milliseconds – *see Activation, 05-33, and Inactivation, 05-37)* and can be mimicked by 5-HT, potentiated by a 5-HT uptake inhibitor, and blocked by selective 5-HT$_3$ receptor antagonists. The **anti-emetic, anxiolytic**, and possibly **anti-psychotic** actions of 5-HT$_3$ antagonists might result from blockade of such synapses[18].

05-14-02: Pre-synaptic 5-HT$_3$ channels mediate **neurotransmitter release** in the CNS. Some examples of known stimulatory, facilitatory or inhibitory effects of endogenous transmitter release involving 5-HT$_3$R subtypes are summarized in *Table 1*.

5-HT$_3$ receptor–channels also mediate neurotransmitter release in the periphery
05-14-03: At locations in the **periphery**† (e.g. **enteric**, **sympathetic** and **para-sympathetic autonomic** and **primary sensory neurones**) 5-HT$_3$-mediated excitation may evoke **neurotransmitter release**[1, 33] and play direct roles in initiation of **intestinal muscular contraction**. Because of its diverse roles in mediating brain function and behaviour, defects in serotonergic signalling systems may contribute to the aetiology of **pain reception** in sensory nerve fibres[34], migraine, depressive conditions and obsessive–compulsive behaviours.

Tachycardic responses due to 5-HT$_3$R-linked transmitter release
05-14-04: 5-HT$_3$ receptor stimulation resulting in **tachycardia**† and **inotropic**† actions is mediated by noradrenaline release from the postganglionic **cardiac**

sympathetic nerves. This response is mimicked by 5-HT agonists and in some cases can be blocked by autonomic antagonists (e.g. propranolol) and 5-HT_3 antagonists (e.g. MDL72222 or ICS205930) *(see Receptor antagonists, 05-51).*

Table 1. *Examples of effects of endogenous transmitter release involving 5-HT_3R subtypes (From 05-14-02)*

5-HT_3 receptors involved in	Example location(s)	Refs
Release of **dopamine**	Rat striatum and nucleus accumbens	19–21
Release of **cholecystokinin**	Cerebral cortex and nucleus acumbens	22
Release of **noradrenaline**	Hippocampus	23
Release of **GABA**	Hippocampus	24
Release of **5-HT**	Frontal cortex and hippocampus	25, 26
Depression of evoked release of **noradrenaline**	Rat hypothalamic slices	27, 28
Depression of evoked release of **acetylcholine**	Cerebral cortex or synaptosomes from cerebral cortex	29–31 but see 32

Reflex bradycardic responses due to 5-HT_3R-linked transmitter release

05-14-05: In the cardiovascular system, the main initial response to a bolus of 5-HT is a short intense **bradycardia**† and **hypotension**, mediated via a Bezold–Jarisch-like reflex†, initiated by stimulation of 5-HT_3 receptors present on afferent vagal nerve endings. Intravenous, intracoronary or local epicardial administration of 5-HT, 1-phenylbiguanide or 2-methyl-5-HT elicits the transient bradycardia, which is effectively antagonized by 5-HT_3-selective blockers[35, 36]. *(See also Receptor/transducer interactions under INR K G/ ACh, 31–49.)*

Protein distribution

5-HT_3 receptors are ubiquitous in the CNS and PNS

05-15-01: On the basis of functional and radioligand-binding studies, 5-HT_3 receptor proteins appear to be **ubiquitously distributed** throughout the **peripheral and central nervous systems**. In central nervous tissue, autoradiographic mapping of 5-HT_3 shows expression associated with cortical, limbic and brainstem structures. In the periphery, 5-HT_3 receptors are found on enteric, autonomic (sympathetic and parasympathetic) and **primary sensory neurones**. Functional 5-HT_3-binding sites have been found in several mammalian cell lines, including neuroblastomas *(see Cell-type* *expression index, 05-08).* A full description of receptor distribution patterns has appeared in a book dedicated to central and peripheral 5-HT_3R *(see Laporte, in Hamon, 1982, pp. 157–87, under Related sources & reviews, 05-56).*

05-15-02: The **anti-emetic† properties** of 5-HT_3 antagonists *(see Receptor antagonists, 05-51)* may correlate with blockade of receptors in the area postrema and solitary tract nucleus of the brainstem, areas which display high densities of the 5-HT_3 subtype. Blockade of peripheral receptors present on **vagal afferents**† may also play a role.

Subcellular locations

5-HT_3 receptors are located on both pre- and post-synaptic membranes

05-16-01: Both pre-synaptic and post-synaptic 5-HT_3 receptors have been studied in brain slice preparations, e.g. those containing the nucleus tractus solitarius (NTS)[37]. Generally, the pre-synaptic 5-HT_3 channels regulate **neurotransmitter release**, while the post-synaptic channels regulate **ionotropic neurotransmission** *(see Phenotypic expression, 05-14, and Receptor/transducer interactions, 05-49).*

SEQUENCE ANALYSES

Encoding

05-19-01: Open reading frame† sizes for cloned subunit cDNAs are listed under *Database listings/primary sequence discussion, 05-53.*

Gene organization

A putative splice variant of the 5-HT_3R carrying a six amino acid deletion

05-20-01: 5-HT_3R-A_S: A shorter form of the prototypic 5-HT_3R has been isolated from the neuroblastoma cell line N1E-115 using reverse transcriptase polymerase chain reaction (RT-PCR†) techniques[3]. The 5-HT_3R-A_L and 5-HT_3R-A_S forms share ~98% amino acid sequence identity, but a contiguous **six amino acid segment (GSDLLP) is deleted** in 5-HT_3R-A_S *(Fig. 1).* The high degree of identity and the fact that both variants can be detected in the N1E-115, NCB-20 and NG108-15 hybridoma cells[17] indicate the transcripts arise as alternative RNA transcripts from a single gene[85] *(see also mRNA distribution, 05-13, and Receptor agonists, 05-50).*

Homologous isoforms

05-21-01: A 461 amino acid 5-HT_3R variant[38] isolated from a rat superior cervical ganglion cDNA library shows ~95% sequence homology with the mouse 5-HT_3R-A_S, and represents a species homologue† of the same splice variant.

Protein molecular weight (purified)

Purified 5-HT$_3$R complexes display molecular masses typical of ELG channels

05-22-01: Polyacrylamide gel electrophoresis of purified 5-HT$_3$ receptor from N1E-115 neuroblastoma cells reveals a **single protein band** of 54.7 ± 1.3 kDa[39]. The oligomeric form of the 5-HT$_3$ receptor solubilized from NG108-15 cells has been estimated as ~600 kDa by gel filtration[40]. This value is consistent with the range reported for the 5-HT$_3$R complex of rabbit small bowel (443–669 kDa)[41] although receptor/detergent complexes and sucrose density gradient techniques estimate an M_r of ~300 kDa[40, 42–44]. When the effects of detergent binding are taken into account, a value of 249 kDa (more typical of other members of the ELG channel family) can be derived[43].

Evidence for heterogeneous molecular sizes of 5-HR$_3$R

05-22-02: On the basis of SDS–PAGE† analysis, a number of studies suggest **heterogeneity** of receptor subunit sizes making up the 5-HT$_3$R complex (typically in the ranges ~50–54 kDa and ~36–38 kDa). The smaller component size matches that of a site labelled by [^{3}H]-zacopride in NG108-15[40] and native cerebral cortex of rat[45].

Protein molecular weight (calc.)

05-23-01: 5-HT$_3$R-A$_L$/5-HT$_3$R-A$_S$: The mouse NCB-20 cell line receptor cDNA sequence[2] predicts a molecular weight of the protein of 55 966 Da including the signal† peptide, or 53 509 Da as a mature protein. The cDNA sequence of the short-splice form isolated from the mouse neuroblastoma cell line N1E-115 predicts a molecular weight of 53 178 including the signal peptide[3].

Sequence motifs

05-24-01: 5-HT$_3$R-A: Isolate 5-HT$_3$R-A[2] displays a signal peptide cleavage motif between residues 23 and 24 of the precursor protein† chain. Motifs for *N*-glycosylation are present at extracellular amino acids 108, 174 and 190 (positions 86, 152 and 168 in the mature 5-HT$_3$-R-A$_S$ protein) *([PDTM], Fig. 1)*. A disulphide bond motif predicts the formation of an S–S link between amino acids 161 and 175 (positions 139 and 153 in the mature protein) *(for relative positions, see [PDTM], Fig. 1)*.

STRUCTURE & FUNCTIONS

In-press updates *(see criteria under Introduction & layout of entries, entry 02): See note above field 07-26 for important new information relating to ionotropic+glutamate receptor structures.*

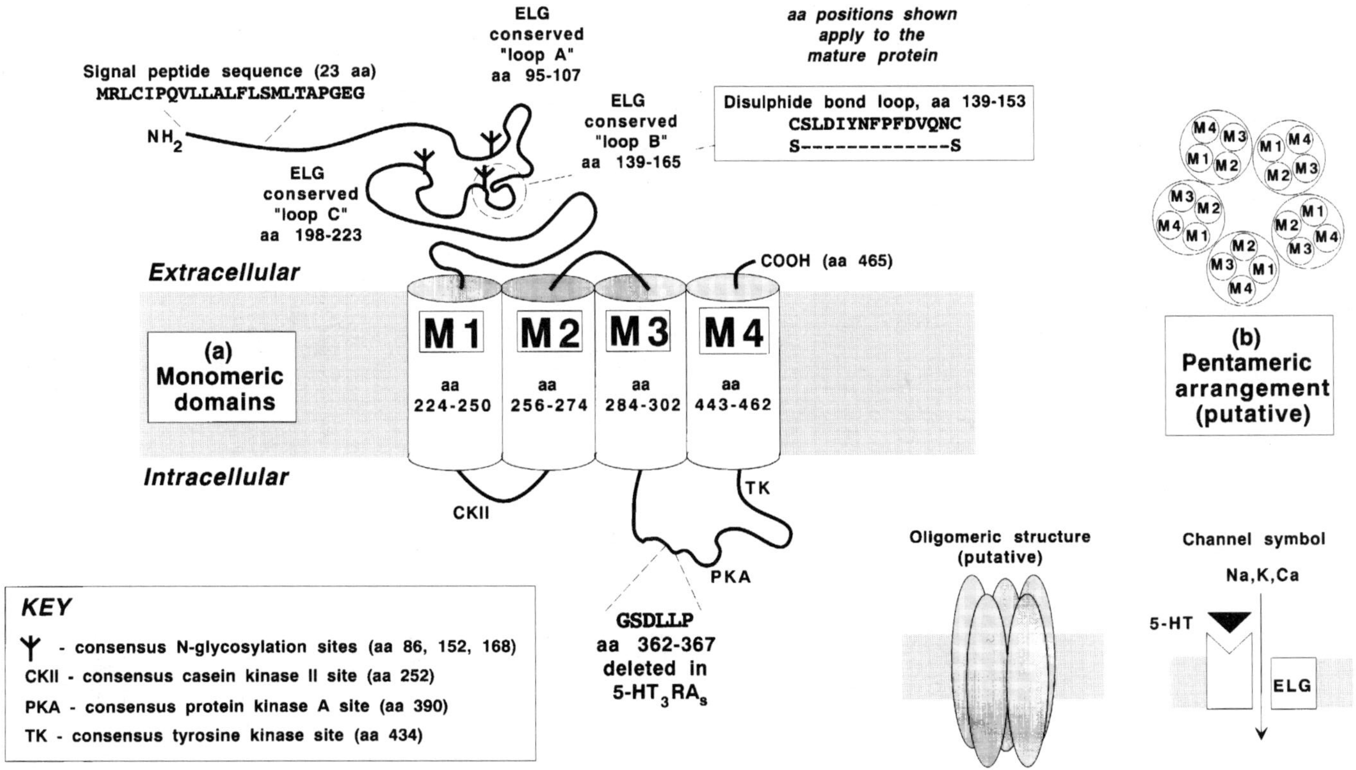

Figure 1. *Monomeric protein domain topography model for the murine 5-hydroxytryptamine-gated receptor–channel (5-HT$_3$R-A) (long splice form).* Note: *All relative positions of motifs, domain shapes and sizes are diagrammatic and are subject to re-interpretation. (From 05-26-01)*

Amino acid composition

05-26-01: 5-HT_3R-A: Hydrophobicity† analysis of the prototypic isolate 5-HT_3R-A[2] reveals a pattern of hydrophobic domains (M1–M4) typical of the ligand-gated ion channel superfamily *([PDTM], Fig. 1) (compare other [PDTMs] of ELG entries).*

Domain arrangement

Structural homologies with other ELG channels suggest pentameric subunit assemblies

05-27-01: In comparison to the homologous domain structure of the nicotinic receptors *(see ELG CAT nAChR, entry 09)*, it is predicted that the intact receptor may be assembled from five monomer subunits (*see [PDTM], Fig. 1*), with the α-helix† of each M2 region lining the channel pore[44] *(see Selectivity, 05-40).*

Many properties of native 5-HT_3R are reproduced on expression of homomultimeric 5-HT_3R-A

05-27-02: Notably, the 5-HT_3R-A homomeric† isolate[2] displays many of the physiological and pharmacological properties of native 5-HT_3 receptors *in vivo* (e.g. in cell lines used as the mRNA sources for cDNA library construction). By comparison with other members of the superfamily, different subunit compositions of 5-HT_3 may be expected *(see other ELG entries)*, but a comparable subunit diversity has not been found (to the end of 1993 – see updates on this entry, accessible via the 'home page' of the CSN, as described in *Feedback & CSN access, entry 12* and Resource J – Search criteria & CSN development, entry 65).

Domain conservation

Amino acid sequence identities with other ELG receptor–channels

05-28-01: 5-HT_3R-A_L: The deduced amino acid sequence of the 5-HT_3R-A isolate[2] *(see Database listings/primary sequence discussion, 05-53)* shows 30% sequence identity with the α subunit of chick brain nAChR *(see ELG CAT nAChR, entry 09)* and both consist of four putative membrane domains. Sequence conservation is greatest in the N-terminal (nAChR ligand-binding) and M2 (nAChR pore-lining) domains. The 5-HT_3R-A isoform shares ~22% amino acid identity with $GABA_A$ β_1 subunit *(see ELG Cl $GABA_A$, entry 10)* and the 48 kDa subunit of glycine receptor–channels *(see ELG Cl GLY, entry 11).*

5-HT_3R residues conserved at equivalent positions in other ELG gene family members

05-28-02: 5-HT_3R-A_L: The relative positions of two **cysteine residues** are conserved in the 5-HT_3R, $GABA_A$R and nAChR. These residues are likely to form part of a small **disulphide bond loop** which may be important for the tertiary† folding of the protein *(see [PDTM], Fig. 1)*. A hydrophobic leucine residue (Leu286) within the M2 domain of the 5-HT_3R-A isolate[2] is conserved within all nicotinic, glycine and $GABA_A$ receptor proteins[46].

Evidence for subunit heterogeneity amongst native 5-HT$_3$ receptor complexes

05-28-03: Native 5-HT$_3$ complexes vary greatly in their single-channel conductance *(see Single-channel data, 05-41)*, suggesting that heterogeneity exists among their subunit assemblies *(see also Domain functions, 05-29)*.

Domain functions (predicted)

Domain functions likely to be common across ELG superfamily members

05-29-01: As shown in the [PDTM] *(Fig. 1)*, a typical 5-HT$_3$R monomeric subunit consists of four hydrophobic transmembrane regions (M1–M4), a long cytoplasmic loop between M3 and M4, and a relatively large N-terminal domain. By analogy with other ELG channels, the N-terminal domain probably forms part of the **agonist-binding site** and the site for some antagonists, while the M3–M4 loop contains potential sites for **channel regulation** with the M2 region lining the **channel pore**.

Direct evidence for structural and functional division of receptor and channel domains

05-29-02: Chimaeric receptor molecules, consisting of the N-terminal domain of the nicotinic acetylcholine receptor subunit α7 *(see ELG CAT nAChR, entry 09)* and the putative transmembrane and C-terminal regions of 5-HT$_3$R-A have been constructed[47]. Chimaeras display *receptor* properties characteristic of the nAChR and *channel* properties similar to those of the 5-HT$_3$R when expressed as homomultimers in *Xenopus* oocytes[47]. These studies represent direct evidence for 5-HT **agonist-binding sites** being formed by assemblies comprising parts of the hydrophilic N-terminal domain.

Single amino acid substitutions affecting desensitization properties of the 5-HT$_3$R

05-29-03: Recombinant 5-HT$_3$A receptor–channels expressed in *Xenopus* oocytes (at −60 mV holding potential) conduct an inward current which declines during the continued presence of 5-HT with a half-time of about 2 s; this **desensitization**† is ~20 times slower in calcium-free solution[48]. Desensitization is markedly faster in channels when Leu286 (near the middle of the M2 segment) is changed to Phe, Tyr or Ala. Desensitization is slower with Thr. Changes at equivalent positions of the nicotinic acetylcholine receptor have similar effects on desensitization, suggesting that the underlying protein **conformational change** might be a common feature of ligand-gated channels[48].

Chemical modification of extracellular tryptophan residues affects agonist/antagonist binding

05-29-04: Pre-treatment of NG108-15 membranes with ***N*-bromosuccinimide** (a selective oxidant of tryptophan residues) greatly reduces the density of sites labelled with the 5-HT$_3$ receptor antagonist **[^{3}H]-zacopride**[49]. The effects of *N*-bromosuccinimide modification can be prevented by pre-incubation or co-incubation with some 5-HT_3 receptor agonists or antagonists. Other agents which modify aspartate, cysteine, cystine and glutamate residues have little effect on ligand binding, while those selective for arginine, histidine and tyrosine residues have only moderate inhibitory effects on [^{3}H]-zacopride labelling[49].

Predicted protein topography

Predicted multimeric assembly patterns

05-30-01: In common with other members of the extracellular ligand-gated cation channel superfamily†, the 5-HT_3 receptor–channel is likely to be formed from five monomers that contain from two to four homologous subunits[50] *(see [PDTM], Fig. 1)*. Direct evidence for heteromultimeric assemblies has not been reported to date. Electron microscopic examination of 5-HT_3 receptor channels reveals '**rosette structures**' of 8–9 nm diameter with stain-filled central regions of ~2 nm diameter[44]. This 'rosette' arrangement is similar to that observed in similar studies of the *Torpedo* nAChR receptor.

Ion-selective pore size predicted by modelling studies

05-30-02: Modelling studies have suggested that 5-HT_3-gated receptor–channels are large water-filled pores, structurally analogous to the nicotinic acetylcholine receptor of the neuromuscular junction *(see ELG CAT nAChR, entry 09)*. Modelling the channel as a simple cylinder[51] suggests a minimum **pore size** of 7.6 Å, which compares to estimates of 7.4 Å for the nicotinic receptor using similar methods.

Protein interactions

05-31-01: 5-HT_3R-A: RNA transcripts synthesized *in vitro* from the 5-HT_3R-A cDNA are sufficient for functional expression of 5-HT-gated ion channels with similar characteristics to endogenous (native†) channels[2].

Protein phosphorylation

05-32-01: Motifs for phosphorylation by protein kinase A, tyrosine kinase and casein kinase II are present in the amino acid sequences of cloned 5-HT_3R subunits[2,3]; the relative positions of these sites are illustrated in the [PDTM] *(Fig. 1)*.

ELECTROPHYSIOLOGY

Activation

5-HT_3 responses display rapid activation and desensitization kinetics

05-33-01: The response of intact cells to 5-HT has a very short **latency**†, typical of an extracellular ligand-gated receptor–channel[8,11,52] *Methodological note:* Substantial receptor desensitization† behaviour *(for details, see Inactivation, 05-37)* necessitates rapid methods of agonist application if 5-HT concentration–effect relationships are to be determined accurately.

Current type

05-34-01: 5-HT_3R-A_L: The 5-HT_3R-A-type ion channels[2] display inward currents which have a divalent cation-mediated negative slope conductance in *Xenopus* oocytes. This pattern is similar to that described for the NMDA subtype of glutamate receptor *(see ELG CAT GLU NMDA, entry 08).*

Current–voltage relation

05-35-01: Most studies of cloned and native† 5-HT_3R *(reviewed in ref.*[53]*)* show current responses to 5-HT reverse polarity at potentials close to 0 mV, yielding outward current at positive holding potentials and inward current at negative holding potentials *(e.g. see ref.*[18]*).* In a number of independent studies on native and immortalized† neuronal cells, the peak single channel I–V relation for the 5-HT_3 receptor shows modest **inward rectification** that is fully developed within <2 ms of the applied voltage step *(e.g. see refs*[9, 11–13, 54, 55]*).*

Dose–response

Evidence for co-operative interaction of agonist-binding sites

05-36-01: A voltage-clamp analysis of the concentration-effect relationship in N1E-115 cells (using rapid applications of 5-HT) has demonstrated that maximal and half-maximal inward currents are evoked by 10 μM and 2 μM 5-HT respectively. The steepness of the dose–response curve (slope = 2.8) indicates **co-operative interaction** between 5-HT-binding sites, requiring at least two agonist molecules *(see ref.*[53]*).*

05-36-02: In ligand-binding assays (where 5-HT_3 recognition sites in NIE-115 neuroblastoma cells are labelled with radiolabelled antagonists), competition curves obtained with the agonists 5-HT and 2-methyl-5-HT have also shown slopes greater than unity, further indicating **co-operativity** within the 5-HT_3 receptor. Patterns of co-operativity are also characteristic of heterologously expressed homo-oligomeric 5-HT_3 receptors[2, 3, 48].

Inactivation

5-HT_3 receptors desensitize in continued presence of agonist

05-37-01: Synaptic depolarizations evoked by iontophoretic† applications of 5-HT show **rapid desensitization**† behaviour when the agonist is continually applied for longer than ~100 ms[8, 18, 56–58]. Furthermore, both synaptic depolarizations and transient responses to **iontophoresis**† of 5-HT are blocked when agonists are added to superfusing solutions[8, 18, 56–58]. The differentiation status of neuronal hybridoma NG108-15 cells alters the desensitization kinetics of their endogenous 5-HT_3R[15]. For a description of amino acid substitutions influencing the rate of development of desensitization, *see Domain conservation, 05-28.*

Pharmacological modifiers of receptor desensitization kinetics

05-37-02: Co-application of 10^{-3}–10^{-2} M **tetraethylammonium ions** (TEA^+) with 5-HT is capable of preventing desensitization in voltage-clamped N1E-

115 neuroblastoma cells, but is incapable of reversing desensitization once it has become established[59]. **5-Hydroxyindole** also slows desensitization of the 5-HT_3 receptor in these cells[60] and can overcome 5-HT receptor desensitization in the continued application of agonist.

A radiolabelled agonist that binds to desensitized conformations of the 5-HT_3R

05-37-03: The binding characteristics of a radiolabelled 5-HT_3 receptor agonist, **[^{3}H]-*meta*-chlorophenylbiguanide** (***m*CPBG**) have revealed two populations of binding sites in membranes of N1E-115 neuroblastoma cells, with $K_d = 0.03 \pm 0.01$ nM and 4.4 ± 1.2 nM and $B_{max} = 11.9 \pm 4.2$ and 897.9 ± 184.7 fmol/mg protein respectively[61]. Competition data suggest that [^{3}H]-*m*CPBG labels high-affinity desensitized states of the receptor.

Kinetic model

05-38-01: A number of studies *(e.g. refs*[10, 52, 62]*)*, have shown that the kinetic properties of the 5-HT_3 receptor-mediated ionic current can only be described by a complex, **co-operative model** *(see Dose–response, 05-36)*.

Rundown

05-39-01: 5-HT_3 receptor–channels in neurons of guinea-pig submucous plexus[8] exhibit high stability. Channel activities are reproducibly evoked by repeated applications of 5-HT even up to 5 h following excision of outside-out† patches, suggesting that neither a G protein nor a diffusible cytoplasmic messenger is necessary for their gating or modulation, *(see also Inactivation, 05-37)*.

Selectivity

5-HT_3 receptors are predominantly Na^+/K^+-selective under physiological conditions

05-40-01: The observed $E_{5\text{-HT}}$ values close to 0 mV *(see Current–voltage relation, 05-35)* are consistent with opening of non-selective cation-permeable channels (with Na^+/K^+ at approximately equal permeability). Under physiological conditions, responses to 5-HT would be predominantly carried by inward movement of Na^+ ions. Selectivity characteristics of native channels are retained following homomeric expression of cDNAs encoding 5-HT_3R and follow the *weak* selectivity sequence amongst monovalent ions of $Cs^+ > K^+ > Li^+ > Na^+ > Rb^+$.

5-HT_3 receptors are Ca^{2+}-permeable in some cell types

05-40-02: In addition to providing a depolarizing stimulus to activate voltage-gated Ca^{2+} channels *(see VLG Ca, entry 42)*, 5-HT_3 receptor stimulation might result in Ca^{2+}-influx through the 5-HT_3 receptor itself. The 5-HT_3 receptor–channel conducts Ca^{2+} and other divalent cations in some cell types studied (e.g. for superior cervical ganglion cells[13] P_{Ca}/P_{Na} is ~0.55, while in N18 neuroblastoma cells[55], P_{Ca}/P_{Na} is ~1.12). The N18 cell 5-HT_3 receptor–channel is equipermeable to Ca^{2+}, Mg^{2+} and Ba^{2+}[55]. By using organic cationic molecules as probes, it has been shown that permeability is inversely related to the geometric **mean diameter** of the permeant molecules (reviewed in ref.[53]) *(see also Blockers, 05-43).*

Ionic permeability ratios and pore size

05-40-03: Ion-substitution experiments in a variety of preparations (superior cervical ganglion cells, N1E-115 neuroblastoma cells, nodose ganglion neurones and N18 neuroblastoma cells) confirm that the 5-HT_3-gated channel has poor discrimination between monovalent metal cations. For example, in PC12 cells, currents induced with nerve growth factor (NGF) or 8-bromo-cAMP (8-Br-cAMP) display a permeability ratio for monovalent cations of $Na^+:Li^+:K^+:Rb^+:Cs^+ = 1:1.19:0.89:0.94:0.91$[14]. Modelling studies suggest a minimum pore size of 7.6 Å for 5-HT_3 channels *(see Predicted protein topography, 05-30).*

Single-channel data

05-41-01: There is considerable variation in reported conductances for 5-HT_3 receptor–channels in different cell types *(see Table 2)*. Recordings have been made under a number of different configurations and the single-channel properties are complex *(reviewed in ref.*[53]*)*. Alternatively, the variable estimates of channel conductances might suggest the existence of several 5-HT_3 molecular subtypes and/or developmentally regulated isoforms (as in NG108-15 hybrid neurones – *see Development regulation, 05-11*).

Table 2. *Variability in reported channel conductances for 5-HT_3-gated receptor–channels (From 05-41-01)*

Cell type/recording method	Estimated channel conductance (pS)	Refs
Coelic ganglion neurones, guinea-pig: WCR/SCR	10	63
Neuroblastoma (N18); WCR/FA	0.59	55
Neuroblastoma (N1E-115); WCR/FA	0.31	11
Neuronal hybridoma NG108-15, differentiated cells; WCR/FA	3.6–4.4	15
Neuronal hybridoma NG108-15, undifferentiated cells; FA/SCR	7.2–12.0	15
Nodose ganglion cells, rabbit; WCR/SCR (chord conductance)	16.5	64
Nodose ganglion cells, rabbit; WCR/SCR (slope conductance)	19.3	64
Submucous plexus neurones, guinea-pig; WCR/SCR	15.0/19.2	8
Superior cervical ganglion cells, rat; SCR	11.1	13
Superior cervical ganglion cells, rat; WCR/FA	2.6	13

WCR, whole-cell recording; FA, fluctuation analysis; SCR, single-channel recording.

Considerations of receptor–channel heterogeneity in single cell types
05-41-02: Fluctuation analysis (FA) of whole-cell recorded (WCR) currents in rat superior cervical ganglion neurons yields an estimate for single-channel (SC) conductance that is ~4 times smaller than that observed in direct outside-out membrane patches. One possible explanation of such variability has been suggested[13] to be **heterogeneity** of 5-HT$_3$ receptors in these cells.

05-41-03: Recordings from excised (outside-out) membrane patches from neurones of guinea-pig submucous plexus[8] show two distinct 5-HT unitary currents of conductances ~15 pS and ~9 pS. Channels opened by 5-HT show characteristic fast openings and bursts† of openings, superficially similar to the features of nicotinic channels in other autonomic neurones *(see ELG CAT nAChR, entry 09).*

Voltage sensitivity

05-42-01: The majority of studies of native and heterologously expressed recombinant 5-HT$_3$R currents show moderate **inward rectification**† but the precise mechanism underlying this is unclear *(see Current–voltage relation, 05-35).*

PHARMACOLOGY

Blockers

(+)-Tubocurarine blockade shows marked species differences
05-43-01: 5-HT$_3$-gated channel currents are potently reduced by **(+)-tubocurarine** *(but see also ELG CAT ATP, ELG CAT nAChR, and VLG Ca, entries 06, 09 and 41).* The high sensitivity typical of the cloned 5-HT$_3$R-A isoform to (+)-tubocurarine[2] has been observed in a number of native tissues (nodose ganglion neurones and superior cervical ganglion cells) and clonal cell lines[53]. In general, 5-HT$_3$ receptor-mediated responses show marked **interspecies differences** – for example, the concentration of (+)-tubocurarine required for 50% inhibition can vary by up to ~10 000-fold across nodose ganglion cells of mouse, rabbit and guinea-pig[65]. In keeping with these observations, mouse and rat homologues of the 5-HT$_3$R-A$_S$ (which only have 16 amino acid differences) exhibit a 100-fold difference in (+)-tubocurarine sensitivity[3,32].

Suppression of inward 5-HT-induced current by extracellular cations
05-43-02: Divalent cations such as Ca^{2+} and Mg^{2+} exert a pronounced modulatory influence on 5-HT$_3$-mediated responses. Both native†[51,55] and cloned[2] 5-HT$_3$ receptor–channels show suppression of 5-HT$_3$-elicited inward current in the presence of the divalent cations Ca^{2+} and Mg^{2+} in the extracellular medium at physiological concentrations. Reduction in external Ca^{2+} concentration augments the amplitude of depolarizing responses to 5-HT in NG108-15 hybrid cells, while both Ca^{2+} and Mg^{2+} (at physiological concentrations) modulate amplitude and duration of 5-HT-induced currents *(reviewed in ref.[9]).*

An apparent difference between cloned and native 5-HT$_3$ receptor–channels

05-43-03: The blocking and/or modulatory effects of **extracellular divalent cations** have been shown to be voltage-sensitive (for both cloned[2] and chimaeric 5-HT$_3$ receptors[47]), with Ca^{2+} and Mg^{2+} introducing a region of negative slope conductance into the I–V relationship upon hyperpolarization[2]. For *native* 5-HT$_3$ receptors, divalent cation blockade appears to be voltage-insensitive[54, 55].

Channel modulation

Potentiation of 5-HT$_3$ current responses

05-44-01: Electrical responses elicited by 5-HT$_3$ receptor activation can be potentiated from submaximal responses by **ethanol** and the dissociative anaesthetic **ketamine**. **Trichloroethanol** enhances 5-HT$_3$ current responses through both native[66, 67] and cloned receptor–channels.

Equilibrium dissociation constant

05-45-01: Saturation analyses of 5-HT$_3$ receptors from N1E-115 neuroblastoma cells with [^{3}H]-GR67330 ligand shows high-affinity binding to homogeneous populations of sites in both solubilized ($K_d = 0.05 \pm 0.02$ nM) and purified ($K_d = 0.10 \pm 0.04$ nM) preparations. Competition experiments indicate that the solubilized and purified receptor preparations retain the characteristics observed in N1E-115 cells *in vivo*[39].

Ligands

05-47-01: 5-Hydroxytryptamine (5-HT; serotonin). Synaptic potentials mediated by 5-HT have been recorded from brain slices *in situ*[18]. Fast neurotransmission in mammalian brain can therefore be mediated by amines as well as amino acids *(see ELG CAT GLU, entries 07 and 08)*.

05-47-02: Available radioligands for the 5-HT$_3$ receptor–channel include **[^{3}H]-GR65630**, **[^{3}H]-ICS205930**, **[^{3}H]-zacopride**, **[^{3}H]-GR65630**, **[^{3}H]-quipazine**, **[^{3}H]-granisetron**, **[^{3}H]-LY278584** and **[^{3}H]-*meta*-chlorophenylbiguanide** (**[^{3}H]-*m*CPBG**).

Receptor/transducer interactions

Role of 5-HT$_3$ channels in modulated neurotransmitter release

05-49-01: There is no indication that 5-HT$_3$ receptors modulate adenylate cyclase activity, although there is evidence that they modulate the release of a variety of CNS neurotransmitters including **acetylcholine**[29], **cholecystokinin**[22], **dopamine**[21], **GABA**[24] and **noradrenaline** *(see also Phenotypic expression, 05-14)*. An analysis of relaxant and contractile effects on cerebrovasculature mediated by 5-HT receptors has been reviewed[68]. A '5-HT$_3$-like' receptor which mediates a non-desensitizing inhibition of rat medial pre-frontal cortical neurons *in vivo* and coupling to phospholipase C has been reported *(cited in ref.[53])*.

Comparison of roles for 5-HT at different classes of receptor
05-49-02: Fast amine neurotransmission via 5-HT_3 receptor–channels can be contrasted with the slow *inhibitory* synaptic potentials mediated by 5-HT agonism through G protein-linked signalling pathways. Thus the majority of known receptor molecules for 5-HT couple to effectors (including ion channels) through G proteins. **G Protein-linked 5-HT receptors** include the subtypes 5-HT_{1A}, 5-HT_{1B} (equivalent to 5-$HT_{1D\beta}$), 5-$HT_{1Ç}$ (now 5-HT_{2C}), 5-$HT_{1D\alpha}$, 5-$HT_{1D\beta}$, 5-HT_{1E}, 5-HT_{1F}, 5-HT_2, 5-HT_{2F} (now 5-HT_{2B}), 5-HT_4, 5-$ht_{5\alpha}$, 5-$ht_{5\beta}$, 5-ht_6 and 5-ht_7 *(for effector pathways, see Resource A, entry 56, and refer to the latest IUPHAR Nomenclature Committee recommendations via the CSN – see Feedback and CSN access, entry 12).*

Receptor agonists (selective)

Present 5-HT_3R agonists are sometimes non-selective and have variable efficacy across species
05-50-01: Derivatives of the endogenous neurotransmitter/agonist 5-hydroxytryptamine such as **2-methyl-5-hydroxytryptamine** act as partial agonists at other sites and agonists such as **1-phenylbiguanide** (PBG) and ***meta*-chlorophenylbiguanide** (*m*CPBG) are of variable efficacy in different species[53, 77]. The compound **SR 57227A** has been reported to act as a full and 5-HT_3-selective agonist[79]. Information about its efficacy on a wide range of preparations was not available at the time of compilation.

Receptor antagonists (selective)

Overview of 5-HT_3-selective antagonists
05-51-01: The **'operational definition'**[4] of 5-HT_3 receptors is based upon (i) resistance to antagonism by compounds acting at receptor classes 5-HT_1, 5-HT_2 and 5-HT_4, while (ii) responses elicited by 5-HT should be mimicked by agonists 'selective' for the 5-HT_3 receptor (as several presently used agonist compounds show **partial agonism** at distinct sites – *see Receptor agonists, 05-50)*, (iii) the receptor should be antagonized by high-affinity competitive antagonists selective for the 5-HT_3R subtype *(see Table 3).*

Table 3. *Common antagonists and their features (From 05-51-01)*

5-HT_3 antagonist and alternative name	Properties	Refs
MDL72222 (Bemestron)		*69*
ICS 205-930 (Tropisetron)	pA_2 value 10–11, 50% block at 50 nM *(Fig. 2)*	*34*
GR38032F (Ondansetron)	pA_2 value 8–10, 50% block at 520 nM	*70*
BRL43694 (Granisetron)	pA_2 value 8–10	*71*
GR 67330	50% block concentration ~3 nM *(Fig. 2)*	*72*
Other antagonists **Quipazine; BRL24924; GR65630**	*see* King, under Jones *et al.*, 1994, pp. 1–44 in *Related sources and reviews, 05-56*	

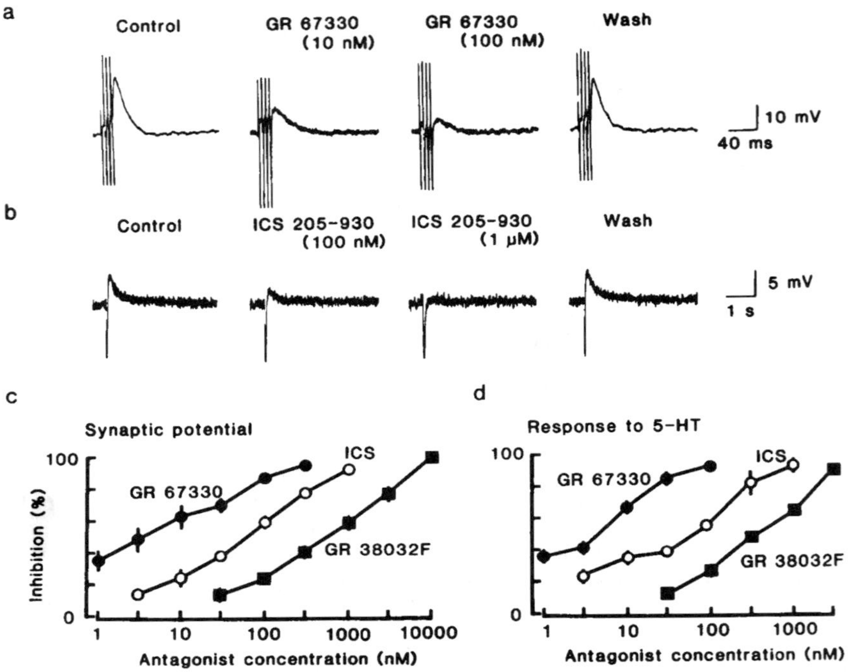

Figure 2. *Examples of reversible blockade of synaptic potentials by 5-HT_3 antagonists. (a) Blockade by* ***GR67330*** *of synaptic responses in the lateral amygdala*[18] *evoked by electrical stimuli consisting of four rapid pulses from a holding potential of −90 mV. (b) Blockade by* ***ICS205-930*** *of depolarization evoked by iontophoretic application of 5-HT. The second and third records were taken 10 min following the change to drug-containing solution, while the fourth was obtained following 20 min washing. (c) Reduction of synaptic potential following application of 5-HT_3 antagonists GR67330, ICS, GR38032F. (d) Reduction of response to iontophoretically applied 5-HT following application of 5-HT_3 antagonists. (Reproduced with permission from Sugita* et al. *(1992)* Neuron *8: 199–203.) (From 05-51-04)*

Many antagonists have been described *(for review, see King, in Jones et al., 1994, pp. 1–44, under Related sources & reviews, 05-56)* and several new compounds are presently being evaluated.

Therapeutic applications of 5-HT_3 receptor antagonists

05-51-02: 5-HT_3R antagonists have selective **anti-emetic properties**, being efficacious in the control of nausea and vomiting reactions to cancer chemotherapy and radiotherapy. Critical reviews on the therapeutic potential and neuropharmacology of 5-HT_3 receptor antagonists have appeared[73], including their use as **anti-emetic** drugs and, in trials, as **anti-psychotic**, **anxiolytic** and **antinociceptive agents**[71, 74–76].

Antagonist binding sites and species differences
05-51-03: The development of selective antagonists for 5-HT_3 receptors has revealed binding sites for this subtype in autonomic neurones and also in the CNS *(see Cell-type expression index, 05-08)*. Notably, 5-HT_3 receptors in guinea-pig ileum, colon and vagus nerve preparations show little heterogeneity in 5-HT_3 antagonist affinities. In comparison, isolated rat vagus nerve preparations show affinities which are ~10- to 100-fold higher than guinea-pig[77, 78]. Significant species differences are summarized in ref.[53].

Reversible blockade of synaptic potentials by 5-HT_3 antagonists
05-51-04: Some examples of reversible blockade by 5-HT_3 antagonists are illustrated in Fig. 2.

INFORMATION RETRIEVAL

Database listings/primary sequence discussion

05-53-01: *The relevant database is indicated by the lower case prefix (e.g. gb:) which should not be typed (see Introduction & layout of entries, entry 02). Database locus names and accession numbers immediately follow the colon. Note that a comprehensive listing of all available accession numbers is superfluous for location of relevant sequences in GenBank® resources, which are now available with powerful in-built* ***neighbouring***† ***analysis*** *routines (for description, see the Database listings field in the Introduction & layout of entries, entry 02). For example, sequences of cross-species variants or related gene family† members can be readily accessed by one or two rounds of neighbouring† analysis (which are based on pre-computed alignments performed using the BLAST† algorithm by the NCBI†). This feature is most useful for retrieval of sequence entries deposited in databases later than those listed below. Thus, representative members of known sequence homology groupings are listed to permit initial direct retrievals by accession number, author/reference or nomenclature. Following direct accession, however, neighbouring† analysis is strongly recommended to identify newly reported and related sequences.*

Nomenclature	Species, DNA source	Original isolate	Accession	Sequence/ discussion
5-HT_3 receptor precursor: 5-HT_3R-A_L ('long form')	Mouse neuroblastoma cell expression library	487 aa; sig: 23 aa (465, mature protein) *see [PDTM]* 55 966 Da (from cDNA)	gb: M74425 pir: P23979 prosite: PS00236	Maricq, *Science* (1991) **254**: 432–7.

continued

Nomenclature	Species, DNA source	Original isolate	Accession	Sequence/ discussion
5-HT_3 receptor precursor: 5-HT_3R-A_S ('short form')	Rat superior cervical ganglion library	461 aa Likely rat variant of mouse 5-HT_3R-A_S	not found	Johnson, *Soc Neurosci Abs* (1992) **18**: 113–15.
5-HT_3 receptor precursor: 5-HT_3R-A_S ('short form')	Rat superior cervical ganglion library	partial cDNA	not found	Isenberg, *NeuroReport* (1993) **5**: 121–4.
5-HT_3 receptor precursor: 5-HT_3R-A_S ('short form')	Mouse neuro-blastoma cell line N1E-115	460 aa sig: 23 aa 53 178 Da (from cDNA)	gb: X72395	Hope, *Eur J Pharmacol* (1993) **245**: 187–92.

Related sources & reviews

05-56-01: Major sources that include subtype definitions[5, 6, 53, 80–83]; other sources with a review element include – the established and potential therapeutic uses for 5-HT_3 receptor antagonists; behavioural pharmacology of 5-HT_3 receptor antagonists[73, 76]; the neuroendocrine pharmacology of serotonergic (5-HT) neurones[84]; molecular cloning and functional expression[2]; advances in electrophysiological characterization of 5-HT_3 receptors[9, 53]; solubilization and physico-chemical characterization of 5-HT_3 receptor-binding sites (Miquel *et al.*, 1993, *Methods in Neurosciences*, Vol. 11 – *see Resource E – Ion channel book references, entry 60*).

Book references

05-56-02:

Andrews, P.L.R. and Sanger, G.J. (eds) (1992) *Emesis in Anti-Cancer Therapy.* Chapman and Hall, London.

Fozard, J.R. (ed.) (1989) *The Peripheral Actions of 5-Hydroxytryptamine.* Oxford University Press, Oxford.

Fozard, J.R. and Saxena, P.R. (eds) (1991) *Serotonin: Molecular Biology, Receptors and Functional Effects.* Birkhauser, Basel.

Hamon, M. (ed.) (1992) *Central and Peripheral 5-HT_3 Receptors.* Academic Press, London.

Jones, B.J., King, F. and Sanger, G.J. (eds) (1994). *5-HT_3 Receptor Antagonists.* CRC Press, Boca Raton.

Saxena, P.R., Kluwer D., Wallis, D.I., Wouters, W.and Bevan P., (eds) (1990) *Cardiovascular Pharmacology of 5-Hydroxytryptamine: Prospective Therapeutic Applications.* Kluwer Academic, Dordrecht.

Stone, T.W. (ed.) (1991) *Aspects of Synaptic Transmission. LTP, Galanin, Opioids, Autonomic and 5-HT.* Taylor and Francis, London, New York.

Feedback

Error-corrections, enhancement and extensions

05-57-01: Please notify specific errors, omissions, updates and comments on this entry by contributing to its **e-mail feedback file** (*for details, see Resource J, Search Criteria & CSN Development*). For this entry, send e-mail messages To: **CSN-05@le.ac.uk,** indicating the appropriate paragraph by entering its **six-figure index number** (xx-yy-zz or other identifier) into the **Subject**: field of the message (e.g. Subject: 08-50-07). Please feedback on only **one specified paragraph or figure per message,** normally by sending a **corrected replacement** according to the guidelines in *Feedback & CSN Access* . Enhancements and extensions can also be suggested by this route (*ibid.*). Notified changes will be indexed via 'hotlinks' from the CSN 'Home' page (http://www.le.ac.uk/csn/) from mid-1996.

Entry support groups and e-mail newsletters

05-57-02: Authors who have expertise in one or more fields of this entry (and are willing to provide editorial or other support for developing its contents) can join its support group: In this case, send a message To: **CSN-05@le.ac.uk,** (entering the words "support group" in the Subject: field). In the message, please indicate principal interests (see *fieldname criteria in the Introduction for coverage*) together with any relevant **http://www site links** (established or proposed) and details of any other possible contributions. In due course, support group members will (optionally) receive **e-mail newsletters** intended to **co-ordinate and develop** the present (text-based) entry/fieldname frameworks into a 'library' of interlinked resources covering ion channel signalling. Other (more general) information of interest to entry contributors may also be sent to the above address for group distribution and feedback.

REFERENCES

[1] Gaddum, *Br J Pharmacol* (1957) **12**: 323–8.
[2] Maricq, *Science* (1991) **254**: 432–7.
[3] Hope, *Eur J Pharmacol* (1993) **245**: 187–92.
[4] Humphrey, *Trends Pharmacol Sci* (1993) **14**: 233–6.
[5] Julius, *Annu Rev Neurosci* (1991) **14**: 335–60.
[6] Peroutka, *J Neurochem* (1993) **60**: 408–16.
[7] Richardson, *Trends Neurosci* (1986) **9**: 424–6.
[8] Derkach, *Nature* (1989) **339**: 706–9.
[9] Peters, *Trends Pharmacol Sci* (1989) **10**: 172–5.
[10] Neijt, *J Physiol Lond* (1989) **411**: 257–69.
[11] Lambert, *Br J Pharmacol* (1989) **97**: 27–40.
[12] Yakel, *Brain Res* (1990) **533**: 46–52.
[13] Yang, *J Physiol London* (1992) **448**: 237–56.
[14] Furukawa, *J Neurophysiol* (1992) **67**: 812–19.
[15] Shao, *J Neurophysiol* (1991) **65**: 630–8.
[16] Tecott, *Proc Natl Acad Sci USA* (1993) **90**: 1430–4.
[17] Werner, *Soc Neurosci Abs* (1993) **19**: 1164.
[18] Sugita, *Neuron* (1992) **8**: 199–203.
[19] Blandina, *J Pharmacol Exp Ther* (1989) **251**: 803–9.

20 Jiang, *Brain Res* (1990) **513**: 156–60.
21 Blandina, *Eur J Pharmacol* (1988) **155**: 349–50.
22 Paudice, *Br J Pharmacol* (1991) **103**: 1790–4.
23 Feuerstein, *Naunyn-Schmiedeberg's Arch Pharmacol* (1986) **333**: 191–7.
24 Ropert, *J Physiol* (1991) **441**: 121–6.
25 Martin, *Br J Pharmacol* (1992) **106**: 139–42.
26 Blier, *Br J Pharmacol* (1993) **108**: 13–22.
27 Blandina, *J Pharmacol Exp Ther* (1991) **256**: 341–7.
28 Goldfarb, *J Pharmacol Exp Ther* (1993) **267**: 45–50.
29 Barnes, *Nature* (1989) **338**: 762–3.
30 Bianchi, *Br J Pharmacol* (1990) **101**: 448–52.
31 Maura, *J Neurochem* (1992) **58**: 2334–7.
32 Johnson, *Soc Neurosci Abs* (1992) **18**: 113–15.
33 Fozard, *Br J Pharmacol* (1976) **57**: 115–25.
34 Richardson, *Nature* (1985) **316**: 126–31.
35 Saxena, *Trends Pharmacol Sci* (1991) **12**: 223–7.
36 Fozard, *Br J Pharmacol* (1982) **77**: 520P.
37 Glaum, *J Physiol* (1990) **438**: 253P.
38 Isenberg, *NeuroReport* (1993) **5**: 121–4.
39 Lummis, *Mol Pharmacol* (1992) **41**: 18–23.
40 Miquel, *J Neurochem* (1990) **55**: 1526–36.
41 Gordon, *Eur J Pharmacol* (1990) **188**: 313–19.
42 McKernan, *Biochem J* (1990) **269**: 623–8.
43 McKernan, *J Biol Chem* (1990) **265**: 13572–7.
44 Boess, *J Neurochem* (1992) **59**: 1692–1701.
45 Gozlan, *Eur J Pharmacol* (1989) **172**: 497–500.
46 Revah, *Nature* (1991) **353**: 846–9.
47 Eiselé, *Nature* (1993) **366**: 479–83.
48 Yakel, *Proc Natl Acad Sci USA* (1993) **90**: 5030–3.
49 Miquel, *Biochem Pharmacol* (1991) **42**: 1453–61.
50 Unwin, *Cell* (1993) **72**: 31–41.
51 Yang, *J Gen Physiol* (1990) **96**: 1177–98.
52 Higashi, *J Physiol* (1982) **323**: 543–67.
53 Peters, *Trends Pharmacol Sci* (1992) **13**: 391–7.
54 Peters, *Eur J Pharmacol* (1988) **151**: 491–5.
55 Yang, *J Gen Physiol* (1990) **96**: 1177–98.
56 Wallis, *Neuropharmacology* (1978) **17**: 1023–8.
57 Yakel, *Neuron* (1988) **1**: 615–21.
58 Yakel, *J Physiol* (1991) **436**: 293–308.
59 Kooyman, *Eur J Pharmacol* (1993) **246**: 247–54.
60 Kooyman, *Br J Pharmacol* (1993) **108**: 287–9.
61 Lummis, *Eur J Pharmacol* (1993) **243**: 7–11.
62 Neijt, *Neuropharmacology* (1988) **27**: 301–7.
63 Surprenant, *Soc Neurosci Abs* (1991) **17**: 601.
64 Peters, *Br J Pharmacol* (1993) **110**: 665–76.
65 Malone, *Br J Pharmacol* (1991) **104**: 68P.
66 Downie, *Br J Pharmacol* (1993) **109**: 53p.
67 Lovinger, *J Pharmacol Exp Ther* (1993) **265**: 771–7.
68 Parsons, *Trends Pharmacol Sci* (1991) **12**: 310–15.
69 Fozard, *Naunyn-Schmiedeberg's Arch Pharmacol* (1984) **326**: 36–44.
70 Butler, *Br J Pharmacol* (1988) **94**: 397–412.
71 Sanger, *Eur J Pharmacol* (1989) **159**: 113–24.
72 Kilpatrick, *Naunyn Schmiedeberg's Arch Pharmacol* (1990) **342**: 22–6.
73 Greenshaw, *Trends Pharmacol Sci* (1993) **14**: 265–70.
74 Russell, *Br J Anaesth* (1992) **69**: Suppl. 1, 63–8S.
75 Costall, *Br J Cancer* (1992) **66**: Suppl. 19, S2–S8.

76 Costall, *Pharmacol Ther* (1990) **47**: 181–202.
77 Ireland, *Br J Pharmacol* (1987) **101**: 591–8.
78 Butler, *Br J Pharmacol* (1990) **101**: 591–8.
79 Bachy, *Eur J Pharmacol* (1993) **237**: 299–309.
80 Tecott, *Curr Opinion Neurobiol* (1993) **3**: 310–15.
81 Brodde, *Clin Physiol Biochem* (1990) **3**: 19–27.
82 Hartig, *Trends Pharmacol Sci* (1989) **10**: 64–9.
83 Hoyer, *J Recept Res* (1991) **11**: 197–214.
84 Van de Kar, *Annu Rev Pharmacol Toxicol* (1991) **31**: 289–320.
85 Uetz, *FEBS Lett* (1994) **339**: 302–6.

Edward C. Conley **Entry 06**

NOMENCLATURES

Abstract/general description

06-01-01: In excitable cells, **extracellular adenosine-5′-triphosphate** (**ATP**) can act as a true neurotransmitter by directly activating **P_{2X} subtype receptors** which possess intrinsic cation channels (designated **$P_{2X}R$–channels** for the purposes of this entry). ATP-activated channels generally confer an excitatory, depolarizing effect on cells through their permeability to Na^+ and K^+ ions and in many cases they mediate agonist-induced Ca^{2+}-entry.

06-01-02: Prior to expression-cloning† of cDNAs encoding two distinct isoforms† of $P_{2X}R$, *functional* similarities with 5-HT_3 and acetylcholine-gated receptor–channels led to a general *expectation* that $P_{2X}R$ would share structural characteristics with these proteins. Surprisingly, the genes encoding P_{2X} receptor channels predict a **distinct subunit structure** atypical of other extracellular ligand-gated receptor channels, consisting of (i) two transmembrane domains, (ii) a large extracellular portion (typically 56–68% of total protein length, probably comprising the **ATP receptor domain**), (iii) a **pore-forming motif**† reminiscent of potassium channels, and (iv) relatively short intracellular **N- and C-terminal domains**.

06-01-03: The physiological roles of P_{2X} receptor–channels include **fast signalling** across synapses of the central and peripheral nervous systems. In the periphery, $P_{2X}R$–channels are involved in control of **effector structures** such as cardiac or smooth muscle. For example, coeliac neurones that innervate the gastrointestinal blood vessels release ATP on to P_{2X} receptor–channels expressed in vascular smooth muscle to constrict the blood vessel. Ca^{2+} influx through ATP-activated channels is also likely to play important roles in **neurosecretory processes**, where the effector may be (for example) an exocrine gland cell.

06-01-04: P_{2X} purinoceptor–channels are present both at nerve terminals and cell bodies of peripheral and central neurones. In the CNS, ATP-gated channels are likely to mediate rapid **sensory**, **motor** and **cognitive functions**. ATP has been directly demonstrated to act as a **fast excitatory synaptic transmitter** at nerve–nerve synapses by activating P_{2X} receptor–channels.

06-01-05: ATP serves as a **co-transmitter** of **acetylcholine** (in postganglionic parasympathetic neurones), **substance P** (in sensory neurones) and **noradrenaline** (in postganglionic sympathetic neurones). In keeping with roles in neurotransmission, $P_{2X}R$–channels activate with short latency† (i.e. in the order of tens of milliseconds). Shortening of both activation and inactivation time constants† with increasing agonist concentrations have been demonstrated in some preparations.

06-01-06: In muscle and other cells, $P_{2X}R$–channels display dual excitatory functions by both direct entry of Ca^{2+} through the channel and, via the depolarizing effects of Na^+ entry, activation of voltage-gated Ca^{2+} channels. In general, activation of P_{2X}R–channels does not affect phosphoinositide hydrolysis and subsequently releases only minor amounts of Ca^{2+} from intracellular stores.

06-01-07: Molecular cloning of the P_{2X}R cDNAs has also revealed a *possible* link with **apoptosis**†. The deduced amino acid sequence of one P_{2X}R isoform is identical to that originally identified (3 years earlier) as the ***RP-2* gene**, activated in thymocytes induced to undergo **programmed cell death**†. Thus in addition to its role in fast signalling, it *may* be possible that the P_{2X}R/RP-2 protein may also function as an 'induced receptor' for ATP or another metabolite released during apoptosis *(see Developmental regulation, 06-11).*

06-01-08: There is **heterogeneity** of P_{2X}R subtypes, as indicated from distinct cDNA sequences, mRNA transcript sizes, marked differences in current–voltage relationships, single-channel properties, agonist selectivity†, desensitization† behaviour and variable permeability† ratios.

Category (sortcode)

06-02-01: ELG CAT ATP, i.e. extracellular ligand-gated cation channels activated by extracellular adenosine-5′-triphosphate. The suggested **electronic retrieval code** (unique embedded identifier or **UEI**) for 'tagging' of new articles of relevance to the contents of this entry is UEI: **P2X-NAT** (for reports or reviews on native† channel properties) and UEI: **P2X-HET** (for reports or reviews on channel properties applicable to heterologously† expressed recombinant† subunits encoded by cDNAs† or genes†). *For a discussion of the advantages of UEIs and guidelines on their implementation, see the section on Resource J under Introduction & layout, entry 02, and for further details, see Resource J – Search criteria & CSN development, entry 65.*

Channel designation

06-03-01: P_{2X}-like purinoceptors; **P_{2X}R–channels**; the P_{2X} purinoreceptor–channel; the purinergic† channel; the ATP-gated channel. *Note: For the purposes of this entry,* characteristics of the two prototype P_{2X}R–channels expressed from cDNA*[1,2]* are distinguished by underlined prefixes denoting their respective open reading frame length, i.e. <u>P_{2X}R (ORF 399aa)</u> and <u>P_{2X}R1 (ORF 472 aa)</u> *(see Gene family, 06-05).* In official nomenclatures, it has been recommended that the 'R' suffix in 'P_{2X}R' designations should be avoided. Its use in this entry is therefore only for convenience.

Current designation

06-04-01: Currents conducted by P_{2X} receptor–channels have been referred to as $I_{ns,ATP}$.

Gene family

P_{2X}R–channel primary structure is <u>distinct</u> from other integral ELG receptor–channels characterized to date

06-05-01: Prototype cDNA sequences encoding two distinct isoforms† of P_{2X}-

Table 1. *Genes encoding P_{2X} purinoceptor–channels (extracellular ATP-activated channels) (From 06-05-01)*

Subunit[a]	Description *(for sequence discussion, see Database listings, 06-53)*	Other distinguishing features	Encoding (ORF)	Molecular weight
$P_{2X}R$ (ORF 399 aa) gb: **X80477**[1] Equivalent to isolate RP-2 *(see below)*	Cation-selective channel; relatively high Ca^{2+}-permeability; functional properties resemble those of native channels expressed in smooth muscle	Two putative transmembrane domains plus **pore-forming motif**; single subunit appears sufficient to form 'fully functional' channels. cDNA isolated from a rat vas deferens cDNA library by expression-cloning†	399 aa Sig[b] Transcript sizes: 1.8, 2.6, 3.6, 4.2 kb *(see mRNA distribution, 06-13)*	~45 kDa (calc.)[c] ~62 kDa (native, by SDS–PAGE†)
P_{2X} R1 (ORF 472 aa) gb: **U14444**[2]	Cation-selective channel; relatively high Ca^{2+}-permeability	As above; cDNA isolated from a rat phaeochromocytoma (PC12) cell cDNA library by expression-cloning†	472 aa Transcript ~2 kb	~52.5 kDa (calc.)
Isolate **RP-2** (partial sequence)[2]	Initially described[3] as 'apoptosis-induced protein' in thymocytes *(for details, see Developmental regulation, 06-11)*	The partial amino acid sequence of RP-2 appears to be the same as the $P_{2X}R$ isolate (ORF 399 aa) listed above. Now considered to be representative of an 'incompletely processed' primary transcript†	(partial, non-coding)	–

[a]In the absence of systematic nomenclature (as going to press), the predicted (non-glycosylated) mol. wt based on (i) the cDNA open reading frame†, (ii) GenBank® (gb:) accession numbers and (iii) trivial names are used here to distinguish isolates reported.
[b]Charged residues in the first 28 aa suggest the absence of a secretion leader peptide in this isoform.
[c]This (calculated) mol. wt agrees with that of an *in vitro* translation product labelled with [^{35}S]-methionine[1]. The higher molecular weight of the $P_{2X}R$ solubilized from rat vas deferens (62 kDa)[4] is likely to be due to glycosylation of the mature protein.

like purinoceptors have been isolated by **expression-cloning**† from rat vas deferens[1] and rat phaeochromocytoma (PC12) cell mRNA[2]. These proteins possess a novel structure, constituting a distinct family of extracellular ligand-gated ion channels. *(For details and distinctions between these isoforms, see Table 1 and information under the appropriate fields in the sections SEQUENCE ANALYSES and STRUCTURE & FUNCTIONS.)*

Subtype classifications

Relationship of P_{2X} receptor–channels to other purinoceptors

06-06-01: Based on pharmacological and functional properties, **purinoceptors** were classified as P_1 or P_2 depending on their preference for adenosine or adenine nucleotides, respectively. Extracellular ATP exerts its cellular effects via P_2 receptors[5]. P_2 receptors are further subdivided into the **P_{2X} subtype** (refs[6,7] and this entry) possessing an integral ionic channel, while the purinoceptor subtypes P_{2Y} and P_{2U} are **G protein-linked receptors**[8] *(see Resource A, entry 56)*. Two further subtypes, the **P_{2T} receptor** and the **P_{2Z} receptor** may also possess intrinsic cation channels. P_{2X} subtypes have received most attention, and in several studies no subtypes are defined beyond the P_2 category.

06-06-02: *Note:* The classification of P_2 series subtypes has not been finalized and in due course will benefit from more specific pharmacological blockers/activators and primary sequence information provided by molecular cloning. For developments in **purinoceptor nomenclature**, refer to the latest IUPHAR Nomenclature Committee recommendations via the CSN *(see Feedback & CSN access, entry 12)*.

Features of P_{2Z} and P_{2T} purinergic receptor subtypes

06-06-03: A **P_{2Z} receptor** subtype specific for ATP is found predominantly on mast cells and other immune cells. Activation of the P_{2Z} receptor by high ATP concentrations appears to be linked to opening of an integral channel. The **P_{2T} receptor** subtype of platelets has been reported to inhibit adenylyl cyclase and to stimulate intracellular Ca^{2+} release. Notably, ATP and AMP are antagonists for this receptor type while ADP is a potent agonist.

Functional evidence for multiple ATP receptor–channel subtypes

06-06-04: Differing functional properties in a range of preparations predict the existence of **purinergic**† **channel subtypes**[6]. Subtypes may be differentiated by characteristic agonist selectivity, desensitization† behaviour and permeability† ratios. In particular, distinct distributions of channel open times in single-cell preparations suggest the existence of multiple ATP-gated channel subtypes *(see Inactivation, 06-37)*.

Despite their structural differences, the functions of P_{2X}R–channels resemble those of other ELG channel types

06-06-05: Functionally, currents through P_{2X}-like purinoceptors show closer resemblance to nicotinic cation channels *(see ELG CAT nAChR, entry 09)* and AMPA-selective glutamate receptor channels *(see ELG CAT GLU AMPA/KAIN, entry 07)* than they do to NMDA-selective glutamate receptors.

Trivial names

Unsystematic names for several 'ATP-activated' channels should not be confused with P_{2X} receptors

06-07-01: ATP-gated cation channels of the P_{2X}-like purinoreceptor subclass *(this entry)* should not be confused with **similarly named channel currents** which may be *indirectly* activated by several subtypes of G protein-linked purinoceptors *(see clarification in Subtype classifications, 06-06)*. Similarly, ATP-gated $P_{2X}R$ channels *(this entry)* should not be confused with novel *intracellular* ATP-activated K^+ channels, such as those described in pancreatic β-cells isolated from a type-2 diabetic human[9] or with the ATP-inhibited K^+ channel family *(see INR K ATP-i, entry 30)*. Putative 'ATP-activated Na^+ channels' recorded in de-folliculated *Xenopus* oocytes[10] are also likely to be activated by a separate (non-integral) G protein-linked ATP receptor. 'ATP-activated' inwardly rectifying K^+ channels of calf atrial cells[11] *(see INR K G/ACh, entry 31)* are also unrelated to $P_{2X}R$–channels.

Trivial names presently in use

06-07-02: Extracellular ATP-gated cation channels; fast-activated ATP-gated channels; directly activated ATP channels; ATP receptor–channels. *Note:* Fast channel activation (in the millisecond range) is the most immediate distinguishing characteristic of the P_{2X}-like purinoreceptor–channels.

EXPRESSION

Cell-type expression index

06-08-02: Consistent with their role in **neurotransmission**, **neurosecretion** and **effector**† **coupling**, $P_{2X}R$–channels are expressed at several synapses of the CNS and PNS, together with a number of characterized secretory and muscle cell types. Some of the important preparations that have been used for the study of $P_{2X}R$ currents are as follows.

acinar cells, e.g. lacrimal and parotid types[12, 13]
blood vessels (various)[15]
cardiac atrial cells, bullfrog[17]
liver, hepatoma cells[20]
cardiac parasympathetic neurones[26]
medial habenula neurones, CNS[28]
cochlea hair cell neurones[31, 31]
cultured hippocampal neurones[35]
dorsal horn neurones[38]
nucleus-solitarii neurones[14]
coeliac ganglion neurones, PNS[16]
sensory ganglia neurones[18, 19]
phaeochromocytoma-derived PC12 cell lines (resembles sensory neurone $P_{2X}R$)[21–25]

locus coeruleus neurones, rat[27]
skeletal muscle fibres, adult rat[29]
vascular smooth muscle[32–34]
vas deferens smooth muscle[36, 37]
bladder smooth muscle[39]

Matching native tissue-specific properties of $P_{2X}R$ with properties of cloned $P_{2X}R$

06-08-02: $P_{2X}R1$ (ORF 472 aa): Native extracellular ATP-gated cation channels have been largely characterized on smooth muscle cells and autonomic sensory neurones. The $P_{2X}R$ type cloned from PC12 cells (472 aa[2]) resembles native $P_{2X}R$ on PC12 cells and some sensory and autonomic neurones[24, 26, 40, 41]. The properties of the 472 aa isoform† differ from those of $P_{2X}R$ in vascular smooth muscle, vas deferens and some CNS neurones. $P_{2X}R$ (ORF 399 aa): The prototype receptor expression-cloned† from vas deferens has properties consistent with the native $P_{2X}R$ in this preparation (e.g. where α, β-methylene-ATP acts as a potent agonist[37]), vascular (arterial) smooth muscle[32], and other smooth muscle preparations[6, 7]. (For further properties of the cloned/native $P_{2X}R$ isoforms, *see the respective fields within the ELECTROPHYSIOLOGY and PHARMACOLOGY sections.*)

Cloning resource

Prototype $P_{2X}R$–channel cDNAs isolated by expression-cloning† protocols

06-10-01: Oocytes injected with poly(A)$^+$ mRNA from rat vas deferens and urinary bladder support ATP-evoked membrane cationic currents which do not appear in uninjected oocytes[1]. Unidirectional cDNA libraries constructed with poly(A)$^+$ RNA from vas deferens and rat phaeochromocytoma (PC12) cells were used as sources to isolate the prototype cDNAs encoding the P_{2X} receptor–channels by expression-cloning† protocols[1, 2]. (For further potential sources of mRNA encoding $P_{2X}R$–channels, *see mRNA distribution, 06-13.*)

Developmental regulation

Possible role of $P_{2X}R$ in apoptosis

06-11-01: In addition to a well-established role in synaptic transmission, the P_{2X} receptor may have a role in the Ca^{2+}-influx associated with induced **apoptosis**†. Following **subtractive hybridization**† procedures designed to enrich mRNAs expressed in thymocytes induced to die (by culture for 8 h in the presence of **dexamethasone** and **cycloheximide**), an earlier study[3] isolated a partial cDNA sequence (***RP-2***) which was later shown[1] to be identical to part of the 399 aa $P_{2X}R$ isoform. The first 37 nucleotides of the reported RP-2 sequence (which did not match the $P_{2X}R$ expressed in vas deferens cDNA) can be attributed to an intronic† sequence not present within spliced $P_{2X}R$ mRNA (thus the original RP-2 sequence was likely to be a fragment of an incompletely processed $P_{2X}R$ gene). Interestingly, ATP (variably) induces **cell**

death in thymocytes[42], hepatocytes[43] and several lymphocytic cell lines[44] (reportedly by action at P_{2Z} or P_{2Y} subtype purinoceptors) with concomitant increases in the intracellular calcium concentration. The identity of RP-2 and $P_{2X}R$ may indicate a role for direct activation of Ca^{2+}-influx following agonism of $P_{2X}R$ during some forms of apoptosis.

ATP-gated conductances are not activated by nerve growth factor in PC12 cells

06-11-02: Nerve growth factor (NGF)-treated rat phaeochromocytoma PC12 cells express an ATP-gated channel *(see Blockers, 06-43)*. NGF stimulates the uptake of radioactive calcium into PC12 cells, but it has been concluded that the NGF-activated influx pathway is independent of both ATP-activated and L-type calcium channels[45].

Cationic- and anionic-permeable channels in developing muscle

06-11-03: Micromolar concentrations of extracellular ATP have been shown to elicit **rapid excitatory responses** in developing chick skeletal muscle. Unusually, these studies concluded that a single class of ATP-gated channels were able to conduct *both* cations and small anions[46] *(see Selectivity, 06-40)*.

mRNA distribution

Strong expression of $P_{2X}R$ mRNA in smooth muscle

06-13-01: $P_{2X}R$ (ORF 399 aa): Northern† blot analysis shows radiolabelled bands with estimated sizes of 1.8, 2.6, 3.6 and 4.2 kb[1]. The representation and intensity of hybridization of each of these transcripts according to preparation, with the strongest signals in vas deferens and urinary bladder (*as summarized in Table 2*). mRNA distributions judged by *in situ* hybridization show prominent signals in the **smooth muscle layer** of the urinary bladder and the smooth muscle layers of small arteries and arterioles[1].

06-13-02: $P_{2X}R1$ (ORF 472 aa): The representation and intensity of hybridization of the 2 kb transcript in various preparations are shown in Table 3.

Phenotypic expression

ATP acts as a transmitter in the CNS and the periphery

06-14-01: A wide range of central and peripheral functions are mediated by extracellular ATP acting at P_{2X}-type receptor–channels. Physiological roles of $P_{2X}R$–channels frequently involve 'fast' signalling, e.g. **synaptic transmission** between central neurones[16,28,48] underlying rapid **sensory**, **motor** and **cognitive functions** and **fast responses of smooth muscle**[5] following sympathetic nerve stimulation. ATP has been confirmed to act as a **co-transmitter** with **noradrenaline** in the sympathetic nervous system[15,36,49]. In keeping with its role as a true neurotransmitter, the stimulus-evoked release of ATP has been demonstrated. Ca^{2+}-influx through ATP-activated channels is also likely to play important roles in **neurosecretory processes**. Table 4 summarizes functional roles in a number of different preparations which are likely to involve P_{2X}-type signalling. Further examples are described elsewhere in this entry.

Table 2. *P_{2X} (ORF 399 aa) mRNA distribution: summary of data derived from Northern analyses[a] (From 06-13-01)*

Preparation	~1.8 kb transcript[b]	~2.6 kb transcript	~3.6 kb transcript[c]	~4.2 kb transcript
Brain	(±)	±	(±)	+
Coeliac ganglia	(±)	+	(±)	++
Lung	+++[d]	++[d]	(±)	+
PC12 cells[e]	(±)	+	(±)	+++
Retina	(±)	(±)	(±)	++
Spinal cord	(±)	+	(±)	++
Spleen	+[d]	+[d]	(±)	(±)
Thymus	++ (bandshift)[f]	+ (bandshift)[f]	(±)	+
Urinary bladder[g]	+	+++++	+++	+
Vas deferens[g]	+++	+++++	+++++	+

[a]The 'relative abundance' of the various transcripts are shown on an arbitrary scale (+ to +++++ and (±) or (−) for low or zero detectable expression in single trials) as judged from the published data[1,2].
[b]The 1.8 kb transcript corresponds in size to that encoding the full-length (post spliced) ORF†.
[c]The 3.6 kb transcript may serve as a precursor for P_{2X} mRNA as it can be detected using a P_{2X} intron†-specific probe *(cited in ref.[1])*.
[d]These signals *may* result from smooth muscle in these organs.
[e]Phaeochromocytoma cells differentiated with NGF (nerve growth factor).
[f]A bandshift to 'slightly larger' transcript sizes are apparent for these bands, although this does not appear to affect the 4.2 kb transcript.
[g]The 'predominant' 2.6 kb mRNA has a 3′ untranslated extension as determined by sequencing.

Protein distribution

Co-distribution of acetylcholine and ATP-release sites

06-15-01: Distributions of ATP-release sites in the CNS are closely associated with those of **acetylcholine**, suggesting ATP may be a **co-transmitter** at **cholinergic† synapses** as shown to be the case in the periphery *(see Receptor agonists, 06-50).*

Subcellular locations

Sites of ATP-induced cation influx in PC12 cells

06-16-01: Activation of $P_{2X}R$ channels in rat phaeochromocytoma (PC12) cells induces a 'mixed' Na^+/Ca^{2+} inward current, but does *not* release Ca^{2+} from internal stores *(see Selectivity, 06-40).* This depolarizing current raises voltage-gated calcium channels on the cell surface to their firing threshold†. PC12 cell ATP-gated influx sites are variable in cell bodies but more homogeneous in growth cones[57].

Local depolarization affecting ATP agonist release at regions of synaptic contact

06-16-02: P_{2X} purinoceptors are present at both **nerve terminals** and **cell bodies** of peripheral and central neurones. Local K^+ depolarization of the ends of coeliac neurites of the guinea-pig evoke single-channel currents characteristic of $P_{2X}R$ in outside-out patches when patches are positioned near the region of apparent **synaptic contact**. This effect is not observed when patches are positioned at remote regions[50].

Transcript size

06-17-01: *See mRNA distribution, 06-13.*

Table 3. *$P_{2X}R1$ (ORF 472 aa) mRNA distribution: summary of data derived from Northern analyses[a] (From 06-13-02)*

Preparation	~2 kb transcript
Adrenal	+
Brain	+
Heart	(−)[b]
Intestine, large	+
Intestine, small	+
Kidney	(±)
Liver	(±)
Lung	(±)
Neurones, superior cervical ganglia	++[c]
Ovary	(±)
PC12 cells	+++++
Pituitary	+++++[d]
Skeletal muscle	(−)[b]
Spinal cord	++
Spleen	(±)
Testis	+
Urinary bladder	+
Vas deferens	++++

[a]The 'relative abundance' of the various transcripts are shown on an arbitrary scale (+ to +++++ and (±) or (−) for low or zero detectable expression in single trials) as judged from the published data[1,2].

[b]The absence of transcripts in these tissues (where native P_{2X} receptors have been characterized[11,46]) *probably* indicates the 472 aa $P_{2X}R1$ isoform does not underlie responses in these preparations.

[c]ATP-gated ion channels have been characterized in sensory ganglia[47].

[d]Although 'strong expression' of mRNA encoding this isoform has been demonstrated in both the intermediate and anterior lobes of the pituitary (both neurosecretory cells and stellate support cells, *cited in ref.*[2]), a physiological role for extracellular ATP has not been described in these cell types to date.

Table 4. *Functional roles of $P_{2X}R$–channels in various preparations (From 06-14-01)*

Preparation	Features/functional roles of purinoceptor–channels	Refs
Heart, neurotransmission	Extracellular ATP-activated cation channels in smooth muscle generally produce contractile (excitatory) responses by direct admission of Ca^{2+}. In cardiac neurones these channels may contribute to **non-adrenergic non-cholinergic** (NANC) neurotransmission and mediate, in part, the **vagal innervation** of the mammalian heart	*32* *26*
Phaeochromocytoma PC12 cells	Ca^{2+}-influx through ATP-activated channels in PC12 cells (but *not* voltage-gated Ca^{2+} channels) contribute to ATP-evoked **noradrenaline release** and at the same time inactivate the Ca^{2+}-selective channels in these cells. The PC12 $I_{ATP.ns}$ resembles the $P_{2X}R$ expressed in sensory neurones. A number of ATP analogues are effective in stimulating catecholamine release, and the receptor antagonists **suramin** and **Reactive blue 2** inhibit the nucleotide-induced noradrenaline release *(see Receptor antagonists, 06-51, and Receptor agonists, 06-50)*	*23* *24*
Coeliac ganglion neurones	ATP acts as a neurotransmitter at several junctions between autonomic nerves and visceral muscle. Features of excitatory junction† currents in the **coeliac ganglion**, e.g. reversal potential†, time course and I–V relation† *(as shown in Current–voltage relation, 06-35)* can be mimicked by application of exogenous ATP. Synaptic currents measured *in situ* possess similar current–voltage relationships to currents produced by ATP, are increased in frequency by K^+ depolarization (in a 'Ca^{2+}-dependent' manner), and are reduced by ATP antagonists *(see Receptor antagonists, 06-51)*	*16, 50*
Hippocampal neurones	In cultured hippocampal neurones, ATP directly activates small sustained currents, and indirectly induces the transient currents by evoking glutamate release	*35*

Table 4. *Continued*

Preparation	Features/functional roles of purinoceptor–channels	Refs
Smooth muscle	ATP acts as a (co-)transmitter at several junctions between autonomic nerves and vascular smooth muscle. The ATP-activated channels provide a distinct mechanism for excitatory synaptic current and Ca^{2+}-entry in smooth muscle. P_{2X} receptors mediate **sympathetic vasoconstriction** in small arteries and arterioles. ATP may also initiate smooth muscle relaxation (vasodilation) by indirect agonism at endothelial cell ATP receptors coupled to second messenger systems *(see Appendix A, entry 56)*	*32*; *51, 52*
Smooth muscle, bladder, non-human	The **bladder** of most non-human species receives dual purinergic and cholinergic excitatory innervation. Activation of P_{2X} purinoceptors depolarizes the cells, increases the spike frequency and causes contraction. Addition of agonists rapidly activates non-selective cation channels, which underlie the excitatory junction potentials seen on stimulation of the intrinsic nerves	*39*; *53*
Skeletal muscle fibres, adult rat	Extracellular ATP (50–100 μM) has been shown to activate junctional and extrajunctional currents similar to those of acetylcholine receptor–channels in isolated adult rat skeletal muscle fibres, but exhibit a shorter open time	*29*
Liver, hepatoma cells	Calcium-permeable channels expressed in rat hepatoma cells are activated by extracellular nucleotides	*20*
Parotid acinar cells (P_{2Z}R channels)	In rat parotid acinar cells, extracellular ATP increases influx of Ca^{2+} across the plasma membrane, in contrast to receptor-mediated responses to carbachol (which also elevates $[Ca^{2+}]_i$ ~3–5-fold, but primarily by release of Ca^{2+} from intracellular stores). Within 10 s, ATP (1 mM) and carbachol (20 μM) reduce the cellular Cl^- content by 39–50% and cell volume by 15–25%. Both stimuli significantly reduce (by ~57–65%) the cytosolic K^+ content of the parotid acinar cell through multiple types of K^+-permeable channels. ATP and carbachol also stimulate the rapid entry of Na^+ into the parotid cell, and elevate the intracellular Na^+ content to ~4.4 and 2.6 times the normal level, respectively – part of this flux is due to **P_{2Z}R–channels** *(see also Receptor antagonists, 06-51)*	*54–56*

SEQUENCE ANALYSES

Note: The [PDTM] symbol denotes an illustrated feature on the channel protein domain topography model (Fig. 1).

Encoding

Predicted sizes of proteins encoded by $P_{2X}R$ genes

06-19-01: The predicted protein encoded by the $P_{2X}R$ cDNA synthesized from vas deferens mRNA[1] has an open reading frame† of 399 amino acids (~45 kDa without glycosylation†); the open reading frame of the $P_{2X}R$ cDNA derived from mRNA from phaeochromocytoma cells[2] predicts a protein of 472 amino acids (~52.5 kDa without glycosylation†).

Gene organization

Evidence for RNA splicing within the protein coding region of $P_{2X}R$ genes

06-20-01: $P_{2X}R$ (ORF 399 aa): The existence of several 'high molecular weight bands' on Northern† hybridizations *(see mRNA distribution, 06-13)* and the observation of **unprocessed forms** of the $P_{2X}R$ gene represented by isolate RP-2[3] *(see Developmental regulation, 06-11)* are indicative of an **RNA splicing mechanism** within the protein coding region of the sequence encoding the 399 aa $P_{2X}R$ isoform.

Homologous isoforms

06-21-01: See the section matching native tissue-specific properties of $P_{2X}R$ with properties of 'cloned' $P_{2X}R$ under *Cell-type expression index, 06-08.*

Protein molecular weight (purified)

06-22-01: *See Table 1 under Gene family, 06-05.*

Protein molecular weight (calc.)

06-23-01: *See Table 1 under Gene family, 06-05.*

Sequence motifs

ATP-binding site motifs

06-24-01: A motif *similar* to the **Walker type A phosphate-binding site** G(X4)GK(X7)(I/V)[58] is found between residues 131 and 144 of the P_{2X} purinoceptor–channel[1]. Walker type A motifs[59] have also been indicated on the reported primary sequence of the cDNA encoding the $P_{2X}R1$ isoform isolated from rat PC12 cells[2] *(see below)*. The extracellular locations of these consensus elements are presumed to form part of the ATP-binding site *(see [PDTM], Fig. 1).*

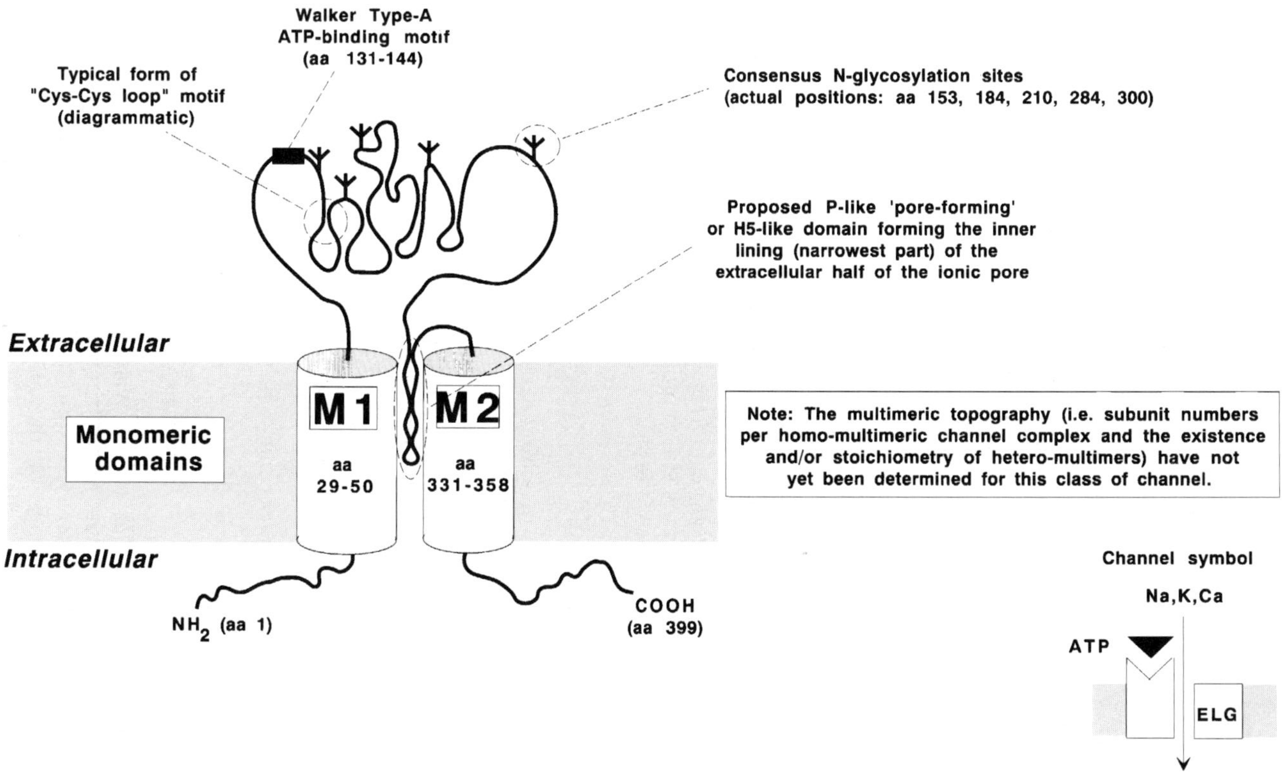

Figure 1. *Monomeric protein domain topography [PDTM] model for the rat extracellular ATP-gated receptor–channel ($P_{2X}R$) exemplified for the 399 amino acid isoform isolated from vas deferens. Note: All relative positions of motifs, domain shapes and sizes are diagrammatic and are subject to re-interpretation. (From 06-24-01)*

$P_{2X}R$/aa residue	131	132	133	134	135	136	137	138	139	140	141	142	143	144
ORF 399 aa (rat vas deferens)	G	C	T	P	G	K	A	E	R	K	A	Q	G	I
$P_{2X}R1$/aa residue	169	170	171	172	173	174	175	176	177	178				
ORF 472 aa (PC12) motif 1	G	T	S	D	N	H	F	L	G	K				
$P_{2X}R1$/aa residue	320	321	322	323	324									
ORF 472 aa motif 2 (partial)	G	Q	A	G	K									

N-Glycosylation sites and putative 'Cys–Cys loop' motifs

06-24-02: $P_{2X}R$ (ORF 399 aa)[1]: The N287 aa hydrophilic (extracellular) region between the two hydrophobic (putative membrane-spanning) domains in this isoform shows five potential sites for *N*-linked glycosylation† (10 cysteine residues). $P_{2X}R1$ (ORF 472 aa)[2]: The ~270 aa extracellular region of this isoform displays three potential *N*-linked glycosylation motifs and several regularly spaced cysteine residues which resemble '**Cys–Cys loop**' motifs† found in the nAChR and other members of the extracellular ligand-gated channel family. Note: Cys–Cys loop motifs are proposed to be important in stabilizing the structure of extracellular ligand-binding pockets *(see also [PDTM], Fig. 1 and Protein phosphorylation, 06-32).*

Apparent lack of secretory leader (signal) peptide

06-24-03: $P_{2X}R$ (ORF 399 aa)[1]: The presence of charged residues in the first 28 amino acids of the open reading frame† of the prototype $P_{2X}R$ suggest the *absence* of a secretion leader. This supports a model in which both the N- and C-termini are in the cytoplasm of the cell *(see [PDTM], Fig. 1).*

STRUCTURE & FUNCTIONS

Amino acid composition

Prediction of a novel structure for $P_{2X}R$–channel subunits

06-26-01: By hydropathicity† analysis, P_{2X} receptors exhibit only *two* hydrophobic† segments 'sufficiently long' to exist as transmembrane domains *(see [PDTM], Fig. 1).* These hydrophobic segments are separated by a large hydrophilic† segment of ~270 aa (for the 472 aa $P_{2X}R1$[2]) or ~287 aa (for the 399 aa $P_{2X}R$[1]) which is cysteine-rich (10 cysteines for the 472 aa $P_{2X}R1$[2] and 9 for the 399 aa $P_{2X}R$[1]). The structure and topography of the receptor therefore differs markedly from those of other extracellular ligand-gated channels *(compare other ELG entries),* which is contrary to what was predicted from electrophysiological and pharmacological properties alone. *(See also Domain arrangement, 06-27, Domain conservation, 06-28, Domain functions, 06-29, and Predicted protein topography, 06-30.)*

Domain arrangement

Expression of cloned $P_{2X}R$ subunits display properties resembling native channels

06-27-01: The electrophysiological properties of heterologously† expressed recombinant P_{2X} receptors cloned 'to date' closely resemble native channels, suggesting that assemblies of single subunits form fully functional channels *(see Cell-type expression index, 06-08).* The stoichiometry and the arrangement of the protein subunits comprising native or recombinant P_{2X} receptors is presently unclear, and heteromeric† assemblies have not been ruled out *(see also Predicted protein topography, 06-30).* Variability of results in Hill plot† analyses for native P_{2X} receptors suggest the ATP-gated channels are composed of multiple subunits in common with other extracellular ligand-gated channels *(see Dose–response, 06-36).*

Domain conservation

Putative pore-forming domain

06-28-01: Amino acid sequence alignments of portions of the prototype $P_{2X}R$-channels in the region immediately preceding the M2 transmembrane domain indicate some similarity with the H5 region typical of '**pore-forming domains**' of voltage-gated and inward rectifier-type K^+ channels *(see the entries beginning VLG K and INR K).* For example, it is possible to align residues 331–338 of the 399 aa $P_{2X}R$ isoform[1] (TMTTIGSG) against the 'pore region' sequences of a range of ion channels (including voltage-gated K^+ channels *(see VLG K entries)*, Ca^{2+}-activated K^+ channels *(see ILG K Ca, entry 27)*, inward rectifier K^+ channels *(see INR K entries)*, K_{ATP} channels *(see INR K ATP-i, entry 30)* and cyclic-nucleotide-gated cation channels *(see ILG CAT cAMP, entry 21, and ILG CAT cGMP, entry 22).* This structural feature is *not* found in other 'extracellular ligand-gated' channels (i.e. those described in other ELG entries).

Potential amphipathic α-helical regions associated with the $P_{2X}R$ M2 domain

06-28-02: The 'topographical similarities' between the $P_{2X}R$ and other ion channels *(see Predicted protein topography, 06-30)* has led to predictions that the $P_{2X}R$ M2 domain is able to form an **amphipathic**† **α-helix**† whose polar residues project into the ion pore. Thus comparative '**helical wheel**'† plots for the two prototype $P_{2X}R$ isoforms clearly illustrate the potential of this domain to form such a structural motif *(see Fig. 2).* Such a model is consistent with a multimeric channel complex in which the lumen of the channel is surrounded by M2 domains from different subunits[2].

Domain functions (predicted)

The large extracellular putative ATP-binding domain

06-29-01: A major portion of both $P_{2X}R$ isoform proteins cloned to date[1,2] appears to be a large hydrophilic extracellular domain which is likely to comprise the **ATP-binding site** *(see Sequence motifs, 06-24, [PDTM], Fig. 1). Note:* For evidence supporting the existence of a 'pore-forming' domain, *see Domain conservation, 06-28.*

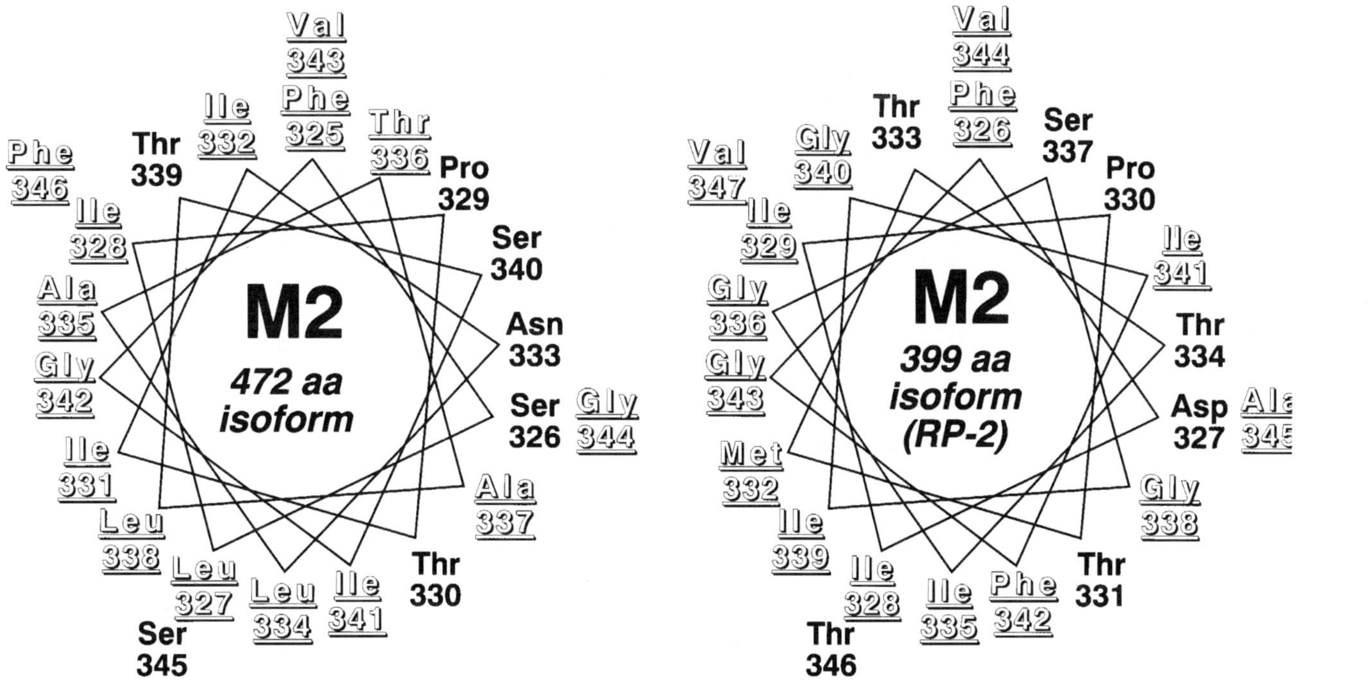

Figure 2. *Comparative 'helical wheel' representations of the amphipathic M2 transmembrane domain of cloned* P_{2X} *receptor–channels showing amino acid residues 325–346 of the 472 aa isoform in comparison with the corresponding region of the 399 aa isoform (equivalent to isolate RP-2). The similar character of homologous amino acids in the two proteins is emphasized by hydrophobic and non-polar residues appearing in outline font. Polar (hydrophilic) residues are predicted to project into the pore. (Based on representations in Brake* et al. *(1994)* Nature ***371:*** *519–23.) (From 06-28-02)*

Predicted protein topography

$P_{2X}R$ topography is atypical of other 'ELG channel' family members

06-30-01: The majority of extracellular ligand-gated receptor–channels share the same overall topography, i.e. four predicted hydrophobic (presumably transmembrane) domains *(see ELG Key facts, entry 04)*. $P_{2X}R$ monomers, however, appear to consist of a **two-transmembrane-domain motif** with a large hydrophilic† extracellular (receptor) domain and a region apparently contributing to the **ion-selective pore**.

Resemblance of predicted $P_{2X}R$ protein topography to other ion channels

06-30-02: The overall protein $P_{2X}R$ monomeric protein *topography* as predicted from primary sequence has some similarity to that proposed for the **mechanosensitive channels** of *Caenorhabditis elegans*[60–62] (see protein domain topography models for *MEC (mechanosensitive), entry 36*, and the rat amiloride-sensitive epithelial sodium channel[63]). There is also some topographical resemblance to cloned inward rectifier[64] and pH-sensitive K^+ channels[65]. Note there is no primary *sequence* homology between the $P_{2X}R$ and these channel types. A two-transmembrane-domain topography is also typical of the inward rectifier potassium channel family *(see the entries beginning INR K)*.

Protein interactions

Functional interactions of P_{2X} receptors with voltage-gated Ca^{2+} and calcium-activated K^+ channels

06-31-01: In single cells isolated from guinea-pig **urinary bladder**, rapid application of ATP (threshold ~100 nM) depolarizes the cell membrane with superimposition of action potentials, followed by transient hyperpolarization[66]. Following addition of the voltage-gated calcium channel blocker **D600**, the amplitude of the ATP-induced depolarization becomes a function of the ATP concentration (EC_{50} ~0.5–1 μM)[66].

Coupling of $P_{2X}R$-mediated Ca^{2+}-influx to activation of potassium channels

06-31-02: The interaction of ATP-gated Ca^{2+}-influx preceding the activation of **K_{Ca} channels** in rat phaeochromocytoma (PC12) cells has also been described[67]. Similarly, the action of ATP upon K_{Ca} channels in bovine aortic endothelial cells is related to second messenger-mediated release of Ca^{2+} from internal stores (coupled to ATP-dependent calcium influx which is also abolished at depolarizing voltages[68]). In cultured aortic smooth muscle cells from rat, ATP-gated channels elevate internal free Ca^{2+} levels to subsequently activate both Ca^{2+}-dependent K^+ and Cl^- currents[69] *(see ILG Cl Ca, entry 25 and ILG K Ca, entry 27.*

Protein phosphorylation

Potentiation of native P_2R–channel responses by cAMP-dependent protein kinase

06-32-01: An ATP-gated channel activation in mouse lacrimal acinar cells is potentiated by stimulation of cAMP-dependent protein kinase activity resulting in an increased responsiveness to external ATP. cAMP-dependent potentiation can be induced by **β-agonists** such as **isoprenaline**[12]. In this preparation, extracellular ATP (1 mM) promotes Ca^{2+}-influx (possibly through a P_{2Z} channel subtype – *see Receptor antagonists, 06-51*), which in turn activates K^+ channels resulting in a delayed, outward current component[12] *(see ILG K Ca, entry 27)*. These ATP-induced responses, and similar ones in rat parotid cells[70], occur in the *absence* of phosphoinositide† hydrolysis (unlike those following ACh application). *Supplementary note:* cAMP elevation following stimulation of **β-adrenergic** receptors is a key second messenger promoting the synthesis and secretion by **exocytosis** of proteins stored in secretory granules. Potentiating effects of GTP and GTP-γ-S on ATP-activated currents have also been characterized in mouse lacrimal cells[12].

Regions of $P_{2X}R$ susceptible to modulation by protein phosphorylation

06-32-02: $P_{2X}R1$ (ORF 472 aa): The ~118 residue C-terminal segment of the $P_{2X}R1$ isoform has 20 proline and 13 serine residues[2] and contains putative sites at which channel activity could be modulated.

ELECTROPHYSIOLOGY

Activation

Activation kinetics for native P_{2X} receptor–channels

06-33-01: As is typical for *integral* extracellular ligand-gated receptor–channels, ATP-gated receptor–channels[32] activate in the **millisecond time range** following agonist application. For example, in single cells isolated from guinea-pig urinary bladder, ATP activates a 'dose-dependent' inward current with a **short latency** (18 ms with 10 μM ATP, measured as the time between the start of application and 10% of the peak)[66]. Similarly, the ATP-stimulated Ca^{2+} conductance in rat hepatoma cells is transient in nature, commencing immediately after ATP addition and reaching a peak at 1 min or less[20].

06-33-02: In keeping with a role in neurotransmission, ATP-gated currents in dorsal root ganglion neurones from rats and bullfrogs activate rapidly upon application of ATP and quickly decay when ATP is removed. Short latencies (< 20 ms) between ATP application and activation of membrane currents are also typical for P_{2X} receptors in rat parasympathetic cardiac ganglia.

Dependence of channel gating kinetics on agonist concentration

06-33-03: The ATP-gated current in dissociated rat nucleus solitarii neurones show time constants† of activation (and inactivation) that are dependent on the extracellular ATP concentration, with *both* parameters becoming faster at higher ATP concentrations[14]. In rat sensory neurones, $P_{2X}R$ activation kinetics are also faster at higher ATP concentrations, with time constants decreasing from ~200 ms at 0.3 μM ATP to ~10 ms at 100 μM ATP. Deactivation kinetics ($\tau \sim$ 100–200 ms) are *independent* of the ATP concentration[18].

Endogenous modulation of activation time constants
06-33-04: In rat phaeochromocytoma PC12 cells, **dopamine** has been shown to shift dependence of activation rate constants† on the concentration of ATP toward a lower concentration range by approximately two-fold[21] *(for details, see Channel modulation, 06-44).*

Potential 'stretch-sensitivity' of ATP-gated currents
06-33-05: The current–voltage relationship of the ATP-activated channel in rat hepatoma cells[20] is 'identical' to that of a previously characterized **stretch-activated channel** in the same preparation[71]. *Note:* The $P_{2X}R$ topographical structure *resembles* that described for **'mechanosensitive' channels** of *Caenorhabditis elegans*[60–62] *(see MEC (mechanosensitive), entry 36).*

Current–voltage relation

Both native and cloned P_{2X} receptors incorporate an inwardly rectifying non-selective cation channel
06-35-01: $P_{2X}R$ (ORF 399 aa/472 aa): The current–voltage relationships of the prototype $P_{2X}R$ isoforms expressed in oocytes and mammalian cells shows a **marked inward rectification**† reminiscent of native P_{2X} receptors as described below *(see also Selectivity, 06-40).*

Comparison of I–V relations for autonomic excitatory synaptic potentials and ATP agonism
06-35-02: Over 95% of **coeliac neurones** display fast inward currents in response to exogenously applied ATP (in comparison to ~45% responding to **acetylcholine** and ~60% which respond to **5-hydroxytryptamine**)[16]. Current–voltage relationships for excitatory post-synaptic currents† (EPSC) recorded in cultured coeliac ganglion neurones show high similarity to currents evoked by exogenously applied ATP *(Fig. 3)*. The EPSC at this synapse are 'blocked' by **suramin** and 'desensitized' by **α,β-methylene-ATP** but are unaffected by the range of antagonists for other candidate receptor channels (e.g. nAChR, NMDAR, non-NMDAR, $5\text{-}HT_3R$, $GABA_AR$). These and additional data led to the identification of ATP acting on the $P_{2X}R$ as the predominant neurotransmitter/receptor combination at this synapse[16].

Functional implications of 'instantaneous' inward rectification in sensory neurones
06-35-03: The current–voltage relationship† for single ATP-activated channels in rat sensory neurones and other preparations is highly **non-linear** and demonstrates **inwardly-directed rectification**[19]. Functionally, inward rectification allows Ca^{2+}-influx to be maximal at hyperpolarized potentials without triggering post-synaptic action potentials. In rat and bullfrog sensory neurones, strong inward rectification is maintained even in symmetric solutions of divalent-free caesium glutamate. Examined at microsecond resolution, the inward rectification occurs within ~10–40 μs (i.e. is instantaneous)[72]. *Note:* I–V curves obtained for single ATP-gated channels in a number of preparations are identical to those for macroscopic† current, but exceptions exist *(see below).*

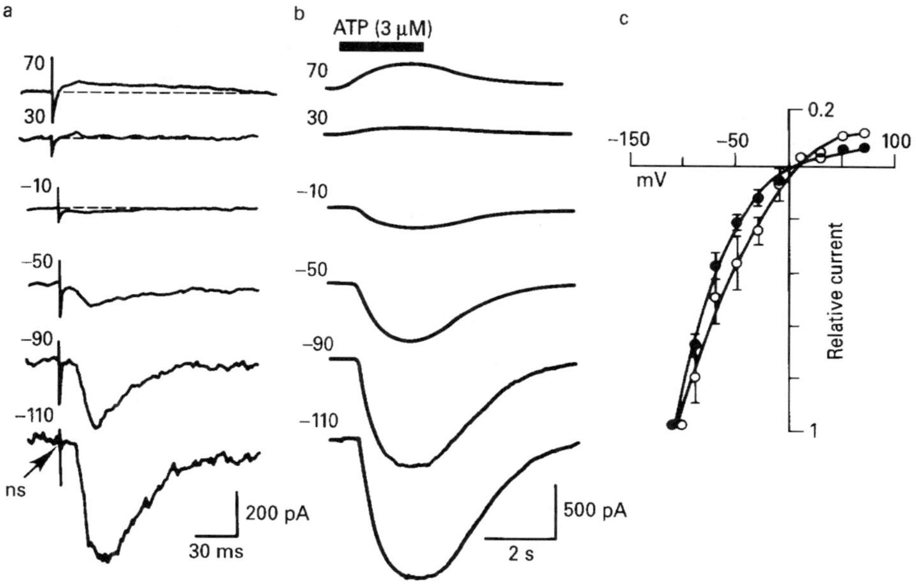

Figure 3. *(a) Amplitude of the coeliac ganglion EPSP recorded from a single neurone at the membrane potentials shown (caesium gluconate in the pipette). The point of nerve stimulation is marked ns (0.1 Hz, 0.5 ms per 20 V). (b) ATP-evoked current under the same conditions as (a). (c) Derived current–voltage relations for data in (a) and (b) (normalized so that current evoked in single cells at −110 mV was equivalent to −1 on the scale). (Reproduced with permission from Evans (1992)* Nature ***357:*** *503–5.) (From 06-35-02)*

Differences in I–V relations for unitary and macroscopic recordings of ATP-gated channels

06-35-04: The current–voltage relationship for macroscopic ATP-evoked currents in rat parasympathetic cardiac ganglia also shows non-ohmic† (inwardly rectifying†) behaviour in the presence *and absence* of external divalent cations with a reversal potential of +10 mV (NaCl outside, CsCl inside)[26]. However, unitary ATP-activated currents in cell-attached membrane patches in this preparation exhibit a linear (ohmic) I–V relationship with a slope conductance of approximately 60 pS[26].

Alteration of I–V relationships by a channel blocker

06-35-05: ATP-activated currents in rat phaeochromocytoma PC12 cells (which resemble the P_{2X} receptors expressed in sensory neurones) show 'less' inward rectification in the presence of the non-selective channel blocker **(+)-tubocurarine**, whereas the receptor antagonists **suramin** or **Reactive blue 2** do *not* affect the voltage-dependency *(see also Receptor antagonists, 06-51).*

Current–voltage relations of cloned P_{2X} receptors expressed in oocytes and HEK-293 cells

06-35-06: $P_{2X}R$ (ORF 399 aa): ATP-evoked currents expressing recombinant $P_{2X}R$ in oocytes display current–voltage relations† (and other properties) which closely resemble those of native channels expressed in smooth muscle[6, 7, 32].

Dose–response

Positive and negative agonist co-operativity in P_{2X} channel gating

06-36-01: ATP-gated channels generally require more than one agonist molecule to bind to the channel in order to open it. Hill plots† of the concentration–current response for low concentrations of external ATP agonist (0.3–1.2 μM ATP) applied to a bullfrog dorsal root ganglion *(see Fig. 4)* show a Hill slope of 3, suggesting a stoichiometry† of (at least) 3 ATP molecules/channel. **Negative co-operativity** between these binding

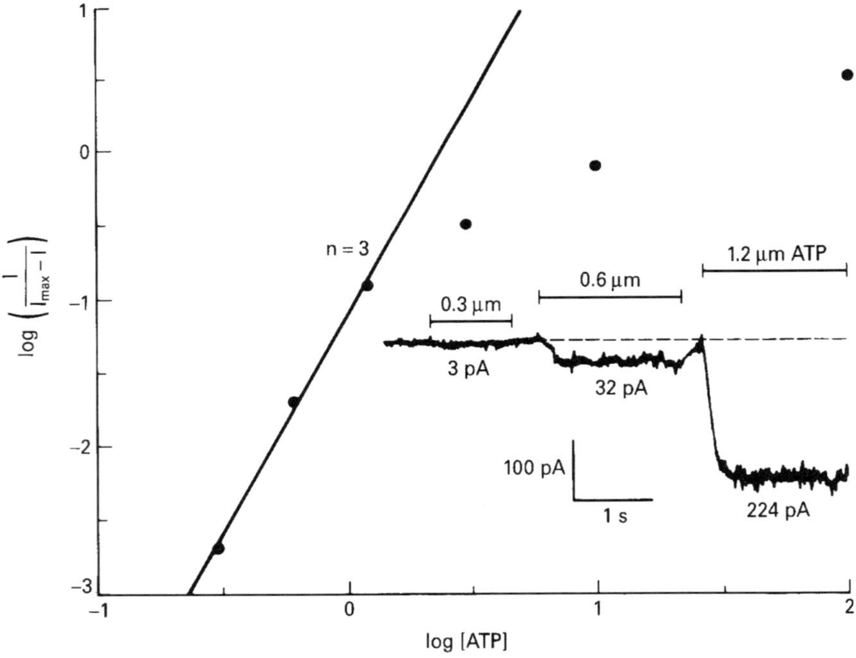

Figure 4. *Responses of a single bullfrog dorsal root (sensory) ganglion voltage-clamped at −80 mV with inward current activated by application of 0.3 μM, 0.6 μM and 1.2 μM ATP (I_{max} activated by 600 μM ATP). At high concentrations the slope becomes less than 1, indicating negative agonist co-operativity. (Reproduced with permission from Bean (1992)* Trends Pharmacol Sci ***13:*** *87–90.) (From 06-36-01)*

sites is suggested by the Hill slope becoming 'less than 1' at high agonist concentrations *(see Fig. 4)*. ATP-gated current in dissociated rat nucleus solitarii neurones increases in a concentration-dependent manner over the concentration range between 10 μM and 1 mM[14]. *Note:* Channels expressed in vas deferens smooth muscle[36,37] show two 'positively co-operative' binding sites (*reviewed in ref.*[6]) *(see also Domain arrangement, 06-27)*.

06-36-02: In the bullfrog dorsal root ganglion neurone preparation, the ATP-gated conductance is half-maximally activated by ~3 μM ATP[18]. At low concentrations, the conductance increases 3- to 7-fold for a doubling in ATP concentration, further suggesting that several ATP molecules must bind in order to activate the current. A steeper concentration–response relationship than expected from 1 : 1 binding is seen in rat dorsal root ganglion neurones[18].

Dose–response characteristics of cloned $P_{2X}R$

06-36-03: P_{2X}R1 (ORF 472 aa): The half-maximal effective ATP concentration for this isoform expressed in oocytes has been determined as ~60 μM[2] *(see also Hill coefficient, 06-46)*.

Inactivation

Variability in observed receptor desensitization properties

06-37-01: Typically, ATP-gated ion channel currents desensitize† with maintained application of ATP ($t_{1/2}$ ~ several seconds). Notably, **desensitization**† is slower in bullfrog sensory neurones than in rat[18] and is not observed at all in some preparations (e.g. guinea-pig hair cells[30,31] and rat parasympathetic cardiac ganglia[26]) *(see also Voltage sensitivity, 06-42)*.

Rates of $P_{2X}R$ desensitization and 'resensitization'

06-37-02: In smooth muscle cells of rat vas deferens, the ATP-induced current disappears within 2 min even in the continuous presence of ATP[73]; cells of the same preparation recover from desensitization in the absence of ATP with a **resensitization** half-time of 2 min[37]. ATP-activated currents in dorsal root ganglion neurones from rats and bullfrogs are similar, except that currents in rat neurones desensitize at a faster rate[18]. In single cells isolated from guinea-pig urinary bladder, the time course of rapid desensitization is a function of the ATP concentration and can be fitted by two exponentials[66].

Minor effects of P_{2X} agonism on release of Ca^{2+} from intracellular stores

06-37-03: In the rat neurosecretory phaeochromocytoma (PC12) cell line, ATP evokes a rise in $[Ca^{2+}]_i$ which rapidly inactivates[74]. The minority of the total response to ATP in these cells (< 20%) is due to intracellular Ca^{2+} redistribution, consistent with a small increase in inositol 1,4,5-trisphosphate level. The majority (> 80%) can be accounted for by ATP-activated cation channels and voltage-gated Ca^{2+} channels.

Selective modulation of inactivation properties

06-37-04: The inactivation of the ATP-activated current in rat phaeochromocytoma (PC12) cells is accelerated by the non-selective ATP channel blocker **(+)-tubocurarine** but not by the ATP receptor antagonists **suramin** or **Reactive blue 2**[25] *(see Blockers, 06-43, and Receptor antagonists, 06-51).* Cell-free patches of PC12 cells show channel inactivation with a half-time of about 5 s[75]. Acceleration of inactivation and the deactivation of ATP-gated channels by dopamine has been reported in PC12 cells (as determined from the current decay upon washout of ATP) *(for further details, see Channel modulation, 06-44).*

Functional evidence for ATP-gated channel subtypes

06-37-05: Open time distributions for ATP-activated channels in rat sensory neurones can be described by a sum of two exponentials, probably reflecting the existence of two channel subtypes within the preparation[19].

Effects of P_{2X}-mediated Ca^{2+}-influx on inactivation of other channels

06-37-06: Ca^{2+}-influx through ATP-gated channels in guinea-pig urinary bladder increments the intracellular Ca^{2+} concentration, and this has been proposed to inactivate voltage-gated Ca^{2+} channels in the same cell[39]. Thus, Ca^{2+}-influx through the $P_{2X}R$ non-selective cation channels has been proposed to be the main determinant of **intracellular Ca^{2+} concentration** in these cells under physiological conditions[39].

Kinetic model

The complex opening and closing kinetics for $P_{2X}R$ have not been analysed in detail. A model for the rapid activation and deactivation kinetics for ATP-activated currents in dorsal root ganglion neurones from rats and bullfrogs requires ATP binding to three identical, non-interacting sites for channel activation[18] *(see Dose-response, 06-36).*

Selectivity

ATP-gated channels as a pathway for calcium influx

06-40-01: P_{2X} purinoceptors possess integral non-selective cation channels[6] possessing significant permeability to **Ca^{2+}**, **Na^+**, **K^+** and **Cs^+** ions. Direct entry of Ca^{2+} into the cell appears to be relatively large in rabbit ear arterial smooth muscle *(see Table 5)*, urinary bladder smooth muscle[39] and cochlear hair cells[31]. Measurement of variable permeability ratios in different membrane preparations may indicate an underlying **heterogeneity** of channel subtypes *(see examples in Table 5).*

Selectivity characteristics of cloned $P_{2X}R$ expressed in oocytes

06-40-02: $P_{2X}R$ (ORF 399 aa): Ion-substitution experiments show the ratio of Ca^{2+} to Na^+ permeability is 4.8 ± 0.4 for this isoform when heterologously expressed in oocytes[1]. The selectivity exhibited by this isoform (and other electrophysiological properties) closely resemble those of native P_{2X} receptors of smooth muscle.

[illegible]5. *Selectivity characteristics of P_{2X} receptor–channels (From 06-40-01)*

[illegible]ation	Selectivity characteristics	Refs
[illegible] [illegible]mpathetic [illegible]es	The amplitude of the ATP-evoked current in cardiac **parasympathetic neurones** is dependent on the extracellular Na^+ concentration. The direction of the shift in reversal potential when NaCl is replaced with mannitol indicates that the purinergic† receptor channel is cation-selective. In this preparation, the cation permeability relative to Na^+ follows the ionic selectivity sequence Ca^{2+} (1.48) > Na^+ (1.0) > Cs^+ (0.67), with anions being 'not measurably permeant'	*26*
Sensory neurones	ATP-gated inward current in rat and bullfrog **sensory neurones** is greatly reduced when *N*-methyl-D-glucamine is substituted for external Na^+. ATP-activated inward currents can be recorded with Ca^{2+} as the *sole* external cation. From reversal potentials, the ratio of Ca^{2+} to Na^+ permeability is ~0.3 : 1 in this preparation *(cf. rabbit ear artery, below)*	*72*
Coeliac neurones	ATP-activated channels which mediate excitatory synaptic transmission between **coeliac neurones** of the guinea-pig show a Na^+ to Cs^+ permeability ratio $P_{Na}/P_{Cs} = 0.6$	*50*
Nucleus solitarii neurones	Calculated relative permeability ratios† for the ATP-gated current in dissociated rat **nucleus solitarii neurones** are $P_{Na}/P_{Cs} = 1.64$ ($[Na^+]_o = 30\text{–}150\,\text{mM}$), $P_{Ca}/P_{Cs} = 2.17$ ($[Ca^{2+}]_o = 2\text{mM}$). Anions are not measurably permeable in this preparation	*14*
Vascular smooth muscle	Although the major cation entering through the ATP-gated channels in **vascular smooth muscle cells** is Na^+, P_{2X} purinoceptor activation also generates subtle, localized increases in calcium concentration. In rabbit ear arterial smooth muscle, ATP activates channels with approximately 3 : 1 selectivity for Ca^{2+} over Na^+ at 'near-physiological' concentrations *(cf. sensory neurones, above)*	*33*, *76*

Table 5. *Continued*

Preparation	Selectivity characteristics	Refs
Skeletal muscle	Permeation of *both* cations and small anions has been shown to occur through a class of ATP-activated ion channels in developing **chick skeletal muscle**. This conclusion is based on fluctuation analysis† about the mean current induced by ATP. At both +40 and −50 mV, ATP elicits a clear increase in noise, but at the reversal potential of the ATP current (−5 mV), no increase in noise above background was observed, indicating that only a single class of non-selective excitatory ATP-activated channels was present. Based on analysis of noise spectra†, the conductance of individual channels is estimated to be 0.2–0.4 pS in this preparation. Calcium is the most permeant ion for this class of channels, with NO_3^- and I^- calculated to be of equal permeance to Na^+	*46*

Selectivity characteristics of cloned $P_{2X}R$

06-40-03: P_{2X}R1 (ORF 472 aa): Ion-substitution experiments indicate the 472 aa isoform incorporates a **non-selective cation channel** equally permeable to Na^+ and K^+, allowing conduction of even large cations. The P_{2X} reeceptor expressed in oocytes displays inward rectification with a reversal potential of approx. −5 mV. Replacement of all extracellular monovalent cations by large organic cations (e.g. Tris) shifts the reversal potential to −45 mV and reduces the current amplitude significantly. Replacement of extracellular Ca^{2+} with Ba^{2+} has no effect on the ATP-induced current[2].

Single-channel data

Heterogeneity of $P_{2X}R$ subtypes is evident at the single-channel level

06-41-01: In general, single-channel currents are significantly larger in neurones, smooth muscle and PC12 cells than they are in cardiac and skeletal muscle, where estimates of small single-channel conductances (typically less than 1 pS) are difficult to obtain by patch-clamp and require techniques of noise analysis†.

Flickering behaviour of unitary $P_{2X}R$–channels

06-41-02: Single-channel open states of ATP-gated currents in a number of preparations display prominent **fluctuations**†. For example, in sensory neurones, fluctuation amplitudes in the 0–4 kHz frequency band display approximately 30% of the mean current amplitude (measured as a double rms). Autocorrelation functions for fluctuations in an open channel can be approximated by a single exponential (time constant ~0.4 ms) and do not depend on the presence of divalent cations in the external medium[19].

06-41-03: Single ATP-gated channels in dorsal root ganglion neurones also display **'flickery' open-channel gating**† behaviour when activated, producing a mean current of about 0.5 pA at −100 mV[18]. High ATP agonist concentrations lengthen the periods when the activated channel flickers, while lower concentrations of ATP produce periods of flickering interspersed with closures[72]. Whole-cell current fluctuations display the expected characteristics if such flickering channels underlie the macroscopic currents[72].

Measured values of ATP-gated unitary conductances

06-41-04: ATP-gated unitary channel activity can be recorded when agonists are applied to outside-out† patches or in the cell-attached† mode. Although a number of different recording configurations and conditions have been employed to measure unitary $P_{2X}R$ channel activity, variations in the values obtained provide support for the existence of **channel subtypes**. Further examples of studies reporting measurements of unitary currents are shown in Table 6.

Voltage sensitivity

Voltage-dependence of ATP agonist potency

06-42-01: ATP-activated currents in dorsal root ganglion neurones from rats and bullfrogs display potency and kinetics of ATP action that are voltage-dependent, with **hyperpolarization** slowing deactivation and increasing ATP's potency. *Note:* In this preparation, deactivation kinetics are sensitive to the concentration of external Ca^{2+} (becoming faster in higher Ca^{2+}). ATP-activated channels in rabbit ear arterial smooth muscle can be opened even at very negative potentials and are characteristically (i) resistant to inhibition by cadmium or nifedipine *(cf. VLG Ca, entry 41)* and (ii) are relatively insensitive to extracellular Mg^{2+} block in the range 1–5 mM *(cf. ELG CAT GLU NMDA, entry 08)*.

Differing voltage-dependence of external versus internal Ca^{2+} block of $P_{2X}R$-channels

06-42-02: External Ca^{2+} block of ATP-activated channels in rat PC12 cells *(see Blockers, 06-43)* does not exhibit voltage dependence between −100 and −210 mV. However, inhibitory effects of *internal* Ca^{2+} are voltage-dependent, with the inhibition being relieved with **hyperpolarization**[22].

Weak voltage-dependence of desensitization properties

06-42-03: The rate of ATP receptor–channel desensitization† *(see Inactivation, 06-37)* is weakly voltage-dependent in cardiac muscle[77].

Dependence on maintained membrane potential

06-42-04: The ATP-stimulated Ca^{2+} conductance in rat hepatoma cells is inhibited by 70% upon dissipation of the membrane potential using the K^{+} ionophore† **valinomycin**[20].

Depolarizing effects of P_{2X}R–channel activation

06-42-05: ATP-activated channels in several preparations[76, 78] serve a dual excitatory function involving (i) **direct entry of Ca^{2+}** through the integral ATP receptor–channel and (ii) **indirect activation of Ca^{2+}-entry** through voltage-gated channels (following Na^{+}-entry and depolarization). Both actions promote action potential discharge. The ATP-activated rise in

Table 6. *Single-channel characteristics of P_{2X} receptor–channels (From 06-41-04)*

Preparation	Single-channel characteristics	Refs
Sensory neurones	In rat sensory neurones, the mean conductance of single ATP-activated channels is ~17 pS (in saline containing 3 mM Ca^{2+} and 1 mM Mg^{2+}; holding potential −75 mV). Subconductance† levels have been detected in this preparation. These channels resemble the P_{2X}R expressed in phaeochromocytoma cells *(see below)*	*19*
Coeliac neurones	ATP-activated channels which mediate excitatory synaptic transmission between coeliac neurones of the guinea-pig display inward rectification† and a mean single-channel conductance† of 22 pS at −50 mV	*50*
Phaeochromocytoma	ATP-activated currents in the phaeochromocytoma PC12 cell line display a unitary conductance of ~13 pS (outside-out, cell-free patch with 140 mM Na^{+} in the external solution)	*75*
Arterial smooth muscle	ATP-activated channels in rabbit ear arterial smooth muscle cells have a unitary ionic conductance of ~5 pS in 110 mM Ca^{2+} or Ba^{2+}	*32*
Vas deferens smooth muscle	In smooth muscle cells of rat vas deferens, an elementary current (mean conductance ~20 pS, zero current potential ~0 mV) is observed in cell-attached patch clamp when the intra-pipette solution is changed to ATP-containing solution	*73*
Cloned P_{2X}R (399 aa isoform) expressed in oocytes	ATP (30 nM) evokes unitary currents in approximately 64% of outside-out membrane patches, while the effect of ATP declines with subsequent applications to the same patch. The chord conductance† between −140 mV and −80 mV has been determined as 19 pS	*1*

$[Ca^{2+}]_i$ within rabbit ear arterial smooth muscle cells[76] is voltage-dependent as *outward* currents evoked by ATP (at positive membrane potentials) are not associated with a change in $[Ca^{2+}]_i$[76] (under these conditions, approximately 10% of the ATP-gated current is carried by Ca^{2+} ions).

PHARMACOLOGY

Blockers

(See also common use of P_{2X} receptor antagonists in 'blocking' extracellular ATP-dependent responses under Receptor antagonists, 06-51).

Block of $P_{2X}R$ channels by external Ca^{2+} ions at high millimolar concentrations

06-43-01: Although it is permeant, relatively high concentrations of **external calcium ions** can reduce inward currents carried by Na^+ ions[19,22,32,41,67,75]. In PC12 cells, the block is concentration-dependent with a Hill coefficient of 1 and a half-maximal concentration of approximately 6 mM. A similar block is observed with other divalent cations, with an order of potency of $Cd^{2+} > Mn^{2+} > Mg^{2+} \approx Ca^{2+} > Ba^{2+}$. Characteristically, high concentrations of Ca^{2+}, Mg^{2+} and Ba^{2+} do *not* block completely, probably because these ions can also carry current in the channel *(see also Voltage sensitivity, 06-42).*

Internal Ca^{2+} ion block of P_{2X} receptors

06-43-02: The amplitude of inward channel currents obtained with 150 mM external Na^+ are reduced by increased internal Ca^{2+} in the inside-out patch-clamp configuration. This reduction is observed at lower concentrations than that by external Ca^{2+}. Internal Ba^{2+} and Cd^{2+} induce similar reductions in current amplitude[22]. A simple one-binding-site model with **symmetric energy barriers** is insufficient to explain bidirectional Ca^{2+} block in PC12 cells[22] *(see also Voltage sensitivity, 06-42).*

Ionic blockers of ATP-gated channels

06-43-03: An effective blocker of the ATP-stimulated Ca^{2+} conductance in rat hepatoma cells is **gadolinium ion**[20]. Voltage-gated calcium channel blockers such as nifedipine and verapamil fail to inhibit $^{45}Ca^{2+}$ uptake in these cells[20]. High concentrations of **zinc ions** reduce and prolong ATP-activated currents in rat sympathetic neurones (consistent with open-channel block) while lower (micromolar) concentrations potentiate $I_{ns.ATP}$[40] *(for details, see Channel modulation, 06-44).*

Pharmacological blockers

06-43-04: In the presence of the non-selective blocker (+)-tubocurarine, maximal responses of rat phaeochromocytoma (PC12) cells to ATP are decreased, but ATP concentrations producing half-maximal responses are unchanged[25]. *Note:* The blocking action of **(+)-tubocurarine** affects influx through other extracellular ligand-gated channels and voltage-gated channels possessing distinct structures. *(For example, see ELG CAT 5-HT_3, entry 05, ELG CAT nAChR, entry 09, and VLG Ca, entry 42).*

Separable P_{2X} purinergic and nicotinic receptor–channel responses
06-43-05: ATP-activated and nicotine-activated influx currents in nerve growth factor (NGF)-treated rat phaeochromocytoma (PC12) cells show many similar properties, and tentatively, the possibility that the channels underlying the currents were *identical* was offered[79]. However, inward currents mediated through ATP-activated channels in these cells can be selectively antagonized by **suramin**[79] *(see Receptor antagonists, 06-51)*. Furthermore, ATP-gated current is *not* affected by ~100 μM **hirsutine** (an alkaloid that produces a potent ganglion blocking effect by potently blocking nicotinic receptor channels and partially inhibiting voltage-gated Ca^{2+} and K^+ channels)[80].

Channel modulation

Positive and negative modulation of native $P_{2X}R$ channels by Zn^{2+} ions
06-44-01: Two distinct modulatory sites of action for **Zn^{2+} ions** have been proposed for ATP-activated channels in rat sympathetic neurones[40]. First, there is evidence for a **positively acting allosteric site** that enhances current amplitude (~five-fold with micromolar Zn^{2+})[47]. By modulation at this site, Zn^{2+} ions can increase membrane depolarization and action potential firing elicited by ATP agonists[47]. Thus low concentrations of extracellular Zn^{2+} rapidly and reversibly potentiate both $I_{ATP.ns}$ and the intracellular Ca^{2+} rise. The potentiation by 10 μM Zn^{2+} is dependent on agonist concentration[40]. Zn^{2+} ions increase the sensitivity of activation without potentiating the maximum response (i.e. possibly by increasing the affinity of $P_{2X}R$ for agonist[40, 47]). Secondly, there is evidence for a **negatively acting modulatory site** for Zn^{2+} (possibly within the pore) that blocks conductance through the $P_{2X}R$–channel[40].

Positive modulation of cloned $P_{2X}R$ expressed in oocytes
06-44-02: $P_{2X}R1$ (ORF 472 aa): Addition of 10 μM Zn^{2+} to the bathing solution shifts the EC_{50} for ATP from 60 μM to 15 μM[2].

Dopaminergic modulation
06-44-03: In phaeochromocytoma (PC12) cells, ATP-activated channel currents are enhanced by **dopaminergic mechanisms**, although this modulation has not been attributed to any single class of **dopamine** receptors[81]. In these cells, 10 μM dopamine enhances an inward current activated by 100 μM ATP. Similar enhancements are produced by 10 μM **apomorphine**, a non-selective dopamine receptor agonist, 10 μM **(+)-SKF-38393** (a selective dopamine D_1 receptor agonist), and 10 μM **(−)-quinpirole** (a selective dopamine D_2 receptor agonist). Moreover, 30 μM **(+)-SCH-23390** (a dopamine D_1 receptor antagonist, and 30 μM **(−)-sulpiride** (a dopamine D_2 receptor antagonist) also enhance the ATP-activated current[81].

Mechanism of dopaminergic modulation of $P_{2X}R$–channels
06-44-04: In PC12 cells, the **'dopamine effect'** *(see above)* has been shown to shift the dependence of **activation rate constants**† on the concentration of

ATP toward a lower concentration range by approximately two-fold[21]. Dopamine also accelerates the inactivation and the deactivation (as determined from the current decay upon washout of ATP). Thus dopamine augments the ATP-activated inward current by **facilitating association of ATP to its binding site**. This augmentation may be mediated through some protein kinase which is different from cyclic-nucleotide-dependent protein kinases or protein kinase C[21].

Comparative note: ATP as a multiple modulator of other cellular proteins
06-44-05: In addition to direct and indirect (**G protein-linked**) gating of ion channels, *intracellular* ATP is a candidate for multiple modulation on other cell-signalling molecules *(for example, see ILG Ca Ca RyR-Caf, entry 17, ILG Ca InsP$_3$, entry 19, and Protein interactions, 06-31).*

Equilibrium dissociation constant

06-45-01: In single cells isolated from guinea-pig urinary bladder, the relationship of the peak current versus ATP concentration is well-fitted by a Michaelis–Menten equation with a K_d of 2.3 μM. For ATP-gated current in dissociated rat nucleus solitarii neurones, the half-maximum concentration is 31 μM ATP[14]. Substitution of ATP with α,β-methylene-ATP shifts the K_d to ~10.4 μM *(see also Inactivation, 06-37, and Hill coefficient, 06-46)*[66]. Dopaminergic modulation can increase association of ATP for its binding site *(see Channel modulation, 06-44).*

Hill coefficient

06-46-01: ATP responses in single cells isolated from guinea-pig urinary bladder display a Hill coefficient of 1.7. Substitution of ATP for α,β-methylene-ATP *(see Equilibrium dissociation constant, 06-45)* does *not* significantly affect the Hill coefficient ($n \sim 1.6$)[66]. For ATP-gated current in dissociated rat nucleus solitarii neurones the Hill coefficient was calculated as 1.2[14] *(see also Dose–response, 06-36).* P$_{2X}$R1 (ORF 472 aa): The Hill coefficent for this isoform expressed in oocytes has been derived as 2.0 for ATP[2].

Ligands

06-47-01: ATP as a **free anion** (**ATP**$^{4-}$) can activate the receptor (i.e. ATP can activate the channel in the absence of divalent cations as well as in their presence). Ca^{2+} ions decrease both macro- and microscopic ATP-activated currents with a concentration-dependence that can *not* be fitted with a single site binding isotherm[19].

Photoaffinity-labelling studies
06-47-02: In an attempt to identify the ATP receptor protein in phaeochromocytoma (PC12) membranes, cells were photoaffinity-labelled with

the radioligand [^{32}P]-3′-*O*-(4-benzoyl-benzoyl)ATP (**[^{32}P]-BzATP**)[24]. SDS–PAGE† analysis revealed that labelling of a 53 kDa protein was inhibited by ATP and its derivatives, as well as by the P_2 antagonists **suramin** and **Reactive blue 2**. *Note:* These antagonists also inhibit **nucleotide-induced noradrenaline release** in these cells[24].

Receptor/transducer interactions

Role of P_{2X} receptors in non-adrenergic, non-cholinergic excitatory transmission

06-49-01: In vas deferens[82] and bladder smooth muscle[39], ATP-gated cation channels produce a component of fast, **non-adrenergic, non-cholinergic** (**NANC**) excitatory transmission by sympathetic neurones. In rabbit ear artery smooth muscle cells, entry of Ca^{2+} ions through the channel is directly associated with contractile events[32, 76].

Discrimination of direct versus G protein-linked Ca^{2+}-influx by ATP agonists

06-49-02: *Indirect* activation of cationic currents by extracellular ATP and the modulation of calcium current through **G protein transducers** have also been described in rabbit portal vein[83] *(see also Resource A, entry 56)*. Notably, the P_{2X} subtype receptor–channel of rabbit ear artery smooth muscle cells is *unaffected* by **SK&F 96365** (a novel inhibitor of receptor-mediated calcium entry[74, 84]). By contrast, this compound reduces G protein-mediated ATP-stimulated currents by about 80% in human neutrophils[85].

Receptor agonists

ATP evokes inward currents through P_{2X} receptors with short latency and fast rise time

06-50-01: ATP, as well as the related molecules **α,β-methylene-ATP**, **2-methylthioATP** and **ADP** evoke inward currents with short latency† (typically <2 ms minimum) and fast rise time (typically <10 ms for 10–90% rise) following application. The P_{2X} receptors display variable desensitization† kinetics with ATP and ATP-derived agonists (with some voltage-dependence), dependent on the preparation and (presumably) molecular subtype *(see Inactivation, 06-37, Voltage sensitivity, 06-42, and Table 7)*.

Agonism at cloned P_{2X} receptors expressed in oocytes and HEK-293 cells – comparisons with native $P_{2X}R$

06-50-02: $P_{2X}R$ (ORF 399 aa): 10 μM **ATP**, **α,β-methylene-ATP**, **2-methylthioATP** and **ADP** evoke 'typical' inward currents with latency† of <2 ms and a rise time of ~7 ms[1]. For this isoform the order of agonist potency is 2-methylthioATP ⩾ ATP > α,β-methylene-ATP >> ADP. $P_{2X}R1$ (ORF 472 aa): ATP, ATP-γ-S and 2-methylthioATP are equipotent as agonists, whereas α,β-methylene-ATP and β,γ-methylene-ATP are inactive as agonists or antagonists. *Note:* The 472 aa isoform displays agonist sensitivity that resembles native $P_{2X}R$ on PC12 and certain sensory and

Table 7. *Common agonists and their features (From 06-50-01)*

Agonist	Characteristics	Refs
P_2 receptor agonism by **2-methylthioATP** on cardiac sympathetic neuronal $P_{2X}R$	ATP receptor–channels described in rat cardiac sympathetic neurones show an order of agonist potency of 2-methylthioATP = ATP > ADP > AMP > adenosine = α,β-methylene-ATP > β,γ-methylene-ATP (a sequence also consistent with the G protein-linked P_{2Y} receptor subtype). ATP and AMP are *antagonists* of the P_{2T} receptor channel subtype expressed on platelets *(see Subtype classifications, 06-06). Note:* ATP-evoked currents in this preparation are *attenuated* by α,β-methylene-ATP ($IC_{50} \sim 10\ \mu M$, *see below*) and reversibly inhibited in a dose-dependent manner by Reactive blue 2 ($K_d = 1\ \mu M$)	*26*
α,β-Methylene-ATP and **ANAPP3**	α,β-Methylene-ATP is a metabolically stable ATP analogue that characteristically activates then desensitizes P_{2X} receptors. Arylazidoamino-propionyl ATP (ANAAPP3) covalently binds tp P_{2x} receptors following irradiation, and is also capable of inducing receptor activation then blockade	*89*
Induction of desensitization by **α,β-methylene-ATP** on neuronal $P_{2X}R$	Fast-synaptic currents within medial habenula central neurones (part of a well-characterized cholinergic pathway) which are blocked by suramin desensitize† following application of α,β-methylene-ATP. Miniature post-synaptic currents observed following spontaneous release of transmitter in this preparation are also desensitized by this agonist	*28*
Induction of desensitization by **α,β-methylene-ATP** on smooth muscle $P_{2X}R$	Desensitization by α,β-methylene-ATP also blocks the response to ATP in single cells isolated from guinea-pig urinary bladder. α,β-Methylene-ATP is ~50–100 times more potent than ATP at eliciting a contractile response of strips of detrusor smooth muscle.	*66*
	Similar desensitization behaviour to ATP ($\geqslant 1\ \mu M$) is observed in cells expressing recombinant $P_{2X}R$ (ORF 399 aa)	*1*
CTP and **dATP**	$P_{2X}R1$ (ORF 472 aa): Among several nucleotide and nucleoside derivatives examined, only CTP and dATP elicit small but detectable current responses from this isoform	*2*
ATP (example of tissue-dependent, endogenous agonist responses	In excitatory synaptic transmission between coeliac neurones of the guinea-pig, ATP evokes inward currents with greater potency† and efficacy† than acetylcholine (ACh)	*50*

autonomic neurones[24,26,40,41] (this pattern differs from that observed for $P_{2X}R$ on vascular smooth muscle, vas deferens and some CNS neurones, where α,β-methylene-ATP acts as a potent agonist[32,37]).

Multiplicity of modulatory and agonist roles of ATP

06-50-03: ATP has multiple actions on other proteins, including (i) indirect gating of ion channels through **G protein-linked receptor**s *(see Resource A – G protein-linked receptors, entry 56)* and (ii) as a modulatory factor on associated signalling components, including ion channels activated by other neurotransmitters or second messengers *(see, for example, ILG Ca Ca RyR-Caf, entry 17, and ILG Ca InsP$_3$, entry 19)*. Partly because of this **multiplicity of targets**, highly selective agonists for P_{2X} receptors are currently not available. Table 7 lists some applications of those presently in use.

Receptor antagonists

06-51-01: The (non-subtype-selective) ATP receptor antagonists **Reactive blue 2** (RB2) and **suramin** reversibly block binding of ATP to P_2 receptors (cf. **(+)-tubocurarine**, a potent antagonist of acetylcholine- and serotonin-gated channels acts as a non-selective blocker of ion permeability through the ATP-activated channel – *see Blockers, 06-43*). RB2 and suramin can distinguish $P_{2X}R$ responses from other extracellular ligand-gated channels[25]. Comparative studies of electrophysiological effects of the above compounds show all three of these compounds inhibit ATP-gated current in rat phaeochromocytoma (PC12) cells in a concentration-dependent manner (order of potency RB2 > suramin > (+)-tubocurarine)[25]. Unlike for suramin or RB2, blockade induced by (+)-tubocurarine is *not* reversed after a 5-min washout period[25]. Further characteristics of these antagonists are listed in Table 8. In general, there is a need for more selective antagonists acting at P_{2X} purinoceptors.

Antagonism at cloned $P_{2X}R$ expressed in oocytes and HEK-293 cells

06-51-02: $P_{2X}R$ (ORF 399 aa): Currents evoked by ATP, α,β-methylene-ATP, 2-methylthioATP and ADP are reversibly blocked by suramin (3–100 μM) and by pyridoxalphosphate-6-axophenyl-2′,4′-disulphonic acid (**PPADS**, 10–30 μM) but not by amiloride (100 μM). These properties served to identify the expressed receptors as of the P_{2X} purinoceptor subtype[1,5]. $P_{2X}R1$ (ORF 472 aa): Both suramin and Reactive blue 2 reversibly antagonize ATP-evoked responses of this isoform by >95%; (+)-tubocurarine only partially blocks (~50%) ATP-evoked responses of this isoform[2].

Relative potency for P_2-mediated release of noradrenaline from PC12 cells

06-51-03: The relative potency of ATP and a number of analogues for eliciting noradrenaline release from rat phaeochromocytoma (PC12) cells in the presence of extracellular Ca^{2+} has been shown to follow the order **adenosine 5′-*O*-(3-thiotriphosphate)** > ATP > adenosine 5′-*O*-(1-thiotriphosphate) = 2-methylthioadenosine 5′-triphosphate (MeSATP) > 2′- and 3′-*O*-(4-benzoylbenzoyl)ATP (BzATP) > ADP > 5-adenylylimidodiphosphate[24].

Table 8. *Characteristics of ATP antagonists (From 06-51-01)*

Antagonist	Application (examples)	Refs
Reactive blue 2	The ATP-activated channel in rat hepatoma cells is inhibited in the presence of **Reactive Blue 2** (RB2), suggesting that channel activation is dependent on purinergic† receptor interaction. In coeliac neurones of the guinea-pig, the antagonists Reactive blue 2 (and suramin, *see below*) reduce the effects of ATP ($IC_{50} \sim 1$–$10\,\mu M$) but not acetylcholine agonists. RB2 is a slowly-acting antagonist of the ATP-gated channel. In rat phaeochromocytoma (PC12) cells noradrenaline release is inhibited by RB2 ($IC_{50} \sim 1$–$10\,\mu M$)	*20*, *50*, *24*
Suramin	The ATP receptor antagonist **suramin** lacks selectivity for P_2 purinoreceptor subtypes, but can discriminate between P_{2X} responses and those of other fast neurotransmitters. Suramin is a competitive blocker of both endogenous transmitter and ATP-evoked currents in coeliac ganglion preparations *(see Cell-type expression index, 06-08)*. In rat phaeochromocytoma (PC12) cells, noradrenaline release is also inhibited by suramin ($IC_{50} \sim 30\,\mu M$)	*16*, *24*
Stilbene isothiocyanate analogues at P_{2Z} purinoceptors *(comparative note only)*	Several **stilbene isothiocyanate** analogues of the calcium-activated chloride current inhibitor DIDS (dihydro-DIDS, SITS but not DNDS – *see Appendix C, entry 58*) can block both the binding of [^{32}P]-ATP to intact parotid cells and the activation of the P_{2Z} purinoceptor–channel[56]. The potency of the **stilbene disulphonates** is related to the number of isothiocyanate groups on each compound (e.g. **DIDS**, $IC_{50} \sim 35\,\mu M$, **SITS** $IC_{50} \sim 125\,\mu M$; DNDS lacks isothiocyanate (SCN^-) groups). **Eosin-5-isothiocyanate** (EITC) and **fluoroscein-5-isothiocyanate** (FITC), non-stilbene isothiocyanate compounds with single SCN^- groups, also block the response to ATP but are less potent than DIDS. **Trinitrophenyl-ATP** (TNP-ATP), an ATP derivative that is not an effective agonist of the parotid acinar cell $P_{2Z}R$, blocks the covalent binding of DIDS to the plasma membrane, suggesting that ATP and DIDS bind to the same site. The distilbene DIDS and 2′3′-dialdehyde-ATP irreversibly inhibit the skeletal muscle ATP-gated channel. DIDS also irreversibly blocks ATP-induced Ca^{2+}-entry in parotid acinar cells of rat	*70*, *56*, *86*, *87*

Table 8. *continued*

Antagonist	Application (examples)	Refs
Adenosine derivatives without agonist activity	These types of compound usually act as weak competitive inhibitors of $P_{2X}R$ (e.g. **adenosine 5′-(β,γ-dichloromethylene)-triphosphonate**, $IC_{50} \sim 21\ \mu M$ at neuronal $P_{2X}R$)	*88*
α,β-Methylene-ATP	**α,β-Methylene-ATP** possesses agonist activity in some preparations *(see Receptor agonists, 06-50)* and may therefore act via a desensitization mechanism. α,β-Methylene-ATP inhibits agonism by ATP in parasympathetic neuronal and cardiac atrial preparations, but *not* in vas deferens, skeletal muscle or sensory neurones	*26, 37, 77, 86*

Effect of ATP analogues on $P_{2X}R$ expressed in smooth muscle from rat vas deferens

06-51-04: In smooth muscle cells isolated from the rat vas deferens, the analogues α,β-methylene-ATP and **AMP-PNP** (**β,γ-imido ATP**) each produce a small, relatively sustained inward current (not resembling the ATP current). The analogue AMP-PCP (β,γ-methylene-ATP) has little or no effect in this preparation[37].

Agonism by other adenosine derivatives and ATP-γ-S

06-51-05: By definition, P_2 receptors are selectively agonized by ATP over adenosine. ADP is a weak agonist *(see above)* and CTP may elicit some current in neurones[88], but GTP and UTP are not effective as agonists. ATP-γ-S (adenosine 5′-O-3-thiotriphosphate, *see above*) is apparently equipotent with ATP in PC12 cells[41], cardiac muscle[11], skeletal muscle[86], and neurones[26].

Inactive agonists at cloned P_{2X} receptors

06-51-06: $P_{2X}R$ (ORF 399 aa): UTP (100 μM), GTP (100 μM), acetylcholine (100 μM) and 5-hydroxytryptamine (50 μM) are ineffective as agonists[1]. $P_{2X}R1$ (ORF 472 aa): ADP, AMP, 5′-adenylylamido-diphosphate, adenosine, GTP, UTP, cAMP, cGMP, acetylcholine, glutamate, glycine, γ-aminobutyric acid (GABA) and 5-hydroxytryptamine (serotonin) do not activate this isoform when expressed in oocytes[2].

INFORMATION RETRIEVAL

Database listings/primary sequence discussion

06-53-01: *The relevant database is indicated by the lower case prefix (e.g. gb:), which should not be typed (see Introduction & layout of entries, entry 02). Database locus names and accession numbers immediately follow the colon. Note that a comprehensive listing of all available accession numbers is superfluous for location of relevant sequences in*

GenBank® resources, which are now available with powerful in-built ***neighbouring† analysis*** *routines (for description, see the Database listings field in the Introduction and layout of entries, entry 02). For example, sequences of cross-species variants or related gene family† members can be readily accessed by one or two rounds of neighbouring† analysis (which are based on pre-computed alignments performed using the BLAST† algorithm by the NCBI†). This feature is most useful for retrieval of sequence entries deposited in databases later than those listed below. Thus, representative members of known sequence homology groupings are listed to permit initial direct retrievals by accession number, author/reference or nomenclature. Following direct accession, however, neighbouring† analysis is strongly recommended to identify newly reported and related sequences.*

Nomenclature	Species, DNA source	Original isolate	Accession	Sequence/ discussion
$P_{2X}R$	Rat, vas deferens cDNA library	**ORF 399 aa**	gb: X80477	Valera, *Nature* (1994) **371:** 516–19.
$P_{2X}R1$	Rat, phaeochromocytoma (PC12) cDNA library	**ORF 472 aa**	gb: U14444	Brake, *Nature* (1994) **371:** 519–23.
Isolate **RP-2** (partial cDNA sequence)	Derived from a subtractive† hybridization† library	Originally described as an 'apoptosis-induced protein' – *see Developmental regulation, 06-11*	gb: M80602	Owens, *Mol Cell Biol* (1991) **11:** 4177–88.

Related sources and reviews

06-56-01: Pharmacological and electrophysiological characteristics of ATP-activated ion channels[6]; ATP-activated channels in excitable cells[7] and vascular smooth muscle cells[33, 90]; ATP as a co-transmitter with noradrenaline in the sympathetic nervous system[15, 49]; ATP receptor classifications and nomenclature[91, 92]. Features of cloned $P_{2X}R$ isoforms[1, 2].

Feedback

Error-corrections, enhancement and extensions

06-57-01: Please notify specific errors, omissions, updates and comments on this entry by contributing to its **e-mail feedback file** (*for details, see Resource J, Search Criteria & CSN Development*). For this entry, send e-mail messages To: **CSN-06@le.ac.uk,** indicating the appropriate paragraph by entering its **six-figure index number** (xx-yy-zz or other identifier) into the **Subject**: field of the message (e.g. Subject: 08-50-07). Please feedback on only **one specified paragraph or figure per message,** normally by sending a **corrected replacement** according to the guidelines in *Feedback & CSN Access* . Enhancements and extensions can also be suggested by this route (*ibid.*). Notified changes will be indexed via 'hotlinks' from the CSN 'Home' page (http://www.le.ac.uk/csn/) from mid-1996.

Entry support groups and e-mail newsletters

06-57-02: Authors who have expertise in one or more fields of this entry (and are willing to provide editorial or other support for developing its contents) can join its support group: In this case, send a message To: **CSN-06@le.ac.uk,** (entering the words "support group" in the Subject: field). In the message, please indicate principal interests (see *fieldname criteria in the Introduction for coverage*) together with any relevant **http://www site links** (established or proposed) and details of any other possible contributions. In due course, support group members will (optionally) receive **e-mail newsletters** intended to **co-ordinate and develop** the present (text-based) entry/fieldname frameworks into a 'library' of interlinked resources covering ion channel signalling. Other (more general) information of interest to entry contributors may also be sent to the above address for group distribution and feedback.

REFERENCES

[1] Valera, *Nature* (1994) **371**: 516–19.
[2] Brake, *Nature* (1994) **371**: 519–23.
[3] Owens, *Mol Cell Biol* (1991) **11**: 4177–88.
[4] Bo, *J Biol Chem* (1992) **267**: 17581–7.
[5] Burnstock, *Ann NY Acad Sci* (1990) **603**: 1–17.
[6] Bean, *Trends Pharmacol Sci* (1992) **13**: 87–90.
[7] Bean, *Ion Channels* (1990) **2**: 169–203.
[8] Barnard, *Trends Pharmacol Sci* (1994) **15**: 67–70.
[9] Williams, *Pflugers Arch-Eur J Physiol* (1993) **423**: 265–73.
[10] Kupitz, *Science* (1993) **261**: 484–6.
[11] Friel, *Pflugers Arch* (1990) **415**: 651–7.
[12] Sasaki, *J Physiol* (1992) **447**: 103–18.
[13] Vincent, *J Physiol* (1992) **449**: 313–31.
[14] Ueno, *J Neurophysiol* (1992) **68**: 778–85.
[15] von-Kiigelgen, *Trends Pharmacol Sci* (1991) **12**: 319–24.
[16] Evans, *Nature* (1992) **357**: 503–5.
[17] Friel, *J Gen Physiol* (1987) **91**: 1–27.
[18] Bean, *J Neurosci* (1990) **10**: 1–10.
[19] Krishtal, *Neuroscience* (1988) **27**: 995–1000.
[20] Bear, *Am J Physiol* (1991) **261**: C1018–24.
[21] Nakazawa, *Pflugers Arch-Eur J Physiol* (1993) **422**: 458–64.
[22] Nakazawa, *J Gen Physiol* (1993) **101**: 377–92.
[23] Nakazawa, *J Neurophysiol* (1992) **68**: 2026–32.
[24] Majid, *Biochim Biophys Acta* (1992) **1136**: 283–9.
[25] Nakazawa, *Pflugers Arch* (1991) **418**: 214–19.
[26] Fieber, *J Physiol* (1991) **434**: 239–56.

27 Harms, *Neuroscience* (1992) **48**: 941–52.
28 Edwards, *Nature* (1992) **359**: 144–7.
29 Mozrzymas, *Neurosci Lett* (1992) **139**: 217–20.
30 Nakagawa, *J Neurophysiol* (1990) **63**: 1068–74.
31 Ashmore, *J Physiol* (1990) **428**: 109–31.
32 Benham, *Nature* (1987) **328**: 275–8.
33 Benham, *Ann NY Acad Sci* (1990) **603**: 275–85.
34 Stjarne, *Trends Neurosci* (1986) **9**: 547–48.
35 Inoue, *Neurosci Lett* (1992) **134**: 215–18.
36 Sneddon, *Science* (1982) **218**: 693–5.
37 Friel, *J Physiol Lond* (1988) **401**: 361–80.
38 Jahr, *Nature* (1983) **304**: 730–33.
39 Schneider, *J Physiol* (1991) **440**: 479–96.
40 Cloues, *Pflugers Arch-Eur J Physiol* (1993) **424**: 152–8.
41 Nakazawa, *J Physiol* (1990) **428**: 257–72.
42 Zheng, *J Cell Biol* (1991) **112**: 279–88.
43 Zoeteweij, *Biochem J* (1992) **288**: 207–13.
44 Spranzi, *Blood* (1993) **82**: 1578–85.
45 Nikodijevic, *J Neurosci Res* (1992) **31**: 591–9.
46 Thomas, *J Gen Physiol* (1990) **95**: 569–90.
47 Li, *Proc Natl Acad Sci USA* (1993) **90**: 8264–7.
48 Silinsky, *Br J Pharmacol* (1994) **106**: 762–3.
49 Burnstock, *Circ Res* (1986) **58**: 319–30.
50 Silinsky, *J Physiol* (1993) **464**: 197–212.
51 Burnstock, *Trends Pharmacol Sci* (1988) **9**: 116–19.
52 Evans, *Br J Pharmacol* (1992) **106**: 242–9.
53 Brading, *Z Kardiol* (1991) **80**: 47–53.
54 Soltoff, *Am J Physiol* (1992) **262**: C934–40.
55 Soltoff, *J Gen Physiol* (1990) **95**: 319–46.
56 Soltoff, *Biochem Pharmacol* (1993) **45**: 1936–40.
57 Reber, *Pflugers Arch* (1992) **420**: 213–18.
58 Walker, *EMBO J* (1982) **1**: 945–51.
59 Saraste, *Trends Biochem Sci* (1990) **15**: 430–34.
60 Chalfie, *Nature* (1993) **361**: 504.
61 Hong, *Nature* (1994) **367**: 470–73.
62 Huang, *Nature* (1994) **367**: 467–70.
63 Canessa, *Nature* (1994) **367**: 463–7.
64 Kubo, *Nature* (1993) **362**: 127–33.
65 Suzuki, *Nature* (1994) **367**: 642–5.
66 Inoue, *Br J Pharmacol* (1990) **100**: 619–25.
67 Neuhaus, *J Neurosci* (1991) **11**: 3984–90.
68 Sauve, *Pflugers Arch* (1988) **412**: 469–81.
69 vonderWeid, *Br J Pharmacol* (1993) **108**: 638–45.
70 McMillian, *Biochem J* (1988) **255**: 291–300.
71 Bear, *Am J Physiol* (1990) **258**: C421–8.
72 Bean, *J Neurosci* (1990) **10**: 11–19.
73 Nakazawa, *Pflugers Arch* (1987) **409**: 644–6.
74 Fasolato, *J Biol Chem* (1990) **265**: 20351–5.
75 Nakazawa, *Neurosci Lett* (1990) **119**: 5–8.
76 Benham, *J Physiol Lond* (1989) **419**: 689–701.
77 Friel, *J Gen Physiol* (1988) **91**: 1–27.
78 Christie, *J Physiol Lond* (1992) **445**: 369–88.
79 Nakazawa, *J Physiol Lond* (1991) **434**: 647–60.
80 Nakazawa, *Jpn J Pharmacol* (1991) **57**: 507–15.
81 Inoue, *Eur J Pharmacol* (1992) **215**: 321–4.
82 VonKugelgen, *Trends Pharmacol Sci* (1991) **12**: 319–24.

83 Xiong, *J Physiol Lond* (1991) **440**: 143–65.
84 Merritt, *Biochem J* (1990) **271**: 515–22.
85 Krautwurst, *Biochem J* (1992) **288**: 1025–35.
86 Thomas, *Br J Pharmacol* (1991) **103**: 1963–9.
87 Soltoff, *Ann NY Acad Sci* (1990) **603**: 446–7.
88 Krishtal, *Br J Pharmacol* (1988) **95**: 1057–62.
89 Kennedy, *Arch Int Pharmacodyn Ther* (1990) **303**: 30–50.
90 Benham, *Jpn J Pharmacol* (1992) **58**: P179–84.
91 O'Connor, *Trends Pharmacol Sci* (1991) **12**: 137–41.
92 Humphrey, *Trends Pharmacol Sci* (1993) **14**: 233–6.

ELG CAT GLU AMPA/KAIN

AMPA/kainate-selective (non-NMDA) glutamate receptor–channels

Edward C. Conley **Entry 07**

NOMENCLATURES

Abstract/general description

07-01-01: Excitatory synaptic transmission in the mammalian central nervous system is mediated predominantly by **ionotropic**† **glutamate receptor–channels** (**iGluR**) which are selectively activated by the compound α-amino-3-hydroxy-5-methyl-4-isoxazolepropionate (**AMPA**). These receptor–channels show fast activation and desensitization† kinetics, and in most (but not all) neurones are characterized by having **high Na^+/K^+-permeability** and **low Ca^{2+}-permeability**.

07-01-02: At different types of '**kainate-preferring**' receptors, the neurotoxin kainic acid (**kainate**) activates fast-desensitizing† currents. Kainate also activates currents through '**AMPA-preferring**' receptor–channels, but these persist in the presence of agonist. Similarities between the physiology and distribution of AMPA-selective and kainate-selective glutamate receptors has led to frequent co-classification as the '**non-NMDA receptors**'.

07-01-03: Molecular cloning of genes encoding iGluR subunits has begun to reveal the basis for wide variations in agonist selectivities and electrophysiological behaviour observed in native† cells. 'High-affinity' AMPA receptor–channel subunits (**GluR-A**, **GluR-B**, **GluR-C**, **GluR-D**) are approximately 900 amino acids in length. Their mRNAs occur in **alternatively-spliced**† **forms**, which are under developmental control. Heterologous† expression of homomultimeric† **'AMPA-preferring'** receptor subunits produces channel currents that desensitize† in response to glutamate and AMPA agonists, while kainate activates non-desensitizing currents.

07-01-04: Genes encoding **'kainate-preferring'** iGluR (**GluR5–GluR7**, **KA1**, **KA2**, **δ1**, **δ2**) do not show any apparent variation through alternative splicing of mRNA. However, some kainate receptor subunits (and AMPA subunits) are found in modified forms produced by **RNA editing**†, which is also developmentally sensitive. In some cases, primary sequence variants introduced by RNA splicing/editing have been demonstrated to produce important functional changes (e.g. ionic selectivity† and receptor desensitization† properties).

07-01-05: While some high-affinity kainate subunits do not express functional channels in homomeric† form, others display high-affinity for kainate and support rapidly desensitizing† currents. Co-expression of these 'inactive' subunits with other iGluR gene family members form hetero-multimeric complexes with modified properties. Despite this, there is some evidence that 'AMPA-preferring' and 'kainate-preferring' subunits do *not* **cross-assemble** but rather form **functionally independent** entities which can co-exist within single cells.

07-01-06: Some iGluR subunits appear to 'dominate' the properties of hetero-multimeric† channel complexes. For example, when recombinant channels are formed *in vitro* from monomeric GluR-A, GluR-C or GluR-D subunits, they show **high Ca^{2+}-permeability** with **doubly-rectifying**† (sigmoid) I–V relationships. Introduction of the subunit **GluR-B** *in any combination* (GluR-A/B, GluR-B/C, GluR-B/D) produces recombinant channels which closely match characteristics of **native**† **receptors** (i.e. low Ca^{2+}-permeability and linear I–V relationship). iGluR in native† cell types which *lack* GluR-B (e.g. **Bergmann glia cells**) also display doubly rectifying† I–V relationships† and **high Ca^{2+}-permeability**. These types of studies indicate native receptors to be **heteromultimers**† of at least two different subunits.

07-01-07: A full appreciation of the functional roles of the non-NMDA glutamate receptors can only be made by taking into consideration parallel interactions with other (co-expressed) signalling proteins. For example, synaptic release of glutamate produces an excitatory post-synaptic current† (**EPSC**) which can be resolved into a **fast-onset**† **and decay**† **component** (mediated by **non-NMDA** ionotropic glutamate receptor–channels) and a **slow-rising, slowly-decaying component** (mediated by **NMDA** receptor–channels – *see ELG CAT GLU NMDA, entry 08*). Glutamate can also initiate responses at metabotropic† receptors, which may be indirectly coupled to other ion channel activities *(see Resource A – G protein-linked receptors, entry 56).*

07-01-08: Receptors for ionotropic† glutamate receptor–channels (iGluR) are expressed on **'virtually every' neuronal cell** and some **glial cells** in the CNS. The availability of subunit-specific probes for distribution analyses and *in vitro* mutagenesis† procedures for protein structure–function and transgenic† expression analyses are likely to further clarify their precise roles in the nervous system.

Category (sortcode)

07-02-01: ELG CAT GLU AMPA/KAIN; i.e. extracellular ligand-gated cation channels selective for the glutamate receptor agonists AMPA and kainate. The suggested **electronic retrieval code** (unique embedded identifier or **UEI**) for 'tagging' of new articles of relevance to the contents of this entry is UEI: AMPA-NAT or UEI: KAIN-NAT (for reports or reviews on native† channel properties) and UEI: AMPA-HET or UEI: KAIN-HET (for reports or reviews on channel properties applicable to heterologously† expressed recombinant† subunits encoded by cDNAs† or genes†). *For a discussion of the advantages of UEIs and guidelines on their implementation, see the section on Resource J under Introduction and layout, entry 02, and for further details, see Resource J – Search criteria & CSN development, entry 65.*

Channel designation

Shorthand designations in use

07-03-01: Ionotropic† glutamate receptor–channels (iGluR–channels) selectively activated by the agonist† **α-amino-3-hydroxy-5-methylisoxazole-**

4-propionic acid are generally designated as **AMPA receptors** (AMPA-R). These receptors show relatively high affinity for AMPA (K_d in the nanomolar range) and have relatively low affinity for kainate (K_d in the micromolar range). Conversely, **kainate receptor subtypes** display relatively high affinity for the agonist **kainate** (kainic acid) over AMPA. Some reports use the designation AMPA/KA when referring to responses from the broad class of AMPA/kainate receptors.

Common abbreviations for flip and flop variants

07-03-02: Names for the '**flip**' and '**flop**' **alternative splice variants** of GluR-A, GluR-B, GluR-C and GluR-D subunits *(for details, see Gene organization, 07-20)* are sometimes referred to in abbreviated form: e.g. GluR-A flip variants can be written as GluR-A_i and GluR-A flop variants as GluR-A_o.

Designation of subclass-specific properties within this entry

07-03-03: In order to distinguish information specific for glutamate receptors selective for AMPA, an underlined prefix '**AMPA**:' is used within this entry. Properties specific for high-affinity kainate receptors are designated with the prefix '**KAIN**:' within this entry.

Grouping of AMPA- and kainate receptors as 'non-NMDA' glutamate receptors

07-03-04: Similarities between the physiology and distribution of AMPA-selective and kainate-selective glutamate receptors has led to frequent **co-classification** as the '**non-NMDA receptors**' *(for details, see Receptor agonists, 07-50).*

Alternative nomenclatures for genes encoding AMPA/kainate subunits

07-03-05: Properties of *recombinant* glutamate receptors generally make specific reference to the **subtype** of the gene(s) encoding them. The nomenclature of presently known GluR subunits and associated proteins is, however, under review[1]. **Alternative nomenclatures** for the same subunit genes have been suggested by different laboratories. For example, Heinemann and colleagues[2–4] have used **serial numbers** for glutamate receptor genes running from GluR1 through GluR*n*. Mishina and collaborators[5] have assigned **Greek letters** to subunit families grouped by amino acid sequence homology†, and serial numbers to the members within each family. Bettler *et al.*[4] have described a further nomenclature for alternatively spliced† variants of a given receptor subunit in the form GluRX-Y where X is the running number of the receptor subunit and Y is the running number of the splice variant. Finally, Seeburg and colleagues have used the **alphabetical series GluR-A** to **GluR-D** for naming genes encoding AMPA receptors. Examples of each of these nomenclatures are presented side-by-side in the listings under *Gene family, 07-05*. For an interim discussion of these alternative nomenclatures, see ref.[1] and the **IUPHAR Nomenclature Committee recommendations** via the CSN *(see Feedback & CSN access, entry 12).*

Nomenclature based on immunological properties

07-03-06: A further nomenclature for AMPA/kainate receptors based on **immunohistochemical assays** of Co^{2+}-permeability has been defined to include subtype K1 (activated by kainate alone), subtype K2 (by glutamate and kainate) and subtype K3 (by kainate, glutamate and quisqualate) *(see Selectivity, 07-40).*

Previous nomenclatures

07-03-07: AMPA receptors were formerly designated as the **quisqualate** or **QUIS receptors**. Since quisqualate has been shown to be an agonist† at metabotropic† glutamate receptors, this classification is now less frequently used.

Distinctive nomenclature for G protein-linked glutamate receptors

07-03-08: In shorthand designation, **metabotropic† glutamate receptors**[6] (i.e. seven transmembrane domain glutamate receptors which couple to G proteins and do not contain integral ion channels) are conventionally designated by the abbreviation '**mGlu**' (although they are often referred to by the older designation 'mGluR'). These designations distinguish them from ionotropic† receptors designated by the abbreviation '**iGluR**' *(see also Resource A – G protein-linked receptors, entry 56).*

Current designation

07-04-01: Generally of the form $I_{agonist}$, e.g. I_{AMPA} (I_{QUIS}), I_{KAIN}, etc.

Gene family

AMPA/kainate receptor genes form part of the ELG channel gene superfamily

07-05-01: The genes encoding **AMPA/kainate-selective glutamate receptors** form part of the extracellular ligand-gated gene superfamily† *(described under ELG Key facts, entry 04).* The expression-cloning† of a subunit of the high-affinity AMPA/low-affinity kainate receptor–channel *(reviewed in ref.[7])* was an important step in determining that all currently known ionotropic† glutamate receptors form part of the same gene family† *(reviewed in refs[1,8]).*

Subgrouping of AMPA/kainate receptor genes based on sequence homology and ligand affinity

07-05-02: The large number of **distinct genes** encoding different subunits of functional non-NMDA glutamate receptors can be listed in three **subgroups** according to their **sequence relatedness** *(Tables 1–3)*. Ordering of subunit gene structure in this manner also enables a functional **subdivision** of the encoded protein subunits on the basis of **agonist affinity**. Three broad functional groups have emerged thus far (**high-affinity AMPA/low-affinity kainate** subunits, **high-affinity kainate/low-affinity AMPA** subunits and **high-affinity kainate** subunits *(see Tables 1–3*). Although this classification is not absolute, it provides a framework for understanding of more detailed subunit structure–function relationships.

The present necessity for citation of alternative nomenclatures in parallel
07-05-03: A listing of genes encoding the broad class of AMPA/KAIN iGluR subunits is given under their respective structural and functional subgroups in *Tables 1–3*. Alternative nomenclatures for the same subunits as used in the literature (e.g. GluR1 vs. GluR-A, *see Channel designation, 07-03)* have been preserved within this entry *according to those stated in the original citation*. This was necessary in the absence of a **universal nomenclature**. For updates on agreed nomenclatures, refer to the latest IUPHAR Nomenclature Committee recommendations via the CSN *(see Feedback & CSN access, entry 12)*.

Subtype classifications

Dependence of pharmacological subtypes on separable gene products
07-06-01: Initially, iGluRs were classified into three separate receptor populations, each defined by its selective activation by the naturally occurring excitatory substances **NMDA**, **AMPA** and **kainate** which are structural analogues of glutamate[15,16]. Properties of each of these broad categories depend on the products of **separate genes** and their combination *(for listing of vertebrate AMPA/kainate subunits, see Tables 1–3)*. Several subunit combinations can be distinguished from one another by their distinct **gating† kinetics**, cation **permeabilities†** and **conductances†**, and their susceptibility to a range of organic and inorganic **antagonists**[17] *(see other fields)*. The molecular biology of mammalian glutamate receptor–channel subtypes has been reviewed[7,8,18–20].

Unitary receptors with 'interchangeable' subunits
07-06-02: Two vertebrate excitatory amino acid ionotropic **unitary† receptors** (i.e. having more than one class of excitatory amino acid agonist specificity within one protein oligomer) have been purified from the *Xenopus* central nervous system[21,22]. These receptors consist of (i) unitary† kainate/AMPA and (ii) kainate/AMPA/NMDA receptors with **interchangeable subunits**. Each isolated oligomer contains 42 kDa subunits of the 'non-NMDA' ligand-binding type, but the second type has an additional NMDA-receptor-specific 100 kDa subunit[21]. Channels reconstituted into bilayers can elicit currents (similar to native non-NMDA GluRs) in response to low levels of AMPA or kainate.

Trivial names

07-07-01: *See Channel designation, 07-03.* The names given to native and recombinant GluRs ultimately depends on their **subunit composition**. As described in Tables 1–3, these groups consist of (i) 'AMPA-preferring' iGluR (high-affinity AMPA, low-affinity kainate receptor–channels) and (ii) 'kainate-preferring' iGluR (high-affinity kainate, low-affinity AMPA receptor–channels).

Table 1. *Genes encoding high-affinity AMPA, low-affinity kainate receptor subunits ('AMPA-preferring glutamate receptors') (From 07-05-02)*

Subunit (equivalent nomenclature in brackets)[a]	Description[b]	Other distinguishing features	Encoding	Predicted mol. wt[c] (non-glycosylated)
GluR1 (= **GluR-A**) (**α1** homologue)	AMPA receptor ~ High-affinity iGluR for AMPA ($K_d \sim$ nM), low-affinity iGluR for kainate ($K_d \sim \mu$M) Alternative splice† variants of GluR1 exist *(see Gene organization, 07-20)*	Glutamate and AMPA are potent agonists producing desensitizing† responses from heterologously† expressed channels formed from the GluR-A to -D classes Kainate is a low-affinity† agonist that activates *non-desensitizing* currents similar to those elicited in CNS neurones[9,10] AMPA-preferring iGluR are strongly potentiated by **concanavalin A** but not **cyclothiazide**[11] *(see Inactivation, 07-37)*	889 aa Sig: aa 1–18 ~ 70% sequence identity to GluR2	~ 99.8 kDa (mRNAs ~ 5.2 kb; ~ 3.9 kb; ~ 3.2 kb)
GluR2 (= **GluR-B**) (**α2** homologue)	High-affinity AMPA, low-affinity kainate iGluR Alternative splice variants of GluR2 exist *(see Gene organization, 07-20)*	Glutamate and AMPA are potent desensitizing agonists for GluR-A to -D classes GluR-B confers native channel properties (e.g. *non*-permeability to Ca^{2+}) in heteromeric† recombinant receptors, *(see Protein interactions, 07-31, and Current–voltage relation, 07-35)*	862 aa Sig: aa 1–21 ~ 70% sequence identity to GluR1; ~ 74% with GluR3	96.4 kDa

GluR3 (= **GluR-C**) (α**3** homologue)	High-affinity AMPA, low-affinity kainate iGluR Alternative splice variants of GluR3 exist *(see Gene organization, 07-20)*	Glutamate and AMPA are potent desensitizing† agonists for GluR-A to -D classes	866 aa Sig: aa 1–22 ~ 69% sequence identity to GluR2; ~ 74% with GluR2	98.0 kDa
GluR4 (= **GluR-D**) (α**4** homologue)	High-affinity AMPA, low-affinity kainate iGluR Three alternative splice variants of GluR4 exist: GluR4/flip, GluR4/flop and GluR4c/flop *(see Gene organization, 07-20)*	Glutamate and AMPA are potent desensitizing† agonists for GluR-A to -D classes Transcripts synthesized *in vitro* from GluR4c flop[12] form kainate/AMPA-activated channels showing strong inward rectification when expressed in *Xenopus* oocytes	881 aa Sig: aa 1–21	98.4 kDa

Supplementary information is given under the appropriate fieldname. For retrieval of full sequences for specific analysis, database **accession numbers** and references describing the cloning and molecular features of individual subunit genes, *see Database listings, 07-53.*

[a]Alternative nomenclatures are used interchangeably in this entry, according to the original description or citation. For a short discussion on the **parallel nomenclature** in use for these GluR-channel subunits, *see Channel designation, 07-03, and refs*[3,13].

[b]For primary sequence discussions, *see references in Database listings, 07-53.*

[c]Encoding data show the number of amino acid residues in the specified channel subunit, with signal peptide residues denoted by the prefix 'Sig:'.

Table 2. *Genes encoding high-affinity kainate, low-affinity AMPA receptor subunits ('kainate-preferring glutamate receptors') (From 07-05-02)*

Subunit (equivalent nomenclature in brackets)	Description[a]	Other distinguishing features	Encoding[b]	Predicted mol. wt (non-glycosylated†)
GluR5 (= **$\beta 1$** homologue)	High-affinity kainate receptor subunit Alternative splice variants of GluR5 exist: the longer GluR5-1 has an ORF of 920 aa and derives from insertion of a 45 nucleotide sequence GluR5-2 variants (905 aa) lack the 45 base insert	~ 41% homology to GluR-A to -D; 80% sequence identity to GluR6. Homo-multimers are weakly responsive to L-glutamate[4] Kainate is a high-affinity agonist for recombinant GluR5–GluR7 channels, producing strongly desensitizing† responses. AMPA is inactive or acts with low potency† Kainate-preferring iGluR are strongly potentiated by cyclothiazide but not concanavalin A[11] *(see Inactivation, 07-37)*	GluR5-1: 920 aa Sig: aa 1–30 GluR5-2: 905 aa	100.9 kDa

GluR6 (= β**2** homologue)	High-affinity kainate receptor subunit	Shows high affinity for kainate, and does not bind AMPA. The competitive antagonist CNQX is substantially less potent† in blocking kainate responses in GluR6 compared to GluR-A. Kainate causes rapid desensitization†	864 aa Sig: aa 1–31	96.2 kDa
GluR7 (= β**3** homologue)	High-affinity kainate receptor subunit	Agonist-elicited current responses are *not* observed in homomeric configurations of GluR7 subunits. Combined with GluR5 and GluR6, KA-1 and KA-2 *(below)*, may form heteromultimeric† high-affinity kainate receptors *(see Predicted protein topography, 07-30)*	956 aa	

Supplementary information is given under the appropriate fieldname. For retrieval of full sequences for specific analysis, database **accession numbers** and references describing the cloning and molecular features of individual subunit genes, *see Database listings, 07-53.*

[a]For primary sequence discussions, *see references in Database listings, 07-53.*

[b]Encoding data show the number of amino acid residues in the specified channel subunit, with signal peptide residues denoted by the prefix 'Sig:'.

Table 3. *Genes encoding high-affinity kainate subunits (Note: homomultimers*† *do not appear to form functional channels) (From 07-05-02)*

Subunit (equivalent nomenclature in brackets)	Description[a]	Other distinguishing features	Encoding[b]	Predicted mol. wt (non-glycosylated†)
KA-1 (= $\boldsymbol{\gamma}$**1**)	Recombinant high-affinity kainate receptor subunit	~ 30% homnology to GluR-A to -D; high-affinity for kainate/low-affinity for AMPA (K_d kainate ~ 5 nM). Homo-mulutimers† do not form functional channels. Combination with GluR5 or GluR6 yields properties distinct from GluR5 or GluR6 homomeric channels	936 aa Sig: aa 1–20	~ 105.2 kDa
KA-2 (= $\boldsymbol{\gamma}$**2**)	Recombinant high-affinity kainate receptor subunit	Highly sequence-related to KA-1, with similar pharmacological profile but a more widespread distribution in the CNS. Homomultimers do not form functional channels[14]. Combination with GluR5 or GluR6 yields properties distinct from GluR5 or GluR6 homomeric† channels	979 aa	~ 109 kDa

δ1	Recombinant high-affinity kainate receptor subunit	Subunit of unknown function	1009 aa	
δ2	Recombinant high-affinity kainate receptor subunit	Subunit of unknown function		
KBP-c **KBP-f** chick/frog brain kainate-binding proteins	Structurally related KBPs from brain of chick (464 aa) and frog (487 aa) (no channel activity)	KBPs may be truncated forms *(see Protein molecular weight (purified), 07-22, Domain conservation, 07-28, and Database listings, 07-53)*	chick, 464 aa Sig: aa 1–23 frog, 487 aa Sig: aa 1–17	~ 51.8 kDa ~ 52.5 kDa (mRNAs ~ 3.9 kb; ~ 6.0 kb)

Supplementary information is given under the appropriate fieldname. For retrieval of full sequences for specific analysis, database **accession numbers** and references describing the cloning and molecular features of individual subunit genes, *see Database listings, 07-53.*

[a]For primary sequence discussions, *see references in Database listings, 07-53.*

[b]Encoding data show the number of amino acid residues in the specified channel subunit, with signal peptide residues denoted by the prefix 'Sig:'.

EXPRESSION

Cell-type expression index

Ubiquity of iGluR expression in the CNS

07-08-01: Receptors for ionotropic glutamate receptor–channels (iGluRs) are expressed on **virtually every neuronal cell** and some glial cells in the CNS. Preparations which display unusual homogeneity or notable absence of responses to identified classes of iGluR are briefly described below. *Note:* Electrophysiological studies in native cells may not always permit clear delineation of receptor subunit types, since **co-expression of subtypes** within single cells is common *(see Subcellular locations, 07-16, and Protein interactions, 07-31).*

Evidence against cross-assembly of AMPA- and kainate-preferring receptor subunits

07-08-02: Evidence from experiments employing **cyclothiazide** blockade of rapid desensitization (having absolute selectivity for GluR1–GluR4 (AMPA) subunit receptors) suggests **independent assembly** of AMPA-preferring and kainate-preferring receptor–channels expressed in the same cell, with little evidence for **cross-assembly**[11] *(for details, see Protein interactions, 07-31, and Inactivation, 07-37).*

Advantages of subunit gene-specific probes for complex distribution studies

07-08-03: The availability of **subunit-specific probes** based on single-cell RT-PCR†, *in situ* hybridization† and immunocytochemistry permit detailed mapping of subunit expression. The results from these approaches can confirm expected **subunit distributions**, but also provide direct evidence for *absence* of expression and can reveal patterns of overlapping distribution between any number of subunits. Table 4 summarizes patterns of expression, co-expression and absence of expression for AMPA/kainate iGluR subunits in a range of neuronal cell preparations. Further details on these **expression patterns** and the contributions of subunit-specific properties to specific cell-type functions are described in other fields.

Differential expression of GluR-B in distinct hippocampal neuronal populations

07-08-04: 'Morphologically heterogeneous' hippocampal type I neurones express AMPA receptors with linear or outwardly rectifying† I–V curves† and have **low Ca^{2+}-permeability** ($P_{Ca}/P_{Cs} \sim 0.2$)[39]. A population of hippocampal neurones in culture which express AMPA receptors with inwardly rectifying† I–V curves and high Ca^{2+}-permeability ($P_{Ca}/P_{Cs} \sim 2.3$)[39] have been referred to as type II neurones. Morphologically, type II neurones are of relatively small size and ellipsoid shape. Type II neurones do not express GluR2 (GluR-B) subunits, which may explain their functional properties[40] *(see Selectivity, 07-40)*. Since the GluR-B gene product is an important determinant of native properties of heteromeric† iGluR (e.g. impermeability to Ca^{2+}) it is to be expected that its gene is subject to **'tight' regulatory control**[18].

Table 4. *Summary of AMPA/kainate subunit expression in selected neuronal preparations (For further information, see also mRNA distribution, 07-13, and Protein distribution, 07-15) (From 07-08-03)*

Subunit gene	Bergmann glia	CA1 pyramidal cells	CA3 pyramidal cells	Dentate granule cells	Cerebellar granule cells	Purkinje cells	Spinal cord motor neurones
GluR-A	+	+	+	+		+	
GluR-A flip?		low	yes				
GluR-A flop?		yes	low			yes	
GluR-B		+	+	+	+	+	+
GluR-B flip?		low	yes		yes	yes	yes
GluR-B flop?		yes	low			yes	yes
GluR-C		+	+	+		+	+
GluR-C flip?		yes	yes			yes	yes
GluR-C flop?		yes					yes
GluR-D	+	+			+		+
GluR-D flip?							yes
GluR-D flop?		yes		yes	yes		
GluR5						+	
GluR6		+	+	+	+		
GluR7				+			
KA1			+	+			
KA2		+	+	+	+		
δ1		±	±	±			±
δ2						+	
References[a]	*23–25*		*2, 4, 23, 26–29*		*30–33*	*2, 30–36*	*37*

This table lists patterns of expression in adult rat neuronal cell types for the principal AMPA/kainate iGluR subunit genes, based on available subunit-specific probes *(see above)*. + indicates unambiguous or abundant expression of the named subunit gene, whereas ± indicates low-abundance expression. The expression of **flip/flop alternative splice**† **variants** of GluR-A to -D are listed separately where these have been identified; otherwise, immunocytochemical, DNA or RNA probes were used which did not discriminate between splice variants.

[a]The arrangement of adult rat neuronal cell types is based on the review by P. Seeburg[18] who with W. Wisden collated the localization data from the indicated references.

Inhibitory and excitatory neurones express different types of non-NMDA receptor

07-08-05: Direct comparison of iGluR channel activities in ***excitatory*** **cortical neurones** (spiny, pyramidal neurones) versus ***inhibitory*** **cortical neurones** (aspiny interneurones) indicate that different types of non-NMDA receptor–channels contribute to spontaneous excitatory post-synaptic currents (sEPSCs) in these preparations[41]. Variable parameters include (i) the sEPSC decay time constant – at 2.5 ms, this is faster in inhibitory aspiny interneurones than the 4.6 ms for excitatory pyramidal neurones; (ii) the rate of desensitization† in patches, which is faster in inhibitory interneurones (3.4 ms) compared with the 12.0 ms of excitatory pyramidal neurones; and (iii) single-channel conductance, which is larger in the inhibitory aspiny interneurones (27 pS) compared with the excitatory pyramidal neurones (9 pS)[41].

Non-NMDA iGluR subtype co-expression – an example

07-08-06: Native cerebellar granule cells commonly express *more than one type* of non-NMDA receptor–channel subunit[42] *(see examples in Table 4).* High-conductance responses (activated by glutamate, AMPA and kainate) mediate most of the synaptic current in granule cells and originate from a GluR channel complex displaying conductance levels of 10, 20 and 30 pS. There is also a distinguishable, low-conductance kainate response in these cells of approximately 1.5 pS[42].

Possible equivalence of cloned and native GluR channel currents

07-08-07: KAIN: Notably, the pharmacology and channel behaviour of GluR5 receptors has been reported to be 'virtually identical' to those of the native **high-affinity kainate receptor** on dorsal root ganglia[43,44]. Mammalian spinal cord C-fibre afferents have been described as possessing 'pure' kainate-selective GluR receptor populations which are insensitive to AMPA[45].

Functional homomeric kainate-selective iGluRs in cultured hippocampal neurones

07-08-08: KAIN: Combined patterns of distribution for the five kainate-selective subunit mRNAs (GluR5–GluR7 plus KA-1 and KA-2) are similar to patterns of [^{3}H]-kainate binding determined by autoradiography[46] *(see Protein distribution, 07-15).* Although currents typical of **'high-affinity' kainate** subunits have been detected in sensory ganglia[44], recordings in central neurones were not reported until 1993[47]. In these studies, high proportions of rat hippocampal neurones expressed functional kainate-selective iGluR early after *in vitro* culturing. The kainate receptors displayed features of pronounced desensitization† with fast onset and very slow recovery, and were activated by **quisqualate** and **domoate**, but not by AMPA[47]. The recordings obtained in this preparation were consistent with the expression of native homomeric† GluR6 receptors.

Cloning resource

Vertebrate brain cDNA libraries

07-10-01: Because of the **ubiqitous expression** of the iGluRs in the CNS, cDNA libraries constructed from total brain poly(A)$^+$ mRNA have been used as the main resource for expression cloning† strategies. Conventional cross-homology†, low-stringency† screening of brain cDNA libraries has resulted in isolation of most of the known GluR subtypes. For possible cloning resources *outside* the CNS, *see mRNA distribution, 07-13.*

Developmental regulation

iGluR genes typically display distinct temporal patterns of expression throughout development

07-11-01: AMPA: In addition to the well-defined **spatial patterns** of expression characteristic of iGluR genes *(see mRNA distribution, 07-13)* developing embryonic and post-natal brain tissues display **dynamic elevations and reductions** characteristic of independent **developmental gene regulation**. For example, studies of GluR1–GluR3 developmental gene expression show GluR1 to be prominent in cortex and caudate-putamen at early ages, but these fall to background levels in the adult[48]. GluR2 mRNA levels in P4–P21 rat brain does not exceed adult levels at any stage, except in cerebellum at P14 and in thalamus (where it diminishes to background levels after P4). The GluR3 gene shows notably high levels of expression in striatum (at P4) and in cerebellum (at P7)[48].

Functional switching of GluR gene exons by alternative splicing during brain development

07-11-02: AMPA: AMPA receptors expressed in the early brain differ in molecular type and functional properties to those in the adult brain: '**Developmentally early**' (**pre-natal**) GluR-A–GluR-D receptors carry mainly 'flip' modules, and these forms persist throughout development into the adult *(for description of 'flip/flop modules', see Gene organization, 07-20)*. Receptors containing flip modules appear to have slower desensitization† kinetics than adult receptors. The 'flop' modules increase only during early post-natal development, while the flip module is expressed in an invariant pattern, leading to frequent co-expression of flip and flop in adult cells. Glutamate activation of flip versions produces more current than those composed of flop[26]. For each receptor variant, the alternatively spliced† messenger RNAs show distinct expression patterns in rat brain, particularly in the CA1 and CA3 fields of the hippocampus. These results identify a '**genetic switch**' in the molecular and functional properties of glutamate receptors operating through alternative splicing†.

'Editing' of GluR mRNA during development

07-11-03: AMPA: The GluR-B subunit, which is dominant in determining functional properties of heteromeric AMPA receptors, exists in **'edited'** and **'unedited'** forms which are developmentally distinct *(for details of **RNA editing processes**, see Gene Organization, 07-20, and Domain functions, 07-29).*

A conserved sequential pattern of ion channel gene expression in spinal cord development

07-11-04: In developing populations of cells from several regions of embryonic (> day 13) rat spinal cord, functional **sodium channels** appear prior to **GABA$_A$ receptors**, which in turn emerge prior to **kainate-activated iGluR**. This **stereotypical pattern** of sequential channel expression during development occurs individually on most cells in all spinal cord regions[49]. Ventrodorsal† and rostrocaudal† gradients of expression reflect known patterns of spinal cord neurogenesis[49].

Multiple co-ordinated iGluR gene expression control in developing CNS and PNS

07-11-05: As with other iGluR receptor genes, GluR4 and GluR5 show striking patterns of **gene induction**† and **gene silencing**† in the course of embryonic and post-natal CNS and PNS development. These patterns have been tabulated for a number of brain regions derived from a systematic *in situ* study of GluR4 and GluR5 gene expression *(see ref.[4])*.

Intrinsic or highly localized signals guide GluR4 gene expression during development

07-11-06: AMPA: Transcripts for three variants of the GluR4 gene (GluR4 flip, GluR4 flop, GluR4c flop *(see Gene organization, 07-20)* are much more abundant in the cerebellum than in other brain areas, and their levels are increased during **cerebellar development**. Maximal increases are observed between post-natal days 1 and 20, ages corresponding to the division and maturation of granule neurones. Granule cells already express GluR4c flop in the pre-migratory zone of the external granular layer, indicating that intrinsic or **highly localized cues** induce GluR4c expression before these cells reach their final position[12].

GluR4 and GluR5 are expressed in areas of neuronal differentiation and synapse formation

07-11-07: The GluR4 and GluR5 genes are expressed in subsets of neurones throughout the developing and adult **central and peripheral nervous systems** (see the systematic *in situ* hybridization study in ref.[4]). Generally, the distribution and expression level of GluR5 mRNA is different from and quantitatively much lower than that of the GluR1–GluR4 (AMPA receptor) mRNAs[4]. During embryogenesis, GluR5 transcripts are detected in areas of **neuronal differentiation** and **synapse formation** (synaptogenesis). '**Phylogenetically old**' brain structures are the first to accumulate GluR4 and GluR5 transcripts[4].

Stage- and cell-type-selective agonist sensitivities within glia

07-11-08: Comparative studies of ionotropic† glutamate receptors in cultured **cerebellar glial cell types** (type 1 and type 2 astrocytes, oligodendrocytes, O-2A progenitors) show agonist responses that are stage- and cell-type-specific[50]. Multiple subunits can be expressed in glial cells – for example, the O-2A lineage display rapidly desensitizing† responses to kainate and expression of mRNAs for GluR6, GluR7, KA-1 and KA-2. They also display rapidly desensitizing† responses to AMPA, and mRNAs for GluR-B, -C and

-D. Thus two receptor populations are present in these cells, with high- and low-affinity for kainate and different sensitivity for potentiation by **concanavalin A** and for block of desensitization by **cyclothiazide**[51].

Developmental regulation in (atypical) calcium permeability of kainate receptors

07-11-09: KAIN: The calcium permeability† of kainic acid-gated receptor–channels is dependent on **subunit composition**[52], and the genes encoding these may be developmentally regulated[53]. *Note:* High Ca^{2+}-permeability is usually associated with the NMDA-selective ionotropic glutamate receptor–channels *(see ELG CAT GLU NMDA, entry 08).*

Differential expression of kainate receptor subunits during brain development

07-11-10: KAIN: From embryonic day 14 (E14) onward, subunit KA-1-expressing cells can be observed in populations of putative **neuronal precursor cells** lining the dorsal† aspect of the brain[14]. **Subunit KA-2** mRNA is ubiquitously expressed at a high level throughout the embryonic CNS. From E14 through to post-natal day 1, KA-2 transcripts are abundant in spinal cord, brain and some areas of the PNS. At E17, KA-2 mRNA is present in all layers of the **spinal cord**, **mesencephalon** and **telencephalic structures**. The embryonic olfactory neurones, nasal epithelium, dorsal root ganglion and pituitary express KA-2 at a high level.

Neural cell competition, survival and death during development

07-11-11: Activity-dependent, **competitive synaptic interactions** (which stabilize some axon branches and dendrites and remove others) involve glutamate receptor expression *(see ELG Key facts, entry 04, and reviews*[54–56]*).* In terms of brain development, these phenomena are fundamental to the ordered **progression of neuronal phenotypes** and demonstrate important interactions between membrane signal transduction and specific patterns of gene expression. The physiological and pathophysiological roles of excitatory amino acids during development have been reviewed[57] *(see also Phenotypic expression under ELG CAT GLU NMDA, 08-14).*

AMPA/kainate iGluR activation in regulation of growth factor gene transcription

07-11-12: Non-NMDA receptor activation in hippocampal pyramidal neurones has been shown to regulate mRNA levels of two **growth factors** essential to neuronal survival: **brain-derived neurotrophic factor**[58] (BDNF) and **nerve growth factor**[58, 59] (NGF).

Isolation probe

07-12-01: The first examples of cloned glutamate receptors were isolated using **expression-cloning**† protocols *(for references, see Database listings, 07-53).* Generally, other subunit genes were isolated using cDNAs as probes in low-stringency† hybridization† protocols (e.g. GluR1 cDNA probe for GluR2 and GluR3[3]).

mRNA distribution

Methodological notes for distribution studies

07-13-01: Immunocytochemical determination of subunit distribution is generally more sensitive than methods based on hybridization† of radiolabelled probes to low-abundance RNA transcripts. Although problems of antibody cross-reactivity might exist, immunological methods do not depend on **protein turnover** (unlike RNA-based methods, which effectively measure the **'synthesis minus degradation'** component of specific molecules in the cytoplasm. The use of antibodies may report the position of **proteins *in transit***, and it is common for large pools of receptor or channel protein to be present in the cytoplasmic vesicles where it has no (apparent) function. *In situ* mRNA hybridization is well-suited for reporting regional and temporal variations in expression level but is not ideal for detection of low-abundance mRNA transcripts (which may be the case for many ion channel types). *Note:* **Single-cell PCR**† methods have the greatest sensitivity and highest resolution for RNA distribution studies, and can be expected to make increasing impact for these types of study. Differences in sensitivity and resolution in these commonly used procedures for localization of ion channel gene expression may account for some **lack of consensus** between independent studies. For details on computer-based 'expression atlases' designed to consolidate gene expression patterns in brain and other tissues, *see Resource J – Search criteria & CSN development, entry 65.*

Generalized patterns of subunit expression at the mRNA level

07-13-02: The AMPA/kainate-selective receptors GluR1–GluR6 exhibit two patterns of mRNA expression: most neurones express GluR1, R2 and R6, whereas only ~20% express significant levels of GluR3, R4 and R5 *(see also Table 4).*

Restricted distribution of 'flip' variants

07-13-03: AMPA: In general, AMPA-selective iGluR subunit genes display **differential spatial expression**[13]. Tissue-specific expression for **'flip' and 'flop' variants** in the AMPA family is of central importance in the function of these receptors[26] *(for a description of 'flip/flop' variation, see Gene organization, 07-20).* In the hippocampus, CA3 neurones preferentially express the 'flip' version of GluR-A, -B and -C, while CA1 neurones preferentially express the 'flop' versions of these receptors *(see summary in Table 4 under Cell-type expression index, 07-08).*

Differential activation and putative silencing of AMPA subunit gene expression

07-13-04: AMPA: Independent studies[60] have shown that *combinations* of GluR-A to -D mRNA expression patterns vary considerably according to location. For instance, GluR-B mRNA is expressed strongly within most brain areas *except* the **Bergmann glia**[61]. GluR-D mRNA is also widely distributed, although the expression level is relatively low. Several brain areas can be observed which lack (or express very little) GluR-A mRNA (e.g. some nuclei in the **motor** and **auditory systems**). Similarly, some nuclei in the **hypothalamus** and general **somatosensory system** lack or express very little GluR-C mRNA[60].

mRNA encoding functional GluR–channels in astrocytes
07-13-05: Messenger RNAs encoding functional acetylcholine and glutamate receptors with similar properties to those in neurones have been detected in human **astrocytoma** cells[62]. In contrast, human **glioblastoma cells** lack these mRNAs[63]. *Note:* 'Non-NMDA'-type glutamate receptors play an important part in determining the resting potential of visual streak astrocytes *in situ* and may be of general importance for the functions of astrocytes *in vivo*[64].

Contrasting distribution patterns of KA-1 and KA-2 subunit gene expression
07-13-06: KAIN: High-affinity kainate receptor KA-1 mRNA expression is high in the CA3 region and dentate gyrus of the hippocampus but is 'virtually silent' in CA1 cells[65]. The expression level of GluR5 mRNA is generally much lower than that of the AMPA or GluR6 receptor mRNAs, but areas of high expression correlate with areas of high-affinity kainate binding in the dorsal root ganglia[44] *(see below).* Subunit KA-2 transcripts are abundant in the cerebral cortex (particularly layers II/III and V/VI), pyriform cortex, caudate-putamen, hippocampal complex (all subfields), medial habenula and granule cell layer of the cerebellum. Thus, the highly selective KA-1 subunit mRNA distribution contrasts with the wide distribution of KA-2 subunit mRNA[14, 65].

Combined kainate receptor mRNA distributions match high-affinity [^{3}H]-kainate-binding sites
07-13-07: The observed patterns of high-affinity [^{3}H]-kainate sites in rat brain overlap with the *combined* patterns of the GluR5, GluR6, GluR7, KA-1 and KA2 mRNAs[18]. These sites include the layer I and the inner laminae of the neocortex and cingulate cortex, caudate-putamen, the CA3 region of the hippocampus, the reticular thalamus, the hypothalamic median eminence and the cerebellar granule cell layer[18, 46].

Co-distribution of GluR6 mRNA and kainate ligand-binding sites
07-13-08: KAIN: The GluR6 subunit mRNA is not expressed in the same *overall* pattern as that for KA-1 *in vivo*, but it is enriched in patterns similar to high-affinity binding sites of kainate located by autoradiographic techniques (the CA3 hippocampal region, the caudate-putamen and cerebellar granule cells).

GluR family mRNA distribution patterns – the retina as an example
07-13-09: *In situ* hybridization studies of **retinal transverse sections** show that mRNAs for seven receptor subunits (GluR1–GluR7) are expressed in both cat and rat **retinal tissue**[66]. Glutamate receptor subunits are employed at many of the retinal synapses, including the photoreceptor input to the **outer plexiform layer** and the contacts of **bipolar cells** with the processes at the **inner nuclear layer** (INL). Probes for GluR1 and GluR2 mRNA produce labelling over the entire INL and ganglion cell layer (GCL). GluR3–GluR7 mRNA has a limited distribution, indicative of expression by only a subset of neurones.

All of the subunits are expressed by the cells at the inner edge of the INL (where amacrine cells are located) though the layers containing the horizontal, bipolar and ganglion cells contain different subsets of subunits. By hybridizing adjacent semi-thin (1 μm) sections of the cat retina with probes for GluR1–GluR3, co-expression was shown of all three subunits (or of pairs of these subunits) in cells within the INL and GCL.

Differential expression of iGluR mRNAs in adrenal gland
07-13-10: AMPA: In *in situ* hybridization studies of the rat **adrenal gland**[67], mRNA encoding all GluR1–GluR4 subunits are found to be expressed in the medullary ganglion cells. Patterns of hybridization suggest that different cell populations of the adrenal gland may express homomeric† forms of different receptor subtypes. The four mRNAs are preferentially expressed as follows: GluR1 (zona glomerulosa of the cortex), GluR3 (remaining parts of the cortex), GluR2 (adrenal medullary cells) and GluR4 (zona glomerulosa at low abundance). Of RNA populations analysed, the 'flip' splice variants of GluR2 and GluR3 are highly represented while GluR2 mRNA is present in the arginine-encoding form[67] *(for significance, see Gene organization, 07-20). Note:* Expression of ionotropic† glutamate receptor genes outside of the nervous system has also been reported in p19 embryonal carcinoma cells[68].

GluR4 and GluR5 expression in the peripheral nervous system
07-13-11: In the **peripheral nervous system**, gene transcripts for GluR4 and GluR5 are strongly expressed in the mural ganglia of the intestinal organs, and variably in the dorsal root ganglia (high for GluR5) and cranial ganglia (e.g. the trigeminal ganglion and acoustic ganglia)[4].

Restricted expression of AMPA subunits in optic nerve white matter
07-13-12: RT-PCR† studies have determined that only GluR1 and GluR3 subunits are expressed in the white matter of rat **optic nerve**[69]. GluR1 'flip' shows a relatively higher (but diffuse) expression pattern as detected by Northern† and *in situ* hybridization† analyses[69].

Phenotypic expression

Phenotypic roles of non-NMDA GluR–channels
07-14-01: In the CNS, the principal mediators of **fast excitatory neurotransmission** are GluRs responsive to **AMPA**. *Intrinsic* channel properties are primary determinants of the kinetics of the fast excitatory post-synaptic currents†[70], as opposed to the rate of neurotransmitter clearance *(see fields under the STRUCTURE & FUNCTIONS and ELECTROPHYSIOLOGY sections).*

Correlation of kainate-selective GluR expression with regional neurotoxicity
07-14-02: KAIN: **Kainic acid** is a potent† **neurotoxin** for certain neurones. The high selective expression of KA-1 messenger RNA in the CA3 region of the

hippocampus *(see mRNA distribution, 07-13)* closely corresponds to high-affinity kainate binding sites defined by autoradiography. This correlation, as well as the specific patterns of **neurodegeneration** observed *in vivo*, suggests that KA-1 subunits participate in receptors mediating one mechanism for kainate sensitivity[65]. *Note:* A clinical syndrome characterized by seizures and brain damage has been linked to ingestion of a kainate receptor agonist (**domoate**) found in contaminated mussels[71].

'Down-regulation' of GluR2 (GluR-B) gene expression prior to neuronal degeneration

07-14-03: Severe, transient **global ischaemia** of the brain induces delayed damage to specific neuronal populations. *Sustained* Ca^{2+}-influx through glutamate receptor–channels is thought to play a critical role in **post-ischaemic cell death**. Following severe, transient forebrain ischaemia GluR2 subunit gene expression is preferentially reduced in CA1 hippocampal neurones at a time point that *precedes* their degeneration[72]. Timing of this change in expression of kainate/AMPA receptor subunits coincides with reported increases of Ca^{2+}-influx into CA1 cells, and may therefore indicate a 'causal' role in post-ischaemic cell death[72]. (For significance of GluR2 (GluR-B) subunit loss on Ca^{2+} conductance, *see Selectivity, 07-40.*) *Note:* Many enzymes are Ca^{2+}-activated and may therefore contribute to excitatory amino acid toxicity *(see Phenotypic expression under ELG CAT GLU NMDA, 08-14).*

An in vitro *model of induced post-ischaemic damage*

07-14-04: AMPA/kainate receptor activation *contributes* to ischaemic damage induced by 5 min of oxygen and glucose deprivation in the CA1 region of the rat hippocampal slice, a treatment that causes **long-term synaptic transmission failure** (LTF)[73]. However, it has been determined that whatever the effects on the AMPA/kainate iGluR during this process, they do *not* involve enhancement of Ca^{2+}-entry through the channel[73]. Notably, total cell Ca^{2+}-*prior to ischaemia* was determined to be adequate to cause complete LTF. Ca^{2+}-influx during the 'first $2\frac{1}{2}$ min' of ischaemia depends entirely on NMDA channels *(see ELG CAT GLU NMDA, entry 08)*, but NMDA channel blockers have no effect during the 'second $2\frac{1}{2}$ min' (an effect probably linked to NMDAR dephosphorylation). The ischaemia-induced Ca^{2+} influx during the second $2\frac{1}{2}$ min of ischaemia could be attenuated 25% by nifedipine (50 μM, *see VLG Ca, entry 42*) and an additional 35% by the Na^+/Ca^{2+} exchange inhibitor benzamil (100 μM)[73].

Changes in GluR-B flip mRNA levels accompanying kainate-induced epilepsy and ischaemia

07-14-05: Kainate-induced **epilepsy**† induces a rapid but transient increase (50%) of GluR-B flip mRNA levels in 'all subregions' of the hippocampus (CA1, CA3, dentate gyrus)[74]. **Seizure**†-resistant CA1 dentate gyrus neurones show a subsequent, persistent GluR-B flip increase while the 'seizure-susceptible' CA3 area displays a 35% decrease in GluR-B flip message levels. In the subiculum and CA1 (areas hypersensitive to ischaemic insult) the levels of GluR-B flip and flop variants are substantially reduced (90–100%) following global ischaemia, and this reduction takes place long before morphological signs of **cell death**[74].

Potentiation of zinc neurotoxicity by AMPA receptor activation
07-14-06: In contrast to its effect on *attenuating* NMDA receptor-mediated excitation and **neurotoxicity**, extracellular Zn^{2+} *increases* AMPA receptor-mediated toxicity[75]. High extracellular K^+ concentrations or kainate also potentiates Zn^{2+} toxicity, however toxicity arising from AMPA plus Zn^{2+} is *attenuated* by raising extracellular Ca^{2+}, or by use of Ca^{2+} channel blockers. Exposure to AMPA plus Zn^{2+} induces an increase in fluorescence from neurones loaded with the **Zn^{2+}-sensitive dye TS-Q** and increases subsequent $^{45}Ca^{2+}$ accumulation[75]. These features may be of significance in understanding mechanisms of **neuronal death** associated with 'intense activation' of glutamatergic† pathways.

Long-term effects of glutamate toxicity in pathological conditions
07-14-07: Prolonged receptor-mediated depolarization (and associated elevated Ca^{2+}-influx) can result in irreversible disturbances in ionic homeostasis[76] which may promote or inhibit gene expression. Activation of non-NMDA receptor subtypes has been shown sufficient to induce the rapid and dramatic increase of **immediate-early**† gene products *(see also ELG Key facts, entry 04, and ELG CAT GLU NMDA, entry 08).*

LTP and synaptic depression phenotypes associated with non-NMDA receptors
07-14-08: While Ca^{2+}-influx through NMDA receptor–channels is usually associated with the initation of **long-term potentiation**† (LTP) in hippocampal area CA1, no NMDA receptors are required to trigger LTP in hippocampal areas CA2/CA3. Similarly there is no apparent requirement for NMDA for **long-term depression**† (LTD) phenotypes in cerebellar Purkinje cells. In the parallel fibre–Purkinje neurone (PF–PN) synapse model, LTD induction requires activation of both AMPA and metabotropic receptors, together with PN depolarization. Post-synaptic Na^+-influx through the AMPA-associated channel is necessary for LTD. While a portion of the Na^+-influx is provided by voltage-gated channels, the AMPA-associated ion channel provides the most important route[77]. *(For illustrations of the mechanisms underlying LTP phenotypes, see Phenotypic expression under ELG CAT GLU NMDA, 08-14).*

'Differential enhancement' of the non-NMDA EPSC component during induced LTP
07-14-09: Pharmacological modulation of transmitter concentrations in CA1 pyramidal cells of guinea-pig hippocampal slices (using the $GABA_B$ agonist **baclofen** or by the adenosine antagonist **theophylline**) result in parallel changes of NMDA and non-NMDA receptor-mediated components of

EPSCs over a 16-fold range[78]. Induction of long-term potentiation by delivery of low-frequency synaptic stimulation in conjunction with depolarization to +30 mV leads to differential enhancement of the non-NMDA receptor-mediated component of the EPSC. Stimuli inducing LTP do *not* cause a sustained enhancement of isolated NMDA receptor-mediated EPSCs† evoked in the presence of the AMPA receptor antagonist CNQX[78].

GluR activation linked to secretory events

07-14-10: AMPA: L-Glutamate has been shown to induce immediate, transient and concentration-dependent **glucagon release** by activating a receptor of the AMPA subtype in isolated rat pancreatic cells[79]. *(See also iGluR-associated* ***growth hormone release*** *in chromaffin cells under Receptor/transducer interactions, 07-49).*

Specialized iGluR's mediating complex auditory information

07-14-11: The fast kinetics of response to glutamate and kainate in **cochlear neurones** enable perception of sound direction by discrimination of **microsecond differences** in the arrival of sound at the two ears. The ability to fire action potentials precisely correlated with synaptic input or '**phase-locking**' can occur at high frequencies (~9000 Hz) in species such as owls. In avian cochlea, AMPA/kainate receptors expressed on neurones in the **nucleus magnocellularis** (nMAG) exhibit unusually rapid onset and termination. For example, in chicken nMAG, 10 mM glutamate-evoked current in patches rises (10–90%) in 0.33 ± 0.18 ms ($N = 13$ patches) and desensitizes biphasically (< 1% peak current) with a fast time constant of 960 μs at 22°C, decreasing to 570 μs at 33°C[127].

iGluR involvement in modulation of inhibitory post-synaptic current

07-14-12: Monosynaptically evoked *inhibitory* post-synaptic currents in hippocampal pyramidal slices diminish in the presence of **CNQX** (an AMPA receptor-selective antagonist – *see Receptor antagonists, 07-51*) and **APV** (an NMDA-selective antagonist) following a train of action potentials. Responses to GABA applied by iontophoresis† however, do not change significantly *(cf. Channel modulation under ELG Cl $GABA_A$, 10-44).*

Expression pattern of iGluR subunits and signalling functions in glial cells

07-14-13: KAIN: **Bergmann glial cells** in mouse cerebellar slices are unusual in that they do *not* express GluR-B subunits although they do express GluR-A and GluR-D subunits[13,23,61]. Bergmann glial cells express native *kainate-type* iGluR with a characteristic (sigmoid) current–voltage relation[27]. These channels are (atypically) **Ca^{2+}-permeable** and can be blocked by CNQX *(see Receptor antagonists, 07-51).* Entry of calcium also leads to a marked reduction in the resting (passive) potassium conductance of the glial cells. Purkinje cells, which are closely associated with Bergmann glial cells, may provide an important stimulus through their glutamergic synapses[27,80] *(see Current–voltage relation, 07-35). Note:* iGluR in **hippocampal astrocytes** can propagate calcium waves in cellular networks, a mechanism which may form part of a 'long-range' glial signalling system[81].

Protein distribution

Radioligand and immunocytochemical distribution studies

07-15-01: KAIN: High-affinity sites for kainate binding ($K_d \sim 5$ and 50 nM) have been shown in the CA3 area of the hippocampal formation[82] and the peripheral neurones of dorsal root ganglia[44]. The **kainate-binding protein** from chick cerebellum is *exclusively* localized on **Bergmann glial membrane** (in close proximity to established glutamatergic synapses). Monoclonal antibodies raised against peptide sequences of the chick KBP (KBP-c) sequence[83] *(see Database listings, 07-53)* cross-react with a 49 kDa protein in cerebellar membranes[83].

07-15-02: Tabulated summaries of ionotropic GluR expression in selected cell types of the rat CNS have appeared[17, 18] *(see also Table 4 and the methodological note under mRNA distribution, 07-13).*

Subcellular locations

Segregation and clustering of ionotropic GluRs in rat hippocampal neurones

07-16-01: Clusters of AMPA-selective GluR1 and GluR2/3 channels **co-localize** in cultured rat hippocampal neurones and are restricted to a subset of post-synaptic sites[84]. GluR1 and GluR2/3 segregate to the **somatodendritic domain** within the first week in culture, even in the absence of synaptogenesis. Glutamate receptor-enriched spines develop later and are present only on presumptive pyramidal cells, not on GABAergic interneurones. There is preliminary, but there is direct evidence for dendritic location of GluR5–GluR7 subunits in primate cortex and hippocampus *(cited in ref.[17])*.

General pattern of iGluR protein family distribution

07-16-02: AMPA and NMDA GluRs are generally post-synaptic[85, 86] and are closely apposed ($\leq 2\,\mu$m) to **transmitter-release sites** or expressed on glial cells. Kainate receptor–channels have been localized to pre-synaptic sites.

Orientation of iGluR N- and C-terminal domains in the post-synaptic face

07-16-03: GluR-A anti-N-terminal antibodies show immunoreactivity to the synaptic face of the plasma membrane, while **anti-C-terminal antibodies** immunolocalize to the intracellular part of the post-synaptic membrane[87]. *Comparative note:* Some protein domain topology models place the C-terminus of iGluR subunits on the extracellular face *(see [PDTM], Fig. 3).*

Immunocytochemical distribution studies

07-16-04: GluR-B and GluR-C subunits are abundantly expressed on **Purkinje cell bodies** and **spines**. However, no AMPA receptor immunoreactivity is detected at the parallel fibre synapse that mediates excitatory post-synaptic currents from granule cells on to Purkinje cells[88].

SEQUENCE ANALYSES

Note: The symbol [PDTM] within this and the next section denotes an illustrated feature on the channel protein domain topography model (Fig.3).

Chromosomal location

Predicted phenotypes of mutant (dysfunctional) iGluR genes

07-18-01: Although no definitive linkage† of iGluR gene defects has been made to any known inherited disease, mutations in iGluR genes would be *predicted* to lead to phenotypes such as **epilepsy**, **cell death**, defective **neural development** or psychiatric symptoms *(see Phenotypic expression, 07-14).* The human AMPA/low-affinity kainate receptor subunit genes have been mapped to specific chromosomes[89], as summarized in Table 5.

Table 5. *Chromosomal locations of the human AMPA/low-affinity kainate receptor subunit genes[a] (From 07-18-01)*

Gene encoding non-NMDA iGluR subunit	Location
GluR-A gene homologue	Human chromosome 5
GluHI (~ 97% identity with rat GluR1)	Human chromosome 5q33[b]
GluR-B gene homologue	Human chromosome 4q32–33[c]
GluR-C gene homologue	Human chromosome X
GluR-D gene homologue	Human chromosome 11[d]
GluR5 gene homologue	Human chromosome 21q21.1–22.2[e]
HBGR1[f]	Human chromosome 5q31.3–33.3 *(from GenBank® entry)*
HBGR2[f]	Human chromosome 5q31.3–33.3[90]
HBGR2[f]	Human chromosome 4q25–34.3[90]

[a]For determination of chromosomal locations of AMPA subunit genes, see ref.[89].
[b]As determined by fluorescence *in situ* hybridization (FISH).
[c]This location excludes GluR-B as a candidate gene involved in Huntington's disease.
[d]The *region* containing the GluR-D gene has been associated with linkage to schizophrenia and major depression.
[e]The GluR5 gene is in the vicinity of the gene linked to familial amyotrophic lateral sclerosis[91, 92].
[f]Isolate name, *see Database listings, 07-53.*

Encoding

07-19-01: For predicted open reading frame† lengths determined from cDNA sequences, see the respective subunit gene name in Tables 1–3 *under Gene family, 07-05.*

Gene organization

GluR–channel diversity via the alternative splice variants 'flip' and 'flop'

07-20-01: AMPA: In the GluR-A to GluR-D family, a 115 bp segment preceding the gene region encoding the predicted M4 transmembrane domain *(see [PDTM], Fig. 3)* has been shown to exist in two versions with different amino acid sequences. These **38 amino acid modules**, designated '**flip**' and '**flop**', are encoded by adjacent exons† of the receptor genes and impart different pharmacological and kinetic properties on currents evoked by L-glutamate or AMPA, but not those evoked by kainate *(for details, see Receptor agonists, 07-50)*. 'Flip' and 'flop' differ in only a few (9–11) of the 38 amino acids. A pentapeptide occurring at a comparable position in each subtype is consistently different in the 'flip' and 'flop' versions *(see Fig. 1, and [PDTM], Fig. 3)*.

Comparison of cDNA and genomic sequences encoding AMPA receptors

07-20-02: The switched versions of 'flip' and 'flop' are generated by alternative splicing† of the two regions which are on adjacent exons†[26] separated by an intron† of ~900 bp *(see Fig. 1)*. For each receptor, the **alternatively spliced**† mRNAs† show **distinct expression patterns** in rat brain, particularly in the CA1 and CA3 fields of the hippocampus *(see mRNA distribution, 07-13)*. Thus, the **flip/flop modules** enable functional properties of glutamate-activated currents to be controlled by alternative splicing† events. (For the phenotypic consequences of flip and flop modules, *see Inactivation, 07-37.*)

A note on splice variant nomenclature

07-20-03: The names 'flip' and 'flop' may imply that the alternative cassettes are in 'reverse orientation' to each other, but this is *not* the case *(see Fig. 1 for clarification)*.

Existence of further alternative splice variants

07-20-04: A third type of transcript (**GluR4c flop**) derived from the GluR4 gene by **differential RNA processing**† has been isolated following screening a rat cerebellar cDNA library[12]. GluR4c transcript encodes a protein with a 'flop' module between transmembrane regions 3 and 4, but with a C-terminal segment of 36 amino acids different from the previously described GluR4 flip/flop cDNAs. Transcripts synthesized *in vitro* from GluR4c flop[12] form kainate/AMPA-activated channels showing strong inward rectification† when expressed in *Xenopus* oocytes.

Alternative splicing in the GluR5 gene

07-20-05: Alternative splice† variants of the high-affinity kainate receptor subunit **GluR5** have also been described[4]. The longer GluR5-1 variant has an open reading frame† (ORF) of 920 aa and derives from insertion of a 45 nucleotide sequence within the first third of the N-terminal (extracellular) glutamate receptor domain. A full open reading frame† for the shorter splice variant (GluR5-2, lacking the 45 base insert for a predicted 905 aa ORF) was not isolated in the original study[4].

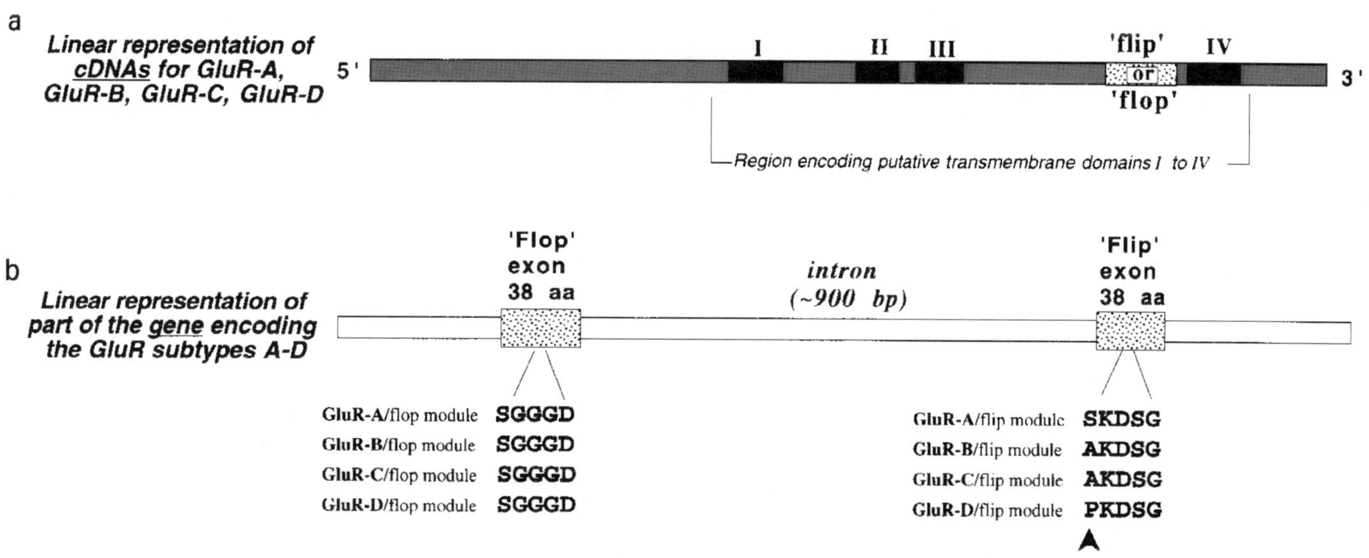

Figure 1. *'Flip' and 'flop' alternative splice variants in AMPA receptor genes. (a) Sequencing of* **cDNAs** *for the four subtypes of the AMPA receptor revealed a 115 base sequence encoding a* ***38 aa*** *segment which existed in two sequence versions designated 'flip' and 'flop'. (b) Sequencing in the region of* **genomic DNA** *preceding that encoding domain IV revealed the 'flip' and 'flop' segments to be encoded by two* **separate exons***, flanking an intron, as shown.* **Alternative use** *of 'flip' and 'flop' exon sequences in AMPA receptor genes confers different kinetic properties of currents evoked by glutamate or AMPA on GluR-A to GluR-D subunits. For details, see Gene organization, 07-20, and Receptor agonists, 07-50. Within the alternative 38 aa flip/flop module (occupying equivalent residue positions in the cDNAs), flop versions of the subunits contain a conserved SGGGD motif, while flip versions are* **variable at a single residue** *(arrowed) in the subunits. (From 07-20-02)*

GluR subunit structural changes via RNA editing mechanisms

07-20-06: A further possible mechanism for generating heterogeneity in the channel coding regions of GluR transcripts† is **RNA editing**†. In the case of the GluR-B gene, it was found that there had been a nucleotide change (an adenosine-to-guanosine transition†) observed between the sequences obtained from genomic† DNA and cDNA† sources[93] (*note:* normally there is perfect agreement of sequence). It has been proposed that **'editing'** of an **adenosine to inosine** gives rise to an Arg (R in single-letter code) instead of a Gln (Q) residue at the critical **Q/R site** *(for the consequences of this change, see Domain functions, 07-29).*

Subunit RNA transcript selectivity for editing processes

07-20-07: Although the GluR-B subunit transcript appears to undergo editing in $>99\%$ of isolates found, the GluR-A, -C and -D subunit transcripts (sharing identical domain M2 sequences) do *not* appear to be edited. In transcripts for GluR5 and GluR6, edited and non-edited transcripts are found at different **ratios**[43]. The principle of RNA editing† is illustrated in Fig. 2. *Note:* RNA editing has been shown *not* to be a *general* cellular mechanism for GluR diversity *(see mRNA distribution under ELG CAT GLU NMDA, 08-13).*

Base-paired intron–exon sequences are required for GluR-B RNA editing

07-20-08: Low RNA editing efficiency is observed in GluR-B gene constructs modified in sequences at the proximal part of the intron† downstream of the

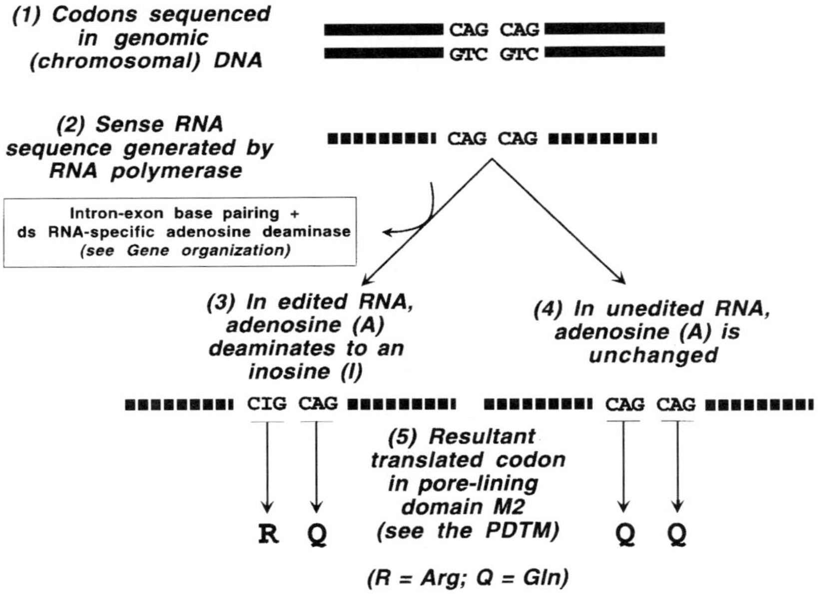

Figure 2. *RNA editing as a mechanism for generating heterogeneity of iGluR transcripts in regions encoding pore-lining domains. (From 07-20-07)*

GluR-B unedited exonic† site[94]. **Perfect intron–exon† sequence base-pairing** of this region is required for efficient Q/R site editing. In the native GluR-B gene, this portion of the intron contains an imperfect inverted repeat preceding a 10 nucleotide sequence with exact complementarity to the exon centred on the unedited codon. Single base substitutions in this short intronic sequence (or its exonic complement) inhibit Q/R site editing, but this can be recovered by restoring complementarity† in the respective partner strand. Base-paired sequences in the GluR-B subunit protein coding region appear to direct base conversion by a nuclear **adenosine deaminase** specific for double-stranded RNA[94].

Editing of GluR6 RNA transcripts – combinatorial variations for edited forms

07-20-09: In addition to transmembrane domain M2 **Q/R site editing**, GluR6 has two additional positions in domain M1 that are modified by RNA editing† *(see below and Domain functions, 07-29).* GluR6 therefore can exist in the genomic† (unedited) form and in seven different edited† forms, depending on the combination of edited or unedited residues. The origin of these combinations and notes on their differing frequencies of occurrence in the CNS are shown in Table 6.

Analytical note: Representation and analysis of complex gene organization

07-20-10: Use of appropriate database accession† numbers and importation of raw data into sequence analysis programs permits a greater appreciation of both marked and subtle variations in **ion channel gene organization** by mechanisms such as alternative splicing† and RNA editing† *(For an example, see the footnote to the table in Database listings, 07-53 describing the organization of the GluR-B gene.)* Standard database entry formats often accommodate data for loci of key **splice sites**† and **alternative exon usage**† and can be used interactively with original journal articles.

Homologous isoforms

07-21-01: For names of **equivalent subunits** isolated in different species and those in the same species by different laboratories, *see Tables 1–3, 7 and Database listings, 07-53.*

Protein molecular weight (purified)

Molecular size of immunoprecipitated recombinant subunits

07-22-01: AMPA: **Subunit-specific antibodies** against pentameric channels containing GluR-A to -D subunits recognize a typical subunit M_r of ~108 kDa[95]. For the AMPA receptors, immunoprecipitation data do not suggest any subunit other than GluR-A to -D within precipitated complexes[95]. **Joro spider toxin** (JSTX) has been used to purify an AMPA-binding protein of $M_r \sim 130$ kDa from native membranes *(see Blockers, 07-43, and Ligands, 07-47).*

Table 6. *Combinatorial variations in Glur 6 introduced by RNA editing. (From 07-20-09)*

	Unedited genome	Edited form 1	Edited form 2	Edited form 3	Edited form 4	Edited form 5	Edited form 6	Edited form 7
Unedited Ile in M1	unedited I	unedited I			unedited I	unedited I		
Edited Ile → Val in M1			edited V	edited V			edited V	edited V
Unedited Tyr in M1	unedited Y		unedited Y		unedited Y		unedited Y	
Edited Tyr → Cys in M1		edited C		edited C		edited C		edited C
Unedited Gln in M2	unedited Q	unedited Q	unedited Q	unedited Q				
Edited Gln → Arg in M2					edited R	edited R	edited R	edited R
Frequency in CNS	~ 10%	remaining 25% of occurrences						~ 65%

Footnotes: The M2 Q/R position can only influence **Ca^{2+}-permeability** when the M1 residues are in their edited form *(see Domain functions, 07-29)*. GluR6 (R-edited) channels show a higher Ca^{2+}-permeability than GluR6 (Q-unedited) channels. Co-expression of GluR6 (R-edited) and GluR6 (Q-unedited) forms with forms 'fully-edited' in M1 produces channels with low Ca^{2+}-permeability, and the influence of the Q/R switch is low when the M1 domain positions remain unedited *(see also Domain functions, 07-29)*. GluR5 also shows a diversity with respect to 'edited' and 'unedited' versions of its Q/R site[43].

Comparisons of native channel sizes by immunoprecipitation and radioligand affinity-labelling
07-22-02: AMPA: When synaptic plasma membranes are solubilized, high-affinity AMPA binding and GluR1 immunoreactivity co-migrate at a native glycoprotein of M_r of ~610 kDa[86].

Cross-reactivity of antibodies to kainate-binding proteins
07-22-03: KAIN: Antibodies against the frog **kainate-binding protein** *(KBP-f, see Database listings, 07-53)* which has a predicted M_r of 48 kDa from its cDNA sequence), cross-react with a native protein of M_r ~99 kDa in rat brain preparations. These antibodies recognize this cross-reactive material in the same brain regions where GluR-K1[96] (a GluR-A homologue) has been localized[97].

Sequence motifs

Sequence motifs for N-*glycosylation*
07-24-01: The number of potential ***N*-glycosylation**† sites per subunit in each cDNA cloned thus far are shown in brackets: Rat GluR1 (6); GluR2 (4); GluR3 (5); GluR4 (5); GluR5 (6); GluR6 (6); GluR7 (6); KBP-f (2); KBP-c (2); KA-1, KA-2 (8–10). The positions of these *N*-glycosylation† motifs† in the individual subunit *see the references and accession numbers given under Database listings, 07-53*. Differing extents of *N*-glycosylation† may account for the selectivity of certain lectins such as **concanavalin A** for desensitization† at kainate-preferring receptors *(for further details, see Inactivation, 07-37).*

N-terminal clustering of N-*glycosylation motifs*
07-24-02: In isolate HBGR-1 (a human homologue of rat GluR1)[90] the *N*-terminal segment preceding M1 contains all potential *N*-glycosylation sites (amino acids 45, 231, 239, 345, 383 and 388).

Universal presence of signal sequences
07-24-03: Signal peptides† of 17–31 aa residues are present at the N-terminus of all glutamate receptor–channel subunits cloned thus far *(see [PDTM], Fig. 3 and Table 1).*

Southerns

07-25-01: It has been verified by Southern† analysis (performed on DNA of the *same* mouse)[93] that multiple genes do *not* exist for GluR-B, GluR5 and GluR6 subunits. This supports the conclusion that **nucleotide exchanges**† found between genomic and cDNA sequences are due to **RNA editing**† processes *(see Gene organization, 07-20).*

STRUCTURE & FUNCTIONS

Note: The symbol [PDTM] denotes an illustrated feature on the channel protein domain topography model (Fig. 3).

In Press updates *(see criteria under Introduction & layout of entries, entry*

02): 1. Hollmann, *Neuron* (1994) **13:** 1331–43: Experimental approaches based on **N-glycosylation**† **site tagging**† as reporter sites for extracellular locations of protein domains in **GluR1** show: (i) the N-terminus is *extracellular*; (ii) the C-terminus is *intracellular*; (iii) only *three* transmembrane domains are present, designated TMD A, TMD B and TMD C (corresponding to M1, M3 and M4 as described in the 'finalized' entry); (iv) contrary to earlier models, the putative channel-lining domain M2 does *not* span the membrane, but lies in **close proximity to the intracellular face** of the plasma membrane or loops into the membrane without transversing it; (v) the region *between* **M3 and M4**, in previous models believed to be intracellular, is an entirely **extracellular domain**. 2. A refined **structural model** of the **glutamate-binding site** of ionotropic† glutamate receptors, based on exchanging portions of the AMPA receptor subunit GluR3 and the kainate subunit GluR6 has also been published (Stern-Bach, *Neuron* (1994) **13**: 1345–57). These new interpretations underline the difficulties of accurate domain topography modelling from indirect data.

Amino acid composition

07-26-01: By hydrophobicity†/hydrophilicity† criteria, all GluR subunits of ~900 aa residues in length are *predicted* to contain four membrane-spanning regions. These are variously referred to as domains TMI–TMIV, M1–M4, TM1–TM4. The 'four transmembrane domain' model should be seen as **tentative** *(see ELG Key facts, entry 04). Note:* The version of the iGluR monomeric protein domain topography model *(Fig. 3)* shows an **additional transmembrane domain** interposed between M3 and M4 (partly replacing the large **M3–M4** (**putative**) **intracellular loop** typical of other models for ELG-gated channels, cf. *Fig. 1 under ELG CAT 5-HT$_3$, entry 05).*

Domain arrangement

General arrangement of subunits

07-27-01: The four predicted membrane-spanning regions *(predicted in some models, see above)* determine the **monomeric subunit architecture** of iGluR. Structurally, the channel subunits are arranged '**barrel-like**' around a central **conductive pore**. In common with other members of the extracellular ligand-gated cation channel superfamily†, native channels are likely to be formed from **five monomers**[98] *(see [PDTM], Fig. 3).*

Subunit arrangements in native receptor–channels

07-27-02: In common with similar studies on other ELG-type receptors *(e.g. see ELG Cl GABA$_A$, entry 10, and ELG CAT nAChR, entry 09)* comparison of current–voltage relations† for native† channels versus those expressed from co-expressed recombinant† subunit combinations predict native† receptors to be **heteromultimeric**†, composed of at least two different subunits *(summarized under Current–voltage relations, 07-35).* Immunochemical studies with GluR subunit-specific antibodies suggest a pentameric, hetero-oligomeric† assembly of subunits in native channels[95] *(but see Protein interactions, 07-31).*

Properties of KA-1/KA-2 subunits within heteromeric complexes
07-27-03: KAIN: Although the KA-1 or KA-2 subunits *fail* to form functional channels when expressed as homomultimers[14,65], they do exhibit a high-affinity for kainate. Heteromeric† expression of KA-2 with the distantly related GluR5 or GluR6 subunits leads to formation of functional channels with **novel properties** – e.g. AMPA activates GluR6/KA-2 channels but does *not* activate homomeric GluR6 receptors[14].

Domain conservation

Conservation of signal sequences and subunit arrangements
07-28-01: All glutamate receptor–channel subunits show a similar *linear* arrangement of protein domains. At the N-terminus, all subunits display a **signal sequence** of between 17 and 31 aa residues *(see 'Sig' on the [PDTM], Fig. 3 and Sequence motifs, 07-24)*. Based on published sequence alignments[4], the extent of the amino acid homology between members of the GluR family is given in *Table 7*.

Conservation of the ELG channel 'core sequence'
07-28-02: The **core sequence** is defined as the region of the extracellular ligand-gated channels which shows the highest **sequence similarity**. Generally, the 'core sequence' includes the domains M1–M4 *([PDTM], Fig. 3)*. The subunits differ most in their N-terminal ~470 residues, e.g. where GluR-A to GluR-D share ~60% and GluR5 and GluR6 share ~75% sequence identity. Pairwise comparisons between these two groups of receptors reduce identity to ~25% in the N-terminal extracellular receptor domain *([PDTM], Fig. 3)*.

Positions of amino acid changes due to alternative splicing
07-28-03: Splice variants† of ionotropic† glutamate receptor can generate insertions or substitutions in several regions *(see [PDTM], Fig. 3 and Gene organization, 07-20)*. The '**flip**' and '**flop**' forms of the AMPA receptor subunits are generated by **alternative splicing**† between regions encoding transmembrane domains M3 and M4. Variant subunit forms also derived from alternative splicing† exist in the N-terminal domain (e.g. the GluR5[4] and the NR-1 subunits). *(For the phenotypic consequences of flip and flop modules, see Inactivation, 07-37)*.

Sequence homology between the iGluR and glutamine-binding proteins
07-28-04: The isolates GluR-K2 and GluR-K3 (corresponding to the 'flip' version of GluR-B and GluR-C) have been reported[99] as showing significant sequence conservation with the glutamine-binding component of the **glutamine permease** of *E. coli*.

Kainate-AMPA subunit similarities
07-28-05: KAIN: KA-1, the first high-affinity kainate subunit to be cloned[65] has a 30% sequence similarity with the AMPA receptor subunits GluR-A to -D. *Note:* The frog and chick brain kainate-binding proteins *(see*

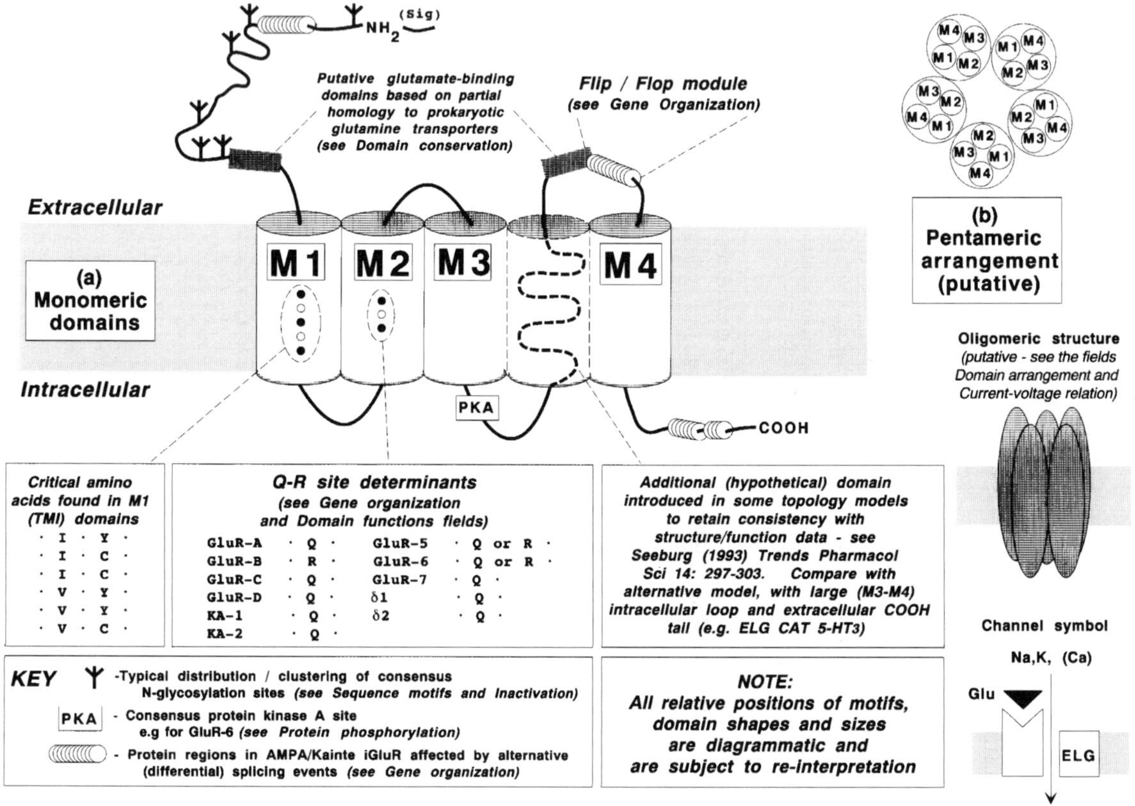

Figure 3. *Monomeric protein domain topography model [PDTM] for AMPA/kainate-selective ionotropic glutamate receptors (iGluR). In press updates (see criteria under Introduction & layout of entries, entry 02): A modified* ***3-transmembrane domain*** *model based on N-glycosylation site tagging data has appeared in press. For brief details, see the insert before Amino acid composition (07-26). For further details, refer to the entry update pages via the CSN. (From 07-28-01)*

Table 7. *Extent of amino acid sequence identity (%) amongst members of the cloned ionotropic GluR subunit family and kainate-binding proteins (From 07-28-01)*

	GluR-A	GluR-B	GluR-C	GluR-D	GluR5	GluR6	KBP-f	KBP-c	KA-1	KA-2
GluR-A	100	70	69	68	40	41	38	37	35	35
GluR-B		100	73	72	40	41	—	—	—	—
GluR-C			100	73	41	42	—	—	—	—
GluR-D				100	41	40	—	—	—	—
GluR5					100	81	42	38	42	—
GluR6						100	43	40	44	44
KBP-f							100	56	35	—
KBP-c								100	34	—
KA-1									100	70
KA-2										100

—: information not found.

Database listings, 07-53) appear to lack the first ~350 aa of the GluR-A protein and possess only ~25% amino acid identity when compared to the N-terminal half of GluR-A.

Domain functions (predicted)

The Q/R site in the M2 domain is critical for regulation of ion permeability and I–V relationships

07-29-01: The **M2 domain** sequence shows variability at a position known as the glutamine/arginine site or **Q/R site** *(see the sequence alignment below and [PDTM] Fig. 3 inset).* GluR-B possesses an **arginine (R)** at this site compared to a **glutamine (Q)** being present in the homologous position in GluR-A, GluR-C and GluR-D. Significantly, the arginine codon is *not* found in the GluR-B gene, but is introduced by an **RNA editing** process *(see Gene organization, 07-20).* The Q/R site also determines the **Ca^{2+}-permeability**† of the channel *(for details, see Selectivity, 07-40*[100,101]*).* **Site-directed mutagenesis**† of the Q/R site has shown that the single amino acid difference in the GluR-B subunit also determines the I–V relationship† of heteromeric† channels[102] *(for details, see Fig. 4).* GluR6 also occurs in two forms with respect to the amino acid residue occupying the Q/R site[93]. For comparison, the sequence alignment below also lists the equivalent (aligned) amino residues of the NMDA receptor subunits NR1 and NR2A *(see ELG CAT GLU NMDA, entry 08).*

Sequence alignments in iGluR subunits surrounding the Q/R site in the M2 domain

GluR-A	F	G	I	F	N	S	L	W	F	S	L	G	A	F	M	[Q]	Q	G	C	D	I	S	P
GluR-B	F	G	I	F	N	S	L	W	F	S	L	G	A	F	M	[R]	Q	G	C	D	I	S	P
GluR6	F	T	L	L	N	S	F	W	F	G	V	G	A	L	M	[Q]	Q	G	S	E	L	M	P
KA-2	Y	T	L	G	N	S	L	W	F	P	V	G	G	F	M	[Q]	Q	G	S	E	I	M	P
δ1	A	T	L	H	S	A	I	W	I	V	Y	G	A	F	V	[Q]	Q	G	G	E	S	S	V
NR1	L	T	L	S	S	A	M	W	F	S	W	G	V	L	L	[N]	S	G	I	G	E	G	A
NR2A	F	T	I	G	K	A	I	W	L	L	W	G	L	V	F	[N]	N	S	V	P	V	Q	N

↑ The Q/R site

Control of Ca^{2+}-permeability of kainate receptors determined by RNA editing

07-29-02: KAIN: Ca^{2+}-permeability of kainate receptor–channels can vary depending on 'editing' of RNAs encoding *both* M1 and M2 transmembrane domain sequences[103]. In addition to editing at the **critical Q/R site** in M2 *(see above and ref.*[104]*)*, sequences forming the GluR6 putative transmembrane domain M1, are also diversified by **RNA editing** *(see [PDTM], Fig. 3 and the* *seven possible edited combinations under Gene organization, 07-20).* This process can generate either isoleucine or valine in one and tyrosine or cysteine in the other M1 domain position. In GluR6 channels the presence of Q (glutamine) at the domain M2 Q/R site forms channels with low Ca^{2+}-permeability (in contrast with AMPA receptor–channels). An arginine at this position determines a higher Ca^{2+}-permeability of GluR6 channels if domain M1 is **'fully-edited'**. In the 'unedited' form of GluR6 domain M1, Ca^{2+}-permeability is less dependent on the presence of either glutamine or arginine in domain M2[103]. These results raise the possibility that RNA editing† may modulate glutamate-activated Ca^{2+}-influx through GluR6 *in vivo.*

Agonist-binding site

07-29-03: The region *preceding* putative transmembrane segment M1 of GluRs is well-conserved among subunits and has been proposed to constitute a part of the **agonist-binding site** *(see also other ELG entries).* Introduction of point mutations into charged residues of the mouse AMPA-selective α1 subunit (Glu398 → Lys398; Lys445 → Glu445) are associated with changes in the EC_{50} values† with different agonists, indicating their involvement for agonist-selective interactions of the GluR channel[105].

Supplementary note for above

07-29-04: L-Glutamate, kainate and AMPA bind to *different* **receptor substructures** on recombinant AMPA receptors[106] *(see Equilibrium dissociation constant, 07-45, and Ligands, 07-47).*

Predicted protein topography

Common assumptions of hydrophobic sequences as structural domains

07-30-01: Like other members of the extracellular ligand-gated (ELG) family, all glutamate receptor–channels display four *predicted* membrane-spanning hydrophobic regions (M1–M4), although some models *(e.g. ref.*[18]) propose an additional **'conjectural' transmembrane domain** between M3 and M4 *(see [PDTM], Fig. 3).* However, the lack of direct structural data does not yet allow any firm conclusions to be made regarding 'true' transmembrane protein topography *(see ELG Key facts, entry 04). In-press updates: See note above field 07-26.*

Protein interactions

Common subunit associations with GluR-B

07-31-01: AMPA: **GluR-B subunits** dominate properties of ionic flow in heteromeric† iGluR complexes[52,102]. For example, co-expression of the **GluR-B** subunit with either GluR-A, GluR-C or GluR-D forms recombinant channels which closely match characteristics of native† receptors[9], implying that most native† receptors are heteromultimeric† and incorporate the GluR-B subunit. GluR-B and GluR-C immunoreactivity has been shown to co-localize with the metabotropic† mGlu1α receptor at the climbing fibre synapse in cerebellum[88].

Evidence against cross-assembly of 'AMPA-preferring' and 'kainate-preferring' subunits
07-31-02: The 'absolute selectivity' of **cyclothiazide** and **concanavalin A** for respective block of fast desensitization† of **AMPA-preferring** and **kainate-preferring receptors** *(see Inactivation, 07-37)* has been used to monitor **subunit assembly patterns** in functional recombinant iGluRs[11] . In all cases, assembly of co-expressed subunits from the two different families suggest **independent assembly** of functional AMPA and kainate receptors without any evidence for **cross-family assembly** of subunits.

Common functional interactions at synapses
07-31-03: Diverse classes of extracellular ligand-gated channels **commonly interact** to shape depolarizing post-synaptic potentials (DPSPs) in the CNS. For example, DPSPs of granule cells in the dentate gyrus is part-mediated by **AMPA**, **$GABA_A$** and **NMDA receptor proteins** *(see Protein interactions under ELG Cl $GABA_A$, 10-31, and Fig. 4 under ELG CAT GLU NMDA).*

Postulation of inhibitory proteins associated with native AMPA receptors
07-31-04: The existence of an **inhibitory protein** (with a possible 'negative regulatory' function) has been invoked to explain the observation that AMPA K_d values become much lower (i.e. AMPA affinity increases) upon various membrane treatments and upon purification[20].

Protein phosphorylation

iGluRs display multiple putative phosphoregulatory motifs
07-32-01: A number of potential **consensus**† **phosphorylation sites** for protein kinases (e.g. protein kinase A, protein kinase C, tyrosine kinase and casein kinase II) have been found in AMPA- and kainate-selective GluR subunit cDNA sequences. Only a minority of these have been shown to have *functional* roles to date. *Note:* Locations of each putative regulatory site can be traced by reference to citations under *Database listings, 07-53.*

Potentiation of Ca^{2+}-fluxes
07-32-02: Ca^{2+}-fluxes can pass through non-NMDA glutamate receptor–channels composed of the subunits GluR-A and GluR-C (i.e. in the absence of GluR-B) *(see Selectivity, 07-40)*. Calcium flux through open KA/AMPA receptor–channels can be **potentiated**† **by phosphorylation** mediated through protein kinase A as part of a cAMP-dependent second messenger system[107].

Enhancement of native iGluR responses through protein kinase A phosphorylation
07-32-03: Non-NMDA channels expressed in cultured hippocampal pyramidal neurones are subject to neuromodulatory regulation through the **adenylate cyclase** cascade. The whole-cell current response to glutamate and kainate is enhanced by **forskolin** (an activator of adenylate cyclase). Single-channel analysis has shown that protein kinase A increases the opening frequency† and the mean open time† of non-NMDA-type glutamate receptor–channels.

Forskolin, acting through PKA, increases the amplitude and decay time of spontaneous excitatory post-synaptic currents[108].

Potentiation of recombinant iGluR responses by protein kinase A
07-32-04: KAIN: Channels expressed from GluR6 subunits (when transiently expressed in mammalian cells) have been shown to be directly phosphorylated by PKA. Application of intracellular PKA increases the amplitude of the glutamate response[25]. Site-directed mutagenesis† of the serine residue (Ser684) representing a PKA consensus site completely eliminates PKA-mediated phosphorylation of this site as well as the potentiation† of the glutamate response[25] *(see [PDTM], Fig. 3).*

Postulated functional roles of iGluR phosphorylation
07-32-05: Protein phosphorylation of glutamate receptors by protein kinase C and cAMP-dependent protein kinase has been suggested to regulate their function in synaptic transmission *(see ref.[25])*, possibly playing a prominent role in **long-term potentiation**†[109,110] and **long-term depression**†[111]. For additional notes on the broad roles of protein phosphorylation in the ELG channel family, *see ELG Key facts, entry 04.*

ELECTROPHYSIOLOGY

Activation

AMPA-selective iGluRs mediate 'fast' excitatory signalling in the CNS
07-33-01: AMPA receptors mediate the **most rapid synaptic excitatory neurotransmission** and conduct mainly **Na^+ currents**. For example, brief (~1 ms) applications of glutamate on membrane patches excised from neurones in the rat visual cortex produce a rapid response that mimicks the time course of miniature excitatory post-synaptic currents† (e.g ~2.4 ms for AMPA-evoked EPSPs[112]). The rate of onset of desensitization† is much slower than the decay rate of the response *(see Inactivation, 07-37)*, implying that the decay of miniature EPSCs† reflects channel closure into a state readily available for re-activation[113].

Rise times for iGluR current activation
07-33-02: Brief pulses (≤ 1 ms) of glutamate (1 mM) on AMPA/kainate receptors in granule cells of dentate gyrus and pyramidal cells of CA3 and CA1 hippocampal regions activate patch currents which rise and decay rapidly[114]. The 20–80% rise time† of these GluR-mediated currents is typically ~0.2–0.6 ms. At −50 mV, peak currents vary from 10 to 500 pA in different patches.

Kainate also activates 'AMPA-preferring' receptor–channels
07-33-03: Kainate also activates non-desensitizing† currents similar to those elicited in CNS neurones through 'AMPA receptor–channels' formed from subunits GluR-A to -D[9,10].

Current–voltage relation

Functional evidence for heteromultimers based on shapes of I–V relations

07-35-01: AMPA: The majority of native† neurones exhibit AMPA receptor-mediated inward currents with linear or outwardly rectifying† current–voltage relationships†[9]. When *Xenopus* oocytes are injected with RNA encoding GluR-A alone, AMPA agonists evoke a smooth inwardly rectifying† current *unlike* the linear I–V relationship seen *in vivo*. However, when the combinations GluR-A/B or GluR-B/C are co-expressed[99], a non-rectifying inward current (i.e. linear and ohmic†) is evoked by kainate or AMPA (closely resembling current types seen in native neuronal cells). In common with similar studies on other ELG-type receptors *(e.g. see ELG Cl* $GABA_A$*, entry 10, and ELG CAT nAChR, entry 09)* these results indicate the native receptor to be a heteromultimer† of at least two different subunits.

Dominant characteristics of GluR-B subunits

07-35-02: KAIN: **Bergmann glial cells** display a kainate-type glutamate receptor with a sigmoid (**doubly-rectifying**†) current–voltage relation[27] and are permeable to the divalent cations Mg^{2+} and Ca^{2+}[52, 102]. *Note:* Bergmann glial cells are unusual in that they do *not* express GluR-B subunits[61] *(see Cell-type expression index, 07-08). Homomeric* GluR-B channels (or heteromeric channels containing GluR-B as described in the previous paragraph) exhibit near linear I–V relations and have low divalent cation permeabilities.

Localization of key amino acids determining I–V relationships

07-35-03: A **single amino acid difference** in the **GluR-B subunit** determines the I–V relationship† of heteromeric† AMPA-selective NMDA receptor–channels[102]. The putative transmembrane domain M2 sequence is identical in each of the GluR-A to -D subtypes, with the exception that GluR-B has a positively charged arginine (Arg, R) residue in aa position 586 (cf. the neutral glutamine (Gln, Q) residue at aa 586 in GluR-A, GluR-C and GluR-D. Exchange of Arg586 → Gln586 in GluR-B and a corresponding Gln → Arg exchange in GluR-D by site-directed mutagenesis† reverses the shapes of the I–V curves† evoked by glutamate in these channels formed by the homomeric† subunits[101] *(see Fig. 4).*

Dose–response

Routes of Ca^{2+}*-influx dependent on agonist concentrations in Purkinje cells*

07-36-01: Ion channels integral to non-NMDA receptors on immature Purkinje cells (3–10-day-old rats) are permeable to Ca^{2+}, Mn^{2+} and Co^{2+}[117]. Increases in $[Ca^{2+}]_i$ induced by *relatively lower agonist concentrations* are largely dependent on **Ca^{2+}-influx** through voltage-sensitive Ca^{2+} channels, which are themselves activated by a large **Na^+-influx**. *Higher concentrations* of agonists dose-dependently increase $[Ca^{2+}]_i$ (under conditions in which activation of voltage-dependent Ca^{2+} channels and NMDA channels are blocked), indicating a Ca^{2+}-influx through the non-NMDA receptor–channel[117].

Differential sensitivity to ethanol dependent on applied agonist concentration

07-36-02: KAIN: Responses produced by low or high concentrations of kainate are differentially inhibited by acute exposure of kainate receptors to **ethanol** when expressed from rat hippocampal mRNA in oocytes[118]. For example, 50 mM ethanol inhibits 12.5 μM kainate responses by 45% compared to only 15% inhibition of 400 μM kainate responses. By contrast, acute ethanol exposure inhibits responses stimulated by low and high concentrations of *N*-methyl-D-aspartate to a similar degree[118]. *Note:* For an illustration of the relative **potentiating effects** on $GABA_A$-mediated Cl^--flux

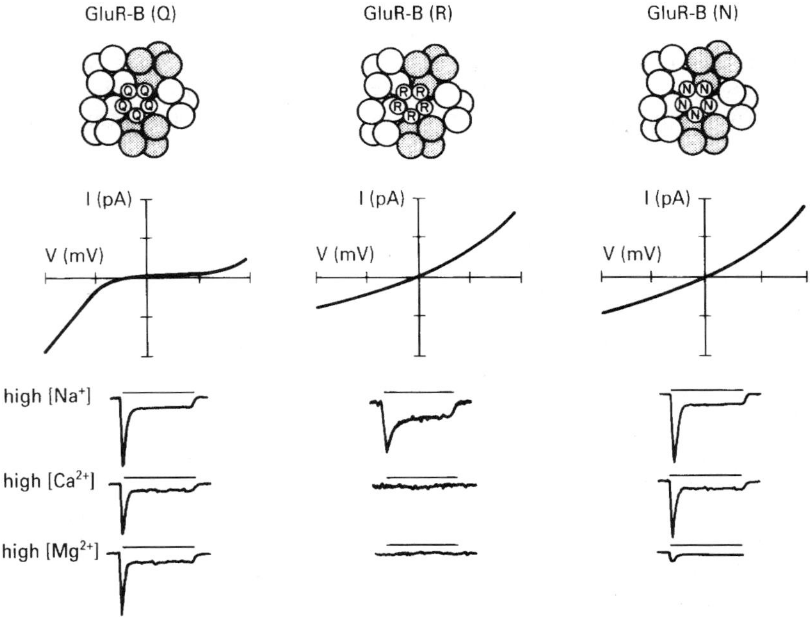

Figure 4. *Comparison of electrophysiological properties from homomeric iGluR-channels containing Q, R or N residues at the Q/R site. Top row: Plan view of Q/R site position in pentameric arrangement of channel subunits. Middle row: Typical I–V relationships. Bottom row: Typical whole-cell currents elicited by glutamate under the stated extracellular ionic conditions. Bar* ~ 300 *ms.* Note: *The experimental substitution of an* ***asparagine (N)*** *into the* **glutamine/arginine (Q/R) site** *of AMPA receptor subunits generates channels displaying a selective permeability for* Ca^{2+} *over* Mg^{2+}*, similar to the NMDA channel*[115]. *Alignments of the NMDA receptor isolate NMDAR-1 in the M2 transmembrane domain identifies the placement of an N residue at the Q/R site (see Selectivity under ELG CAT GLU NMDA, 08-40, and Selectivity under this entry, 07-40). (Reproduced with permission from Sommer and Seeburg (1992)* Trends Pharmacol Sci ***13:*** *291–6; original data from refs*[102, 115, 116]*.) (From 07-35-03)*

and the depressive effects on NMDA-, kainate- and voltage-gated Ca^{2+}-flux, *see Fig. 5 under ELG Cl GABA$_A$.*

Selective responses from iGluR populations dependent on agonist dose

07-36-03: In dorsal horn neurones, where mixed subtypes of glutamate receptors are expressed, glutamate responses at concentrations *less than 3 μM* are due *exclusively* to NMDA receptor activation[119]. *At higher glutamate concentrations,* intracellular responses are mediated by both NMDA and non-NMDA receptors[119].

Inactivation

Physiological roles of desensitization

07-37-01: Ionotropic receptor **desensitization**† properties help govern the **strength** of fast excitatory synaptic transmission in the brain. Under equilibrium conditions, >90% of available receptors are desensitized†, although their affinity to glutamate is much higher than that measured prior to desensitization†[120]. Desensitization at AMPA/kainate receptors has been proposed to contribute to the **fast decay** of excitatory† synaptic currents.

Desensitization kinetics

07-37-02: Kainate receptors *generally* desensitize† only 'extremely slowly', whereas AMPA receptors (with rare exceptions, see ref.[121]) undergo this transition relatively rapidly and in a concentration-dependent manner *(see examples below). Note:* Kainate has also been shown to cause rapid desensitization† of homomeric† channels expressed from subunit **GluR6**.

Utility of 'absolute selective block' of rapid desensitization for AMPA channels by cyclothiazide

07-37-03: Potentiation by **cyclothiazide** (**CYZ**) of recombinant glutamate receptor responses via an allosteric block of rapid desensitization shows **absolute selectivity** for GluR1–GluR4 (AMPA) receptors when expressed in *Xenopus* oocytes[11]. Rapid desensitization in HEK-293 cells transfected with AMPA receptors is also blocked by CYZ, but is only weakly attenuated by **concanavalin A** (**Con A**). Conversely, desensitization† at kainate receptors (GluR5–GluR7 and KA-1/KA-2) is blocked by Con A but unaffected by CYZ[11]. *Note:* Cyclothiazide is a benzothiadiazine diuretic and antihypertensive drug structurally related to diazoxide.

CYZ/Con A-sensitivity phenotypes can report iGluR subunit assembly patterns in vivo

07-37-04: Native† cell types shown to predominantly express **kainate-preferring subunits** (e.g. dorsal root ganglion neurones, mainly expressing GluR5) show an expected Con A-sensitive/CYZ-insensitive phenotype[11]. Conversely, native† cell types which preferentially express AMPA-preferring subunits (e.g. hippocampal neurones) display a CYZ-sensitive/Con A-insensitive phenotype. Table 8 summarizes findings of comparative studies on native† cell preparations.

Table 8. *Cyclothiazide (CYZ) versus concanavalin A (Con A) sensitivity phenotypes for AMPA-selective iGluR–channels expressed in native neurones (From 07-37-04)*

Preparation	Cyclothiazide phenotype	Refs
Hippocampal spiny 'mossy cells' versus aspiny hilar interneurones	A greater sensitivity to cyclothiazide in hippocampal spiny 'mossy cells' versus aspiny hilar interneurones has been reported (with half-maximal removal of desensitization being 90 mM and 200 mM, respectively)	*122*
Hippocampal slices, glutamergic neurones responding to glutamate	Cyclothiazide (CYZ) reduces rapid desensitization, enhancing the steady-state and peak current produced by 1 mM quisqualate with EC_{50} values of 14 and 12 μM respectively. CYZ causes glutamate to induce long bursts of channel openings, and greatly increases the number of repeated openings. At 10 μM CYZ does not have measurable effects on the fast component of deactivation nor does it have statistically significant effects on the distribution of the faster components of glutamate-induced burst duration	*123*
Hippocampal neurones responding to kainate	Responses of hippocampal neurones to kainate are strongly potentiated† (300%) by cyclothiazide, which is considerably more effective ('complete block of desensitization') than aniracetam in reducing desensitization evoked by glutamate	*124*
Dorsal root ganglion neurones compared with hippocampal neurones	Cyclothiazide completely blocks desensitization produced by 5-chlorowillardiine in hippocampal neurones and strongly potentiates responses to kainate (the action of aniracetam is similar but much weaker). In DRG neurones, cyclothiazide and aniracetam has no effect on desensitization but produces weak inhibition of responses to kainate	*125*
Cultured cerebellar glial cells (oligo-dendrocyte lineage, O-2A progenitors)	Two receptor populations are present in these cells, with high and low affinity for kainate showing different sensitivity for potentiation by concanavalin A and for block desensitization of cyclothiazide	*51*

A mechanistic basis for kainate subunit-selective phenotypes of concanavalin A

07-37-05: Concanavalin A (Con A) is a **lectin** which can bind to **glycosylated**† **membrane proteins** with high affinity. Expression of recombinant iGluR subunits which display a higher proportion of glycosylated† N-termini (extracellular) might be therefore expected to bind Con A with greater affinity. Although both AMPA and kainate cDNAs show *N*-glycosylation motifs *(see Sequence motifs, 07-24)* treatment of glycosylated GluR6 (kainate-preferring) with **N-glycosidase**[25] induces a 13 kDa shift in M_r compared with a shift of only 5–6 kDa for AMPA-preferring subunits[18, 126]. These analyses suggest kainate-preferring iGluRs are glycosylated to a greater extent than AMPA subunits, and may therefore explain the higher sensitivity of the kainate subunits to lectins like Con A[11]. *Note:* The flip/flop variants *(see below)* show no apparent differences in sensitivity to concanavalin A.

Desensitizing effects of cyclothiazide vary in flip/flop splice variants of AMPA receptors

07-37-06: Although the molecular basis for '**absolute selective block**' of rapid desensitization in AMPA subunit channels is unknown, the **'flop' splice variants** show much less potentiation by **cyclothiazide** (22 ± 4-fold for glutamate responses, 4.2 ± 0.7-fold for kainate responses) than their **'flip' variants** (130 ± 30-fold for glutamate responses, 12.4 ± 2.7-fold for kainate responses). Similar properties are observed with flip/flop variants in heteromeric combinations (e.g. GluR-A_i + GluR-B_i versus GluR-A_o + GluR-B_o). These results are consistent with a role for the **flip/flop locus** in regulating **desensitization**† *(see below and refs*[11, 175]*)*.

Desensitization plateaux in alternative splice variants flip and flop

07-37-07: AMPA: Upon fast application, glutamate elicits currents at AMPA receptor–channels which exhibit a **fast rise time**† and then decay to a **plateau value** in the continued presence of agonist. The plateau is more pronounced with flip- than flop-containing GluRs *(see Gene organization, 07-20)*. Thus the differing desensitization kinetics shown by receptors containing flip and flop modules affect the '**peak : steady state**' component of iGluR formed from AMPA-preferring GluR-A–GluR-D subunits. *Note:* Kainate evokes identical **non-desensitizing**† currents in both flip- and flop-containing GluR channels formed from the GluR-A to -D classes.

Modulation of desensitization

07-37-08: The nootropic† drug **aniracetam**, wheat germ **agglutinin**, and **concanavalin A** act via separate mechanisms to reduce desensitization evoked by L-glutamate in rat hippocampal neurones[120]. The decay of excitatory synaptic currents, and miniature excitatory post-synaptic currents† (EPSCs) evoked by sucrose are slowed 2- to 3-fold by aniracetam. Aniracetam also increases the magnitude of glutamate-evoked EPSCs 1.9-fold, probably via a *post-synaptic* mechanism of action. Aniracetam increases the **burst length**† and **peak amplitudes**† of L-glutamate-activated single-channel responses[128]. Simulations suggest that aniracetam either slows entry into a desensitized state† or decreases the closing rate constant† for ion channel gating†. *Comparative note:* Wheat germ agglutinin and concanavalin A reduce EPSC amplitude via a *presynaptic* mechanism[120] *(see below)*. **Diazoxide** can also reduce desensitization† of hippocampal AMPA receptors to AMPA, glutamate and quisqualate.

Different neuronal cells show different rates of recovery from desensitization

07-37-09: In native dentate gyrus, hippocampal CA3 and CA1 cell patches, applications of 1 mM glutamate of 100 ms duration show time constants† for desensitization of 9.4 ± 2.7, 11.3 ± 2.8, and 9.3 ± 2.8 ms respectively[114]. Desensitization time constants† are only weakly dependent on glutamate concentration (200 μM and 1 mM) for the three cell types. Double pulse applications of glutamate indicate that 1 ms pulse of 1 mM glutamate cause partial GluR channel desensitizations (~60%). The time course of *recovery* from desensitization is slower in dentate gyrus granule cell patches than in CA3 or CA1 pyramidal cell patches[114]. *Note:* Specialized iGluRs mediating **complex auditory information** in cochlear neurones show unusually **rapid onset and termination** of responses *(for details, see Phenotypic expression, 07-14).*

Decay time constants are independent of membrane potential and agonist concentration

07-37-10: Offset decay time constants† of the currents following 1 ms pulses of 1 mM glutamate show mean values of 3.0 ± 0.8, 2.5 ± 0.7, and 2.3 ± 0.7 ms for granule cells of dentate gyrus, hippocampal CA3 and CA1 cell patches, respectively[114]. Offset time constants† are independent of both membrane potential and glutamate concentration (200 μM and 1 mM) for all three cell types[114].

Selectivity

Principal ionic selectivity differences between iGluR–channels

07-40-01: AMPA/kainate receptor–channels in native tissues have usually been classified as mediating an influx of **monovalent cations** (in contrast to NMDA receptors which are thought to mediate their physiological response mainly through the influx of extracellular calcium). However, Ca^{2+}-selectivity is now known to be influenced by subunit composition *(see below)*. The permeation pathways of a range of neurotransmitter-gated ion channels has been reviewed[129].

The GluR-B subunit dominates properties of ionic flow

07-40-02: AMPA: Heteromeric† AMPA receptors containing the GluR-B subunit display **low divalent ion permeabilities**. However, recombinant† AMPA receptors ***lacking* the GluR-B subunit are Ca^{2+}-permeable** at physiological Ca^{2+} concentrations (e.g. GluR-A, GluR-C or GluR-A + GluR-C in combination[52, 107]). Heterologous† expression of GluR-B mRNA pre-mixed in **different molar ratios** with mRNAs encoding GluR-A, GluR-C or GluR-D show a large range of glutamate-activated Ca^{2+}-permeabilities[100].

Side-chain size and charge affect divalent permeability in recombinant GluRs

07-40-03: AMPA: Following expression in HEK-293 cells, *homomeric*† assemblies of GluR-B(586R) channels *(see Domain functions, 07-29)* display a **low divalent permeability**†, whereas homomeric GluR-B(586Q) and GluR-D channels exhibit a **high divalent permeability**†[100]. Mutational analysis has shown both the positive charge and the side-chain size of the amino acid located at the Q/R site (position 586) control the divalent permeability of homomeric channels. Changes in **GluR-B(586R) expression** are therefore capable of regulating the AMPA receptor-dependent divalent permeability of a cell[100, 101].

Ca^{2+}-influx through non-NMDA receptors in native tissues

07-40-04: Ca^{2+} currents through native mammalian non-NMDA channels have been reported in retinal bipolar cells[130], hippocampal neurones[131], Bergmann glial cells (GluR-A/GluR-B)[27, 61] and type-2 astrocytes (when activated by kainate, *see refs in*[132]). *(See also Cell-type expression index, 07-08, and Dose–response, 07-36.)*

Differences in Ca^{2+}-permeability determinants for kainate receptor–channels

07-40-05: <u>KAIN</u>: In general, the Ca^{2+}-permeability of kainic acid-gated receptor–channels is governed by their **subunit composition**[52] and this may be **developmentally regulated**. In GluR6 homomeric channels, the presence of Q (glutamine) in the domain M2 Q/R site produces channels with low Ca^{2+}-permeability (in contrast with AMPA receptor–channels – *for details, see also Domain functions, 07-29*).

Separate determinants of rectification and divalent ion mobility in selectivity filters

07-40-06: Different amino acid residues control (i) the ability to pass outward current (rectification† properties) and (ii) divalent ion mobility (permeability† properties) in the **selectivity filter**† of GluR-C and GluR6 non-NMDA channels[100, 132]. Mutagenesis at (or near) the **Q/R site** in GluR-C indicates that (i) the position of the arginine is critical to function and (ii) the ability to pass outward current is not necessarily linked to low barium permeability.

Relative Ca^{2+} fraction through native NMDA and non-NMDA receptor–channels

07-40-07: The **Ca^{2+} fraction** of the ion current flowing through glutamatergic NMDA and AMPA/kainate receptor–channels has been compared directly in forebrain neurones of the medial septum[133]. A fractional Ca^{2+} current of 1.4% was determined for the linearly conducting AMPA/kainate receptor–channels found in these neurones[133]. In comparison, at negative membrane potentials (extracellular free Ca^{2+} concentration of 1.6 mM) the Ca^{2+} fraction of the current through the NMDA receptor–channels is ~6.8%, or ~2-fold lower than previously estimated from reversal potential† measurements.

Novel immunochemical assays detecting divalent ion-permeable kainate or AMPA receptors

07-40-08: Glutamate analogues stimulate uptake of **cobalt ion** into neuronal cells in cell culture or tissue slices. Since Co^{2+}-permeable channels are also Ca^{2+}-permeable, the **precipitable Co^{2+}** has been used to quantitate agonist

Table 9. *General features relevant to the conductance states of AMPA- and kainate-selective iGluRs (From 07-41-01)*

iGluR type	Features	Refs
AMPA channels	Distinct *single*-channel conductances were determined for non-NMDA receptors for (inhibitory) aspiny interactions (27 pS) compared with (excitatory) pyramidal neurones (9 pS) *(see Cell-type expression index, 07-08)*. In cerebellar neurones, multiple conductance states are observed with quisqualate and kainate agonists (mainly below 20 pS with the predominant conductance state for AMPA channels being ~ 8 pS). A rapidly inactivating, high-conductance state (~ 3 ms, ~ 35 pS) associated with the fast (quisqualate) receptor-mediated EPSC† has been described.	***41, 135–138***
Kainate channels	Kainate gates primarily low-conductance channels in neurones of the hippocampus, spinal cord, cortex and cerebellum, e.g. outside-out patches from these membranes display a principal conductance of ~ 4 pS with an open time of ~ 0.5–3 ms for kainate agonists. Noise analysis indicates that kainate may activate a 140 fS channel in some preparations	***135, 137, 139***
High-conductance channels in granule cells	High-conductance non-NMDA channels, such as the 10–30 pS glutamate receptor–channel previously characterized in granule cells, carry the majority of the fast component of the EPSC† at the cerebellar mossy fibre–granule cell synapse. Low numbers (~ 10) of non-NMDAR–channels appear to be activated by a single packet of transmitter. Conductance estimates for the non-NMDA receptor component of synaptic currents activated during EPSCs at this synapse show a mean single-channel conductance of approximately 20 pS	***140***
Low-conductance channels in goldfish retinal horizontal cells	Analysis of whole-cell noise in the presence of Mg^{2+} induced by glutamate, quisqualate and kainate in the retinal horizontal cells of the goldfish indicate all three agonists activate channels with a conductance of 2.5–3 pS. *Note:* Vertebrate retinal horizontal cells appear to lack NMDA receptors	***141***
Affinity-purified, reconstituted iGluRs	Excitatory amino acid receptor proteins purified from *Xenopus* central nervous system using domoate affinity columns followed by reconstitution into lipid bilayers exhibit variable single open channel conductance levels depending on agonists used to elicit current. These have been measured as ~ 6 pS with AMPA, ~ 9 pS with kainate, and ~ 50 pS with NMDA. Occasionally, unitary† channel openings of up to 400 pS are observed, suggesting that reconstituted receptors may form functional aggregates	***142***

affinities. Three types of Co^{2+}-permeable kainate receptors have been defined using these assays: **K1** (activated by kainate alone), **K2** (by glutamate and kainate) and **K3** (by kainate, glutamate and quisqualate)[134].

Single-channel data

AMPA- and kainate-activated channels have distinct conductances

07-41-01: A number of general features relevant to the conductance states of AMPA- and kainate-selective iGluRs are summarized in Table 9. A minority of studies have concentrated on determination of *single-channel* properties under defined conditions, so only a few examples are quoted in the table.

PHARMACOLOGY

Blockers

Open-channel blockers isolated from spider venom

07-43-01: A component of the spider venom from *Argiope lobata*, **argiotoxin**, has been characterized as an antagonist of homomeric† and heteromeric† glutamate-activated receptor–channels. Argiotoxin acts as an open-channel blocker in a voltage-dependent manner and discriminates between AMPA receptors[143]. A determinant in the M2 domain for divalent cation permeability also determines argiotoxin sensitivity *(see Selectivity, 07-40)*. Subunit-specific differences in time courses of argiotoxin block and recovery demonstrate that heteromeric AMPA receptors can assemble in variable ratios[143]. Notably, the spider venom toxins argiotoxin and **Joro spider toxin** *(see below)* have higher potency at NMDA receptors[144].

Joro spider toxin binds to the pore domain of iGluR subunits at a glutamine residue

07-43-02: Joro spider toxin (JsTx) is a potent non-NMDA receptor antagonist which can exert a subunit-specific block at submicromolar concentrations[145]. Receptor subunits with **rectifying I–V relationships** (GluR1, GluR3, GluR4 and GluR1/3) are reversibly blocked by JsTx. Receptor subunits forming a receptor–channel with a linear I–V relationship (GluR1/2, GluR2/3 and GluR6) are not affected. JsTx binds close to the **central pore region** of the channel – a single amino acid position (a Glu-586 at the Q/R site) appears critical for the JsTx block[145] *(see Current–voltage relation, 07-35 and Selectivity, 07-40). Note:* **Philanthotoxin** is also known to block a *Drosophila* kainate-selective glutamate receptor–channel[146].

Channel modulation

Modulation of GluR agonist responses by nootropic drugs

07-44-01: Micromolar concentrations of **piracetam**, **aniracetam** and **oxiracetam** enhance AMPA-stimulated Ca^{2+} influx in primary cultures of cerebellar granule cells[147]. Such **nootropic† drugs** increase the efficacy† but not the potency† of AMPA, and their action persists in the presence of the voltage-sensitive calci[illegible]r **nifedipine**. Piracetam, aniracetam and oxiracetam increase the maximal density of the specific binding sites for [^{3}H]-AMPA in synaptic membranes from rat cerebral cortex[147]. Aniracetam (1-*p*-anisoyl-2-pyrrolidinone) allosterically† potentiates ionotropic **quisqualate (iQA)** responses induced in *Xenopus* oocytes expressed from rat brain mRNA in a reversible manner[148] *(see also Inactivation, 07-37).*

07-44-02: *Note:* Further examples of iGluR channel modulation are described under *Phenotypic expression, 07-14 and Protein phosphorylation, 07-32.*

Equilibrium dissociation constant

High-affinity kainate binding sites in native tissues

07-45-01: KAIN: Kainate-binding sites that differ from high-affinity AMPA-binding sites have been identified by ligand-binding studies[149]. 'Classical' **high-affinity kainate sites** (K_d ~ 5 and 50 nM kainate) exist in the CA3 area of the hippocampal formation.

Similar agonist affinities for native and recombinant kainate receptors

07-45-02: The 'pharmacological profile' of expressed recombinant KA-1 (determined in binding experiments with [^{3}H]-kainate) differs from that of the cloned AMPA receptors, but is similar to the mammalian high-affinity kainate receptor (kainate > quisqualate > glutamate ≫ AMPA, where $K_{d\,(kainate)}$ is ~5 nM[65]). In comparison, the inhibitory constant (K_i) values for quisqualate, L-glutamate and AMPA are ~ 18 200 and > 5000 nM respectively[65]. *Note:* Recombinant KA-2 subunits do *not* form channels in homomultimers† but exhibit high affinity for kainate (K_d ~ 15 nM)[14].

Increased agonist affinities observed for purified iGluR

07-45-03: AMPA K_d values have been observed to become 'much lower' upon various membrane treatments and following protein purification *(see Protein interactions, 07-31).*

K_d values for unitary† receptors

07-45-04: In *Xenopus* brain, kainate- and AMPA-binding sites co-exist in a **1:1 ratio** and *cannot be separated* by physical and chemical fractionations[21, 22] *(see Subtype classifications, 07-06).* In these proteins, AMPA and kainate are mutually and **fully competitive**†, with K_i values identical to the K_d values for the radioligand (AMPA, 34 nM; kainate, 15 nM). Channels reconstituted in lipid bilayers can elicit currents (similar to native non-NMDA GluRs) in response to low levels of AMPA *or* kainate.

Ligands

Glutamate-, AMPA- and kainate-binding sites

07-47-01: From pharmacological considerations, L-glutamate, kainate and AMPA bind to different **receptor substructures** on recombinant AMPA receptors[106]. Receptor binding/autoradiographic approaches to characterization of excitatory amino acid receptors have been reviewed[149].

Available radioligands

07-47-02: AMPA: **[^{3}H]-AMPA** and **[^{3}H]-CNQX**. Use of AMPA or CNQX can define the non-specific binding of [^{3}H]-glutamate. *Note:* CNQX also binds to the **glycine site** and possibly the NMDA iGluR. Radioligands for kainate sites include **[^{3}H]-kainate** and **[^{3}H]-domoate** (but note the heterogeneity of domoate-affinity purified products – *see Single-channel data, 07-41*).

Affinity purification of AMPA receptor–channels using [^{3}H]-AMPA and Joro spider toxin

07-47-03: A glutamate receptor has been purified from Triton X-100-solubilized bovine cerebellum membranes by affinity chromatography using a spider toxin (**Joro spider toxin**; JsTx, immobilized on a lysine–agarose column followed by a Mono Q anion exchange column)[150]. The active fraction purifies an **AMPA-binding protein** of $M_r \sim 130$ kDa. Lineweaver–Burk† plots indicate the protein to have a K_d of 12.7 nM [^{3}H]-AMPA in the purified fraction. In reconstituted liposomes, the purified protein yields a glutamate-activated channel which can be inhibited with JsTx[150] *(see Blockers, 07-43)*.

Receptor/transducer interactions

Involvement of second messengers in excitatory amino acid signal transduction processes

07-49-01: Excitatory amino acids (EAA) activate second messenger† systems via metabotropic† receptors *in addition to* the direct gating of 'integral' (ionotropic†) receptor–channels *(reviewed in ref.*[151]*)*. Through these 'indirect' metabotropic pathways, EAAs are capable of activating both **adenylate cyclase** and **guanylate cyclase** and also to induce phosphoinositide† (PI) turnover *(see, for example, Fig. 4 under ELG CAT GLU NMDA and tables in Resource A – G protein-linked receptors, entry 56*.

Calmodulin-dependent inhibition of post-synaptic voltage-gated Ca^{2+} currents

07-49-02: Glutamate-evoked Ca^{2+}-influx through both NMDA and non-NMDA receptor–channels in rat hypothalamic neurones inhibits high-voltage-activated (HVA) Ca^{2+} channels *(see VLG Ca, entry 42)* via a calmodulin-dependent mechanism[152]. A pre-synaptic glutamate receptor agonist (**L-2-amino-4-phosphonobutyric acid**) and a selective metabotropiç agonist (**trans-ACPD**) are ineffective in mimicking the HVA Ca^{2+} current inhibition promoted by glutamate. Inhibition is also dependent on the presence of extracellular Ca^{2+}, and can be blocked by internal perfusion of the cells with **BAPTA**. The calmodulin antagonists **trifluoperazine** and **çalmidazolium** completely prevent the inhibition[152].

Links between GluR1 agonism and hormone secretion by heterologous gene expression

07-49-03: Co-expression† of a plasmid† construct encoding **growth hormone** and a plasmid encoding a non-NMDA glutamate receptor, GluR1, yields chromaffin cells in which Ca^{2+}-dependent growth hormone secretion can be stimulated by kainate.

Receptor agonists (selective)

07-50-01: *Note:* Availability of selective antagonists+ and agonists have been central to the recognition of receptor subtypes underlying native† iGluR responses. The basic features of these and other agonists are listed in Table 10.

Agonist affinities of recombinant GluRs
07-50-02: Receptors generated from the GluR1 to GluR4 cDNAs† have a higher apparent affinity for AMPA than for kainate. When homomeric† receptors of the GluR6 class are expressed in *Xenopus* oocytes, the receptors are activated by kainate, quisqualate and L-glutamate, but not by AMPA. Furthermore, the apparent affinity for kainate is higher than for receptors from the GluR1–GluR4 class[2].

EC_{50} values for iGluR agonists
07-50-03: Typical EC_{50} values for recombinant† GluR-A/B receptors expressed in oocytes are 3.31 μM (AMPA); 6.16 μM (glutamate); 57.5 μM (kainate)[106]. EC_{50} values for recombinant GluR-B/D receptors expressed in oocytes are 5.01 μM (AMPA); 32.3 μM (glutamate); 64.6 μM (kainate)[106].

Agonist affinities and potencies at GluR–channel splice variants
07-50-04: Gene expression control of the alternative splice variants '**flip**' and '**flop**' *(see Gene organization, 07-20)* can confer **different kinetic properties** on the currents evoked by the agonists glutamate or AMPA, but *not* on those evoked by kainate. For both 'flip' and 'flop' versions of GluR-A to GluR-D, kainate evokes a non-desensitizing† current whereas both glutamate and AMPA cause an initial fast-desensitizing† current followed by a **steady-state plateau**. Whereas glutamate, AMPA and kainate evoke currents of similar amplitude in 'flip'-expressing cells, kainate-evoked currents are much larger than those evoked by glutamate or AMPA in 'flop'-expressing cells. Glutamate activates channels 4–5-times more effectively when acting at the 'flip' version of the GluRs[26].

Receptor antagonists

07-51-01: Several studies have indicated that antagonism at NMDA receptors *(see ELG CAT GLU NMDA, entry 08)* is only partially protective in some models of **focal ischaemia** and may be ineffective in **global ischaemia** (see *Phenotypic expression, 07-14 and discussion in ref.*[153]). Development of AMPA/kainate-selective antagonists (particularly NBQX and those of the 2,3-benzodiazepine class) have indicated their value as **'neuroprotective'** and **anti-convulsant agents**. The basic properties of these antagonists are listed in Table 11 in comparison with other (less-selective) antagonists in use.

Table 10. *Common agonists of AMPA/kainate receptors and their features (From 07-50-01)*

Agonist	Features	Refs
Principal agonists **AMPA**	For constitution of 'AMPA-preferring' versus 'kainate-preferring' iGluRs, *see Gene family, 07-05.* For reported activities of **AMPA** (α-amino-3-hydroxy-5-methylisoxazole-4-propionic acid), see paragraphs below this table and other fields prefixed with 'AMPA:'	*161*
Kainate	Kainic acid (kainate) is a full, non-desensitizing† agonist at subsets of iGluR assemblies *(see 07-05).* A functional kainate receptor has been cloned which is insensitive to quisqualate/AMPA. Conductance responses evoked by kainate at the GluR channels are competitive with those evoked by AMPA and are not additive. *Note:* **Domoate** has also been used as a kainate receptor agonist and affinity ligand *(e.g. see Table 9).* For further activities of these agonists, see paragraphs below this table and other fields prefixed with KAIN:	*20*
Willardiines and bromo-/chloro-derivatives	The (*S*)- but not (*R*)-isomers of the naturally occurring heterocyclic excitatory amino acid **willardiine** and **5-bromowillardiine** are potent agonists for AMPA/kainate receptors. Willardiine [(*S*)-1-(2-amino-2-carboxyethyl)pyrimidine-2,4-dione] produces rapidly but incompletely desensitizing responses. At equilibrium, (*S*)-**5-fluorowillardiine** (EC_{50}, 1.5 μM) is ∼ 7 times more potent† than (*R, S*)-AMPA (EC_{50}, 11 μM) and ∼ 30 times more potent† than willardiine (EC_{50}, 45 μM)[162]. *Note:* Willardiines are the first compounds characterized in which simple substituent changes in molecular structure are associated with marked differences in the ability of agonists to produce desensitization of AMPA/kainate receptors	*162*
Mixed agonists **L-Glutamate** **L-Aspartate** **L-Proline**	**L-Glutamate** and **L-aspartate** are mixed agonists† of the AMPA-, kainate- and NMDA-selective receptors and their effects are partially inhibited by all selective antagonists. In dorsal horn neurones of the rat spinal cord, millimolar concentrations of **L-proline** elicit an inward current that is partially antagonized by strychnine, APV and CNQX. Thus, L-proline is a weak agonist at strychnine-sensitive glycine receptors and at both NMDA and non-NMDA glutamate receptors. The ability of L-proline to stimulate CNQX-sensitive Ca^{2+}-entry following activation of excitatory amino acid receptors implicates L-proline as a potential **endogenous excitotoxin**	*163*
Quisqualate	**Quisqualate** (QUIS) was formerly used as a principal agonist for AMPA receptors, but it has also been shown to be an agonist at metabotropic† glutamate receptors.	
Amino toxins of plant origin **BOAA**	The amino acid toxin β-*N*–oxalylamino-L-alanine (**BOAA**) is associated with incidence of **neurolathyrism** (a spastic disorder with acute and chronic onset) associated with consumption of the chick pea, *Lathyrus sativus*. BOAA appears to act as an excitant on spinal neurones via agonist activity at AMPA receptors	*82, 174*

Table 11. *Common antagonists and their features (From 07-51-01)*

Antagonist	Features	Refs
Competitive selective antagonist **NBQX**	Competitive antagonists with differential selectivity for recombinant iGluRs include 6-nitro-7-sulphamobenzo-*f*-quinoxaline-2,3,-dione (**NBQX**, pA_2 7.1). The potency of NBQX for blocking currents mediated by GluR-A/B receptors changes depending on the agonist used to activate the receptors (pA_2 values: 7.23 ± 0.01 for block of kainate responses; 6.78 ± 0.02, for block of L-glutamate responses; 6.95 ± 0.02 for block of AMPA responses). Differences between agonists are less marked in cells expressing GluR-B/D receptors (pA_2 values: 7.28 ± 0.01 for block of kainate responses; 7.30 ± 0.02 for block of L-glutamate responses; 7.35 ± 0.01 for block of AMPA responses). NBQX acts as a potent and selective antagonist at recombinant receptors, but its action can be overcome by increasing agonist concentration (i.e. it is competitive†). NBQX has anticonvulsive properties in several seizure† models and is neuroprotective in brain ischaemia† models. In native membranes NBQX is $>$ 500-fold more selective for AMPA over NMDA receptors (compare the relative non-selectivity of CNQX and DNQX at recombinant receptors below)	*106*
Non-competitive selective antagonists **2,3-Benzo diazepines GYKI-52466 GYKI-53655**	The homophthalazine (+/−)-**GYKI-52466** is a highly-selective AMPA/kainate antagonist (IC_{50} for kainate ~ 7.5 μM) and does not significantly affect NMDA, mGlu (metabotropic† glutamate), or $GABA_A$ responses. GYKI-52466 binds to both open and closed receptor–channels, and is voltage-independent in its antagonistic effects. The action of GYKI-52466 cannot be overcome by raising agonist concentrations (i.e. it is non-competitive†). GYKI-52466 is a broad-spectrum anticonvulsant. The methyl-carbamoyl derivative **GYKI-53655** is several-fold more potent than GYKI-52466. The **2,3-benzo-diazepine** class of AMPA/kainate receptor non-competitive antagonists may have therapeutic applications in epilepsy, ischaemia, neurodegeneration and Parkinson's disease	Review[1] *53*

Table 11. *Continued*

Antagonist	Features	Refs
'Low-selectivity' antagonists **CNQX**	In addition to their specific antagonistic effects on neuronal GluRs, quinoxalinediones (e.g. **CNQX**) have also been shown to block glutamate-induced responses mediated by recombinant AMPA/KA receptor–channels when expressed in heterologous† systems (e.g. GluR-A/B and GluR-B/D receptors). Antagonism occurs irrespective of the particular subunit composition and displays little selectivity between AMPA and kainate receptors). 6,7-Dinitroquinoxaline-2,3-dione (**DNQX**) also shows relatively low selectivity	*106*
Lipophilic competitive antagonists **DDHB** and derivatives	A class of glutamate receptor antagonists that show competitive† action, significant potency at multiple sites, and a high degree of lipophilicity are the **substituted benzazepines**. 2,5-Dihydro-2,5-dioxo-3-hydroxy-^{1}H-benzazepine (DDHB) and three substituted derivatives, 4-bromo-, 7-methyl- and 8-methyl-DDHB, block the activation of non-NMDA receptors by kainate and L-glutamate *(see also Receptor antagonists under ELG CAT GLU NMDA, 08-51)*	*154*
Competitive antagonists **GAMS** **GDEE** **DGG**	γ-Glutamylaminomethyl sulphonate (**GAMS**) has been used as a partially selective non-NMDA receptor antagonist which offers protection against audiogenic seizures†. Other competitive antagonists at the AMPA-binding site are **GDEE** and **DGG**	*34, 155, 156*
Inhalational anaesthetics **Enflurane**	**Enflurane** at anaesthetic concentrations (1.8 mM) inhibits AMPA-, kainate- and NMDA-induced currents expressed in oocytes by 29–40%, 30–33% and 20–27%, respectively. Inhibition by enflurane is independent of the concentrations of the agonists (NMDA, AMPA and kainate) or the NMDA-co-agonist (glycine) suggesting that enflurane inhibition does *not* result from a competitive interaction at glutamate- or glycine-binding sites	*157*
Biphenyl derivative of NDSA **Evans blue**	Selective blockade of a subset of AMPA/KA receptors has been reported for the non-competitive† antagonist **Evans blue**[158]. This biphenyl derivative of naphthalene disulphonic acid blocks (at low concentrations) kainate-mediated responses of the subunits GluR-A, GluR-A,B, GluR-A,C, and GluR-B,C expressed in *Xenopus* oocytes but *not* responses of GluR-C or GluR6	*158*

Table 11. *Cont*

Antagonist		Refs
	(IC ... or the subunit combination Glu... B). The blocking action of Evans blue is partially reversible and does *not* compete with the kainate for the agonist-binding site	
Non-competitive antagonist **Riluzole**	Responses from kainic acid-evoked currents in *Xenopus* oocytes injected with mRNA from rat whole brain or cortex can be non-competitively† blocked by the anticonvulsant and 'neuro-protective' compound **riluzole** ($IC_{50} \sim 167\ \mu M$ – cf. CNQX: $IC_{50} \sim 0.21\ \mu M$ ands NBQX: $IC_{50} \sim 0.043\ \mu M$). Riluzole is more potent at blocking responses to NMDA ($IC_{50} \sim 18.2\ \mu M$ – cf. the competitive NMDA receptor antagonist 2-APV: $IC_{50} \sim 6.1\ \mu M$)	*159*
Spider and wasp venom toxins	For the non-selective actions of certain spider venom toxins (argiotoxin and Joro spider toxin) and the digger wasp toxin philanthotoxin, *see Blockers, 07-43*	*160*

INFORMATION RETRIEVAL

Database listings/primary sequence discussion

07-53-01: *The relevant database is indicated by the lower case prefix (e.g. gb:), which should not be typed* (see Introduction & layout of entries, entry 02). *Database accession numbers immediately follow the colon. Note that a comprehensive listing of all available accession numbers is superfluous for location of relevant sequences in GenBank® resources, which are now available with powerful in-built* ***neighbouring† analysis*** *routines (for description, see the Database listings field in the Introduction and layout of entries, entry 02). For example, sequences of cross-species variants or related gene family† members can be readily accessed by one or two rounds of neighbouring† analysis (which are based on pre-computed alignments performed using the BLAST† algorithm by the NCBI†). This feature is most useful for retrieval of sequence entries deposited in databases later than those listed below. Thus, representative members of known sequence homology groupings are listed to permit initial direct retrievals by accession number, author/reference or nomenclature. Following direct accession, however, neighbouring† analysis is strongly recommended to identify newly reported and related sequences.*

Nomenclature (non-systematic)	Species, DNA source	Original isolate	Accession	Sequence/ discussion
GluR-4c flop	Rat Sprague–Dawley; alternatively spliced, cerebellum, cDNA	885 aa	gb: S94371	Gallo, *J Neurosci* (1992) **12**: 1010–23
GluR-A	Rat forebrain cDNA/ expression-cloned in oocytes. Equivalent to GluR-K1 (now renamed GluR1 or GluR-A)	889 aa (clone GluR-K1)	gb: X17184	Hollmann, *Nature (1989)* **342**: 643–8. Hollmann, *Cold Spring Harbor Symp Quant Biol* (1990) **55**: 41–55
GluR1 (human isoform)	Human hippocampal cDNA library	888 aa (mature)	gb: X58633 gb: S40299	Potier, *DNA Seq* (1992) **2**: 211–18
HBGR-1 (human isoform)	Full-length human homologue of GluR1 (or the flop version of the GluR-A clone)	888 aa	gb: M81886	Sun, *Proc Natl Acad Sci USA* (1992) **89**: 1443–7
GluHI (= GluR1 homologue)	Human brain cDNA library; chromosome 5	907 aa $M_r \sim 100$ kDa 97% homologous to rat GluR-A	gb: M64752	Puckett, *Proc Natl Acad Sci USA* (1991) **88**: 7557–61
'Glutamate receptor 1'	Mouse, brain cDNA	908 aa	gb: X57497 sp: P23818	Sakimura, *Neuron* (1992) **8**: 267–74
GluR-A **GluR-B** **GluR-C** **GluR-D**	Rat brain cDNA library	889 aa 862 aa 866 aa 881 aa	gb: M36418 sp: P19490 gb: M36419 sp: P19491 gb: M36420 sp: P19492 gb: M36421 sp: P19493	Keinanen, *Science* (1990) **249**: 556–60

Nomenclature (non-systematic)	Species, DNA source	Original isolate	Accession	Sequence/ discussion
GluR-A to GluR-D: Flip and flop variants	GluR-A flop GluR-B flop GluR-C flop GluR-D flop GluR-A flip GluR-B flip GluR-C flip GluR-D flip	Flip and flop; cell-specific functional switch *(see Gene organization, 07-20)*	gb: M36418 gb: M36419 gb: M36420 gb: M36421 gb: M38060 gb: M38061 gb: M38062 gb: M38063	Sommer, *Science* (1990) **249**: 1580–5 Koehler, *J. Biol Chem.* (1994) In press.
GluR2 (= GluR-B)	Adult rat forebrain cDNA library	Original published sequence name: RATGLUR2; 884 aa	gb: M85035 sp: P19491	Boulter, *Science* (1990) **249**: 1033–7
'Glutamate receptor 2'	Mouse, brain cDNA	884 aa	gb: X57498 sp: P23819	Sakimura, *Neuron* (1992) **8**: 267–74
GluR2 (= GluR-B) genomic[†a]	*Mus musculus* (strain BALB/c, subspecies *domesticus*) male liver DNA	exon[†] 2	gb: L32190 gb: L32151	Koehler, M., Kornau, H.-C. and Seeburg, P.H. unpublished (1994).
GluR2 (= GluR-B) genomic[†a]	*Mus musculus* (strain BALB/c, subspecies *domesticus*) male liver DNA	exon[†] 3	gb: L32191 gb: L32151	Koehler, M., Kornau, H.-C. and Seeburg, P.H. unpublished (1994).
GluR2 (= GluR-B) genomic[†a]	*Mus musculus* (strain BALB/c, subspecies *domesticus*) male liver DNA	exon[†] 1 and promoter region	gb: L32189 gb: L32151 gb: L32152	Koehler, M., Kornau, H.-C. and Seeburg, P.H. unpublished (1994).

Nomenclature (non-systematic)	Species, DNA source	Original isolate	Accession	Sequence/ discussion
GluR-B to GluR-6 genomic sequences	Mouse genomic GluR-B, genomic GluR-C, genomic GluR-D, genomic GluR-5, genomic GluR-6, genomic		 gb: M76437 gb: M76438 gb: M76439 gb: M76440 gb: M76441	Sommer, *Cell* (1991) **67**: 11–19
HBGR2 (human glutamate receptor 2)	Human brain cDNA library		gb: L20814	Sun, *Neuroreport* (1993) **5**: 441–4.
GluR3 (= GluR-C)	Adult rat forebrain cDNA library	original published sequence name: RATGLUR3; 889 aa	gb: M85036 sp: P19492	Boulter, *Science* (1990) **249**: 1033–7.
hGluR3 flip (human)	Stratagene cDNA libraries 936205 and 936206	895 aa	gb: U10301	Rampersad, V. unpublished (1994).
hGluR3 flip (human)	Stratagene cDNA libraries 936205 and 936206	895 aa	gb: U10302	Rampersad, V. unpublished (1994).
GluR4a (= GluR-D)	*Rattus norvegicus* (strain Sprague–Dawley)	903 aa	gb: M85037	Bettler, *Neuron* (1990) **5**: 583–95.
The long (putatively) **intracellular loop** of AMPA-selective iGluR	N.B. Subcloned for phosphorylation studies	nucleotide coding sequence segments	gb: S56679 gb: S56890	Wright, J *Recept Res* (1993) **13**: 653–65
*β*2 (= **GluR6** homologue)	Mouse, cDNA	864 aa	gim: 405312	Morita, *Mol Brain Res* (1992) **14**: 143-6.

Nomenclature (non-systematic)	Species, DNA source	Original isolate	Accession	Sequence/ discussion
GluR5 (= **β1** homologue)	GluR5 has two splice variants, GluR5-1 (920 aa) and GluR5-2 (905 aa) *(see Gene organization, 07-20)*	GluR5-1: 920 aa (a full open reading frame for the shorter GluR5-2 splice variant was not found)	gb: M83552	Bettler, *Neuron* (1990) **5**: 583–95.
GluR6 (= **β2** homologue)	Rat cerebellum cDNA library	884 aa	not found	Egebjerg, *Nature* (1991) **351**: 745–8.
GluR7	not found	956 aa	not found	not found
δ1	Mouse delta-1 GluR chain precursor	1009 aa	PIR: JH0266	Yamazaki, *Biochem Biophys Res Commun* (1992) **183**: 886–92.
δ2	Mouse delta-2 GluR chain precursor		not found	Lomeli, *FEBS Lett* (1993) **315**: 318–22.
γ2 (= KA-2 homologue)	Mouse GluR gamma 2 subunit selective for kainate	979 aa	gb: D01273	Sakimura, *FEBS Lett* (1990) **272**: 73–80.
GluR-K2 (= flip version of GluR2)	Rat hippocampus and cerebral cortex cDNA	888 aa	gb: X54655	Nakanishi, *Neuron* (1990) **5**: 569–81.
GluR-K3 (= flip version of GluR3)		883 aa	gb: X54656	
humEAA2	Human hippocampal cDNA library; structurally related, though not identical to KA-1	962 aa plus 18 aa signal sequence $M_r \sim 107\,176$ kDa	not found	Kamboj, *Mol Pharmacol* (1992) **42**: 10–15.

Nomenclature (non-systematic)	Species, DNA source	Original isolate	Accession	Sequence/ discussion
KA-1 (= γ1 homologue)	Rat brain cDNA, high-affinity kainate; homomeric assemblies do not form channels	956 aa	gb: X59996	Werner, *Nature* (1991) **351**: 742–4.
KA-2 (= γ2)	Rat brain cDNA, high-affinity kainate; homomeric assemblies do not form channels	979 aa	em: X59996 em: Z11581 gb: X59996 gb: Z11581	Herb, *Neuron* (1992) **8**: 775–85.
KBP-c	Chick cerebellum cDNA kainate binding protein	KA binding (channel inactive) 464 aa	not found	Gregor, *Nature* (1989) **342**: 689–92.
KBP-f	Frog brain cDNA/kainate-binding protein initially purified by domoic acid affinity chromatography	KA binding (channel inactive) 487 aa	not found	Wada, *Nature* (1989) **342**: 684–9.

[a]The complex **alternative splicing**† patterns observed in the GluR-B gene (Koehler, M., Kornau, H-C. and Seeburg, P.H. unpublished, 1994) can be traced from splice site joining data* accompanying certain database entries. Importing sequences in a standard file format can be interpreted by some sequence analysis programs and incorporated in an interactive feature table. *For example, the GluR-B short-splice form entry (gb: L32204/gb: L32151) contains the following 'joining' protocols:
mRNA join (L32189:1075..1594,L32190:1..141,L32191:1..240,L32192:1..197, L32193:1..54, L32194:1..162,L32195:1..168,L32196:1..105,L32197:1..111, L32198:1..207, L32199: 1..371,L32200:1..199,L32201:1..248,L32202:1..115, L32203:2..115,1..249) and
CDS join (L32189:1507..1594,L32190:1..141,L32191:1..240,L32192:1..197, L32193: 1..54,L32194:1..162,L32195:1..168,L32196:1..105,L32197:1..111, L32198:1..207,L32199:1..371,L32200:1..199,L32201:1..248,L32202:1..115, L32202:2..115,1..246)
which relate the contents of database entries *('L-numbers' in the example above)* to designated splice points.

Sources of information on other glutamate receptors
07-53-02: *Note:* The genes and cDNAs tabulated above encode subunits forming **ionotropic**† **glutamate receptors** (**iGluR**). In database searches, these should not be confused with the gene nomenclature used to describe the **metabotropic**† glutamate receptors such as the $mGlu_1$–$mGlu_7$ series[6, 164–166]. mGlu receptor functions through G proteins and inositol phosphate turnover[167] are characterized by selective activation with 1-amino-cyclopentyl-1,3-dicarboxylate (ACPD). By gating of K^+ currents via metabotropic glutamate receptors, excitatory amino acids can also act as slow neuromodulatory transmitters[168] *(see Receptor/transducer interactions under ILG K Ca, 27-49).*

Related sources & reviews

07-56-01: Major quoted sources[1, 7, 17–19, 169]; GluRs in hippocampal neurones[170]; molecular neurobiology of GluRs[7, 8, 18–20]; molecular biology of ionotropic glutamate receptors in *Drosophila*[171]; permeation pathways of neurotransmitter-gated ion channels[129]; physiological and pathophysiological roles of excitatory amino acids during development[57]; single-channel recording from iGluR[172]; therapeutic potential of selective AMPA/kainate receptor antagonists[153]; non-NMDA glutamate receptors in glial cell signalling[80]; roles of GluRs in CNS function[55, 56, 135]; excitatory amino acids as endogenous functional neurotransmitters[173]; excitatory amino acid activation of second messenger systems in addition to a direct gating of ion channels[151]. *See also the Resource E, – Ion channel book references, entry 60.*

Feedback

Error-corrections, enhancement and extensions
07-57-01: Please notify specific errors, omissions, updates and comments on this entry by contributing to its **e-mail feedback file** (*for details, see Resource J, Search Criteria & CSN Development*). For this entry, send e-mail messages To: **CSN-07@le.ac.uk,** indicating the appropriate paragraph by entering its **six-figure index number** (xx-yy-zz or other identifier) into the **Subject**: field of the message (e.g. Subject: 08-50-07). Please feedback on only **one specified paragraph or figure per message,** normally by sending a **corrected replacement** according to the guidelines in *Feedback & CSN Access* . Enhancements and extensions can also be suggested by this route (*ibid.*). Notified changes will be indexed via 'hotlinks' from the CSN 'Home' page (http://www.le.ac.uk/csn/) from mid-1996.

Entry support groups and e-mail newsletters
07-57-02: Authors who have expertise in one or more fields of this entry (and are willing to provide editorial or other support for developing its contents) can join its support group: In this case, send a message To: **CSN-07@le.ac.uk,** (entering the words "support group" in the Subject: field). In the message, please indicate principal interests (see *fieldname criteria in the Introduction for coverage*) together with any relevant **http://www site**

links (established or proposed) and details of any other possible contributions. In due course, support group members will (optionally) receive **e-mail newsletters** intended to **co-ordinate and develop** the present (text-based) entry/fieldname frameworks into a 'library' of interlinked resources covering ion channel signalling. Other (more general) information of interest to entry contributors may also be sent to the above address for group distribution and feedback.

REFERENCES

[1] Sommer, *Trends Pharmacol Sci* (1992) **13**: 291–6.
[2] Egebjerg, *Nature* (1991) **351**: 745–8.
[3] Boulter, *Science* (1990) **249**: 1033–7.
[4] Bettler, *Neuron* (1990) **5**: 583–95.
[5] Sakimura, *Neuron* (1992) **8**: 267–74.
[6] Schoepp, *Trends Pharmacol Sci* (1993) **14**: 13–20.
[7] Gasic, *Annu Rev Physiol* (1992) **54**: 507–36.
[8] Nakanishi, *Science* (1992) **258**: 597–603.
[9] Jonas, *J Physiol* (1992) **455**: 143–71.
[10] Patneau, *Neuron* (1991) **6**: 785–98.
[11] Partin, *Neuron* (1993) **11**: 1069–82.
[12] Gallo, *J Neurosci* (1992) **12**: 1010–23.
[13] Keinanen, *Science* (1990) **249**: 556–60.
[14] Herb, *Neuron* (1992) **8**: 775–85.
[15] Mayer, *Prog Neurobiol* (1987) **28**: 197–276.
[16] MacDermott, *Trends Neurosci* (1987) **17**: 280–4.
[17] Wisden, *Curr Opin Neurobiol* (1993) **3**: 291–8.
[18] Seeburg, *Trends Pharmacol Sci* (1993) **14**: 297–303.
[19] Nicoll, *Physiol Rev* (1990) **70**: 513–65.
[20] Barnard, *Trends Pharmacol Sci* (1990) **11**: 500–7.
[21] Henley, *Proc Natl Acad Sci USA* (1992) **89**: 4806–10.
[22] Henley, *New Biol* (1989) **1**: 153–8.
[23] Monyer, *Neuron* (1991) **6**: 799–810.
[24] Shen, *J Biol Chem* (1993) **268**: 19070–5.
[25] Raymond, *Nature* (1993) **361**: 637–41.
[26] Sommer, *Science* (1990) **249**: 1580–5.
[27] Müller, *Science* (1992) **256**: 1563–6.
[28] Bettler, *Neuron* (1992) **8**: 257–65.
[29] Lomeli, *FEBS Lett* (1992) **307**: 139–43.
[30] Lambolez, *Neuron* (1992) **9**: 247–58.
[31] Monyer, *Science* (1992) **256**: 1217–21.
[32] Meguro, *Nature* (1992) **357**: 70–4.
[33] Ishii, *J Biol Chem* (1993) **268**: 2836–43.
[34] Moriyoshi, *Nature* (1991) **354**: 31–7.
[35] Lomeli, *FEBS Lett* (1993) **315**: 318–22.
[36] Wisden, *J Neurosci* (1993) (cited as in press in source).
[37] Tölle, *J Neurosci* (1991) (cited as in press in source).

[38] Kuhse, *FEBS Lett* (1991) **283**: 73–7.
[39] Iino, *J Physiol* (1990) **424**: 151–65.
[40] Bochet, *Neuron* (1994) **12**: 383–8.
[41] Hestrin, *Neuron* (1993) **11**: 1083–91.
[42] Wyllie, *J Physiol* (1993) **463**: 193–226.
[43] Sommer, *EMBO J* (1992) **11**: 1651–56.
[44] Huettner, *Neuron* (1990) **5**: 255–66.
[45] Evans, *Br J Pharmacol* (1987) **91**: 531–7.
[46] Monaghan, *Annu Rev Pharmacol Toxicol* (1989) **29**: 365–402.
[47] Lerma, *Proc Natl Acad Sci USA* (1993) **90**: 11688–92.
[48] Pellegrini-Giampietro, *Proc Natl Acad Sci USA* (1991) **88**: 4157–61.
[49] Walton, *J Neurosci* (1993) **13**: 2068–84.
[50] Wyllie, *J Physiol* (1991) **432**: 235–58.
[51] Patneau, *Neuron* (1994) **12**: 357–71.
[52] Hollmann, *Science* (1991) **252**: 851–3.
[53] Ogura, *Neurosci Res* (1992) **12**: 606–16.
[54] Lipton, *Trends Neurosci* (1989) **12**: 265–70.
[55] Collingridge, *Trends Pharmacol Sci* (1990) **11**: 290–6.
[56] Bliss, *Nature* (1993) **361**: 31–9.
[57] McDonald, *Brain Res Rev* (1990) **15**: 41–70.
[58] Zafra, *EMBO J* (1990) **9**: 3545–50.
[59] Gall, *Mol Brain Res* (1991) **9**: 113–23.
[60] Sato, *Neuroscience* (1993) **52**: 515–39.
[61] Burnashev, *Science* (1992) **256**: 1566–70.
[62] Moran, *FEBS Lett* (1992) **302**: 21–5.
[63] Matute, *Proc Natl Acad Sci USA* (1992) **89**: 3399–403.
[64] Clark, *J Neurosci* (1992) **12**: 664–73.
[65] Werner, *Nature* (1991) **351**: 742–4.
[66] Hamassakibritto, *J Neurosci* (1993) **13**: 1888–98.
[67] Kristensen, *FEBS Lett* (1993) **332**: 14–18.
[68] Ray, *Biochem Biophys Res Commun* (1993) **197**: 1475–82.
[69] Jensen, *J Neurosci* (1993) **13**: 1664–75.
[70] Tang, *Science* (1991) **254**: 288–90.
[71] Perl, *N Engl J Med* (1990) **322**: 1775–8.
[72] Pellegrini-Giampietro, *Proc Natl Acad Sci USA* (1992) **89**: 10499–503.
[73] Lobner, *J Neuroscience* (1993) **13**: 4861–71.
[74] Pollard, *Neuroscience* (1993) **57**: 545–54.
[75] Weiss, *Neuron* (1993) **10**: 43–9.
[76] Olney, *Exp Brain Res* (1971) **14**: 61–76.
[77] Linden, *Neuron* (1993) **11**: 1093–1100.
[78] Perkel, *J Physiol* (1993) **471**: 481–500.
[79] Bertrand, *Eur J Pharmacol* (1993) **237**: 45–50.
[80] Teichberg, *FASEB J* (1991) **5**: 3086–91.
[81] Cornell, *Science* (1990) **247**: 470–3.
[82] Meldrum, *Trends Pharmacol Sci* (1990) **11**: 379–87.
[83] Gregor, *Nature* (1989) **342**: 689–92.
[84] Craig, *Neuron* (1993) **10**: 1055–68.
[85] Petralia, *J Comp Neurol* (1992) **318**: 329–54.
[86] Blackstone, *J Neurochem* (1992) **58**: 1118–26.

87 Molnar, *Neuroscience* (1993) **53**: 307–26.
88 Martin, *Neuron* (1992) **9**: 259–70.
89 McNamara, *J Neurosci* (1992) **12**: 2555–62.
90 Sun, *Proc Natl Acad Sci USA* (1992) **89**: 1443–7.
91 Eubanks, *Proc Natl Acad Sci USA* (1993) **90**: 1782–87.
92 Potier, *Genomics* (1993) **15**: 696–7.
93 Sommer, *Cell* (1991) **67**: 11–19.
94 Higuchi, *Cell* (1993) **75**: 1361–70.
95 Wenthold, *J Biol Chem* (1992) **267**: 501–7.
96 Hollmann, *Cold Spring Harbor Symp Quant Biol* (1990) **55**: 41–55.
97 Wada, *Nature* (1989) **342**: 684–9.
98 Unwin, *Cell* (1993) **72**: 31–41.
99 Nakanishi, *Neuron* (1990) **5**: 569–81.
100 Burnashev, *Neuron* (1992) **8**: 189–98.
101 Hume, *Science* (1991) **253**: 1028–31.
102 Verdoorn, *Science* (1991) **252**: 1715–18.
103 Kohler, *Neuron* (1993) **10**: 491–500.
104 Egebjerg, *Proc Natl Acad Sci USA* (1993) **90**: 755–9.
105 Uchino, *FEBS Lett* (1992) **308**: 253–7.
106 Stein, *Mol Pharmacol* (1992) **42**: 864–71.
107 Keller, *EMBO J* (1992) **11**: 891–6.
108 Greengard, *Science* (1991) **253**: 1135–8.
109 Raymond, *Trends Pharmacol Sci* (1993) **147**: 147–53.
110 Madison, *Annu Rev Neurosci* (1991) **14**: 379–97.
111 Linden, *Science* (1991) **254**: 1656–9.
112 Stern, *J Physiol* (1992) **449**: 247–78.
113 Hestrin, *Neuron* (1992) **9**: 991–9.
114 Colquhoun, *J Physiol* (1992) **458**: 261–87.
115 Burnashev, *Neuron* (1992) **8**: 8–20.
116 Burnashev, *Science* (1992) **257**: 1415–19.
117 Cole, *Nature* (1989) **340**: 474–6.
118 Dildy, *J Neurochem* (1992) **58**: 1569–72.
119 Reichling, *J Physiol* (1993) **469**: 67–88.
120 Vyklicky, *Neuron* (1991) **7**: 971–84.
121 Hori, *Brain Res* (1988) **457**: 350–4.
122 Livsey, *J Neurosci* (1993) **13**: 5324–33.
123 Yamada, *J Neurosci* (1993) **13**: 3904–15.
124 Patneau, *J Neurosci* (1993) **13**: 3496–509.
125 Wong, *Mol Pharmacol* (1993) **44**: 504–10.
126 Hullebroeck, *Brain Res* (1992) **590**: 187–92.
127 Raman, *Neuron* (1992) **9**: 173–86.
128 Isaacson, *Proc Natl Acad Sci USA* (1991) **88**: 10936–40.
129 Lester, *Annu Rev Biophys Biomol Struç* (1992) **21**: 267–92.
130 Gilbertson, *Science* (1991) **251**: 1613–15.
131 Ozawa, *J Neurophysiol* (1991) **66**: 2–11.
132 Dingledine, *J Neurosci* (1992) **12**: 4080–7.
133 Schneggenburger, *Neuron* (1993) **11**: 133–43.
134 Pruss, *Neuron* (1991) **7**: 509–18.
135 Collingridge, *Physiol Rev* (1989) **40**: 145–210.

[136] Cull-Candy, *Nature* (1987) **325**: 525–8.
[137] Ascher, *J Physiol* (1988) **399**: 227–45.
[138] Tang, *Science* (1989) **243**: 1474–7.
[139] Sansom, *Int Rev Neurobiol* (1990) **32**: 51–106.
[140] Traynelis, *Neuron* (1993) **11**: 279–89.
[141] Ishida, *Proc Natl Acad Sci USA* (1985) **82**: 1837–41.
[142] Kerry, *Mol Pharmacol* (1993) **44**: 142–52.
[143] Herlitze, *Neuron* (1993) **10**: 1131–40.
[144] Priestley, *Br J Pharmacol* (1988) **97**: 1315–23.
[145] Blaschke, *Proc Natl Acad Sci USA* (1993) **90**: 6528–32.
[146] Ultsch, *Proc Natl Acad Sci USA* (1992) **89**: 10484–8.
[147] Copani, *J Neurochem* (1992) **58**: 1199–204.
[148] Ito, *J Physiol* (1990) **424**: 533–43.
[149] Young, *Trends Pharmacol Sci* (1990) **11**: 126–33.
[150] Shimazaki, *Mol Brain Res* (1992) **13**: 331–7.
[151] Smart, *Cell Mol Neurobiol* (1989) **9**: 193–206.
[152] Zeilhofer, *Neuron* (1993) **10**: 879–87.
[153] Rogawski, *Trends Pharmacol Sci* (1993) **14**: 325–31.
[154] Swartz, *Mol Pharmacol* (1992) **41**: 1130–41.
[155] Lodge, *Trends Pharmacol Sci* (1990) **11**: 81–6.
[156] Chapman, *Neurosci Lett* (1985) **55**: 325–30.
[157] Lin, *FASEB J* (1993) **7**: 479–85.
[158] Keller, *Proc Natl Acad Sci USA* (1993) **90**: 605–9.
[159] Debono, *Eur J Pharmacol* (1993) **235**: 283–9.
[160] Jackson, *Trends Neurosci* (1988) **11**: 278–83.
[161] Hollmann, *Nature* (1989) **342**: 643–8.
[162] Patneau, *J Neurosci* (1992) **12**: 595–606.
[163] Henzi, *Mol Pharmacol* (1992) **41**: 793–801.
[164] Tanabe, *Neuron* (1992) **8**: 169–79.
[165] Sugiyama, *Neuron* (1989) **3**: 129–32.
[166] Miller, *Trends Pharmacol Sci* (1991) **12**:
[167] Sugiyama, *Nature* (1987) **325**: 531–3.
[168] Charpak, *Nature* (1990) **347**: 765–7.
[169] Henneberry, *Bioessays* (1992) **14**: 465–71.
[170] Ozawa, *Jpn J Physiol* (1993) **43**: 141–59.
[171] Betz, *Trends Pharmacol Sci* (1993) **14**: 428–31.
[172] Cull-Candy, *Trends Pharmacol Sci* (1987) **8**: 218–24.
[173] Headley, *Trends Pharmacol Sci* (1990) **11**: 205–11.
[174] Bridges, *J Neurosci* (1989) **9**: 2073–9.
[175] Partin, *Mol Pharmacol* (1994) **46**: 129-38.

ELG CAT GLU NMDA

***N*-Methyl-D-aspartate (NMDA)-selective glutamate receptor–channels**

Edward C. Conley **Entry 08**

NOMENCLATURES

Abstract/general description

08-01-01: Pre-synaptic release of glutamate produces an **excitatory postsynaptic current**† (EPSC†) which can be resolved into (i) a fast EPSC component where the onset and decay is mediated by *non*-NMDA ionotropic† glutamate receptor–channels *(see ELG CAT GLU AMPA/KAIN, entry 07)* and (ii) a **slow** (or '**long-lasting**') **EPSC component** mediated by **NMDA-gated receptor–channels** (**NMDARs**).

08-01-02: Responses of the NMDA receptor–channel are characterized by a **slow rise and decay**, a **large Ca^{2+}-permeability**†, **voltage-dependent Mg^{2+} block** and a requirement for **glycine** (or a glycine-like endogenous molecule) as a **co-agonist**†.

08-01-03: Genes encoding cation channels selective for the glutamate receptor agonist ***N*-methyl-D-aspartate** (referred to as NMDARs in this entry for convenience) form part of the extracellular ligand-gated channel gene superfamily *(described under ELG Key facts, entry 04).* Functional NMDA receptors have been shown to be composed of a **fundamental subunit**, encoded by the **NR1 subunit gene** (or equivalent name) and its **potentiating subunits** (encoded by the **NR2A–NR2D gene** family†). When heterologously expressed†, the NR1 subunits form homomeric receptor–channels whose responses are significantly potentiated† by co-expression with any NR2 subunit. By contrast, the NR2-series do *not* possess homomeric channel-forming activity when expressed alone.

08-01-04: Receptors for ionotropic† NMDA receptor–channels are expressed in virtually **every neuronal cell**. 'Functional diversity' of NMDAR may arise through **co-expression of multiple types** of subunit genes within single cells, which is a common occurrence *(see Cell-type expression index, 08-08, and mRNA distribution, 08-13).* In general, regional variations in NMDAR properties are effected by **differential expression**† of NR2 subunit genes.

08-01-04: NMDA receptor activation is also essential for **neuronal differentiation** processes and establishment or elimination of synapses in developing brain *(see Developmental regulation, 08-11).* Transient increases in NMDA receptor gene expression occur during development, and application of NMDA-receptor-selective antagonists can block **experience-dependent plasticity**† in the adult. NMDAR densities appear to be modulated by **developmental stimuli**, and influx of Ca^{2+} through the NMDAR and L-type Ca^{2+} channels of hippocampal neurones has been shown to regulate **gene transcriptional**† events in the nucleus.

08-01-05: The involvement of NMDA receptors in several **neurophysiological** and **pathophysiological processes** has been demonstrated, including normal **information processing** of auditory and visual signals, activation and maintenance of **motor rhythms** (e.g. in respiratory control), central sensitization during **nociception** (pain reception), cardiovascular pressure control (**baroreception** and control of **vasomotor tone**), development of **motoneurone disorders**, epileptic states and post-ischaemic brain damage (**glutamate toxicity**). There is a large literature relating the function of the NMDAR to roles in **synaptic plasticity**† and by implication, **learning** and **memory**† processes *(see Phenotypic expression, 08-14, and the next paragraph).*

08-01-06: A large number of associated signalling components and mechanisms have been shown to modulate mechanisms of **long-term potentiation**† (**LTP**) a long-lasting increase in the strength of synaptic transmission due to brief, repetitive activation of excitatory afferent† fibres. As a striking example of **synaptic plasticity**†, the experimental induction of LTP and other distinct forms of synaptic potentiation have been shown to be reliant on molecular properties of the NMDAR–channel. Principally, this involves activation of NMDAR by synaptically released glutamate with **concomitant post-synaptic membrane depolarization** which relieves a **voltage-dependent**, **extracellular Mg^{2+}-block** of the NMDA receptor ion channel. Removal of Mg^{2+}-block, present under resting conditions, allows calcium to flow into the dendritic spine *(for details, see Phenotypic expression, 08-14).* Notably, several NMDAR-*independent* pathways for synaptic potentiation have also been characterized.

08-01-07: Comparison of genomic† and cDNA† sequences for genes encoding NMDARs has revealed that **alternative splicing**† processes affecting the fundamental (NR1) subunit can generate a number of functionally distinct NMDA receptors. In common with other ELG receptor–channels, there is strong evidence that **protein phosphorylation** events can modulate NMDAR functional responses *in vivo.*

08-01-08: In addition to their susceptibility to voltage-dependent Mg^{2+} block, a key molecular feature of the NMDAR series is their **Ca^{2+}-permeability**, which is central to their proposed phenotypic roles such as synaptic potentiation, glutamate toxicity and information processing. The **molecular determinants** of both Mg^{2+}-sensitivity and Ca^{2+}-permeability have been 'mapped' using defined mutants, and amino acid residues in a position homologous to the site in the M2 region (**Q/R site**) of AMPA receptors have been shown critical for determination of Ca^{2+}-permeability *(compare the Domain functions, 08-29, in this entry and Domain functions under ELG CAT GLU AMPA/KAIN, 07-29).*

08-01-09: A large number of **endogenous modulators** and **pharmacological antagonists** have been characterized for the NMDA receptor and appear to act at several distinguishable '**sites**' including the **NMDA** (or **glutamate**) site, the **glycine** (**co-agonist**) **site**, the **Zn^{2+}** (**modulatory**) **site**, the **redox modulatory site**, the **polyamine site**, the **Mg^{2+}-block site**, and sites for **H^+ ion modulation** and those involved in binding other **open-channel blockers**† *(see Receptor antagonists, 08-51, including Fig. 6 and Table 10).*

Category (sortcode)

08-02-01: ELG CAT GLU NMDA, i.e. extracellular ligand-gated cation channels selective for the glutamate receptor agonist *N*-methyl-D-aspartate. The suggested **electronic retrieval code** (unique embedded identifier or **UEI**) for 'tagging' of new articles of relevance to the contents of this entry is UEI: NMDA-NAT (for reports or reviews on native† channel properties) and UEI: NMDA-HET (for reports or reviews on channel properties applicable to heterologously† expressed recombinant† subunits encoded by cDNAs† or genes†). *For a discussion of the advantages of UEIs and guidelines on their implementation, see the section on Resource J under Introduction & layout, entry 02, and for further details, see Resource J – Search criteria & CSN development, entry 65.*

Channel designation

Distinction of ionotropic from metabotropic glutamate receptors
08-03-01: NMDA receptor–channels are frequently referred to as NMDARs. Ionotropic† glutamate receptor–channels in general *(see also ELG CAT GLU AMPA/KAIN, entry 07)* are sometimes designated as **'iGluR'** as compared to the metabotropic† GluRs such as the $mGlu_1$–$mGlu_7$ series[1–4] which are linked to **G proteins** and do *not* contain integral ion channels. Dependent on their molecular subtype, mGlu receptors can either activate phosphoinositide† metabolism via phospholipase C activation (subtypes $mGlu_1$, $mGlu_5$), ultimately releasing Ca^{2+} from intracellular stores *(see ILG Ca InsP$_3$, entry 19)*, or inhibit the activity of adenylate cyclase via a pertussis toxin-sensitive G protein (remaining subtypes) *(see Resource A – G protein-linked receptors, entry 56)*. The role of glutamate-receptor-linked second messenger generation in regulating intracellular Ca^{2+} has been reviewed[5].

NMDAR subunit/gene nomenclature
08-03-02: Recombinant† NMDA receptor subunits are designated according to the **name of the gene** that encodes them, e.g. for rat – NMDAR1, NMDAR2A, NMDAR2B, NMDAR2C and NMDAR2D *(see Gene family, 08-05)*. Hyphenated forms of these gene/subunit names are also in common use (i.e. NMDAR-1, NMDAR-2A, etc.). Other nomenclatures in use are listed in Table 1. Mouse equivalents of the NMDAR1 (NR1) and NMDAR2 (NR2) series have been designated by **Greek letters**. *(For further discussion of nomenclatures, see Gene family, 08-05, and the IUPHAR Nomenclature Committee recommendations via the CSN 'home page' (see Feedback & CSN access, entry 12).)*

Current designation

08-04-01: Conventionally of the form $I_{agonist}$, e.g. I_{NMDA}, or by specification of the current following co-expression of named subunits, e.g. $I_{(NMDA\text{-}R1/2A)}$.

Gene family

Distinction between genes encoding 'fundamental' and 'potentiating' NMDAR subunits

08-05-01: The genes encoding NMDA receptors form part of the extracellular ligand-gated channel gene superfamily *(introduced under ELG Key facts, entry 04).* Functional NMDA receptors have been shown to be composed of a **fundamental subunit**, encoded by the NR1 subunit gene (or equivalent name) and its **potentiating subunits** (encoded by the genes NR2A–NR2D).

Alternative nomenclatures for independently isolated NMDAR subunit genes

08-05-02: The NMDA receptor–channel subunit genes have been classified into the **epsilon** (**ε, fundamental**) and **zeta** (**ζ, potentiating**) **families** according to the degree of shared amino acid sequence *(for clarification, see Table 1).* Independent description of the same receptor subtypes has led to the proposal of several **alternative nomenclatures** which are compared in Table 1 and under *Database listings, 08-53.* In the absence of a **universal nomenclature**, species-specific isoform data are quoted using the name derived from the original citation – e.g. NMDAR1, NMDAR2A–2D/NR1, NR2A–2D (generally from rat) and zeta-1 (ζ1), ε1–ε4 (generally derived from mouse). For brevity, descriptions of rat subunits are cited in the abbreviated form (e.g. NR1, NR2A, etc.).

Mechanism for accessing updates on iGluR nomenclature (IUPHAR recommendations)

08-05-03: For updates on the agreed nomenclatures, refer to the latest IUPHAR Nomenclature Committee recommendations via the CSN *(see Feedback & CSN access, entry 12).*

Relatedness of the fundamental subunit NR1 to NR2 series- and 'non-NMDA' subunit genes

08-05-04: NR1 coding sequences† share only ~18% sequence homology† with the 'potentiating' NR2-type subunits. NR2A–2C share 55–70% sequence identity† with each other[6,8]. Typical homologies between the NR1 subunit[10,15] and subunits encoding AMPA- and kainate-selective iGluR are ~22–23% (for GluR-B–GluR-D), ~23–24% (for GluR6 and GluR7) and ~22% (for KA-2). Despite these relatively low sequence homologies the NR1 isolate shows overall *structural* similarities with AMPA and kainate receptors *(see ELG CAT GLU AMPA/KAIN, entry 07, and Domain arrangement, 08-27).*

Sequence variants within populations

08-05-05: Minor differences in sequence between the NR2A-D subunits, possibly due to **polymorphic variation**† between different strains of rat have been reported and discussed[8].

Subtype classifications

08-06-01: For a subtype classification based on similarity of NMDAR protein primary† sequences, *see Gene family, 08-05.*

Table 1. *Distinguishing features of genes encoding NMDA receptor–channels (From 08-05-02)*

Gene/subunit (equivalent nomenclature in brackets)[a]	Description[b]	Encoding[c]	Transcript size	Predicted mol. wt (non-glycosylated)
NR1 = **NMDAR1** Homologue: = **zeta 1** (ζ1)[d]	NMDA 'fundamental' or 'key' subunit. Forms homomeric† channels; significantly potentiated† by co-expression with any NR2 subunit. Different NR1/NR2 series combinations have different properties[6–9] *(see other fields)*	938 aa	Two adjacent hybridizing bands ~ 4.2 and 4.4 kb[10] or ~ 4.5 and 4.8 kb (human[11])	~ 105 kDa[e] ~ 116 kDa by SDS–PAGE†
NR2A = **NMDAR2A** Homologue: = **epsilon 1** (ε1)	NMDA 'potentiating' or 'modulatory' subunit[6, 8]. Does not possess homomeric† channel-forming activity	1464 aa 1445 aa, mature Sig: 19 aa	mRNA ~ 12 kb	~ 180 kDa (glycosylated†) ~ 165 kDa de-glycosylated – *see ref.*[12] and *Protein molecular weight (purified), 08-22*
NR2B = **NMDAR2B** Homologue: = **epsilon 2** (ε2)	NMDA 'potentiating' or 'modulatory' subunit[6, 8]. Does not possess homomeric† channel-forming activity	1482 aa	mRNA ~ 15 kb	

NR2C = **NMDAR2C** Homologue: = **epsilon 3** (ε3)	NMDA 'potentiating' or 'modulatory' subunit[6,8]. Does not possess homomeric† channel-forming activity	1239 aa	mRNA ~ 6 kb	
NR2D = **NMDAR2D** Homologue: = **epsilon 4** (ε4)	NMDA 'potentiating' or 'modulatory' subunit[6,8]. Does not possess homomeric† channel-forming activity	1323 aa	mRNA ~ 7 kb	
Other NMDAR-associated subunits	*See Database listings, 08-53*	*See Database listings, 08-53*	*See Database listings, 08-53*	*See Database listings, 08-53*

[a]For discussion of conventional nomenclatures, *see Channel designation, 08-03.*
[b]For sequence retrieval via accession numbers, *see Database listings, 08-53.*
[c]'Encoding' data show the number of amino acid residues in the specified channel subunit, with signal peptide residues denoted by the prefix 'Sig:'
[d]The ζ symbol for the NMDA receptor subunit cloned in Nakanishi's laboratory[10] is part of the **Greek letter nomenclature** used by Mishina and colleagues for mouse iGluR subunits[13] *(see Gene family under ELG CAT GLU AMPA/KAIN, 07-05).*
[e]Radiation inactivation analysis† previously suggested[14] that the NMDA/glycine-binding site of NMDA receptors was ~ 120 kDa, in reasonable agreement with this predicted (non-glycosylated†) M_r.

Multiple agonist activation of single receptors

08-06-02: Purification and characterization of two *Xenopus* CNS excitatory amino acid ionotropic receptors show that they display **unitary receptor**† **activity (i.e. they have more than one class of excitatory amino acid agonist specificity within one protein oligomer). A subunit-exchange** hypothesis has been proposed to account for the known multiplicity of excitatory amino acid receptor types[16].

Putative pre-synaptic ionotropic GluRs

08-06-03: A cDNA clone encoding a 33 kDa protein (GR33) has been obtained by screening a library with an antibody generated against glutamate binding proteins[17]. The sequence of GR33 is identical to that of the previously reported *pre-synaptic* protein **syntaxin**. When GR33 is expressed in *Xenopus* oocytes, it forms glutamate-activated ion channels that are pharmacologically similar to NMDAR but have distinct electrophysiological properties. *In vivo,* the GR33 may be a *pre-synaptic* glutamate receptor[17].

Trivial names

08-07-01: The NMDA receptor; the NMDA receptor–channel; the Ca^{2+}-permeable glutamate receptor–channel; the NMDAR; the NMDAR–channel.

EXPRESSION

Cell-type expression index

For details of selective NMDAR subunit gene expression, see other fields, particularly mRNA distribution, 08-13, and Protein distribution, 08-15.

Ubiquity of iGluR expression in the CNS

08-08-01: Receptors for ionotropic† NMDA receptor–channels (iGluRs) are expressed on virtually **every neuronal cell**, but are notably *absent* from **glial cells** *(cf. Cell-type expression index under ELG CAT GLU AMPA/KAIN,* 07-08). Functional diversity of NMDAR may arise through **co-expression of multiple types** of subunit genes within single cells, which is a common occurrence[18]. Table 2 gives examples for patterns of expression, co-expression and absence of expression for NMDAR subunits derived from immunochemical and RT-PCR† studies in a range of neuronal cell preparations. In general, regional variations in NMDAR properties are effected by differential expression† of NR2 subunit genes *(see mRNA distribution, 08-13, and Domain functions, 08-29).* Specific examples of well-characterized preparations for studying NMDAR are listed in other fields.

Restrictive expression of NMDAR

08-08-02: Although the neurotransmitter glutamate exerts excitatory† effects in both CNS neurones and glial cells, the latter do *not* express NMDAR[20]. The type 2 astrocyte (one type of macroglial cell) from rat cerebellum possesses AMPA- and kainate-selective GluRs even though it lacks receptors selective for NMDA[21].

Table 2. *Patterns of expression for NMDAR subunits in adult rat neuronal cell types based on available 'subunit-specific' probes (From 08-08-01)*

Subunit gene	CA1 pyramidal cells	CA3 pyramidal cells	Dentate granule cells	Cerebellar granule cells	Purkinje cells	Spinal cord motor neurones
NR1	+	+	+	+	+	+
NR2A	+	+	+	+		
NR2B	+	+	+			
NR2C				+		
NR2D			±			+

Channel density

NMDAR densities modulated by developmental stimuli

08-09-01: Transient **increased densities** of NMDA-binding sites have been reported in the developing rat hippocampus[22]. Increases in NMDA agonist site density can be regulated by the ovarian hormone **oestradiol**[23] *(see also Developmental regulation, 08-11).*

Cloning resource

Expression-cloning

08-10-01: The principal subunit of the rat NMDA receptor–channel (NR1) was first isolated by both **expression cloning**†[10] and subsequently by low-stringency† cross-homology†[15] screening of brain cDNA libraries. Expression-cloning† protocols for ionotropic† NMDAR were based on a search for NMDA-evoked currents measured in Mg^{2+}-free buffer in the presence of glycine *(see Isolation probe, 08-12, and Receptor agonists, 08-50).*

Cell line sources for NMDAR mRNA

08-10-02: The astrocytoma cell line R-111 has been used as a source of mRNA encoding acetylcholine and glutamate receptors supporting the expression of a small number of NMDA receptors when injected into *Xenopus* oocytes[24]. Expression of ionotropic glutamate receptor genes has also been reported in p19 embryonal carcinoma cells[25]. The terminally differentiated post-mitotic human neuronal cell line hNT has been reported to express functional NMDA- and non-NMDA receptor–channels[26].

Developmental regulation

NMDAR control of gene-activation events in hippocampal cells

08-11-01: Influx of Ca^{2+} through the NMDAR and L-type Ca^{2+} channels of hippocampal neurones has been shown to regulate **gene transcriptional**† **events** in the nucleus[27]. Activation of multifunctional Ca^{2+}/calmodulin-dependent protein kinase (CaM kinase) is evoked by stimulation of either

NMDA receptors or L-type Ca^{2+} channels (but appear 'critical' only for propagating the L-type Ca^{2+} channel signal to the nucleus). The NMDAR and L-type Ca^{2+} channel pathways activate transcription by means of different ***cis*-acting regulatory elements**† in the promoter† of the immediate-early† proto-oncogene† c-*fos*†[27] *(see below)*. For roles of NMDAR activation in modulation of synaptic plasticity (resetting of synaptic strength), *see Phenotypic expression, 08-14.* The relationships of neuronal activity to gene expression have been reviewed[28].

NMDAR control of cell migration ('growth cone movement') in developing cerebellum

08-11-02: NMDARs appear to play an early role in the regulation of Ca^{2+}-dependent **cell migration** before neurones reach their targets and form synaptic contacts[29]. Blockade of NMDA receptors by specific antagonists curtails **granule cell migration** in developing mouse cerebellum[29]. Furthermore, enhancement of NMDAR activity (by the removal of Mg^{2+} or glycine application) increases the rate of **cell movement**. Increase of endogenous extracellular glutamate (by inhibition of its uptake) also accelerates cell-migration rates[29].

NMDA channel activation is required for continued growth of cerebellar cells in culture

08-11-03: Cerebellar granule cells in culture develop **survival requirements** which can be offset either by chronic membrane depolarization (25 mM extracellular K^+, *see below*) or by stimulation of ionotropic† excitatory amino acid receptors[30]. This **trophic effect**† is mediated via Ca^{2+} influx, either through dihydropyridine-sensitive, voltage-dependent calcium channels (activated directly by high K^+ or indirectly by kainate) or through NMDAR–channels. **Calmidazolium** (a calmodulin inhibitor) counteracts the trophic effect† of elevated K^+ with high potency ($IC_{50} \sim 0.3\,\mu M$), indicating that the trophic† effects involve a Ca^{2+}/calmodulin-dependent protein kinase II activity[30].

Interrelations between neurotrophic responses and synaptic activity

08-11-04: NMDA **neurotrophic**† **responses** in rat cerebellar granule cells are modified by **chronic depolarization** in culture[31]: Cells cultured in high (25 mM) K^+ conditions only respond with a rise in cytosolic free calcium concentration when external Mg^{2+} is removed. When granule cells are grown in low (5 mM) K^+, NMDA exerts a neurotrophic† effect. At the critical time for this effect, NMDA elicits a $[Ca^{2+}]_i$ rise in 5 mM K^+ cultures even in the presence of Mg^{2+}. Growth in 25 mM K^+ induces the rapid appearance of **Mg^{2+} block** *(see Blockers, 08-43)*. Thus, rises in $[Ca^{2+}]_i$ are associated with the neurotrophic effect of NMDA. *Note:* Neurotrophic factor production (which in turn influences subsequent cell-developmental lineage†) may itself require synaptic activity *(see Development regulation under ELG CAT GLU AMPA/KAIN, 07-11).*

Pharmacological induction of immediate-early gene expression via NMDAR activation

08-11-05: The **proto-oncogenes**† **Fos** and **Jun**† form a non-covalent nucleoprotein† complex that binds to the consensus recognition sequence of the AP-1 transcription factor†. Fos, Jun and other **immediate-early**† gene products have been described as '**third messengers**†' which are regulated by second messengers† such as intracellular calcium[32]. In the brain, **c-*fos***† and **c-*jun***† may be induced by elevated neuronal activity such as occurs during **pentylenetetrazole** (PTZ) seizures. NMDAR-gated Ca^{2+}-influx plays a role in the induction of c-*fos* expression in PTZ seizures†[32].

Transcriptional activation coupled to NMDAR stimulation
08-11-06: NMDA application has been shown to directly stimulate rapid c-*fos* mRNA accumulation in dentate gyrus neurones[33] ('**stimulus–transcription coupling**'[34]). In these cells, Ca^{2+} serves as a second messenger† coupling the GluRs to **transcriptional activation**† of c-*fos* mRNA. The route of Ca^{2+} entry into dentate neurones, however, depends on the excitatory amino acid receptor subtype stimulated. Non-NMDA receptor activation results in the indirect enhancement of Ca^{2+}-influx via voltage-sensitive calcium channels, whereas NMDA receptor activation results in Ca^{2+} influx directly through the NMDA channel itself[33]. *Note:* Inhibition of c-Fos synthesis *fails* to protect against glutamate-induced cell death[35] *(see Phenotypic expression, 08-14).*

Modulation of NMDAR gene expression during brain development
08-11-07: Transient increases in NMDA receptor expression occur during development, and application of antagonists can block **experience-dependent plasticity**† in the adult *(reviewed in ref.*[36]*)*. At early developmental stages, NMDA receptors are 'spontaneously' activated by endogenous neurotransmitter[37], but these receptors are normally lost by an activity-dependent process. Absence of **sensory input** delays the loss of these functions[38]. Notably, *increased* levels of RNA transcripts† for **Ca^{2+}/calmodulin protein kinase**†, **GAP43**† and **glutamic acid decarboxylase**† have been reported under these conditions of sensory deprivation[39], suggesting their genes are differentially regulated† by calcium influx.

Increases in NMDAR expression following treatment with oestradiol
08-11-08: Increases in NMDA agonist sites have been observed following treatment of CA1 hippocampal neurones with the ovarian hormone **oestradiol**[23]. *Note:* Oestradiol has been shown to affect **cognitive function** and lowers the threshold for **seizures**†.

Control of NMDAR properties during development
08-11-09: In general, the molecular properties of NMDA receptors are **developmentally regulated** and may control the ability of synapses to change in early life. The physiological and pathophysiological roles of excitatory amino acids (and their receptors) during development have been reviewed[40,41].

Developmental variations in extracellular Mg^{2+}-block (see also Blockers, 08-43)
08-11-10: Heterologous† NMDAR–channel expression studies have revealed **developmental variations** in **Mg^{2+}-block sensitivity** amongst different

receptor subtypes *(for significance, see Phenotypic expression, 08-14, and Blockers, 08-43)*: Release from Mg^{2+}-block often allows or 'facilitates' the occurrence of **long-term potentiation**† (LTP). Within the immature visual cortex (which is more susceptible to LTP than adult visual cortex) synaptically activated NMDAR have varying but clearly reduced sensitivities to Mg^{2+} block[42]. This variability is *not* observed in the adult brain, and it has been proposed that 'initially expressed, later-eliminated' NMDAR exhibiting 'reduced' Mg^{2+}-block phenotypes may underlie the greater susceptibility to plasticity† in the immature neocortex[42].

Other developmental changes in NMDAR characteristics

08-11-11: NMDA receptors in rat hippocampus appear less voltage-sensitive[43] and less Mg^{2+}-sensitive[44] at early stages of development, allowing greater influx of Ca^{2+} than in the adult *(see also Blockers, 08-43)*. Developmental increases in **glycine-binding sites** associated with the NMDA receptor[22,45] can also enhance Ca^{2+}-influx in mature cells. **Neuronal connectivity patterns** are altered by selective blockade of post-synaptic NMDA receptors in developing visual pathways[41]. *(See also Developmental regulation under VLG Na, 55-11.)*

08-11-12: The duration of evoked NMDAR-mediated excitatory post-synaptic currents (EPSCs†) in the superior colliculus are several times longer at **early developmental stages** compared to that measured in older animals[46]. In contrast, the amplitude of NMDAR-mediated **miniature EPSCs**† do *not* change during development.

NMDAR activity affecting phenotypic development

08-11-13: There are several documented **developmental phenotypic** consequences of altering NMDA-gated channel function. For example, pharmacological blockade of the NMDA receptor in developing retinorectal projections alters retinal ganglion cell **arbor**† **structure**[47–49] and renders striate cortex resistant to the effects of monocular deprivation[50].

Pharmacological modulation of neuronal development

08-11-14: NMDA agonists can *restore* plasticity† of *Xenopus* binocular maps beyond a critical period in development[51].

Environmental 'cues' for control of NMDAR expression

08-11-15: Activity-dependent, **competitive synaptic interactions**, which stabilize some axon branches and dendrites and remove others, involve glutamate receptor expression *(see ELG Key facts, entry 04, and reviews[52–54])*. **Dark-rearing** delays the loss of NMDA receptor function in visual cortex[38] *(see below)*.

NMDA receptor responses underlying developmental reductions in synaptic plasticity†

08-11-16: NMDA receptors are crucial for **experience-dependent synaptic modifications** that occur in the developing visual cortex. NMDA-mediated excitatory post-synaptic currents† (EPSCs) in layer IV neurones of the visual cortex have been shown to 'last longer' in young rats than in adult rats. Furthermore, durations of the EPSCs become progressively shorter, in parallel with the developmental reductions in synaptic plasticity†[55]. This decrease in NMDA receptor-mediated EPSC duration is delayed when the animals are reared in the dark, a condition that prolongs developmental plasticity, and is prevented by inhibition of neural activity with tetrodotoxin *(see Blockers under VLG Na, 55-43)*. Modifications of NMDA receptor gating† properties may account for the 'age-dependent' decline of visual cortical plasticity[55] *(see paragraph below and Fig. 1)*.

NMDAR–channel activation in development of the somatosensory cortex

08-11-17: Changes in NMDA channel activity may explain the **transient plasticity** observed in layer IV during **early post-natal development**. NMDA receptor-mediated currents are prominent in layer IV cells of immature mouse **somatosensory cortex** (thalamocortical synapses) before maturation of inhibition[56]. *Earlier than* post-natal day 9 the majority of responses are monosynaptic and purely excitatory, with both non-NMDAR and NMDAR-mediated glutamatergic components. In *older animals*, **disynaptic inhibitory currents** summate with the excitatory ones and lower the reversal potential of the response to voltages at which the NMDAR conductance is largely blocked[56].

Isolation probe

Expression cloning protocol

08-12-01: A functional cDNA clone for the NMDA receptor was first isolated by **expression cloning**†[10]. Poly(A)$^+$ mRNA prepared from the forebrains of 4-week-old male rats was subjected to centrifugation on 5–25% sucrose density gradients *(see Ref.[57])*. A fraction giving 'potent electrophysiological responses' to 100 μM NMDA (measured in Mg^{2+}-free medium supplemented with 10 μM glycine) following injection into *Xenopus* oocytes was used for construction of a directional cDNA library† in a phage lambda cloning vector†. T7 RNA polymerase† cRNA-runoff† from DNAs prepared from individual sub-pools of this library was used to inject individual *Xenopus* oocytes. Isolation of a single NMDA receptor clone (lambda vector λN60/plasmid vector pN60) was achieved by stepwise fractionation of a 'response-evoking' cDNA mixture[10].

Degenerate PCR probes for isolating related gene family members

08-12-02: An RT-PCR† strategy was employed to isolate probes for the NMDAR subtypes NR2A–NR2D[8]. This method relied on a **conserved amino acid sequence** between NR1 and other ionotropic GluRs (**YTANLAA**) in the vicinity of the putative transmembrane M3 segment of these receptors. **Degenerate† PCR primers** synthesized between this site and a T7 RNA polymerase† promoter† sequence in the library cloning vector† permitted retrieval of DNA fragments which were used as NR2A–NR2D-specific probes.

Immunoprobes for detection of glutamate-binding proteins
08-12-03: A cDNA clone encoding a 33 kDa protein (**GR33**) with similar *pharmacological* properties to NMDAR has been obtained by screening a library with an antibody generated against glutamate-binding proteins[17] *(see Subtype classifications, 08-06).*

mRNA distribution

See the methodological note in mRNA distribution under ELG CAT GLU AMPA/KAIN, 07-13.

'Near-universal' expression of the 'fundamental' NMDAR subunit in the CNS
08-13-01: *In situ* hybridization to adult rat brain sections shows that the NR1 mRNA is distributed in almost **all neuronal cells** throughout the brain. The spatial pattern observed with NR1 mRNA is largely consistent with autoradiographic studies using several radiolabelled ligands† of the NMDA receptor.

08-13-02: 'Prominent' expression of NR1 mRNA is observed in the **cerebellum**, **hippocampus**, **olfactory bulb** and **hypothalamus**. In the latter, large signals are observed in granule cells and CA4 cells of the dentate gyrus and in pyramidal cells throughout the CA1–CA3 regions[10,58]. *See also Table 3.*

NMDAR subunit mRNA distribution by single-cell RT-PCR
08-13-03: Glutamate receptor subunit mRNA composition has been measured using **RT-PCR**† in several **individual neurones** within rat hippocampal CA1 slices[18]. Generally, each CA1 neurone contains varying amounts of *most* glutamate receptor mRNAs. These **single-cell expression** studies have revealed novel, alternatively spliced forms of the NR1 *(see Gene organization, 08-20)* and kainate receptor type 2 subunits[18]. Surprisingly, compared to other GluR subunit message, NMDAR type 1 mRNA is of relatively *low* abundance at the single-cell level *(see below).* Furthermore, RNA editing† was shown not to be a *general* cellular mechanism for GluR diversity *(compare Gene organization under ELG CAT GLU AMPA/KAIN, 07-20).*

Differential gene expression of mRNAs encoding 'potentiating' subunits
08-13-04: In contrast to the wide distribution of the zeta 1 (NR1) and epsilon 1 (NR2A) subunit messenger RNAs in the brain, the **epsilon 2** (**NR2B**) subunit mRNA is expressed only in the **forebrain** and the **epsilon 3** (**NR2C**) subunit mRNA is found predominantly in the **cerebellum**[59]. Other than the ubiquitous NR1, NMDA-selective GluR subunit genes display **differential spatial expression**[10] *(see Table 3).*

Summary of regional mRNA expression patterns for NMDAR in brain
08-13-05: Individual mRNAs for the **NR2 subunit series** (which provide a molecular basis for the 'functional diversity' of the NMDA receptor)

Table 3. *mRNA distribution: summary of reported tissue distributions of NMDA receptor family mRNAs (From 08-13-04)*

Subunit	Distribution
NR1	Expressed ubiquitously (i.e. almost all neuronal cells in all brain regions)[10]. The hypothalamus contains only the NR1 transcript, suggesting the existence of additional NMDA receptor subunits[6]
NR2A	Widely expressed in many brain regions. Abundant in cerebral cortex and hippocampus, internal granule layer of the olfactory bulb, anterior olfactory nuclei, olfactory tubercle, certain thalamic nuclei, inferior colliculus, pontine nuclei, inferior olivary nuclei and cerebellar cortex[8]. NR2A mRNA distribution bears the closest resemblance to that of NR1 and is present in both forebrain and cerebellum[6]. The amygdaloid nuclei express mRNAs encoding NR2A and NR2B but *not* NR2C[6]
NR2B	Widely expressed in many brain regions, but more restricted than NMDR2A. Abundant in cerebral cortex and hippocampus, telencephalic and thalamic regions, but lowly expressed in the hypothalamus, lower brainstem and cerebellum[8]. NR2B is expressed in forebrain, and has a complementary distribution to NR2C mRNA[6]. The amygdaloid nuclei express mRNAs encoding NR2A and NR2B but *not* NR2C[6]. NR2B-specific mRNA is expressed mainly in granule cells[6]
NR2C	Shows localized mRNA distribution in comparison to NMDR2A and NMDR2B. Prominently expressed in the cerebellar granular layer. Moderate expression of this mRNA is seen in the glomerular and mitral cell layers of the main olfactory bulb. Also observed in some of the pontine, thalamic and vestibular nuclei[8]. NR2C is expressed at highest levels in the cerebellum, and has a complementary distribution to NR2B and mRNA[6]. The amygdaloid nuclei express mRNAs encoding NR2A and NR2B but *not* NR2C. Similarly, in the caudate-putamen, there is no NR2C mRNA, but moderate signals are detected with NR2A and NR2B probes[6]. NR2C-specific mRNA is expressed mainly in tufted and mitral cells[6]
NR2D	Highly expressed in the diencephalic and lower brainstem (subcortical) regions. High *in situ* hybridization signals are observed in the glomerular layer of the main olfactory bulb, the ventral pallidum, the majority of the thalamic nuclei, the hypothalamus, superior colliculus, substantia nigra, vestibular nuclei, pontine nuclei, and deep cerebellar nuclei. Low expression is observed in the central cortical regions and the granular layer of the cerebellum[8]

Note: The spatial distribution of channel subunit-specific mRNAs presumably reflects a 'functional specialization' in particular cell types. Mapping of expression patterns is a complex task and has to take many variables into account, such as *in situ* localization, developmental regulation, subunit stoichiometry, and factors regulating overlapping or co-expression. For notes on the integration of computer-based information resources able to cross-reference these diverse factors, *see Feedback & CSN access, entry 12, and Appendix J – Search criteria & CSN development, entry 65.*

overlap in some brain regions but also show **localized expression patterns**[6, 8]. Anatomical distributions which have been described in the literature are given in Table 3 *(see the original* in situ *hybridization studies – refs*[6, 8, 10, 11] *and the footnote to Table 3).*

Phenotypic expression

NMDA receptors contribute to a large number of neurophysiological phenotypes

08-14-01: The demonstrated involvement or 'association' of NMDA receptors in several **neurophysiological** and **pathophysiological** processes has given great impetus for the development of selective **pharmacological modulators** of the NMDAR *(see Receptor antagonists, 08-51)* and for definition of these phenotypes at the molecular and cellular level[60, 61] *(Table 4)*. A more extensive list of **chronic neurodegenerative diseases** in which iGluR dysfunction or 'overstimulation' may play a role is included in ref.[62].

Table 4. *Functional roles of NMDAR–channels (From 08-14-01)*

Physiological function/pathophysiological phenotype involving NMDAR–channels	Selected refs
Developmental synaptic plasticity† *(see below and Protein interactions, 08-31)*	*41, 50, 53, 54, 63–72*
Learning and **memory**	*53, 66, 73, 74*
Trans-synaptic **gene expression control**†	*28, 34*
Normal **information processing** *See also* Collingridge, 1989, *under Related sources & reviews, 08-56*	*75*
Sensory neurotransmission including amplification of **auditory responses** and amplification of excitatory and inhibitory **visual signals**	*38, 75–82*
Anxiogenesis†	*83*
Development and maintenance of **epileptic**† **states**	*84, 85*
Development of olivopontocerebellar **atrophy**†	*86*
Development of **motor neurone disorders**	*87, 88*
Hypoglycaemic† **episodes**	*89*
Post-ischaemic brain damage: **pathophysiology** and **cell death**	*40, 90–93*
Cardiovascular pressure control (**baroreception**† and control of **vasomotor tone**†)	*94–96*
Central sensitization during **nociception**† (**pain reception**) *(see also below)*	*97–99*
Activation and maintenance of **motor rhythms** (e.g. in **respiratory control** NMDAR blockade causes prolonged inspiration)	*Reviewed in refs.*[75; 100]

Concentration- and time-dependent glutamate toxicity

08-14-02: Excess net Ca^{2+}-influx through NMDA receptor–channels is a key step in triggering the **neuronal death** induced by brief, intense glutamate exposure[101]. The amount of $^{45}Ca^{2+}$ accumulation correlates closely with the 'degree of neuronal' death 24 h later[101] over applied glutamate concentrations ranging between 10 and 1000 μM and durations of exposure between 0 and 10 min. *Comparative note:* Relatively little Ca^{2+}-entry (as measured by $^{45}Ca^{2+}$ accumulation) has been observed in similarly treated glial cells.

A large number of signalling components may contribute to glutamate toxicity

08-14-03: As summarized in Table 7 (*see Protein interactions, 08-31*) and reviewed in ref.[102], several **Ca^{2+}-activated enzymes** may contribute to excitatory amino acid toxicity, including **protein kinase C**† [103–105], **phospholipase A_2**† [106–109], **phospholipase C**† [110], various **endonucleases**† [111], **Ca^{2+}/calmodulin-dependent protein kinase II**[112], **nitric oxide synthase**[113–115], **calpain I**† (micromolar Ca^{2+}-sensitive proteolysis) and **calpain II**† (millimolar Ca^{2+}-sensitive proteolysis)[116–120]. Since regulation of Ca^{2+}-influx appears fundamental to the 'ordered progression' of gene expression, dysregulation of Ca^{2+} homeostasis is also likely to have longer term effects on cell phenotype or the initiation of **apoptosis**†.

Relevance of glutamate toxicity to neurodegeneration processes

08-14-04: A number of **pathological/neurodegenerative conditions** can be explained by **glutamate toxicity** as a central mechanism (e.g. when prolonged receptor-mediated depolarization results in 'irreversible disturbances' in ionic homeostasis[40, 62, 92, 102, 121–125]). The correct functioning of **Na^+/K^+ pumps** is also important for prevention of glutamate toxicity *(see Protein interactions, 08-31). Note:* Although NMDAR may be the *predominant* route for Ca^{2+}-entry at toxic levels, other routes may be involved. Other candidate pathways for **neurotoxic**† **Ca^{2+}-influx** include **L- and N-type voltage-gated Ca^{2+} channels** *(see VLG Ca, entry 42)*, **non-NMDA iGluRs** (through AMPA-selective iGluR *lacking* GluR-B subunits – *see ELG CAT GLU AMPA/KAIN, entry 07*), influx channels indirectly opened by **metabotropic**† **glutamate receptor** stimulation, **Na^+/Ca^{2+} exchangers**, and non-specific **membrane leak**[122]. Some of these routes are illustrated in *Fig. 4 under Protein interactions, 08-31.*

Neuropathological conditions associated with 'overstimulation' of ionotropic glutamate receptors

08-14-05: The involvement of iGluR are well-established in the initiation and propagation of seizures† [126, 127] and in the massive **neuronal cell death**† that follows periods of **stroke**†**-induced ischaemia**† and **hypoglycaemia**† [90, 122] *(see below)*. Dysfunction of glutaminergic pathways has been implicated (though not *specifically* associated) with the pathogenesis of a number of **neurodegenerative diseases** (*see below*, full references listed in refs[128, 129]). These disorders include **epilepsy**† [127], **Alzheimer's disease**† (tentative), **Huntington's disease**†, **AIDS encephalopathy**†**/dementia complex**, **amyotrophic lateral sclerosis**† and **lathyrism**†.

Time-dependent changes in synaptic transmission failure in models of 'in vitro *ischaemia*'

08-14-06: Five minutes of oxygen and glucose deprivation ('***in vitro* ischaemia**') causes **long-term synaptic transmission failure** (LTF) in the CA1 region of the rat hippocampal slice. In buffer containing 2.4 mM Ca^{2+}, the extent of LTF largely depends on the average level of 'exchangeable' cell Ca^{2+} in CA1 during this procedure[130]. In this model, **unidirectional Ca^{2+}-influx** during the 'first $2\frac{1}{2}$ min' of ischaemia depends *entirely* on NMDAR–channels; however the NMDAR antagonist MK-801 has no effect during the 'second $2\frac{1}{2}$ min', probably as a result of NMDAR dephosphorylation. **Ischaemia-induced Ca^{2+}-influx** during the second $2\frac{1}{2}$ min of ischaemia can be partially attenuated by ~25% by the voltage-gated channel blocker **nifedipine** (50 μM) and by an additional 35% by the Na^+/Ca^{2+} exchange inhibitor **benzamil** (100 μM). *Note:* The AMPA/kainate antagonist DNQX has *no effect* on the Ca^{2+}-influx in this model[130].

Total cell Ca^{2+} prior to in vitro *ischaemia is adequate to cause complete LTF*

08-14-07: In the hippocampal CA1 slice model of *in vitro* ischaemia *(previous paragraph)*, pharmacological blockade of enhanced Ca^{2+}-entry during ischaemia in 2.4 mM Ca^{2+} has no effect on the LTF phenotype[130]. However, the NMDAR antagonist MK-801 strongly protects against LTF when the buffer contains Ca^{2+} at concentrations closer to physiological levels (1.2 mM). A *combination* of blockers for NMDAR–channels (**MK-801**) and AMPA receptor–channels (**DNQX**) prevents LTF in buffer containing 1.2 mM Ca^{2+}. Thus, AMPA/kainate receptor activation makes some contribution to ischaemic damage, although this appears *independent* of enhanced Ca^{2+}-entry[130].

Longer term NMDAR-associated neurotoxicity

08-14-08: *Slow increases* in NMDA channel P_{open} may also provide an **excitotoxic**† mechanism in that Ca^{2+}-influx can increase markedly in cells subject to **prolonged depolarization**[131] *(see below and Current–voltage relation, 08-35).*

Extended neuronal depolarization following removal of glutamate agonist

08-14-09: 'Physiological responses' of hippocampal pyramidal neurones in primary culture to prolonged (~10 min) exposure to 500 μM glutamate show that the neurones **remain depolarized** (⩾20 mV from rest) following 'washout'. This depolarization is accompanied by a ~57.8% increase in membrane conductance, *initially* by NMDA channel opening and *subsequently* through other Ca^{2+}-influx channels. Since depolarization can be maintained for long periods (30 min to <4 h), the phenomenon has been described as **extended neuronal depolarization** (END)[125]. During END, cells retain both the ability to fire action potentials and the ability to respond to glutamate, and can exclude vital dyes†[125].

Indirect neurotoxic effects

08-14-10: In retinal neurones, **metabolic inhibition** results in the loss of voltage-dependent Mg^{2+} blockade of the NMDA receptor–channels, leading to an increased potency of glutamate agonists[132] *(for significance, see Blockers, 08-43).*

'Protection mechanisms' for excitotoxic phenotypes
08-14-11: Extracellular Zn^{2+} *decreases* NMDA receptor-mediated toxicity[133]. This effect is in contrast to its *potentiating* effect on AMPA receptor-mediated neurotoxicity *(see Phenotypic expression under ELG CAT GLU AMPA/KAIN, 07-14).* **Nitric oxide** may have a role in **protection** of NMDA-mediated excitotoxicity *(for details, see Channel modulation, 08-44).* **Antisense**† **oligonucleotides** to the NR1 sequence have been shown to 'protect' cortical neurones from 'excitotoxic' reactions by reducing (i) NR1 gene expression and (ii) the volume of experimentally induced **focal ischaemic infarctions**[134].

Prevention of 'cell death phenotypes' in cells expressing NR1/NR2A heteromeric NMDAR
08-14-12: Co-expression of NR1 and NR-2A subunits in HEK-293 cells results in cell death†, but viability can be maintained by including the NMDAR antagonist DL-2-amino-5-phosphonopentanoic acid (**AP5**) in the culture medium post-transfection[12] *(see Receptor antagonists, 08-51).*

NMDAR modification in maintenance of epileptic states
08-14-13: Chronic epilepsy induced by **kindling**† can be considered as an NMDA receptor-dependent form of activity-dependent neuronal plasticity† induced *in vivo,* which results in *lasting modifications* in the function of single NMDARs[135]. In control neurones, the amplitude of whole-cell NMDA currents is *not* sensitive to the presence of an intracellular ATP regeneration system†, whereas NMDA currents in kindled† cells show a great variability, with larger amplitudes consistently recorded in the presence of intracellular 'high-energy' phosphates.

NMDA channel properties affected by kindling
08-14-14: Kindling†-induced epilepsy predominantly affects the mean open time†, burst†, and cluster† duration of NMDAR–channels, their sensitivity to intracellular 'high-energy' phosphates, and their block by Mg^{2+}, but *not* rates of receptor desensitization† or single-channel conductance† values[135]. Such alterations may reflect a change in the molecular structure of NMDA channels and may underlie the *maintenance* of the epileptic state.

Pulsatile glutamate release following NMDAR–channel activation
08-14-15: Periodic inward currents have been shown to be generated in neonatal neurones by a synchronous, persistent, **pulsatile glutamate release** from pre-synaptic nerve terminals (secondary to NMDA receptor stimulation and oscillations in intracellular calcium)[136] *(see Fig. 4 under Protein interactions, 08-31).*

NMDAR–channel function in central sensitization during nociception (pain reception)
08-14-16: Glutamate is co-localized with **substance P** in the terminals of C-

fibre inputs to the spinal cord and brainstem[137] and, in addition to **serotonergic**†, **peptidergic**† and **noradrenergic**† mechanisms, may mediate pain receptive input in the **dorsal horn** (*Note:* C-fibre inputs were first associated with **pain reception** pathways over 60 years ago[138]). NMDAR–channel activation mediates three related phenomena associated with pain reception in dorsal horn neurones *(reviewed in ref.*[75]*)*: (i) the **formalin response**, a model of pain reception by subcutaneous injection of formalin, (ii) amplification or **'wind-up'** of dorsal horn neuronal response, and (iii) modulation of the **flexion reflex**, initiated by pinching of the cutaneous surface of the foot (the response can be increased by 'prestimulation' of C-fibre inputs). *Notes:* 1. The **'wind-up' response** has also been characterized as being mediated by L-type Ca^{2+} channels in turtle dorsal horn neurones[139] and can be potentiated† by protein kinase C in isolated mouse trigeminal neurones *(see Channel modulation, 08-44)*. 2. All three responses *(above)* can be blocked by systemic administration of NMDAR antagonists in decerebrate animals. 3. Antagonism at the **glycine site** on the NMDAR reduces **spinal nociception**† in the rat[98].

Role of NMDA receptors in non-vesicular endogenous release of GABA

08-14-17: The glutamate-induced **endogenous release** of the neurotransmitter **gamma-aminobutyric acid** (**GABA**) is mainly mediated by NMDA receptors, and consists of a single, sustained phase[140]. This phase is insensitive to **nocodazole**, partly inhibited by **verapamil** and can be blocked by **Co^{2+} ions** as well as **SKF 89976A**. The action of Co^{2+} has been attributed to a block of NMDA-associated ion channels. *Note:* The majority of GABA release occurs from **vesicles** (whose release can be stimulated experimentally by K^+-induced depolarization) whereas the **glutamate-dependent release** is **non-vesicular**[140]. *See also SYN (vesicular), entry 40.*

NMDAR–channels as 'biosensors' for release of neurotransmitters

08-14-18: Outside-out membrane patches excised from rat hippocampal neurones have been used to detect release of endogenous excitatory amino acids (EAAs) from synaptic terminals of isolated turtle photoreceptors[141]. Electrical stimulation or application of **lanthanum chloride** to **photoreceptors** induces an increase in the opening frequency of 50 pS NMDA receptor–channels[141]. Spontaneous channel activity is also observed near **synaptic terminals**†. *Note:* **Exocytotic release** of endogenous EAAs is an important criterion for classification as an **authentic transmitter**.

Involvement of NMDA receptor–channels in synaptic plasticity† – special note

08-14-19: Within this entry, the descriptions of processes affecting synaptic plasticity† have been limited to phenotypic aspects dependent upon the **molecular characteristics*** of the NMDA receptor *(*see 08-14-26)*. The subject of **neuronal phenotype modulation** is a complex one, and is difficult to describe outside of a **neurophysiological** and **neuroanatomical context**. For comprehensive accounts of these aspects, including the molecular perspective, *see the selected references listed in Related sources & reviews, 08-56.*

General phenotypic changes in long-term potentiation of synaptic transmission

08-14-20: The phenomenon of **long-term potentiation**† (LTP), a long-lasting increase in the strength of synaptic transmission due to brief, repetitive activation of **excitatory afferent**† **fibres**, is a striking example of **synaptic plasticity**†[142]. Brief, high-frequency **tetanic stimulation** of afferent fibres in hippocampal slices induces an LTP of synaptic transmission†, which causes an increase in the size of the synaptic response elicited by low-frequency stimulation of the same synapse. LTP persists for several hours *in vitro* and up to several weeks *in vivo*, and is at present the most **extensively studied form** of activity-dependent synaptic plasticity† *(for comprehensive reviews, see refs*[54, 67, 143–148] *and Related sources & reviews, 08-56).*

Experimental induction and phases of LTP

08-14-21: LTP processes can be induced by delivery of a **tetanus**† (typically 50–100 stimuli at 100 Hz or greater) to the pathway of interest. Alternatively, LTP can be induced by (i) **theta burst stimulation**[149] (typically by multiple bursts of four shocks at 100 Hz delivered at 200 ms intervals) or (ii) **primed burst stimulation**[150] (typically by delivery of a 'priming stimulus' followed by a single burst of four 100 Hz shocks 200 ms later). These experimental procedures appear to simulate **synchronized firing patterns** that occur at similar frequencies in the hippocampus during learning[151].

Methodological note – measurement of synaptic potentiation phenomena

08-14-22: LTP† phenotypes are commonly determined by plotting a graph of the slope of the **field EPSP**† versus time in experiments where stimuli of *specified* frequency are delivered (i.e. low-frequency, ~1–10 Hz, ranging up to tetanic† stimuli *(see Fig. 1e)*. **EPSP slope parameters** (mV/ms or %) can be determined in the presence of pharmacological agonists/antagonists of the NMDAR, or activators/inhibitors of associated signalling components. **Sustained enhancement** of synaptic transmission (i.e. **LTP**) is indicated by a **'non-decremental response'** *(see Fig. 1b)* whereas **short-term potentiation** (i.e. **STP**) is characterized by a **'decremental response'** *(see below and Fig. 1b(ii)).*

Nomenclatures for alternative forms of synaptic potentiation

08-14-23: The use of the phrase 'long-term potentiation' has been *non-systematically* applied to any form of **synaptic enhancement**† lasting more than a few minutes. However, several experimental manipulations can result in potentiation† of synaptic transmission† that declines over ~5–40 min (decremental synaptic transmission or **short-term potentiation**†, STP). STP is likely to be a prerequisite for stable LTP (i.e. non-decremental synaptic enhancement). For further details on the relationships between these multiple forms of potentiation, *see Fig. 1d and refs*[54, 64, 74].

LTP and LTD

08-14-24: The relationship of LTP to **long-term depression** (**LTD**) of excitatory synaptic transmission is discussed in ref.[152]. There is evidence for the involvement of NMDA receptors in the induction of **homosynaptic LTD**†, where post-synaptic depolarization and increases in Ca^{2+} *resemble* those for LTP induction (as described below), however LTP requires a **markedly stronger post-synaptic depolarization** *(see ref.*[152] *for further details).* Development of LTD phenotypes depends on functional **metabotropic**† glutamate receptors *(see Table 7).*

Dependence of LTP on NMDAR activation and depolarization
08-14-25: In the CA1 region of the hippocampus, the induction of LTP requires activation of NMDAR by synaptically released glutamate with **concomitant depolarization** of the post-synaptic membrane[142,153]. This relieves the voltage-dependent magnesium block of the NMDA receptor–channel, allowing calcium to flow into the dendritic spine. A **synaptic model of memory**† involving LTP formation in the hippocampus has been proposed *(reviewed in ref.*[54]*).*

An overview of experimental models used in the study of hippocampal long-term potentiation (LTP)
08-14-26: Within the scope described above *(paragraph 08-14-19), Fig. 1* attempts to summarize aspects of **NMDAR LTP phenotypes**† that ultimately depend on **molecular characteristics**† of the receptor–channels. Supporting information on several subtopics can be found under relevant fieldnames of this entry and the *ELG CAT GLU AMPA/KAIN* entry. *(A broad illustration of the types of pre- and post-synaptic signalling proteins interacting with NMDA receptor–channels is also shown under Protein interactions, 08-31.)*

Direct experimental evidence for NMDAR-mediated Ca^{2+}-influx in synaptic plasticity
08-14-27: Optical measurements of synaptically induced Ca^{2+} transients through NMDA receptors have been observed under conditions which eliminate activation of dendritic voltage-sensitive Ca^{2+} channels in pyramidal cell dendrites within hippocampal slices (i.e. steady post-synaptic depolarization to the synaptic reversal potential)[154]. Ca^{2+}-influx through synaptically activated NMDA receptors are directly implicated in the **induction of LTP**†, since the magnitude of LTP is *diminished* when induced with the post-synaptic membrane held at progressively more positive potentials (with complete suppression at potentials near +100 mV)[154]. *Note:* Induction of LTP can be blocked by injection of **intracellular Ca^{2+} chelators**† – e.g. microinjection of EGTA into post-synaptic neurones blocks the induction of LTP[155] and further implicates a direct role for Ca^{2+} signalling in the induction process.

'Augmentation' of NMDAR-associated calcium transients and other Ca^{2+} sources
08-14-28: The **Ca^{2+}-permeability**† of NMDAR–channels *(see Selectivity, 08-40)* has led to a common *assumption* that NMDA receptors play a direct role in potentiation of synaptic responses. However, **voltage-gated Ca^{2+}-entry** *(see VLG Ca, entry 42)* and Ca^{2+}-release mediated by **neuronal ryanodine receptor–channels** *(see ILG Ca Ca RyR-Caf, entry 17)* and **neuronal $InsP_3$ receptor–channels** *(see ILG Ca $InsP_3$, entry 19)* appear to

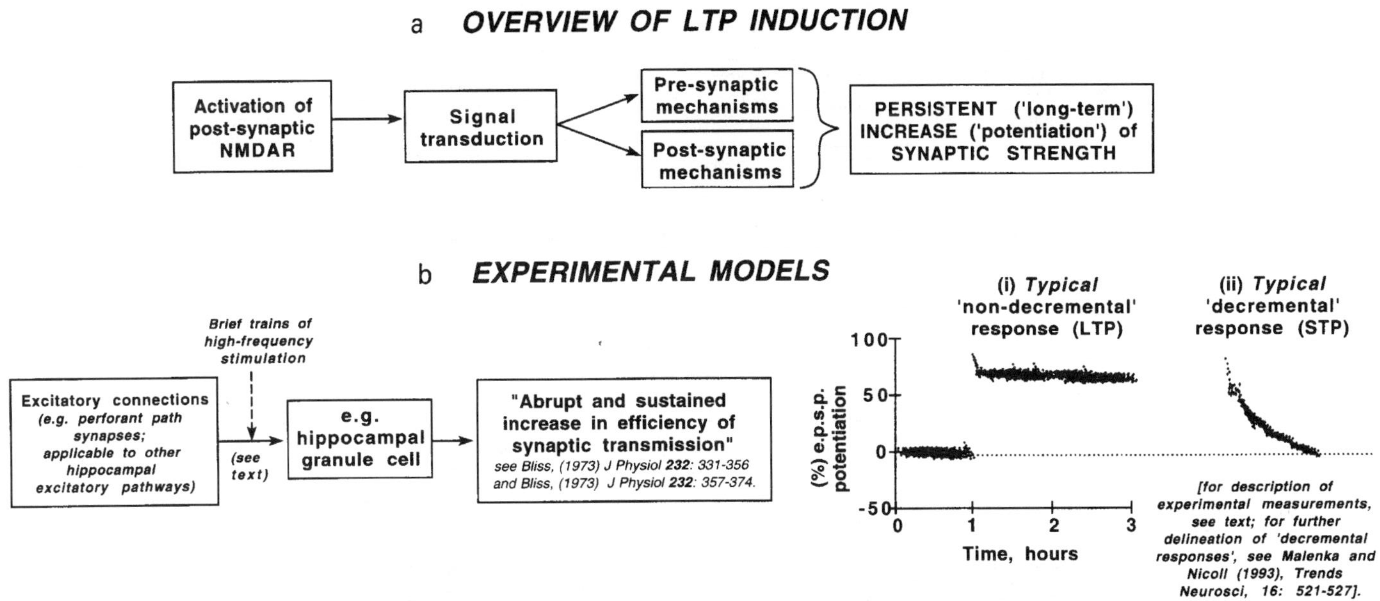

c *TIME COURSE OF LTP-INDUCTION AND PERSISTENCE*

Induction phase:	Persistent phase:		
In vivo and *In vitro*	*In vitro* (hippocampal slice)	*In vivo* (non-anaesthetised)	*In vivo* (anaesthesia)
generally <10 mS	several hours	several days	several hours

Figure 1. *Experimental models for synaptic LTP in the hippocampus. (a) Overview of LTP induction. (b) Experimental models. (c) Time course of LTP induction and persistence. (d) Classification of potentiated synaptic responses. (e) Absolute dependence of synaptic potentiation on 'sufficient' post-synaptic depolarization. (f) Molecular mechanism for NMDAR-dependent induction of LTP. (From 08-14-26)*

d CLASSIFICATION OF POTENTIATED SYNAPTIC RESPONSES

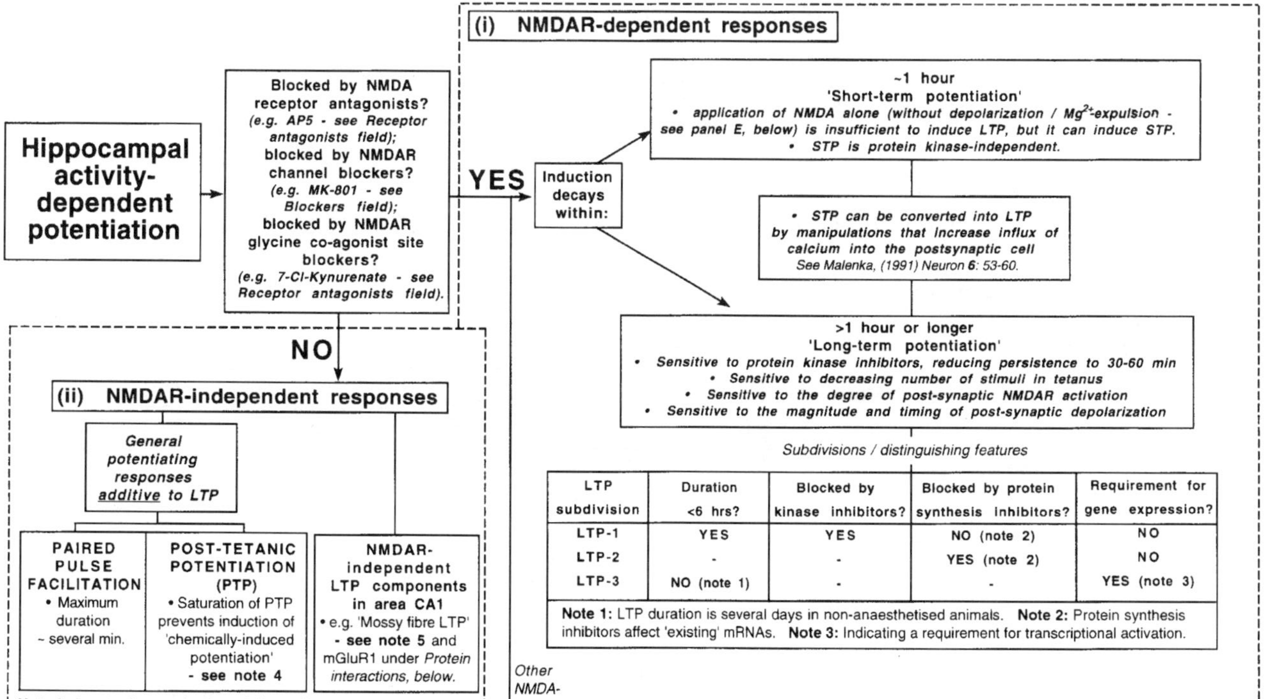
Hippocampal activity-dependent potentiation
Blocked by NMDA receptor antagonists? (e.g. AP5 - see Receptor antagonists field); blocked by NMDAR channel blockers? (e.g. MK-801 - see Blockers field); blocked by NMDAR glycine co-agonist site blockers? (e.g. 7-Cl-Kynurenate - see Receptor antagonists field).
YES
NO
(i) NMDAR-dependent responses
Induction decays within:
~1 hour
'Short-term potentiation'
• application of NMDA alone (without depolarization / Mg2+-expulsion - see panel E, below) is insufficient to induce LTP, but it can induce STP.
• STP is protein kinase-independent.
• STP can be converted into LTP by manipulations that increase influx of calcium into the postsynaptic cell
See Malenka, (1991) Neuron 6: 53-60.
>1 hour or longer
'Long-term potentiation'
• Sensitive to protein kinase inhibitors, reducing persistence to 30-60 min
• Sensitive to decreasing number of stimuli in tetanus
• Sensitive to the degree of post-synaptic NMDAR activation
• Sensitive to the magnitude and timing of post-synaptic depolarization
Subdivisions / distinguishing features
Other NMDA-dependent forms
"Non-Hebbian" Long-term potentiation (see references in Related sources and reviews)
and
e.p.s.p.-spike (E-S) form of activity-dependent potentiation (see references in Related sources and reviews)
(ii) NMDAR-independent responses
General potentiating responses additive to LTP
PAIRED PULSE FACILITATION
• Maximum duration ~ several min.
POST-TETANIC POTENTIATION (PTP)
• Saturation of PTP prevents induction of 'chemically-induced potentiation' - see note 4
NMDAR-independent LTP components in area CA1
• e.g. 'Mossy fibre LTP' - see note 5 and mGluR1 under Protein interactions, below.

LTP subdivision	Duration <6 hrs?	Blocked by kinase inhibitors?	Blocked by protein synthesis inhibitors?	Requirement for gene expression?
LTP-1	YES	YES	NO (note 2)	NO
LTP-2	-	-	YES (note 2)	NO
LTP-3	NO (note 1)	-	-	YES (note 3)

Note 1: LTP duration is several days in non-anaesthetised animals. **Note 2:** Protein synthesis inhibitors affect 'existing' mRNAs. **Note 3:** Indicating a requirement for transcriptional activation.

Note 4: Agents characterised as elicitors of 'chemically-induced potentiation' include arachidonic acid, metabotropic glutamate receptor agonists, the potassium channel blocker TEA, calcium ions and G protein activators (e.g. sodium fluoride / aluminium chloride).
Note 5: Sensitive to calcium channel antagonists; requires stronger tetanic stimulation than NMDAR-dependent LTP.

e *ABSOLUTE DEPENDENCE OF SYNAPTIC POTENTIATION ON 'SUFFICIENT' POST-SYNAPTIC DEPOLARIZATION*

Typical experimental arrangement:

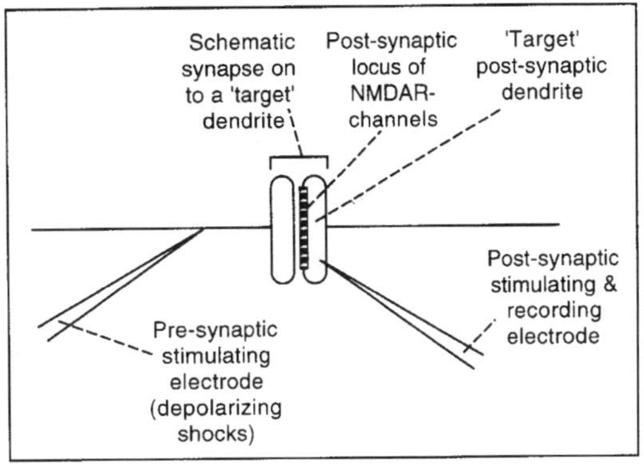

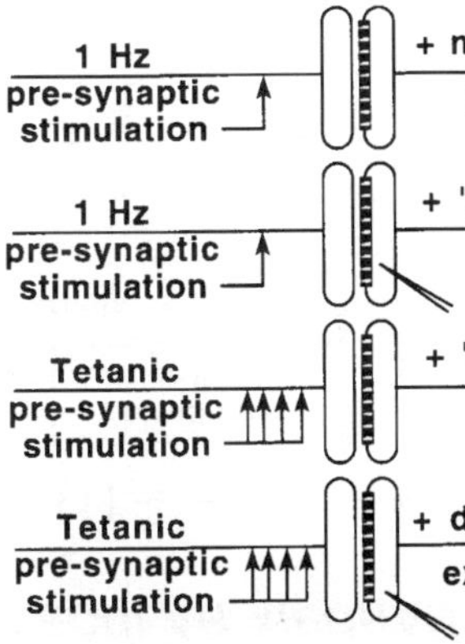

Experimental condition			LTP phenotype
(1)	1 Hz pre-synaptic stimulation	+ no depolarization of post-synaptic dendrites	No LTP induction. Analogous to weak *in vivo* stimuli (i.e. those activating only a few input fibres) not being able to approach a 'threshold' of post-synaptic depolarization.
(2)	1 Hz pre-synaptic stimulation	+ 'paired' depolarization of post-synaptic dendrites	Robust LTP-induction; *simultaneous* depolarization (even at low frequency) facilitates LTP (the NMDAR-channel acts as a 'molecular co-incidence detector').
(3)	Tetanic pre-synaptic stimulation	+ 'sufficient' depolarization of target dendrites	Robust LTP-induction; analogous to 'synaptically-coupled' neurons *in vivo (see sections on temporal summation of synaptic inputs by the NMDAR under Activation, below).*
(4)	Tetanic pre-synaptic stimulation	+ depolarization 'limited' experimentally during tetanus	No LTP induction. There is an absolute dependence for 'sufficient' post-synaptic depolarization for LTP induction - *for underlying molecular mechanisms, see below.*

f MOLECULAR MECHANISM FOR NMDAR-DEPENDENT INDUCTION OF LTP

(1) ----------→ (2) ----------→ (3)

(1) Under 'resting' conditions (or weak stimuli or those produced by synaptic inhibition) (Hyperpolarized)

(2) 'Sufficiently depolarized' (Strong tetanic stimuli inducing Mg^{2+} -expulsion and agonist binding)

(see Panel F, above)

(3) Extracellular ligands (agonists) bound with concomitant post-synaptic depolarizing 'threshold' reached

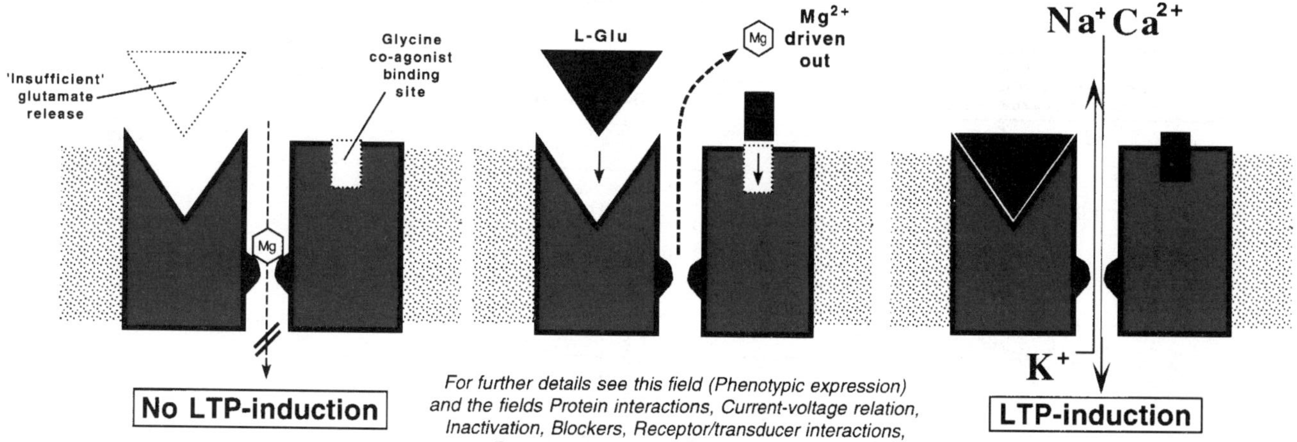

For further details see this field (Phenotypic expression) and the fields Protein interactions, Current-voltage relation, Inactivation, Blockers, Receptor/transducer interactions, Receptor agonists and Receptor antagonists. See also the entries ELG CAT GLU AMPA/KAIN and VLG Ca

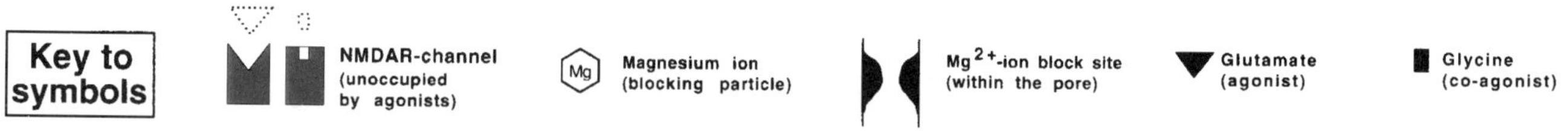

augment the Ca^{2+} transient associated with the synaptic activation of NMDAR. These results, together with observations on the inhibition of LTP by **dantrolene** *(see Phenotypic expression under ILG Ca Ca RyR-Caf, 17-14)* and the Ca^{2+} store Ca^{2+}-uptake pump inhibitor **thapsigargin** *(see Phenotypic expression under ILG Ca CSRC, 18-14, and Channel modulation under ILG Ca InsP$_3$, 19-44)* both inhibit LTP induction[156–158]. A role for Ca^{2+}-influx through AMPA receptors *lacking* the **GluR-B subunit** *(see Selectivity under ELG CAT GLU AMPA/KAIN, 07-40)* has *not* been associated with induction of LTP phenotypes. *Note:* For roles of **metabotropic**† glutamate receptors in LTP induction, see separate paragraphs *(below).*

Examples of NMDAR-independent LTP

08-14-29: LTP events are *not* exclusively associated with NMDA receptor–channel activity *(for a review, see ref.[159])*. For example, **low-threshold Ca^{2+} channels** have been shown to mediate induction of LTP in kitten visual cortex[160]. Although Ca^{2+}-entry via post-synaptic voltage-sensitive Ca^{2+} channels can potentiate excitatory synaptic transmission in the hippocampus transiently ($\geqslant$30 min)[161], other factors may be required for development of **sustained** (i.e. 'true' long-term potentiation) *(see ref.[161] and VLG Ca, entry 42)*. Similarly, a component of LTP associated with **tetanic**† stimulation and opening of voltage-gated calcium channels (but not requiring the activation of NMDA receptors) can be induced in hippocampal area CA1[162]. *Note:* Some forms of **long-term depression**† (LTD) phenotypes (notably **homosynaptic**† **LTD**) require NMDAR activation for their induction *(for reviews, see refs[64, 152])*, but many of these are independent of NMDAR–channel activity *(e.g. mGlu$_1$, see Table 7 under Protein interactions, 08-31)* .

Further distinction of role for NMDAR during initiation but not maintenance of LTP in CA1

08-14-30: Independent studies have shown synaptically activated increases in Ca^{2+} concentration in hippocampal CA1 is primarily due to voltage-gated Ca^{2+} channels[163], although other work has associated *initation* of LTP in CA1 with Ca^{2+}-influx through NMDA receptors. Furthermore, NMDA receptors do *not* appear to be required for development of LTP in **hippocampal area CA2/CA3** or **synaptic depression**† in cerebellar Purkinje cells. Thus in these brain areas, intracellular [Ca^{2+}] changes may also be due to influx of calcium through certain subunit arrangements of **non-NMDA-type Ca^{2+} channels** *(see ELG CAT GLU AMPA/KAIN, entry 07)*, Ca^{2+} channels gated by **intracellular ligands** *(e.g. see ref.[164])* or **voltage-gated channels** *(see above)*. Distinct forms of LTP have been described in guinea-pig hippocampal CA1 region which consist of *both* non-NMDA- and NMDA-mediated components which can be measured both separately and in parallel[165] *(see Fig. 1d).*

Novel NMDA-independent hippocampal LTP by K⁺ channel blockers enhancing glutamate release

08-14-31: In hippocampal CA1, a **transient block** of I_C, I_M and the delayed rectifier (I_K) by the K^+ channel blocker tetraethylammonium produces a

Ca^{2+}-dependent, NMDA-independent LTP, referred to as **LTPK**[166]. This novel form of LTP is induced by a **transient enhanced glutamate release** which generates a depolarization via non-NMDA receptors and the consequent activation of voltage-dependent Ca^{2+} channels[166]. *Comparative note:* NMDAR-dependent post-synaptic mechanisms of LTP induction can also lead to sustained enhancement of pre-synaptic transmitter release[167].

External Mg^{2+}-block of NMDAR 'masks' observation of LTP

08-14-32: Long-term potentiation in the **striatum** (a brain region associated with acquisition of **memory motor skills**) is 'unmasked' by removing the voltage-dependent Mg^{2+}-block of NMDA receptor–channels[168] *(see Blockers, 08-43)*. Under control conditions, NMDA receptor–channels are inactivated† by the voltage-dependent Mg^{2+}-block and repetitive cortical stimulation induces long-term depression† (LTD) which also does *not* require activation of NMDA channels in this preparation. Removal of external Mg^{2+} removes voltage-dependent block and reveals a component of the EPSP† which is potentiated by repetitive (tetanic†) activation[168]. For an illustration of the role of **voltage-dependent Mg^{2+} 'expulsion'** during LTP induction, *see Fig. 1e.*

Synaptic activation of NMDAR can induce 'short-term' potentiation in rat striatum

08-14-33: Maintained activation of NMDAR by synaptically released glutamate in the **corpus callosum** can produce a **sustained enhancement** of the EPSP† which could contribute to basal ganglia-related **motor function**[169]. Mg^{2+}-free cerebrospinal fluid (artificial CSF) increases the duration of pre-tetanus EPSP† together with increases in amplitude and duration of the direct response to tetanic† stimulation. **Post-tetanic† potentiation†** in normal artificial CSF (containing Mg^{2+}) is followed by a long-lasting *depression* of the EPSP. Thus Mg^{2+}-free artificial CSF enables the expression of a **short-term potentiation** of the EPSP amplitude and duration[169].

LTP enhancement by antibody 'agonists' acting at the glycine recognition site

08-14-34: A monoclonal antibody (B6B21) which (i) displaces [^{3}H]-glycine bound specifically to hippocampal NMDA receptors and (ii) enhances the opening of the NMDA integral cation channel in a 'glycine-like' fashion can be competitively antagonized by **7-chlorokynurenic acid**[170] *(see Receptor antagonists, 08-51).* **Antibody B6B21** also enhances LTP in hippocampal slices[170]. Intraventricular infusions of B6B21 significantly enhances acquisition rates in hippocampus-dependent trace eye blink conditioning in rabbits, halving the number of trials required to reach a criterion of 80% conditioned responses *(see also effect of partial agonists at the glycine site under Receptor agonists, 08-50).*

Heterogeneity of synaptic firing patterns associated with NMDAR in different neuronal loci

08-14-35: Ionotropic† NMDA receptors expressed in rat area **CA1 pyramidal cells** (PC) and **interneurones** (IN) differ in their mode of firing and possess different kinetic properties when depolarized by a prolonged current pulse[171]: Whereas PCs fire a single action potential, most INs respond with non-accommodating high-frequency spike firing. *Note:* NMDA EPSCs† in the majority of channels located in the **oriens layer** and **alveus** display unusually slow rise times[171].

Intracellular modulation of NMDA channel activation

08-14-36: A consideration of how each of the distinguishing properties of the NMDA receptor (i.e. glycine co-agonism, voltage-dependent Mg^{2+} block, relatively long open states and Ca^{2+}-permeability) contribute to the modulation of synaptic strength and firing synchronization in **computational neural circuits** has been made[128,172]. *(See also Voltage sensitivity, 08-42, Blockers, 08-43, and Channel modulation, 08-44).*

Parallel 'down-modulation' of inhibitory signalling associated with induction of LTP

08-14-37: During induction of LTP, $GABA_A$ inhibition decreases, causing the NMDAR-mediated excitation to increase, and induce LTP *(see Fig. 4, 08-31).* During *5 Hz* stimulation, **2-OH-Saclofen** (a **$GABA_B$-receptor** antagonist) has been shown to prevent (i) reduction of inhibition, (ii) increase of excitation, and (iii) induction of LTP[173]. *Note:* $GABA_B$ receptor modulation of synaptic plasticity† ('disinhibition') can occur at ~5 Hz which is within the frequency range of the **theta rhythm**† endogenous to the hippocampus and which has been shown to modulate LTP *in vivo.*

Regulation of intracellular Mg^{2+} by NMDA receptor stimulation

08-14-38: The regulation of *intracellular* **Mg^{2+} ion** concentration by neurotransmitters such as glutamate may also be important in controlling neuronal excitability: NMDAR activation by glutamate plus glycine has been shown to raise intracellular Mg^{2+} from 1 to more than 11 mM (compared with the resting concentration of 0.5 mM)[174]. The major component of this glutamate-induced intracellular Mg^{2+} increase is dependent on extracellular Ca^{2+} but is *independent* of extracellular Mg^{2+}. A second (minor) component of the increase is independent of extracellular Ca^{2+}, but requires extracellular Mg^{2+} and can be amplified by extracellular Na^+ removal[174]. *Comparative note:* See also the effects of intracellular Mg^{2+} on inwardly rectifying potassium channel gating within the *INR K series (entries 30 to 33).*

Antagonism of metabotropic† glutamate receptors reduce the duration of LTP

08-14-39: The (relatively non-selective) mGluR antagonists **2-amino-4-phosphonobutanoate** (**AP4**) and **2-amino-3-phosphonopropionate** (**AP3**) reduce the duration of LTP[175]. Conversely, selective mGluR *agonists* such as **1S,3R-aminocyclopentane dicarboxylate** (**ACPD**) can augment tetanus-induced potentiation and can induce NMDA-dependent LTP with 'subthreshold'[176] or low-frequency stimuli[177]. *Note:* ACPD is known to augment responses of hippocampal neurones to NMDA[176] and induce a 'slow-onset' NMDAR-independent mode of LTP[156] *(see also Receptor/transducer interactions, 08-49).*

Comparative note: Role of metabotropic glutamate receptors in parallel release of Ca^{2+} from $InsP_3$-sensitive stores

08-14-40: Synthesis of $InsP_3$ following the activation of mGluRs *(see ILG Ca $InsP_3$, entry 19, and Resource A – G protein-linked receptors, entry 56)* can elevate $[Ca^{2+}]_i$ from $InsP_3$-sensitive stores *(see ILG Ca $InsP_3$, entry 19)* in addition to Ca^{2+}-induced Ca^{2+} release initiated by Ca^{2+}-influx through NMDAR–channels *(see ILG Ca Ca RyR-Caf, entry 17)*. Activation of mGluRs can induce LTP by a **thapsigargin**-sensitive mechanism even in the presence of NMDAR antagonists *(see Phenotypic expression under ILG Ca CSRC, 18-14, and Channel modulation under ILG Ca $InsP_3$, 19-44)*[156]. These results suggest that glutamate-induced release of Ca^{2+} from stores can *substitute* for Ca^{2+}-influx following activation of ionotropic† NMDAR. *Note:* A more general role for Ca^{2+}-store-mediated release in *amplifying* localized transient signals through NMDAR–channels is suggested by the 'insufficiency' of (i) slow depletion of Ca^{2+} stores[158] or (ii) Ca^{2+} currents[103] to induce LTP.

Alternative forms of synaptic plasticity dependent upon the degree of post-synaptic depolarization

08-14-41: Intracellular recordings from **granule cells** of the hippocampal **dentate gyrus** maintained *in vitro* have shown isolated NMDA receptor-mediated synaptic responses to express both LTP *and* LTD[178]. Additionally, the level of post-synaptic depolarization can determine which of the two forms of synaptic plasticity is expressed in response to an identical input. These studies have also shown that post-synaptic Ca^{2+}-influx is essential not only for induction of LTP but also for induction of LTD of NMDA receptor–channel function[178].

Protein distribution

See the methodological note in mRNA distribution under ELG CAT GLU AMPA/KAIN, 07-13.

NR1 protein distribution

08-15-01: The 'fundamental' NMDAR subunit **NR1** is generally confined to the central nervous system. Following subcellular fractionation of the cerebral cortex, NR1 protein 'co-enriches' with **synaptic membranes**. Prominent selective immunostaining for NR1 protein occurs in several layers of the cerebral cortex, in the hippocampus and dentate gyrus, as well as in the cerebellum[58].

Differential NMDAR distributions as determined by displacement of agonists

08-15-02: Under conditions of differential receptor activation, regional differences in NMDA receptor pharmacology can be detected in the CNS using **[^{3}H]-MK-801** binding and quantitative autoradiography[179]. The compounds **CPP** and **7-Cl-Kyn** *(see Receptor antagonists, 08-51)* have distinguished at least three populations of native† NMDA receptors by displacement of [^{3}H]-MK-801[179].

Functional assays for NMDA protein distribution

08-15-03: By use of selective antagonists for NMDA receptors, it has been shown that NMDA receptors contribute only a *small* and variable amount to the EPSPs† in hippocampus, cortex, spinal cord and neostriatum.

Subfields 'devoid' of NMDAR in hippocampus

08-15-04: Mossy fibres† which terminate in the **stratum lucidum** of area CA3 are devoid of NMDAR and LTP is not blocked by NMDAR antagonists in this subfield†. **Mossy fibre LTP** thus forms one class of NMDAR-independent synaptic potentiation in the hippocampus *(see classification in refs*[54, 180] *and the classification of synaptic responses in Phenotypic expression, 08-14).*

Subcellular locations

08-16-01: NMDA receptors are generally located on post-synaptic membranes, closely apposed to transmitter release sites across the **synaptic cleft** ($\leqslant 2\,\mu$m). NMDAR are generally *assumed* to be located on **dendritic spines**; a dendritic location of NMDAR proteins may act to localize Ca^{2+} signals as spines can restrict the diffusion of Ca^{2+}[181]. A comprehensive discussion of pre- and post-synaptic factors affecting LTP phenotypes has appeared[54].

'Clustering' and 'mobility' of conantokin-G-sensitive NMDAR

08-16-02: The *Conus geographus* venom peptide **conantokin-G** (CntkG) has been reported as a 'reliable' probe for determination of NMDAR protein distribution[182]. These studies have shown NMDAR to be clustered and immobilized on dendrites of living cortical neurones. In hippocampal slices, the **CA1 dendritic subfield** is strongly labelled by CntkG, whereas the CA3 mossy fibre region is not labelled. On CA1 hippocampal neurones in culture, dendritic **CntkG-sensitive NMDAR** are **clustered** at sites of **synaptic contacts**, whereas **somatic NMDAR** are distributed **diffusely** and in patches[182]. Notably, NMDAR distribution differs from the distribution of voltage-dependent calcium channels. A significant fraction of labelled NMDAR on somata and dendrites has been found to be highly **mobile**. Rates of mobility are consistent with rapid recruitment of NMDAR to specific synaptic locations[182].

Methodological note

08-16-03: Neurones receive many excitatory synapses (10^3–10^5 per cell) which are located on slender dendritic elements generally far away from the somatic recording site, making *in situ* studies of discrete (localized) synaptic events difficult[187]. Typically, when *populations* of synapses are activated, NMDA receptor-mediated synaptic potentials appear as **slowly rising, long-lasting waves** superimposed on faster, **non-NMDA-receptor potentials**[187]. Excitatory **autaptic**† **currents** (identical to those found *in vivo)* have been shown to be mediated by NMDA receptors in isolated hippocampal neurones maintained in cell culture[188].

Co-localization and co-activation of NMDAR and non-NMDAR channels at single synapses

08-16-04: Monosynaptic excitatory post-synaptic potentials (EPSPs) evoked between pairs of cultured neurones from either mouse hippocampus or spinal cord have demonstrated that two functionally distinct excitatory amino acid receptor–channels can be **simultaneously activated** by transmitter release from a single pre-synaptic neurone[183]. The **co-localization** of NMDA and non-NMDA receptor–channels at single synapses is important for the development of LTP†, which may be differentially expressed at each synapse according to the mix of receptor subtypes at that synapse[184] *(see Protein interactions, 08-31). Note:* As summarized by Daw *et al.*[75] NMDA agonists multiply (amplify) synaptic responses (increasing the slope of the response curve), whereas non-NMDA agonists add to them (moving the response curve upwards). The roles of post-synaptic calcium in the induction of LTP† have been reviewed, e.g. refs[54, 185, 186]. For experimental protocols of LTP† induction, *see Phenotypic expression, 08-14.*

Transcript size

08-17-01: *See Table 1 under Gene family, 08-05.*

SEQUENCE ANALYSES

Note: The symbol [PDTM] denotes an illustrated feature on the channel protein domain topography model (Fig. 2).

Chromosomal location

08-18-01: The human gene encoding the NR1 subunit (**NR1**, **zeta 1**) has been mapped to chromosome 9q34.3[11, 189] with genes encoding potentiating subunits **epsilon 1** and **epsilon 3** being localized to chromosomes 16p13 and 17q35 respectively[189].

Encoding

08-19-01: For open reading frame† lengths of reported cDNAs, *see Database listings, 08-53.*

Gene organization

Alternative splice variants of the fundamental subunit

08-20-01: Alternative splicing† generates **functionally distinct** NMDA receptors[190–193]. Analysis of the gene structure encoding NR1 revealed eight **splice variants** arising from (i) different combinations of a single 5′-terminal **exon insertion** and (ii) three different 3′-terminal **exon deletions**[193] *(for further details, see Domain functions, 08-29).*

Analysis of NR1 gene 'upstream sequences'
08-20-02: Cloning and sequence analysis[194] of a 3.8 kb *Eco*RI fragment of the rat NR1 gene included 3 kb of **promoter**† and **enhancer**† region, exon† 1 and a portion of intron† 1. NR1 possesses sequence motifs characteristic of a **housekeeping gene**† regulated by **immediate-early**† gene products. Two major **transcriptional start sites**† were identified at −276 and −238 from the first nucleotide in codon† one. One GSG and two SP1 motifs, but no TATA box† or CAAT box† exists in the region proximal to the transcriptional start sites†[194].

Homologous isoforms

08-21-01: The human NMDA receptor cDNA hNR1 shares high sequence homology (~99%) with the rat brain NMDA1 and the mouse zeta 1 subunit[195]. The rodent and human homologues diverge near the C-terminus, suggesting that they represent **alternatively spliced**† messages of the same gene *(see Gene organization, 08-20, and Database listings, 08-53).* Of the 7 of 938 amino acids which are different between the rodent and human sequences, three occur in the region of the **signal peptide**† and the others in the extracellular (N-terminal) domain preceding the four putative transmembrane segments[11] *(see also alternative (five transmembrane domain) model shown in [PDTM], Fig. 2).*

Protein molecular weight (purified)

Glycosylated monomeric and oligomeric proteins
08-22-01: Both native and heterologously† expressed rat brain NR1 subunit have an apparent molecular mass of 116 kDa determined by SDS–PAGE†[58]. **Chemical cross-linking** of native synaptic membrane proteins shows that the NR1 protein is part of a **receptor protein complex** with a molecular mass of 730 kDa[58]. The NR1 receptor protein is heavily **glycosylated**† *(see below).*

N-*Glycosylation can account for differences between predicted and purified M_r of NMDAR*
08-22-02: An antibody raised to aa positions 1435–1445 of NR2A recognizes four immunoreactive species with M_r of 180 kDa, 122 kDa, 97 kDa and 54 kDa in rat brain, but only a single band of M_r of 180 kDa in HEK-293 cells transiently expressing NR2A[12]. *De-N*-glycosylation of HEK cell membranes yields a 165 kDa immunoreactive species, which agrees with an M_r predicted from the open reading frame† length of the cDNA sequence for the mature NR2A subunit *(see also Sequence motifs, 08-24).*

Sequence motifs

Multiple N-*glycosylation sites*
08-24-01: The amino acid sequence of the NR1 clone[10, 58] predicts 10 possible ***N*-glycosylation** sites at the extracellular domain. In the isoforms NR2A to NR2D, (under the assumed membrane topography described in *Amino acid*

composition, 08-26) there are 6, 6, 5 and 6 **canonical**† **Asn-X-Ser/Thr sequences** respectively for potential *N*-glycosylation in the extracellular *N*-terminal regions[8]. A large number of possible *N*-glycosylation sites are also present in the extracellular C-terminal regions of NR2A (12 sites) and NR2B (10 sites) but only one in NR2C and none in NR2D. At the C-terminal of these polypeptides, all but NR2D-1 share a **common amino acid sequence** of undetermined function: Ser/Pro–Ser–Leu/Ile–Glu–Ser–Glu/Asp–Val. The cloned NMDAR subunits also possess many phosphorylation motifs† *(see Protein phosphorylation, 08-32).*

STRUCTURE & FUNCTIONS

Note: The symbol [PDTM] denotes an illustrated feature on the channel protein domain topography model (Fig. 2).
In-press updates *(see critera under Introduction & layout of entries, entry 02): See note above field 07-26, for important new information relating to ionotropic glutamate receptor structures.*

Amino acid composition

Amino acid assignments of N- and C-terminal residues
08-26-01: The overall structure of NR1 subunits is similar to that predicted for the AMPA/kainate family of receptor–channels, i.e. a large **N-terminal extracellular domain** followed by four putative transmembrane segments (TMI–TMIV or M1–M4 representing the '**core region**' *(see Domain conservation, 08-28)*[10]. **Hydrophobicity**† **analysis** of the NR1 receptor predicts a hydrophobic N-terminal signal† peptide and four **hydrophobic**† **transmembrane domains** *(but see Domain arrangement, 08-27, and compare the protein domain topography models under ELG CAT nAChR, entry 09, ELG CAT 5-HT$_3$, entry 05, ELG Cl GABA$_A$, entry 10, ELG Cl GLY, entry 11, and the 'cautionary note' in ELG Key facts, entry 04). According to this type of assignment,* the NR1 protein has ~540 N-terminal residues and ~110 C-terminal residues[10].

08-26-02: Sequence relationships between NR1 and the potentiating subunits are outlined under *Gene family, 08-05, and Domain conservation, 08-28.*

Domain arrangement

08-27-01: Native NMDA receptors are composed of a **fundamental** or **key** subunit (NR1) and its **potentiating subunits** (NR2A–NR2D). Variants of **NMDAR-2 subunits** potentiate and functionally differentiate native NMDA receptors by forming different heteromeric† configurations with NR1 *(for further details, see Domain functions, 08-29).*

Alternative domain models predict five transmembrane segments
08-27-02: The subunits NR2A–NR2D[8] also possess large **hydrophilic**† **domains** at both N- and C-terminal sides of the four putative transmembrane

segments, though NR2 subunits clearly display **five hydrophobic segments** consisting of 20 uncharged amino acid residues. NR2 subunits also show a **large C-terminal extension** following the M4 segment which is not seen in other cloned ligand-gated ion channels (cf. >400 aa residues in NR2 versus 50–100 residues typical of other ionotropic† GluRs)[8]. The **'five transmembrane domain' model** (e.g. see ref.[19]) is represented in [PDTM], Fig. 2, with dotted lines representing the 'extra' transmembrane domain.

Domain conservation

Determinants of Mg^{2+}-sensitivity and Ca^{2+}-selectivity

08-28-01: NMDA and AMPA receptor–channels contain **common structural motifs**† in their transmembrane M2 segments that are responsible for some of their **ion selectivity**† and **conductance**† properties. The high Ca^{2+}-permeability and extracellular Mg^{2+}-sensitivity of the NMDAR are imparted by **asparagine residues** in a putative channel-forming segment of the protein, transmembrane domain 2 (M2). In the NR1 subunit, replacement of this **asparagine** by a **glutamine** residue decreases **Ca^{2+}-permeability**† of the channel and slightly reduces **Mg^{2+}-block**[196] (see *Domain functions, 08-29).* The same substitution in NR2 subunits strongly reduces magnesium block and thus increases the Mg^{2+}-permeability but does not significantly affect Ca^{2+}-permeability. These asparagines are in a position *homologous to* the site in the M2 region (**Q/R site**) of AMPA receptors[196] *(cf. Domain functions under ELG CAT GLU AMPA/KAIN, 07-29, and see Domain functions, 08-29).*

Variation of sequence homology in different structural domains of the NMDAR family

08-28-02: Levels of sequence homology are variable across the structural domains of NMDAR. Homology is extremely high within the four **transmembrane segments** (~80–90%, the **'core' region**) but only moderate in the N-terminal portions (~45–60%) and 'very low' in the **C-terminal** portions (~20–30%)[8]. On structural grounds, the NR2A and NR2B sequences closely resemble each other, as do the NR2C and 2D subtypes. Thus, they may be classified into two subgroups[8]. Subunits NR2A–NR2D[8] are only about 15% identical with the key subunit of the NMDA receptor (NR1) but are highly homologous (approximately 50% homology) with one another.

Domain functions (predicted)

Agonist and co-agonist binding site – signal transduction via electron transport

08-29-01: In the NR1 subunit, the **agonist-binding signal** may be carried from Y456 to W590 through an **electron transport chain**, including W480 which is a candidate locus for the **glycine modulatory site**. NMDA channel opening may arise from **repulsion of negatively charged Trp590s**, analogous to Trp435s of the Shaker K^+ channel[197].

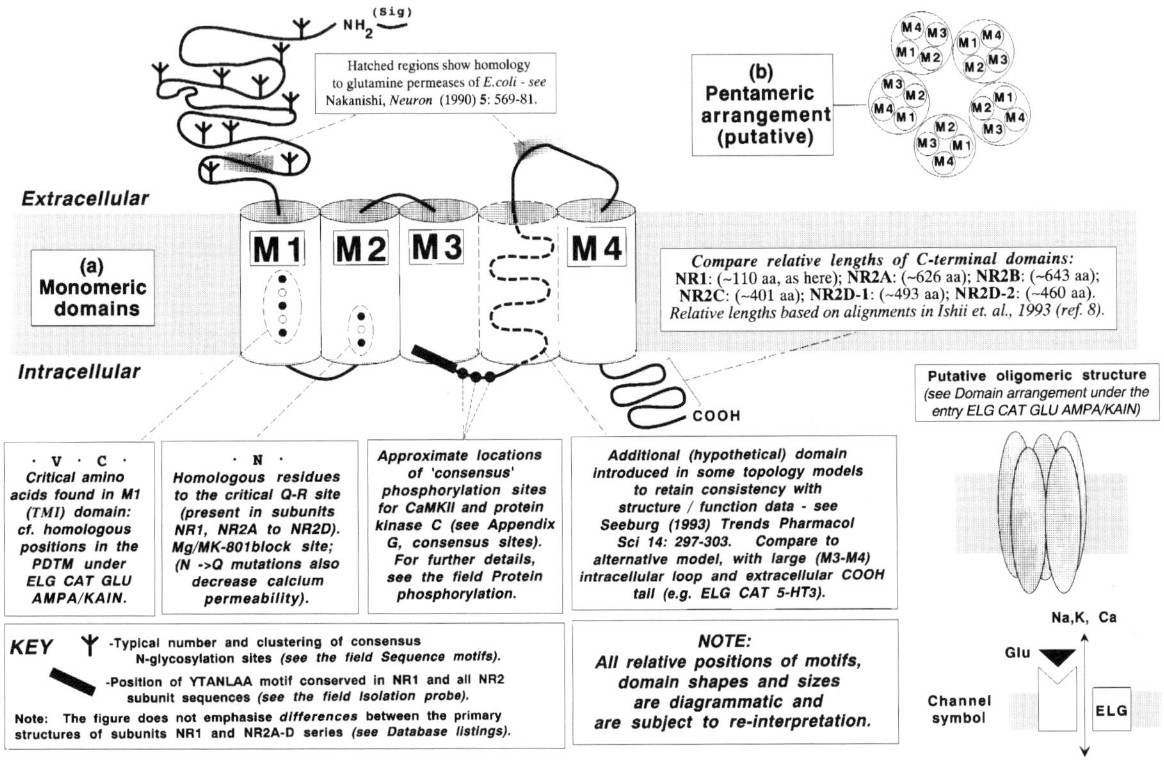

Figure 2. *Monomeric protein domain topography model (PDTM) exemplified for NMDA-selective ionotropic glutamate receptor (iGluR) NR1 subunits. In press updates: (see criteria under Introduction & layout of entries, entry 02): A modified 3-transmembrane domain model based on N-glycosylation site tagging data has appeared in press. For brief details, see note above field 07-26. For further details, see entry update pages via the CSN. (From 08-27-02)*

Amino acids regulating Ca^{2+}-permeability of NR1 homomeric channels

08-29-02: Localised protein sequences contributing to NMDAR Ca^{2+}-permeation† *(see Selectivity, 08-40)* and channel blockade *(see Blockers, 08-43)* have been determined following *in vitro* mutagenesis and expression in *Xenopus* oocytes[198]. The substitution of an **asparagine** (N) into the NR1 equivalent of the **glutamine/arginine (Q/R) site** of AMPA receptor subunits (determined following sequence alignment) generates channels displaying a selective permeability for Ca^{2+} over Mg^{2+} *(the role of the Q/R site is described in detail under the ELG CAT GLU AMPA/KAIN fields Gene organization, 07-20, Domain functions, 07-29, Current–voltage relation, 07-35, Selectivity, 07-40, and Blockers, 07-43).* The permeation pathways of a range of neurotransmitter-gated ion channels has been reviewed[199].

Overlapping blocking sites for extracellular Mg^{2+} ions and dizocilpine (MK-801)

08-29-03: The conserved asparagine residue in segment M2 constitutes a **Mg^{2+}-block site** of the NMDA receptor–channel. The MK-801 site overlaps the Mg^{2+} site[198,200]: Asn598 → Glu598 mutations in segment M2 of the zeta 1 subunit (ζ1-N598Q) strongly reduce the sensitivity of the heteromeric epsilon 2/zeta 1 NMDA receptor–channel to Mg^{2+}-block when expressed in *Xenopus* oocytes *(see Fig. 3).* The '**asparagine ring**' formed in the central part of the channel-forming M2 has been determined to play a critical role in determining the Ca^{2+}-permeability† and the inhibition of open channel blockers[198]: Asp598 → Glu598 or Asp598 → Arg598 mutations alter both the **Ca^{2+}-permeability**† and the sensitivity to **Mg^{2+}-block** and **MK-801**. These mutations also reduce the inhibitory effects of Zn^{2+} and an antidepressant, **desipramine**[198].

NR1 splice variant functions

08-29-04: Properties of seven isoforms of the NMDA receptor generated by alternative splicing† have been described[192]. Combinatorial RNA splicing† has been shown to alter the **surface charge** on the NMDA receptor[191]. Splice variants of NR1 encoded by rat ventral midbrain cDNA differ in their functional properties *(for the location of NMDAR protein regions affected by alternative splicing events, see [PDTM] Fig. 2).* The structural variation in some splice variants is shown in *Table 5.*

Presence of NR2 series 'potentiating' subunits in heteromeric complexes reproduce native NMDAR properties

08-29-05: Homomultimers† formed from expression of NR1[10] (zeta 1)[15] subunits exhibit several features of native NMDA receptors. Significantly, these include pronounced **Ca^{2+}-permeability** *(see Selectivity, 08-40),* a modulatory action of **glycine** *(see Channel modulation, 08-44),* and a negative slope conductance of currents in the presence of **Mg^{2+}** *(see Current–voltage relation, 08-35).* Subunits NR2A, NR2B and NR2C yield prominent, typical glutamate- and NMDA-activated currents only when they are in heteromeric† configurations with NR1. NR1/NR2A and NR1/NR2C channels differ in **gating**† **behaviour** and **magnesium sensitivity**. Heteromeric NMDA receptor subtypes probably exist in native neurones,

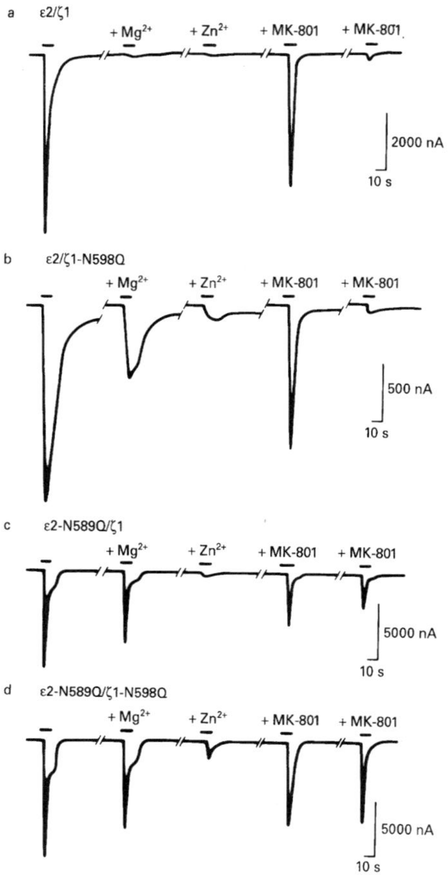

Figure 3. *NMDAR mutations affecting Mg^{3+} block, Ca^{2+}-permeability and sensitivity to Zn^{2+} and open-channel blockers in heteromeric combinations of NMDAR. The figure compares phenotypes for (a) the wild-type† subunit combination* ***ϵ2/ζ1****; (b) the mutant subunit combination* ***ϵ2/ζ1-N598Q****; (c) the mutant subunit combination* ***ϵ2-N589Q/ζ1*** *and (d)* ***ϵ2-N589Q/ζ1-N598Q****. The heteromeric† epsilon 2/zeta 1 NMDA receptor–channel with the mutation on both subunits – shown in panel (d) – displays greatly reduced sensitivity to dizocilpine (MK-801) but is still susceptible to inhibition by Zn^{2+}* ***200***. *The corresponding mutation of the epsilon 2 subunit has a similar effect to panel (d). The figure shows non-competitive† antagonism of inward current responses to 10 μM L-glutamate plus 10 μM glycine (co-agonists) at −70 mV in Ringers† solution. The black bars represent (in order) application of L-glutamate plus glycine, co-agonists plus bath Mg^{2+} (1 mM), co-agonists plus $ZnCl_2$ (100 μM), and co-agonists plus two successive applications of the open-channel blocker (+)-MK-801 (1 μM). (Reproduced with permission from Mori (1992)* Nature ***358****: 673–5.) (From 08-29-03)*

Table 5. *RNA splice variants of NMDA receptor–channel (From 08-29-04)*

Splice variant	Structural variation
NR1a and NR1b	The NR1b splice variant† differs from NR1a by the presence of a 21 aa insert near the amino end of the N-terminal domain and by an alternate C-terminal domain in which the last 75 amino acids are replaced by an unrelated sequence of 22 amino acids[201]. Otherwise, NR1b is virtually identical to NR1a in the remainder of the N- and C-terminal domains, at the 5′ and 3′ non-coding ends, and within the predicted transmembrane domains and extracellular and cytoplasmic loops
NR1c	The NR1c splice variant has been shown to be identical to NR1b in its C-terminus but lacks the N-terminal insert[201] *(see also Protein phosphorylation, 08-32, and Channel modulation, 08-44)*

Note: Features and differences of alternative splice variants designated NMDAR1–1a, NMDAR1–1b, NMDAR1–2a, NMDAR1–2b, NMDAR1–3a, NMDAR1–3b, NMDAR1–4a, NMDAR1–4b and others can be found in the original references shown in the *Database listings, 08-53*

since NR1 messenger RNA is synthesized throughout the mature rat brain, while NR2 messenger RNA show a differential distribution[6] *(for summary, see mRNA distribution, 08-13)*. Further 'potentiating' properties of NR2 series subunits in heteromeric NMDAR complexes are listed in Table 6. *(See also Domain functions under ELG CAT GLU AMPA/KAIN, 07-29).*

Subunit-specific patterns of modulation for recombinant homomeric NMDARs

08-29-06: Homomeric† NMDA channels of the splice variant NR1b possess electrophysiological properties distinct from those of NR1a homomeric channels[201] *(see Gene organization, 08-20)*. NR1b channels exhibit a lower apparent affinity for NMDA and for glutamate. Furthermore, NR1b channels exhibit a lower affinity for D-2-amino-5-phosphonovaleric acid (APV) and a higher affinity for Zn^{2+}. The two receptor variants show 'nearly identical' affinities† for **glycine**, **Mg^{2+}**, and **phencyclidine**. **Spermine potentiation** of NMDA responses, prominent in oocytes injected with rat forebrain mRNA, is also prominent for NR1a receptors, but is greatly reduced or absent for NR1b receptors[201]. For mechanisms of spermine potentiation, *see Channel modulation, 08-44.*

Predicted protein topography

08-30-01: 'True' transmembrane topography has not yet been *directly* determined for any ionotropic† glutamate receptor–channel. For *predicted* arrangements, *see Amino acid composition, 08-26.* ***In-press updates:** See note above field 07-26.*

Table 6. *Functional properties and molecular features contributed to NMDAR complexes by NR2 series subunits (From 08-29-05)*

Property/feature	Description	Refs
NR2 subunits alone do not form functional channels	When expressed individually in *Xenopus* oocytes, NR2A and NR2C show no electrophysiological response to agonists. However, when NR2A and NR2C are co-expressed with the ubiquitous NR1, complexes show marked potentiation† of NR1 activity and produce functional variabilities in the affinity of agonists, the effectiveness of antagonists, and the sensitivity to Mg^{2+} blockade	*8*
'Full reproduction' of native NMDAR–channel properties in NR1/NR2 heteromultimers	Heteromeric NR1/NR2A and NR1/NR2C combinations display *all* of the basic properties characteristic of the native NMDA receptor, including Ca^{2+}-permeability†, glycine modulation, voltage-dependent Mg^{2+} block and selective inhibition by competitive and non-competitive antagonists and open-channel blockers. *Note:* In contrast to its role as a *co-agonist* with glutamate, glycine alone can activate epsilon 1/zeta 1 subunit heteromultimers when co-expressed in oocytes	*8*, *7*
The epsilon 4 (NMDAR4, NR4) subunit forms heteromultimers with distinct properties	The epsilon 4 (NMDAR-D) subunit is distinct in functional properties from the epsilon 1, epsilon 2 and epsilon 3 subunits, and contributes further diversity to the NMDA receptor–channel. The epsilon 4/zeta 1 heteromeric channel exhibits high apparent affinities for agonists and low sensitivities to competitive antagonists when expressed in *Xenopus* oocytes	*202*

Variations in regional expression in patterns of diverse NR2 series subunits	The molecular diversity and different spatial distribution patterns of the NR2 (epsilon) subunit family underlies the functional heterogeneity of the NMDA receptor–channel *(see mRNA distribution, 08-13)*. For example, the heteromeric† epsilon 1/zeta 1, epsilon 2/zeta 1 and epsilon 3/zeta 1 NMDA receptor–channels exhibit distinct functional properties in affinities for agonists and sensitivities to competitive antagonists and Mg^{2+} block	*59*
Conserved residues in NR2 subunits	All NR2-type 'potentiating' subunits contain a positively charged lysine residue within M2. The corresponding residue of NR1 subunits (Thr602) controls ion permeation at homologous positions in the nicotinic acetycholine receptor *(see ELG CAT nAChR, entry 09)* and may explain its conservation in NR2A to -2D subunits	
The 'asparagine ring' predicted for NR1 homomultimers is conserved within heteromultimers	A corresponding asparagine residue forming the putative NR1 'asparagine ring' *(see above)* is conserved in NR2-type subunits, indicating the ring may form and control Ca^{2+}-permeability and channel conductance properties in the pore region within heteromeric† (NR1/NR2) receptors	
Variations in NR2 series extracellular domains	Structural variability in the extracellular domains of the NR2 subunits is responsible for governing different affinities of agonists and antagonists that act at the glutamate-binding site and the glycine-modulatory site. A gating mechanism for NMDAR channels has been proposed which shares features with *Shaker*-type K^+ channels	*197*
Relative length of intracellular domains	NR2 subunits possess relatively long C-termini (typical size ~ 550 aa) which generally exceeds that of extracellular domain preceding M1 (typical length ~ 500 aa in all iGluRs)	

Protein interactions

Co-localization of NMDA and non-NMDA receptor–channels

08-31-01: NMDA and the non-NMDA (AMPA/kainate) channel subtypes are often **co-localized** at individual excitatory synapses[184]. This assumption is critical for models of **long-term potentiation**† *(see Phenotypic expression, 08-14).* The NMDA class, by virtue of its voltage-dependent channel block by magnesium and calcium permeability†, provides the 'trigger' for the **induction** of long-term potentiation†, whereas the actual **enhancement** of synaptic efficacy has been proposed to be contributed by the **non-NMDA class** of GluR channels.

Frequencies of NMDAR/non-NMDAR subtype co-localizations

08-31-02: As described under *Domain arrangement, 08-27, and Domain functions, 08-29),* interaction with NR1 is *necessary* for the functional expression of all other cloned NMDA receptor subunits. Therefore, the NR1 subunit is likely to be a **central component** of all known NMDA receptors in brain *(see ref.*[58]*)*. Measurement of **miniature**† **synaptic currents** in cultured hippocampal neurones has shown ~70% of excitatory synapses to possess both the NMDA and non-NMDA classes of receptor, although to differing extents. Of the remaining excitatory synapses, ~20% contain only the **non-NMDA subtype** and the remainder possess *only* NMDA receptors[184]. The depolarizing post-synaptic potential† (DPSP) in cerebellar granule cells has been determined to be *part-mediated* by NMDA, $GABA_A$ and AMPA receptor proteins *(see Protein interactions under ELG Cl $GABA_A$, 10-31).*

Operation of extracellular Mg^{2+}-block depends on functional Na^+/K^+ ATPase

08-31-03: Central to the prevention of '**neurotoxic**' **influx of calcium** through the NMDAR is the operation of **voltage-dependent block** by extracellular Mg^{2+} *(described under Blockers, 08-43).* Failure of the **Na^+/K^+-ATPases** (for example under conditions of **ischaemia**†) can directly affect NMDAR function in two main ways. First, if *pre-synaptic* Na^+/K^+ pumps fail, the elevated **intracellular sodium concentration** can lead to failure or reversal of '**glutamate (uptake) transporters**' *(see Fig. 4)* thereby increasing glutamate concentration within the synaptic cleft†. Moreover, if *post-synaptic* Na^+/K^+ pumps fail, the neuronal membrane will depolarize, and extracellular Mg^{2+}-block will not operate *(see Fig. 1e and Blockers, 08-43).* Thus 'ambient' levels of glutamate will open the NMDAR channels[132]. This type of **glutamate toxicity** has been demonstrated for NMDAR in retinal ganglion cells maintained *in vitro*[203].

'Functional clustering' of other pre- and post-synaptic signal transduction proteins

08-31-04: Co-localization of a large number of **signal transduction proteins** *(see Fig. 4 and Table 7)* within single synapses permits **rapid, local cross-modulation** through **multiple diffusible factors**. *Figure 4* illustrates a *selection* of these proteins and their interactions. *Note:* The arrangement shown is intended to supplement the text of several fields and does not take into account the important roles of **potassium and chloride channels** which can regulate **rates of neuronal firing**. Likewise, the contribution of **G protein-linked receptor** signalling to modulation of ion channels is largely ignored. Brief descriptions of functional processes dependent upon interactions of synaptic proteins (such as long-term potentiation†) are included under *Phenotypic expression, 08-14,* though the complexity of potential interactions at individual synapses is difficult to generalize into a single model *(see additional references listed under Related sources & reviews, 08-56).*

Properties of other signalling components associated with NMDAR-mediated phenotypes

08-31-05: Neurons **co-ordinate** the expression of a large number of signalling proteins which contribute to phenotypes and functions associated with **synaptic transmission**† (as described in the previous paragraph and partly illustrated in Fig. 4). Table 7 summarizes important features of some of these molecules, *where they are likely to involve interactions with the NMDAR*. For descriptions and classifications of **synaptic phenotypes**† (such as long-term potentiation, LTP) as mentioned in the table, *see Phenotypic expression, 08-14.*

Protein phosphorylation

Several other pre- and post-synaptic signalling proteins contributing to or undergoing 'phosphomodulation' are also described in Table 7 under Protein interactions, 08-31.

Conservation of kinase-modulatory sites in cloned NMDAR subunits

08-32-01: Cloned NMDA receptor subunits display many potential phosphorylation sites† for **Ca^{2+}/calmodulin-dependent protein kinase type II** and **protein kinase C.** (For the locations of these sites, *see references given under Database listings, 08-53*). By analogy with other 'ELG superfamily' channels, kinase sites located on intracellular domains (such as the putative **M3–M4 intracellular loop**, *see [PDTM], Fig. 2*) can act as important modulators of channel function[10].

Examples of homomeric and heteromeric NMDAR current potentiation by PKC activators

08-32-02: The NMDA zeta 1 homomeric† channel activity is positively modulated by treatment with the protein kinase C activator phorbol 12-myristate 13-acetate (**TPA**)[15]. Heteromeric† epsilon 1/zeta 1 and epsilon 2/zeta 1 channels (but not the epsilon 3/zeta 1 channel), are activated by treatment with TPA when expressed in *Xenopus* oocytes[59].

Relative 'susceptibility' of NR1a and NR1b to PKC activators

08-32-03: Treatment with **TPA** *(see above)* potentiates NMDA responses in oocytes injected with mRNA encoding the **splice variant**† **NR1b** by about 20-fold compared to ~4-fold potentiation in **splice variant NR1a**-injected oocytes[201] *(see Gene organization, 08-20).*

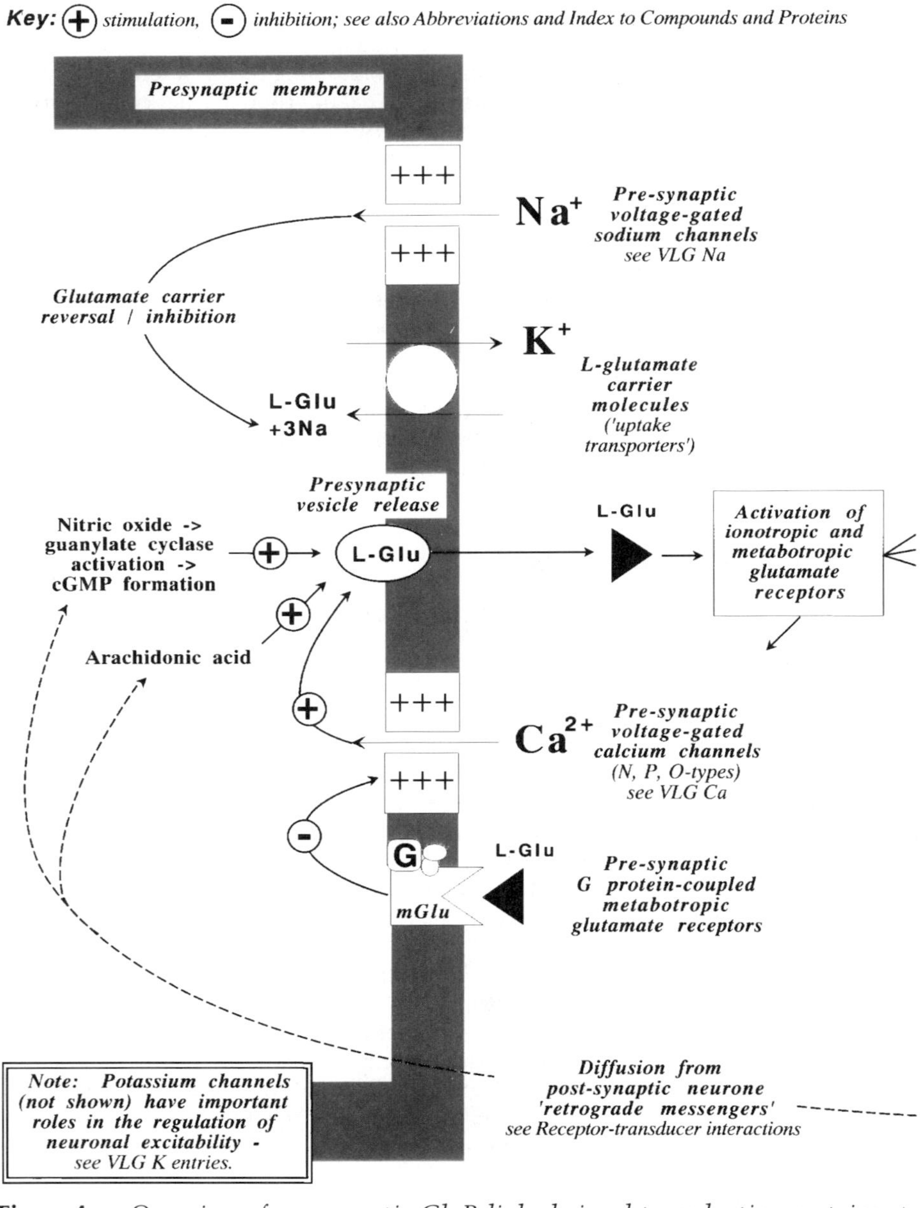

Figure 4a. *Overview of pre-synaptic GluR-linked signal transduction proteins. +, stimulation; –, inhibition. (From 08-31-04)*

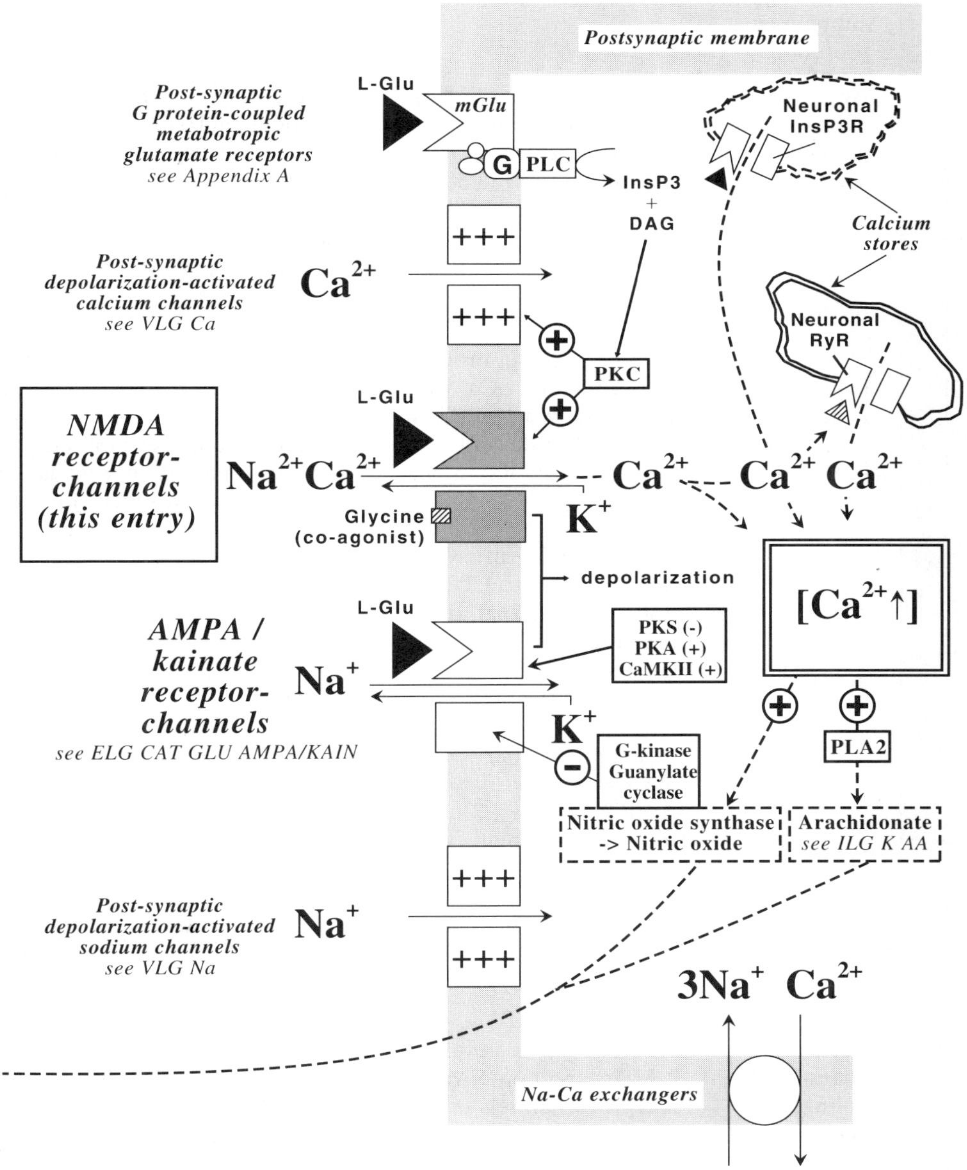

Figure 4b. *Overview of post-synaptic GluR-linked signal transduction proteins. +, stimulation; –, inhibition. (From 08-31-04)*

Table 7. *Summary of NMDAR-associated signal transduction components (From 08-31-05)*

Class and subtype of protein	Key roles/interaction	Regulatory functions/notes
Adenylyl cyclases calmodulin-sensitive	Increased levels of cAMP have been demonstrated following tetanic† stimulation of the Schaffer collateral pathway in the CA1 region[204]	Calmodulin-sensitive adenylyl cyclase in this preparation depends on both activation of the NMDAR and increases in $[Ca^{2+}]_i$[204] *(see also 'Calcium channels, high-voltage activated', this table)*
Ca^{2+}/ calmodulin-dependent protein kinase (CaMKII)[112]	Targeted disruption of the gene encoding αCaMKII (i.e. gene-knockout†) markedly reduces (but does *not* eliminate) induction of LTP in brain slices[205] αCaMKII-null† mice are deficient in both **LTP†-induction** and **spatial learning** ability[205] but show no gross morphological changes in the brain αCaMKII is also likely to 'positively-modulate' AMPA/kainate iGluR receptor–channels *(see 'Non-NMDA-type ionotropic glutamate receptors', this table and ELG CAT GLU AMPA/ KAIN, entry 07)*	αCaMKII is highly expressed in post-synaptic densities. Primary sequences of cloned NMDAR exhibit consensus† phosphorylation motifs for CaMKII *(see [PDTM], Fig. 2 and Protein phosphorylation, 08-32)*
Calcium channels, high-voltage activated *(see VLG Ca, entry 42)*	Calcium influx through both NMDA and non-NMDA receptor–channels in cultured rat hypothalamic neurones activates a calmodulin-dependent inhibition of the high voltage-activated (HVA) Ca^{2+} current[206]	*Comparative note:* NMDA receptor activation has been shown to increase cAMP levels and thereby the fractional open time† of high-threshold Ca^{2+} channels in CA1 pyramidal cells[204]

Table 7. *Continued*

Class and subtype of protein	Key roles/interaction	Regulatory functions/notes
Calcium channels, voltage-gated, type, N pre-synaptic *(see VLG Ca, entry 42)*	NMDA receptor agonists have been reported to selectively and effectively *depress* N-type Ca^{2+} channels which modulate neurotransmitter release from pre-synaptic sites[207] *(see VLG Ca, entry 42).* The *inhibitory* effect is eliminated by the competitive NMDA antagonist D-2-amino-5-phosphonovalerate (APV) and does not require Ca^{2+} entry into the cell	Implies a **'negative feedback'** between liberation of excitatory transmitter and entry of Ca^{2+} into the cell, modulating pre-synaptic inhibition and regulating synaptic plasticity† [207]
Calcium-store release channels: Ryanodine receptors *(see ILG Ca Ca RyR-Caf, entry 17) and* **$InsP_3$ receptors** *(see ILG Ca $InsP_3$, entry 19)*	In cerebellar granule cells a major component of both K^+- and NMDA-induced elevation of Ca^{2+} involves release from **intracellular stores**[208]. The Ca^{2+}-store depletors **thapsigargin** (which blocks the action of the Ca^{2+} ATPase) and **ryanodine** *(see ILG Ca RyR-Caf, entry 17)* display **partial additivity** to K^+- and NMDA-induced responses, showing that these agents affect two **overlapping but non-identical Ca^{2+} pools**	Perfusion of **dantrolene** on to groups of cells during the sustained plateau phase of the $[Ca^{2+}]_i$ response to K^+ or NMDA reduces the response to both agents in a concentration-dependent manner[208]. Note: Dantrolene is used as a clinical antidote for ryanodine receptor-mediated **malignant hyperthermia** *(see ILG Ca Ca RyR-Caf, entry 17)*
Cytoskeletal proteins	A protein interaction between **actin** and NMDA channel regulatory proteins can affect the **'rundown'** phenotype of NMDAR in native cells	*See Rundown, 08-39*

Table 7. *Continued*

Class and subtype of protein	Key roles/interaction	Regulatory functions/notes
Endo-nucleases[111]	Several Ca^{2+}-activated enzymes may contribute to excitatory amino acid **toxicity**, which may include the activation of various endonucleases†	Regulation of Ca^{2+}-influx appears fundamental to the **'ordered progression'** of gene expression; dysregulation of Ca^{2+} homeostasis is also likely to have longer term effects on cell phenotype or the initiation of **apoptosis**† *(see also 'Ca^{2+}-sensitive proteases', this table)*
G protein-coupled GABA receptors, $GABA_B$ subtypes	For role of $GABA_B$ receptors in low-frequency (~ 5 Hz) induction of LTP *(see Receptor/transducer interactions, 08-49)*	
G protein-coupled glutamate receptors (metabotropic† glutamate receptors, mGluR)	mGluR which activate protein kinase C appear to lower the threshold of induction for LTP *(see review, ref.*[143] *and 'Protein kinase C', this table).* Targeted disruption of the gene encoding $mGluR_1$ in mice having severe deficits in **motor co-ordination** and **spatial learning**[209]	mGluR-null† mice have no gross anatomical or basic electrophysiological abnormalities in either the cerebellum or hippocampus, but show impaired cerebellar long-term depression† and hippocampal mossy fibre long-term potentiation[209] *(see Fig. 1d). See also Receptor/transducer interactions, 08-49*
G protein-coupled muscarinic receptors (M)	Muscarinic receptors which activate protein kinase C appear to lower the threshold of induction for LTP *(see review*[143] *and 'Protein kinase C', this table)*	*See Appendix A – Index of G protein-linked receptors, entry 56*
G protein-coupled opioid receptors, μ-subtypes	**μ-opioid** receptor agonists† potentiate† NMDAR-activated currents in trigeminal neurones of rat medullary slices, probably via activation of protein kinase C *(see this table and ref.*[210]*)*	μ-opioid potentiation† may be a feature of synaptic plasticity† observed in central **pain reception pathways**[210] *(see Phenotypic expression, 08-14, and Channel modulation, 08-44)*

Table 7. *Continued*

Class and subtype of protein	Key roles/interaction	Regulatory functions/notes
$GABA_A$ (inhibitory) **receptor–channels** *(see ELG Cl $GABA_A$, entry 10)*	In most cortical neurones the activation of the NMDARs (and hence the induction of LTP† – *see Phenotypic expression, 08-14*) requires a concomitant reduction of GABAergic† inhibition by low doses of the $GABA_A$ antagonist **bicuculline**[211]. This indicates that in the neocortex the **activation threshold**† of the NMDA mechanism and consequently the susceptibility to LTP†, are strongly influenced by **inhibitory processes**	Morphological damage and psychotomimetic† effects induced by NMDA-active drugs can be prevented by other drugs which act at the gamma-aminobutyric acid ($GABA_A$) receptor–channel complex[212], indicating a functional interaction between these channel types *in vivo (see Receptor antagonists, 08-51, and ELG Cl $GABA_A$, entry 10)*
	For interactions of **GABAergic†, cholinergic†** and **glutaminergic**† pathways in the control of NMDAR-mediated **glutamate toxicity**†, *see Fig. 7*	Monosynaptically evoked *inhibitory* post-synaptic currents in hippocampal pyramidal slices diminish in the presence of CNQX (an AMPA receptor-selective antagonist) and APV *(see Receptor antagonists, 08-51)* following a train of action potentials. However, responses to GABA applied by iontophoresis† do *not* change significantly *(cf. Channel modulation under ELG Cl $GABA_A$, 10-44)*
Guanylate cyclase	**Sustained activation** of guanylate cyclase and accumulation of cyclic GMP has been observed following NMDAR activation in primary cultures of cerebellar granule cells[213]	cGMP formation can occur via Ca^{2+}-dependent stimulation of **nitric oxide synthase** and **lipoxygenase** metabolism of **arachidonate** (released by **phospholipase A_2**). For further details of the apparent **neuro-protective** role of cGMP and nitric oxide, *see refs*[113,214,215] *and Channel modulation, 08-44*

Table 7. *Continued*

Class and subtype of protein	Key roles/interaction	Regulatory functions/notes
Nitric oxide synthase (Ca^{2+}-dependent) (NOS)[113–115]	Inhibitors of NOS appear to block induction of LTP†	*See also 'Guanylate cyclase' this table, and Channel modulation, 08-44*
non-NMDA-type ionotropic glutamate receptors (AMPA- and kainate-selective glutamate receptor–channels)	The co-involvement of NMDAR and non-NMDAR in LTP† induction is well-characterized in the Schaffer collateral-CA1 pathway *(see refs[54, 216] for reviews)*. Generally, non-NMDA (AMPA/kainate receptors) provide **'sufficient' depolarization** for removal of Mg^{2+} block from NMDAR, facilitating the induction of LTP *(see Fig. 1e, f)*	In the **maintenance** phase of LTP, both NMDAR and non-NMDAR currents appear to be enhanced[216], possibly as a consequence of (i) increased pre-synaptic glutamate release[217, 218] and (ii) increased post-synaptic responsiveness of iGluR[219, 220] *See also ELG CAT GLU AMPA/KAIN, entry 07*
Phosphatases (various, Ca^{2+}-sensitive) e.g. Ca^{2+}/calmodulin-dependent phosphatase 2B (calcineurin)[221]	In cell-attached recordings of acutely dissociated adult rat dentate gyrus granule cells, application of **okadaic acid** (a non-selective **phosphatase inhibitor**) prolongs NMDAR-channel openings only at a concentration that inhibits the Ca^{2+}/calmodulin-dependent phosphatase 2B (**calcineurin**), and is ineffective when Ca^{2+} entry through NMDA channels is prevented[222]. In adult dentate gyrus granule cells, calcineurin is activated by calcium entry through native NMDAR channels and shortens the duration of channel openings. Simulated synaptic currents are enhanced following phosphatase inhibition. Application of a calcineurin inhibitor (**FK-506**) mimics the effects of okadaic acid[222]	Calcineurin inhibition prolongs the duration of single NMDA channel openings, bursts†, clusters and superclusters† *Note:* Intracellular dialysis with calcineurin does *not* induce rundown† of NMDAR in rat hippocampal neurones[223] *See also Protein phosphorylation, 08-32*

Table 7. *Continued*

Class and subtype of protein	Key roles/interaction	Regulatory functions/notes
Phospholipases, (various, Ca^{2+}-sensitive) e.g. phospholipase A_2 (PLA_2)[106–109] and phospholipase C[110]	Post-synaptic [Ca^{2+}] elevation activates PLA_2. Inhibitors of PLA_2 block LTP[224, 225] Transient application of arachidonic acid (*see next column*) to hippocampal synapses induces a 'slow-onset' potentiation	*See Fig. 4.* PLA_2 releases arachidonic acid (AA) from membrane phospholipids which may serve as a 'retrograde messenger' (i.e. diffusing from post-synaptic to pre-synaptic loci) *(for details, see Receptor/transducer interactions, 08-49)*. AA stimulates phosphoinositide (PI) turnover. PI turnover has been observed pre-synaptically during LTP induction[226]. AA may stimulate pre-synaptic protein kinase C *(see Fig. 4)*
Proteases (various Ca^{2+}-sensitive) e.g. calpain[227]	Conversion of *transient* NMDAR-mediated Ca^{2+} influx into *persistent* modifications of synaptic strength† in LTP† induction	Ca^{2+}-sensitive proteases include **calpain I**† (μM Ca^{2+}-sensitive proteolysis) and **calpain II**† (mM Ca^{2+}-sensitive proteolysis)[116–120] *(see Protein phosphorylation, 08-32)*
Protein kinase C (Ca^{2+}/phospholipid-dependent protein kinase; PKC)[103–105]	NMDAR-mediated Ca^{2+}-influx which activates PKC directly mediates (i) 'positive modulation' of the NMDAR itself (i.e. increased glutamate sensitivity during the maintenance phase of LTP); (ii) positive modulation of AMPA/kainate receptors *(see this table)* and (iii) activation of Ca^{2+}/calmodulin-dependent protein kinase II *(this table)* There is likely to be a **multiplicity of PKC isoforms**, with post-synaptic kinase (PKC-γ?) activating transiently ($\tau_{1/2}$ ~several min) followed by pre-synaptic kinase activation (PKC-β?) with sustained activity ($\tau_{1/2} < 1$ h)[228]	PKC inhibitors can block LTP induction *after* the **tetanic**† **stimulation phase** *(see Fig. 1b)*. PKC inhibitors (at low doses) block induction of long-term potentiation without affecting short-term potentiation *(see Fig. 1d)* PKC activators (e.g. phorbol esters) induce synaptic potentiation[230] when injected post-synaptically or applied extracellularly[231] *Note:* As reviewed[54], PKC *alone* has been judged insufficient to induce LTP but may convert STP to LTP_1 *(see Fig. 1d)*

Table 7. *Continued*

Class and subtype of protein	Key roles/interaction	Regulatory functions/notes
	PKC increases P_{open} and reduces voltage-dependent Mg^{2+}-block of the NMDAR *(see ref.*[97] *and Blockers, 08-43)*	Primary† sequences of cloned NMDAR exhibit consensus† phosphorylation motifs for PKC *(see [PDTM], Fig. 2 and Protein phosphorylation, 08-32)*
	Currents conducted through heterologously† expressed NMDAR in *Xenopus* oocytes can be potentiated by PKC activators[59, 229]	
Tyrosine kinase; TyrK	Experimental induction of LTP can be blocked by co-application of tyrosine kinase inhibitors[232]	NMDAR activation has been linked to tyrosine phosphorylation of **MAP-2 kinase**†[233]

For further information on the roles of NMDAR **protein phosphorylation** and **G protein-mediated signalling**, *see Protein phosphorylation, 08-32, and Receptor/transducer interactions, 08-49.*

Role of phosphomodulation of NMDAR in development of LTP phenotypes

08-32-04: Direct, positive modulation of NMDA currents by **protein kinase C** (PKC) has been observed, and it has been suggested that PKC activity may determine the **threshold of LTP† induction** *(see review, ref.*[143]*)*. Enhanced kinase activity may underlie the central role of the NMDA receptor–channel complex in **neuronal plasticity**† *(see sections on LTP induction under Phenotypic expression, 08-14).*

NMDAR phosphorylation may be associated with the maintenance of epileptic states

08-32-05: Kindling†-induced epilepsy affects the sensitivity of NMDAR channels to intracellular high-energy phosphates[135] *(see Phenotypic expression, 08-14).*

Mechanism of PKC potentiation of NMDAR responses

08-32-06: In a number of neuronal preparations, **protein kinase C** has been shown to potentiate the NMDA response by increasing the probability of channel openings and by reducing the voltage-dependent Mg^{2+}-block of NMDA receptor–channels[97] *(see Channel modulation, 08-44).*

ELECTROPHYSIOLOGY

Activation

Slow rise and decay times of NMDA-activated currents
08-33-01: NMDAR-channels elicit the **slowly rising**, **slowly decaying** (> several hundred milliseconds open state†) component of excitatory postsynaptic currents (EPSCs†) in response to glutamate *(see Inactivation, 08-37)*. In comparison, **AMPA receptors** mediate the most rapid synaptic excitatory neurotransmission† and conduct mainly Na^+ currents *(see ELG CAT GLU AMPA/KAIN, entry 07)*.

Intrinsic properties of NMDAR–channels determine duration of current
08-33-02: Intrinsic NMDAR channel kinetics determine the **time course** of NMDA receptor-mediated synaptic currents[234,235]. For example, **kinetic responses** of NMDA receptors in excised membrane patches from hippocampus and superior colliculus show similarities to that of the NMDA EPSC†, suggesting the time course† of the NMDA EPSC† reflects **slow NMDA channel properties** in this preparation[46].

'Re-binding' of glutamate to the NMDAR from the synaptic cleft does not occur
08-33-03: Brief pulses of glutamate applied to outside-out membrane patches results in openings of NMDA channels that persist for *hundreds of milliseconds,* indicating that glutamate can *remain bound* for this period[234]. Current rise† and decay† is markedly **temperature-dependent**, indicating that changes in rates of **free transmitter diffusion** cannot alone account for its time course[235].

Glutamate and glycine activate the NMDAR as independent co-agonists
08-33-04: Occupation of a separate, **allosteric 'glycine receptor site'** is also an **absolute requirement** for NMDAR activation *(see Receptor agonists, 08-50)*. The *concentration* of glycine at the synaptic cleft† is *below* a saturated level. In trigeminal neurones, external Ca^{2+} contributes to unusually **high glycine affinities** for NMDA receptors (potentiation† occurring when glycine sites are unsaturated – *see Equilibrium dissociation constant, 08-45*). This form of **Ca^{2+} modulation** may have a role in regulating the NMDA receptor–channel activities during **intensive** or **sustained** neuronal stimulation[236]. Analysis of NMDA channel **activation kinetics** in outside-out patches of cultured hippocampal neurones (following rapid steps into high concentrations of glutamate, e.g. ~200 μM Glu) has determined the activation time course† to be concentration-independent and limited by **transitions**† between the shut (but 'fully-liganded') state and the open state†. Kinetic models which can account for **activation kinetics** following glutamate concentration jumps have been derived for NMDAR in hippocampal neurones[237] *(see Fig. 5)*.

'Temporal summation' of synaptic currents relieves Mg^{2+}-block by depolarization

08-33-05: Single, brief applications of glutamate are sufficient to produce an **extended activated state** of *several hundred milliseconds*, whose transition

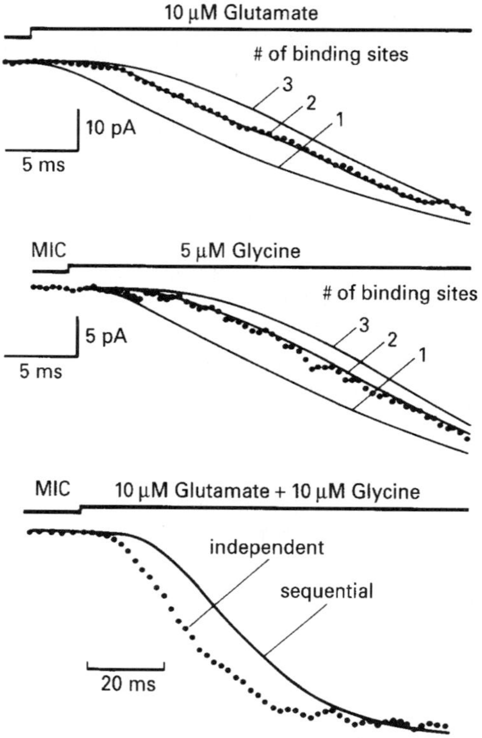

Figure 5. *Models accounting for NMDAR activation kinetics following agonist concentration jumps. At relatively* ***low agonist concentrations*** *(e.g. ~2–10 μM glutamate) a* ***two glutamate-binding site model*** *can account for activation kinetics following glutamate concentration jumps (upper panel)*[237]*. A two glycine-binding site model can also be fitted for channel activation in the* ***continuous presence*** *of glutamate (middle panel)*[237]*. Agonist and co-agonist binding kinetics are better described by an* ***independent co-agonist binding model****, rather than a sequential binding model (lower panel)*[237]*. Data points shown represent approximately the first 30 ms following agonist addition, with fitted model predictions indicated by arrows. MIC denotes addition of 25 mM* ***methoxyindole carboxylic acid*** *used here to eliminate spurious gating due to glycine contamination from solutions (see Receptor antagonists, 08-51). Note: These studies predict the NMDAR to be at least tetrameric, containing* ***four ligand-binding subunits****, assuming a single binding site per subunit*[237]*. (Reproduced with permission from Clements and Westbrook (1991)* Neuron *7: 605–13.) (From 08-33-04)*

to the **closed state** is independent of the conducting state of the channel[234]. This mode of gating causes **temporal summation**† of synaptic inputs† (and consequential depolarization†). *Note:* Summation can explain why 'single shock' stimulation of afferents is not as effective as natural (repetitive) stimulation, since single pulses do not **'sufficiently depolarize'** neurones to relieve the Mg^{2+} block[81] *(for significance, see Fig. 1e and Blockers, 08-43).*

Numbers of NMDA channel openings required to generate an EPSC†
08-33-06: Brief applications of L-glutamate to outside-out† patches from hippocampal neurones in the presence and absence of the open-channel blocker MK-801 have shown that about 30% of the L-glutamate-bound channels are open at the peak of the current[238]. The high probability of opening for NMDA receptor–channels following stimulation by L-glutamate[238] suggests that relatively few channels are required to 'guarantee' a large, localized post-synaptic calcium transient.

Agonists probably remain bound during 'superclustering'
08-33-07: NMDA receptors have an unusual property of binding certain agonists (including glutamate) for a long period of time. This property may partly explain why brief (~1 ms) applications of glutamate (1 mM) produce a **slowly decaying current**, the major component of which has a time constant of ~200 ms. 'Superclustering'† behaviour observed at low glutamate concentrations *(see Single-channel data, 08-41)* may correspond to a single period during which one or more molecules of glutamate are bound to the receptor[240].

Current–voltage relation

Effects of different extracellular ions on I–V relationships close to E_{rev}
08-35-01: The response activated by NMDA agonists† exhibits a **voltage-dependent extracellular Mg^{2+} block**. Ca^{2+}-influx is restricted under resting conditions, but post-synaptic **membrane depolarization** can remove the block. Furthermore, the larger the **driving force**† for Mg^{2+} to penetrate the membrane, the larger the block *(for further details, see Blockers, 08-43).* However, in tight-seal†, whole-cell† recordings of cultured spinal cord and hippocampal neurones, high concentrations (20 mM) of Ba^{2+} and Ca^{2+} display **linear I–V relationships**† (within ±15 mV of the reversal potential) although they do reduce slope conductances. By contrast, extracellular Mn^{2+} ions produces a strong, voltage-dependent block of responses to NMDA, such that even close to the reversal potential†, the NMDA current–voltage relationship† is highly non-linear[241].

The slow voltage-dependence of P_{open} in Mg^{2+}-free solutions
08-35-02: Studies of non-linear† whole-cell† I–V curves† in **Mg^{2+}-free solutions** have shown that NMDA channels in excised patches reversibly shift their P_{open} in a voltage-dependent manner (i.e. they exhibit ~3- to 4-fold greater P_{open} at positive potentials than at rest)[131]. Changes in P_{open} are mainly attributable to shifts in **opening frequency**†: P_{open} changes over a 'very slow' time course (~2–15 min) which underlies the observed

hysteresis† of whole-cell current–voltage curves obtained under non-equilibrium (i.e. non-steady-state†) conditions. The **slow increase in P_{open}** provides a potential **excitotoxic mechanism** in that Ca^{2+}-influx can increase markedly in cells depolarized for prolonged periods of time[131].

Quantitative NMDAR activation models predicting E_{rev} and Ca^{2+}-influx at different voltages and $[Ca^{2+}]_o$

08-35-03: In accordance with a **quantitative model** developed for NMDAR–channels expressed in cultured hippocampal neurones[242], increasing $[Ca^{2+}]_o$ markedly shifts the reversal potential to positive values and *simultaneously* decreases the single-channel conductance at potentials negative to the reversal potential. Using the model, relatively simple quantitative descriptions of calcium permeation† and channel block† by calcium ions can account for observed channel behaviour and accurately predict reversal potentials and magnitudes of calcium influx over a wide range of conditions[242].

Dose–response

Thresholds for activation of glutamate receptor–channel subtypes in mixed populations

08-36-01: In dorsal horn neurones, activation of Ca^{2+}-permeable NMDA receptors evokes intracellular Ca^{2+} transients that are large (~780 nA), rise at a moderate rate, and maximize amplitude† at NMDA concentrations of ~300 μM. Where **mixed subtypes** of glutamate receptors are expressed, glutamate responses at concentrations less than 3 μM are due *exclusively* to NMDA receptor activation[243]. At higher glutamate concentrations, intracellular responses are mediated by *both* NMDA and non-NMDA receptors[243].

Inactivation

Inactivation parameters shape decay times of NMDA receptors

08-37-01: The characteristically **slow decay** of NMDA-mediated EPSCs† appear to be due to (i) **persistence of bound glutamate** and (ii) the **long open state** of the channels. In hippocampal neurones, **repetitive stimulation** of glutamate receptors elicits increasingly smaller ionic currents. For example, in the presence of 2.2 mM $[Ca^{2+}]_o$, repetitive glutamate applications (15 episodes of 4 s/min) elicit progressively smaller currents which stabilize at ~45% of their initial peak value[244]. This '**inter-episode inactivation**' is exacerbated by elevating extracellular Ca^{2+} to 11 mM, and is attenuated by reducing **extracellular Ca^{2+}** to 0.22 mM. Current decay shown during individual stimuli ('**intra-episode inactivation**') is dependent on extracellular Ca^{2+} yet remains stable during repetitive stimulation. Thus, inter- and intra-episode inactivations of NMDAR currents result from two distinct processes triggered by Ca^{2+}. These '**modalities' of inactivation** may arise from Ca^{2+} binding either to the receptor or to closely associated regulatory proteins[244].

Distinctions between NMDA receptor–channel desensitization and inactivation

08-37-02: Two distinct mechanisms have been suggested for modulation of NMDA receptors by **intracellular Ca^{2+}**[245]. *Desensitization* of NMDA receptors is induced when both intracellular Ca^{2+} is increased and NMDA receptors activated by agonist. *Note:* Two types of **steady-state desensitization** for the NMDA agonists aspartate and glycine have been shown in isolated rat hippocampal neurones[246]. *Inactivation* of NMDA receptors is produced by increased levels of intracellular Ca^{2+} but does not require NMDA receptor activation for induction[245] *(see below).*

'Down-regulation' of post-synaptic Ca^{2+}-entry

08-37-03: Studies of **calcium-dependent inactivation** of NMDA channels in cultured rat hippocampal neurones have suggested a mechanism for **down-regulation** of post-synaptic calcium entry during sustained synaptic activity[247]. In normal $[Ca^{2+}]_o$ (1–2 mM) and 10 μM glycine, macroscopic currents evoked by 15 s applications of NMDA (10 μM) inactivate slowly following an initial peak. At −50 mV in cells buffered to $[Ca^{2+}]_i < 10^{-8}$ M with 10 mM EGTA, the inactivation time constant τ_{inact} is ~5 s. **Inactivation** does *not* occur at membrane potentials of +40 mV and is absent at $[Ca^{2+}]_o \leqslant 0.2$ mM, suggesting inactivation results from transmembrane Ca^{2+}-influx[247]. In these studies, the percentage inactivation and τ_{inact} were dependent on $[Ca^{2+}]_o$.

Transient Ca^{2+}-induced inactivation

08-37-04: In cell-attached† patches, transient increases in $[Ca^{2+}]_i$ following cell depolarization† also result in **inactivation** of NMDA channels without altering the single-channel conductance. Ca^{2+} entry through local **voltage-gated Ca^{2+} channels** may substitute for (or aid) this inactivation process, although Ca^{2+}-entry through NMDA channels is more efficient. $[Ca^{2+}]_i$ transients may induce NMDA channel inactivation by binding to either the channel or an 'associated' regulatory protein to alter channel gating[247].

Mechanism of NMDA receptor reactivation following inactivation

08-37-05: Under conditions of brief (1 ms) glutamate application (1–10 mM), neurones in rat visual cortex produce a response that mimicks the time course of **miniature EPSCs**† (mEPSCs). The rate of onset of desensitization is much slower than the decay rate of the response to a brief application of glutamate, implying that the decay of mEPSCs reflects channel closure into a state readily available for **reactivation**[248]. Furthermore, at steady state, MK-801 completely blocks subsequent responses to NMDA, suggesting that 'inactivated' channels *(see above)* can **re-open** at steady state. Inaçtivation is fully reversible in the presence of ATP but is *not* blocked by inhibiting phosphatases or proteases. For a description of the complex closed-time distributions of the NMDAR *see Single-channel data, 08-41.*

Transient Cl^- currents activated by Ca^{2+}-influx through NMDA receptors expressed in Xenopus *oocytes*

08-37-06: NMDA receptors expressed in ***Xenopus* oocytes** injected with rat brain RNA elicit a rapid inward current on NMDA application that decays in several seconds to a relatively stable level ('apparent desensitization'). However, the early Ca^{2+}-dependent transient component can be evoked more than once during single applications of NMDA, suggesting that the receptor does *not* desensitize[249]. A variety of chloride channel blockers 'almost eliminate' the transient component and, in addition, inhibit the plateau current. Thus it has been proposed *for oocyte expression systems* that a significant portion of the NMDA current recorded is carried by a transient inward **calcium-activated chloride current** (Cl_{Ca})[249] *(see ILG Cl Ca, entry 25).*

Rundown

Relationship of cytoskeletal element function to 'rundown'
08-39-01: The ATP- and calcium-dependent **rundown**† of NMDA channels can be prevented when **actin depolymerization** is blocked by **phalloidin**[250]. *Comparative note:* Rundown of AMPA/kainate receptors is unaffected by phalloidin. Application of **cytochalasins** (which enhance actin-ATP hydrolysis) induce NMDA channel rundown, whereas **taxol** or **colchicine** (which stabilize or disrupt **microtubule assembly**) have no effect. **Protease inhibitors** also have no effect. Calcium and ATP can therefore influence NMDA channel activity by altering the state of **actin polymerization**. To explain these findings, a model has been proposed in which actin filaments '**compartmentalize**' an NMDAR channel regulatory protein[250].

Sensitivity of rundown to intracellular Ca^{2+}
08-39-02: Increases in **intracellular calcium** lead to NMDA channel rundown† during whole-cell† recording of rat hippocampal neurones by reducing the **open probability** of the NMDA channel[223].

ATP regeneration retards rundown
08-39-03: Although high concentrations of **ATP** (or the inclusion of **'ATP-regeneration' systems**†) in the patch pipette[247] can prevent or retard rundown†, their action has been suggested *not* to be a direct result of receptor phosphorylation[223].

Selectivity

Activation of NMDA receptors mediates calcium influx
08-40-01: Among the ionotropic† glutamate receptors (iGluR), NMDA receptors are thought to mediate their physiological response mainly through the influx of **extracellular calcium** – cf. AMPA- and kainate-selective iGluR which (with some exceptions) mainly mediates Na^+-influx *(see special cases in Selectivity under ELG CAT GLU AMPA/KAIN, 07-40).* **High Ca^{2+}-permeability**† of recombinant NMDAR has been directly demonstrated[6].

The structural basis of Ca^{2+}-selectivity in iGluR–channels
08-40-02: NMDA receptor–channels display selective permeability† for Ca^{2+}

over Mg^{2+} ions dependent on the placement of an asparagine (N) residue at the equivalent locus to the **glutamine/arginine (Q/R) site** within the M2 domain of 'non-NMDA' channels. For further details of structure/function at this site, *see Domain functions, 08-29, and Gene organization, 07-20, Domain functions, 07-29, Current–voltage relation, 07-35, Selectivity, 07-40, and Blockers, 07-43, under ELG CAT GLU AMPA/KAIN.*

Direct comparisons of Ca^{2+}-influx mediated by NMDA- versus non-NMDA receptor–channels

08-40-03: The **Ca^{2+} fraction** of the ion current flowing through glutamatergic NMDA and AMPA/kainate receptor–channels has been directly compared in forebrain neurones of the medial septum[251]. At negative membrane potentials (extracellular free Ca^{2+} concentration of 1.6 mM) the Ca^{2+} fraction of the current through the NMDA receptor–channels was only 6.8%. *Comparative note:* A high fractional Ca^{2+} current of 1.4% has been determined for the linearly conducting AMPA/kainate receptor–channels found in these neurones[251].

Influence of $[Ca^{2+}]_o$ and $[Na^+]_o$ on the conductance mechanism of the NMDAR

08-40-04: In tight-seal, whole-cell recordings of spinal cord and hippocampal neurones in cell culture, raising the extracellular calcium concentration shifts the reversal potential† of responses to NMDA in the depolarizing direction (calculated $P_{Ca}/P_{Na} \sim 10.6$, with extracellular Na^+ held constant at 105 mM). There is an apparent increase in P_{Ca}/P_{Na} on lowering extracellular Na^+, which may result from interaction of permeant† ions within the channel[241] *(see also Current–voltage relation, 08-35).*

Single-channel data

'Superclustering' properties of NMDA channel openings at low glutamate concentrations

08-41-01: Single-channel recordings of NMDA receptors in adult rat hippocampus (CA1) at **low glutamate concentrations** (20–100 nM glutamate with 1 μM-glycine without extracellular divalent cations) show NMDAR activations consist of **clusters**† of channel openings in this preparation[239]. Sample distributions of the length of these clusters have mean time constants of 88 μs (45% of openings), 3.4 ms (25%) and 32 ms (30%). **Long clusters** contain short-, intermediate- and long-duration openings as well as **subconductance† openings**, with the average open probability† within clusters averaging ~0.62. Three components are evident in distributions of the number of openings per cluster, having mean values of 1.22, 3.2 and 11 openings per cluster[239].

Complexity of NMDAR closed time distributions

08-41-02: Single-channel recordings of NMDA receptors in adult rat hippocampus (CA1) at low glutamate concentrations (20–100 nM glutamate with 1 μM-glycine without extracellular divalent cations) display **complex closed time distributions**, requiring fitting of five exponential components for adequate description[239]. Of these five components, at least three, with time constants of 68 μs, 0.72 ms and 7.6 ms (relative areas of 38, 12 and 17%) represent **gaps** within single activations of the receptor[239].

Sublevel transitions reflect NMDAR subunit composition
08-41-03: Single-channel properties for NMDAR have been described in a number of preparations[21, 187, 252–260]. Following expression of cloned NMDAR, single-channel conductances and characteristic patterns of sublevel transitions can be used as **diagnostic criteria** for subunit composition[9]. Table 8 lists some single-channel characteristics resulting from defined subunit combinations compared to those of native cells.

Voltage sensitivity

Voltage-dependent extracellular Mg^{2+}-block of the NMDAR–channel
08-42-01: The block of NMDA receptor–channels by **extracellular Mg^{2+} ions** exhibit **strong voltage-dependence**†, and allows the NMDAR to function as a **molecular 'coincidence detector'** *(for further details, see Fig. 1e, f and Blockers, 08-43)*. In *Mg^{2+}-free* solutions, slow voltage-dependent changes in channel open-state probability underlie **hysteresis**† of NMDA responses[131] *(see Current–voltage relation, 08-35)*.

Voltage-sensitive block by arcaine
08-42-02: The *voltage-dependent* binding site for the polyamine antagonist **arcaine** *(see Channel modulation, 08-44)* is distinct from either the phencyclidine/MK-801 site or the voltage-dependent channel site for magnesium[261]. The voltage-sensitivity of **arcaine block** indicates the binding site is located in a region of the NMDA receptor ionophore complex capable of **sensing transmembrane potential**[261]. Arcaine has also been shown to act as an NMDAR open-channel blocker†, an action that is *independent* of the polyamine site *(see Receptor antagonists, 08-51)*. For models of voltage- and use-dependent NMDAR blockade by the **dissociative anaesthetics** ketamine, phencyclidine (PCP) and dizocilpine, *see Blockers, 08-43*. For the voltage-dependence of **polyamine modulation**, *see Channel modulation, 08-44.*

PHARMACOLOGY

Blockers

Under resting (hyperpolarized) conditions, NMDAR–channels are blocked by extracellular Mg^{2+}
08-43-01: At negative membrane potentials, **Mg^{2+}-block** curtails transmitter-evoked ion conductance through NMDAR[262, 263]. Thus glutamate release from pre-synaptic sites is unable to activate the channel unless the post-synaptic membrane is **sufficiently depolarized** to remove the block *(for an illustration of this principle, see Fig. 1e, f)*. The larger the **driving force**† for Mg^{2+} to penetrate the membrane, the larger the block. Variant strengths of

Table 8. *Selected single-channel characteristics resulting from defined NMDAR subunit combinations compared to those of native cells (From 08-41-03)*

Subunit combination/ preparation	Description/features	Refs
NR1/NR2A NR1/NR2B	NR1–NR2A and NR1–NR2B combinations both have 50 pS openings, brief 40 pS sublevels (in 1 mM external Ca^{2+}), with similar mean lifetimes and frequencies. NR1–NR2A and NR1–NR2B combinations also show close quantitative resemblance to the channels of hippocampal CA1 and dentate gyrus cells and of cerebellar granule cells, except that the NR1–NR2A combination has a lower glycine sensitivity than the native channels	*9*
NR1/NR2C	The NR1–NR2C combination produces a channel with 36 pS and 19 pS conductances of similar (brief) duration. NR1–NR2C channels closely resemble the 38–18 pS channels that have been observed (together with 50 pS channels, *see below*) in large cerebellar neurones in culture	*9*
NR1 alone	Low current amplitudes observed following expression of subunit NR1 in *Xenopus* oocytes predict that 'natural' NMDA receptors occur in hetero-oligomeric configurations	*10*
Native NMDAR, rat hippocampus	Single-channel recordings of native NMDA receptors in adult rat hippocampus (CA1) at low glutamate concentrations (20–100 nM glutamate with 1 μM glycine without extracellular divalent cations) show two main conductance levels of 50 pS and 40 pS (extracellular Ca^{2+} at 1mM). Approx. 80% of openings are to the 50 pS conductance level. Single-channel conductances increase as extracellular Ca^{2+} is reduced. Single-channel open time distributions can be described by three exponential components of 87 μs, 0.91 ms and 4.72 ms (relative areas of 51%, 31% and 18%) with the majority of long openings being to the large conductance level	*239*
Native NMDAR, Mg^{2+}-free solutions	In Mg^{2+}-free solutions, increases in temperature between 14 and 24°C, increased the NMDAR–channel conductance with a Q_{10} of ~ 1.6 while the mean open times decreased with a Q_{10} of ~ 2[257]	*257*

Mg^{2+}-blockade of recombinant NMDARs has been demonstrated directly[6]. Open-channel block† of the NMDAR by Mg^{2+} may explain its observed **'neuroprotective'** and **anti-epileptic** effects. *Notes:* 1. Voltage-dependent block of the NMDA receptor–channel occurs at concentrations 'well below' the (millimolar) extracellular concentrations found in the CNS. 2. Mg^{2+} is relatively weak as an open-channel blocker because it leaves the channel relatively quickly *(cf. channel occupancy times of memantine and MK-801, below).* 3. Proper functioning of pre- and post-synaptic **Na^+/K^+-ATPases** are crucial for operation of voltage-dependent Mg^{2+}-blockade and prevention of neurotoxicity *(see Protein interactions, 08-31).* For further implications of Mg^{2+}-block, *see also Phenotypic expression, 08-14, Current–voltage relation, 08-35, and Channel modulation 08-44).*

Behaviour of NMDAR in Mg^{2+}-free solutions and with other extracellular ions

08-43-02: In Mg^{2+}-free solutions, the opening and closing of the NMDAR–channels leads to rectangular current pulses, the mean duration of which varies little with membrane potential[259]. Following addition of Mg^{2+}, the single-channel currents recorded at negative potentials appear in bursts of short openings separated by brief closures *(compare with features in Table 8).* While Co^{2+} (and to a lesser extent Mn^{2+}) mimic effects of Mg^{2+} on the NMDAR–channel, Ca^{2+}, Ba^{2+} and Cd^{2+} do not[259] *(see below).*

Divalent cation block versus permeability through the NMDAR–channel

08-43-03: NMDA receptors are distinct from other glutamate receptor–channels because of their **high Ca^{2+}-permeability**† and inhibition by selective cationic channel blockers such as **Zn^{2+} ions** (typically ~5–100 μM)[264, 265], **dizocilpine** (MK-801, typically ~1 μM) and **Mg^{2+} ions** (typically ~1–5 mM, *see above*)[266]. The 'blockage/permeability' distinction between Mg^{2+}-like and Ca^{2+}-like divalent cations may correspond to a difference in the **speed of exchange** of the water molecules surrounding the cations in solutions. Thus, it is possible that permeation† occurs for all the divalent cations, but is slower for those which are **slowly dehydrated**[259].

Differential sensitivity to open-channel blockers by certain NMDAR subunit combinations

08-43-04: Recombinant† NMDAR subunit combinations ζ1/ε1 from mouse (~NR1/NR2A of rat) or ζ1/ε2 (~NR1/NR2B of rat) are more sensitive to the open-channel† blocker† **dizocilpine** (MK-801) than the subunit combinations ζ1/ε3 (~NR1/NR2C of rat) or ζ1/ε4 (~NR1/NR2D of rat)[267]. *Note:* By definition, open-channel blockers such as **ketamine**, **dextrorphan**, **PCP** and **dizocilpine** can only gain access to the channel when agonists have *first activated* (opened) the channel (in consequence, certain open-channel† blockers† may be of value in selective blockade under pathological conditions, where high concentrations of glutamate would inhibit channel populations to a relatively greater extent[129, 267]).

Modelling of NMDAR blockade by dissociative anaesthetics

08-43-05: Voltage- and use-dependent blockade of NMDA receptor currents by the **dissociative**† **anaesthetics ketamine**, **phencyclidine** (**PCP**) and **dizocilpine** (MK-801) has been analysed[268]. Following the assumptions of the 'guarded receptor hypothesis' (a model used to interpret action of these anaesthetics, *see below*)[268], the estimated **reverse rates** for anaesthetic binding are independent of blocker concentration while **forward rates** increase with concentration. Changing the level of positively charged **ketamine** (pK_a 7.5) 10-fold (by changing pH from 6.5 to 8.5) causes a corresponding change in the forward rate but has no effect on the reverse rate. The **voltage-dependence of blockade** can largely be accounted for by reductions of the reverse rates of binding by depolarization†. Differences in potency can be accounted for by differences in the reverse rate constants, which increase at positive potentials[268]. Important assumptions of the 'guarded receptor hypothesis' are (i) that NMDA receptors are maximally activated at the peak of their response with a P_{open} approaching 1; (ii) that there is no receptor desensitization and (iii) that the blocking drug only associates with, or dissociates from, receptor–channels which have been activated by agonist (i.e. are open). *Note:* PCP, dizocilpine and similar drugs also affect several other functions and targets in the CNS – as reviewed in ref.[269], these include inhibition of **noradrenaline** and **dopamine uptake**, inhibition of **acetylcholinesterase**†, antagonism of **muscarinic receptors**, inhibition of activated **nicotinic receptor** and **delayed rectifier K^+ channels**.

'Protective' effects of NMDAR blockers against calcium neurotoxicity

08-43-06: PCP and **ketamine** 'protect' against brain damage in neurological disorders such as **stroke**†. However, these agents have **psychotomimetic**† properties in humans and damage neurones in the cerebral cortex of rats, although morphological damage can be prevented by **anticholinergic drugs** or by **diazepam** and **barbiturates**, which act at the $GABA_A$ receptor[212]. Furthermore, **dizocilpine** and similar drugs show a pattern of **'build-up'** for blockade, such that following binding of antagonist molecules, they leave the channel only very slowly ($t_{1/2}$ > 1 h). Notably, open-channel blockers which are well-tolerated clinically, such as **memantine** *(see below)* leave the channel promptly ($t_{1/2}$ > 5 s at micromolar concentrations)[270]. $GABA_A$-selective drugs have been observed to reduce the psychotomimetic† symptoms caused by **ketamine**.

08-43-07: Memantine, an **adamantane**, derivative related to the anti-viral drug **amantadine** has been shown to be an open-channel† blocker† for NMDA receptors. Adamantane is well-tolerated clinically unlike other open-channel blockers like MK-801 (dizocilpine)[270]. *Note:* The anti-viral drug **amantadine** is also used in the treatment of Parkinson's disease, but is considerably less potent than memantine at clinically tolerated doses.

Non-selective suppression of excitatory amino acid receptor–channel currents

08-43-08: In oocytes expressing mouse brain mRNA, **enflurane** at an anaesthetic concentration (1.8 mM) inhibits NMDA-, AMPA- and kainate-induced currents by 29–40%, 30–33% and 20–27%, respectively, suggesting that all three glutamate ionotropic receptors are susceptible to suppression by **inhalational anaesthetics**[271]. Inhibition by enflurane is *independent* of the concentrations of the agonists (NMDA, AMPA and kainate) or the NMDA co-agonist (glycine), suggesting that enflurane inhibition does *not* result from a competitive interaction at glutamate- or glycine-binding sites. Enflurane also suppresses the oscillation and 'apparent desensitization' of NMDA currents, suggesting an inhibition of Ca^{2+}-influx through the NMDA channel[271].

Dual suppression of ELG-channel and voltage-gated channel currents by morphinans

08-43-09: The morphinan **dextromethorphan** and the related molecule **dextrorphan** block NMDA-induced currents and voltage-operated inward currents in cultured cortical neurones and PC12 cells[272]. Dextromethorphan is at least 100 times more potent ($IC_{50} \sim 0.55$ μM) as a blocker of the current induced by NMDA in cortical neurones than for voltage-gated Ca^{2+} channels in the same preparation ($IC_{50} \sim 80$ μM). Notably, this class of blockers are well-tolerated clinically[129]. The **neuroprotective** effect of dextromethorphan (which occurs in a concentration range of 10–100 μM) may therefore be due to a complete blockade of the NMDA receptor–channel and a partial inhibition of voltage-dependent Ca^{2+} and Na^{+} channels[272].

Non-selective channel block by TEA

08-43-10: The organic cation **tetraethylammonium** (TEA, 1–5 mM) antagonizes NMDA responses in cultured mouse cortical neurones in a concentration-dependent manner. TEA causes voltage-dependent decreases in single-channel conductance and the frequency of channel events is also decreased. These effects are not accompanied by any change in average channel open time†[273].

Channel modulation

Protein kinase C modulation of NMDAR channel currents in pain reception

08-44-01: Phosphorylation by protein kinase C (PKC) has been reported to enhance NMDA receptor-mediated glutamate responses in a number of preparations[97,143,210,229]. For example, in the trigeminal subnucleus caudalis, a centre for processing pain-sensory (**nociceptive**) information from the orofacial areas, a **μ-opioid receptor** agonist causes a sustained increase in NMDA-activated currents by activating intracellular PKC *(see also Phenotypic expression, 08-14)*. Protein kinase C appears to reduce the voltage-dependent Mg^{2+}-block in **wind-up**†[97] (e.g. triggering of a dramatic increase in discharge following tissue injury or repetitive stimulation of small-diameter afferent† fibres *(see Phenotypic expression, 08-14)*. Thus protein kinase C potentiates NMDA responses by increasing P_{open} due to reductions in Mg^{2+}-block[97]. A role for **metabotropic† glutamate receptors** in PKC modulation of NMDA-mediated processes has also been described[229] *(see Channel designation, 08-03, and Protein interactions, 08-31)*.

Modulation of LTP processes by protein kinase activation

08-44-02: In the post-synaptic cell, both activation of **calmodulin** and **kinase**† activity are required for the generation of LTP[216,274] *(see Phenotypic expression, 08-14). Extracellular* application of **protein kinase inhibitors** to the hippocampal slice preparation has been shown to block the induction of LTP. *Intracellular* injection of the protein kinase **inhibitor H-7** into CA1 pyramidal cells also blocks LTP, as does injection of the calmodulin antagonist **calmidazolium**[274]. Continuous influx of Ca^{2+} through the NMDAR can be prevented by prior treatment with the protein kinase C inhibitor **sphingosine**[105]. LTP is also blocked by injection of synthetic peptides characterized as potent **calmodulin antagonists** presumably by inhibiting CaM-KII auto- and substrate phosphorylation *(see Protein phosphorylation, 08-32).*

Negative and positive modulation of NMDAR by oxidizing and reducing agents

08-44-03: NMDA responses in several different neuronal preparations are 'substantially decreased' following exposure to the disulphide **reducing**† **agent dithiothreitol** (DTT, 0.1–10 mM), while **oxidation** with **5,5-dithiobis-2-nitrobenzoic acid** (DNTB, ~500 μM) can potentiate† the magnitude of the response[275,276]. Modification of the NMDA response by either oxidation or reduction at this **redox modulatory site**[275] does *not* appear to affect the *pharmacological* properties of the receptor–channel complex. Since **redox state**† of the native NMDAR varies widely among neurones, regulation of NMDA-activated functions by either **reduction** or **oxidation** may operate *in vivo*[275,276] *(see also below).*

Mechanism of NMDA receptor blockade by nitric oxide at the redox modulatory site

08-44-04: Clinically important **nitroso compounds** that generate **nitric oxide** (NO•, where the dot signifies one free electron) have been shown to inhibit responses mediated by the NMDA-selective glutamate receptor on rat cortical neurones *in vitro*[129,277]. It has been proposed that **free sulphydryl groups** on the NMDAR react to form one or more ***S*-nitrosothiols** in the presence of nitric oxide. For example, reaction of NMDAR free thiol groups with **nitroglycerin** via ***S*-nitrosylation** has been shown to modulate protein function in an analogous manner to protein phosphorylation† or acylation†[278]. If **vicinal**† **thiol groups** react in this way, they may form **disulphide bonds**, constituting the **redox modulatory site**† of the receptor[275]. *Note:* Formation of disulphide bonds would result in reduction of Ca^{2+}-influx in response to NMDA. Thus, reaction with nitric oxide appears to protect cells from NMDA receptor-mediated neurotoxicity[277]. Conversely, reducing conditions favour the formation of free thiol (-SH) groups, and this condition is present following stroke†[279], being associated with both a net increase in Ca^{2+}-influx through NMDAR channels[280–282] and higher apparent neurotoxicity[281] *(see also Phenotypic expression, 08-14).*

'Reversible' sulphydryl redox modulation of NMDAR in rat cortical neurones
08-44-05: Reversal of the effects of **sulphydryl**† **oxidizing agents**† by DTT *(see above)* can occur over a rapid time course ($t_{1/2} \sim 0.6$ min). *Spontaneously oxidized* receptors can be further oxidized with DTNB and 'fully reduced' with DTT. When the **redox modulatory site**[275] or sites are alkylated with ***N*-ethylmaleimide** (NEM, 300–500 μM) following reduction with DTT, responses become '**permanently potentiated**†' and largely insensitive to oxidation by DTNB. Blocking effects of protons and potentiating† actions of glycine are unaffected by **alkylation**†, and Zn^{2+} and Mg^{2+} ions produce a significantly weaker block of the NMDA whole-cell response[283].

Modulation of NMDAR–channels by (endogenous) zinc ions
08-44-06: Both **pro-convulsant**† and **depressant**† actions of **Zn^{2+} ions** have been reported *(see ref.[265])*. Zinc ions are potent non-competitive antagonists of NMDA responses in cultured hippocampal neurones at relatively high (micromolar) concentrations[265]. Unlike Mg^{2+}, the effect of Zn^{2+} is not voltage-sensitive between −40 and +60 mV, suggesting that Zn^{2+} and Mg^{2+} act at distinct sites. Zn^{2+} could also modulate neuronal excitability because it is present at **high concentrations** in brain, especially the synaptic vesicles of **mossy fibres** in the hippocampus and is released with neuronal activity[265]. *Comparative note:* Zn^{2+} ions also antagonize responses to the inhibitory transmitter GABA in the hippocampus *(see ELG Cl GABA$_A$, entry 10)*.

NMDAR-potentiating effects of Zn^{2+} ions
08-44-07: Zinc ions may also be *positive modulators* of NMDA receptors in certain regions of the brain[271]. Analysis of Zn^{2+} modulation of currents through homomeric† receptors assembled from different splice variants of NR1 subunit show that, in addition to its well-characterized inhibitory effect at high concentrations *(see above)*, Zn^{2+} potentiates agonist-induced currents at submicromolar concentrations ($EC_{50} = 0.50$ μM). Potentiation† is observed only with a subset of NR1 splice variants[193]. Zn^{2+} potentiation is rapidly reversible, voltage-independent† and non-competitive† with glutamate or glycine. Zn^{2+} potentiation is mimicked by Cd^{2+}, Cu^{2+} and Ni^{2+}, but not by Mn^{2+}, Co^{2+}, Fe^{2+}, Sn^{2+} or Hg^{2+} [193]. *Note:* For a description of the molecular 'loci' affecting Mg^{2+}-block phenotypes on the NMDAR, *see Domain functions, 08-29.*

Autacoid modulation of NMDAR
08-44-08: The ether phospholipid **platelet-activating factor** (PAF, an ether phospholipid **autacoid**†) induces a stable, concentration-dependent long-term potentiation† (LTP) in hippocampal slices[284]. PAF-induced LTP is blocked by antagonists of the PAF and NMDA receptors, but the former antagonists do not block LTP induced by high-frequency stimulation. Facilitation† induced by PAF cannot be reversed by PAF receptor antagonists[284]. For an introduction to long-term potentiation phenotypes associated with the NMDAR, *see Phenotypic expression, 08-14.*

Complex modulation of NMDAR-channels by polyamines

08-44-09: Several biochemical and electrophysiological studies *(reviewed in refs.*[285–287]*)* suggest that a recognition site exists on the NMDAR for **endogenous polyamines** (e.g. normal catabolic intermediates of **arginine**† like **putrescine**, **spermidine** and **spermine**). The **polyamine site** is distinct from binding sites for glutamate, glycine, Mg^{2+}, Zn^{2+}, and open-channel blockers. **Polyamines** increase the binding of open-channel blockers but can also increase NMDA-elicited currents in cultured neurones, suggesting they may play a role in '**excitotoxic**†' responses following neuronal damage. **Polyamine segments** are also associated with several **neurotoxic venoms** from invertebrates *(see Receptor antagonists, 08-51).* Spermine and spermidine have been described as 'agonists' at the **polyamine recognition site**, which may modulate excitatory synaptic transmission[285]. Further features of polyamine modulation of NMDA receptors are described in *Table 9.*

Modulation of NMDAR by the co-agonist glycine

08-44-10: Glycine-evoked potentiation† of NMDA receptor activity is accompanied by **reduced desensitization**†[293]. Dose–response analysis for the glycine-sensitive activation of NMDA receptors at +60 mV reveals a 3–5-fold increase in apparent affinity for glycine in the presence of 1 mM spermine. This increase in affinity for glycine is accompanied by a 3.3-fold decrease in the rate of development of **glycine-sensitive desensitization**, and a 2.4-fold decrease in the rate of dissociation of glycine from NMDA receptors (while the rate constant for dissociation of NMDA is not reduced[290]).

Alcohol-inhibition of NMDAR currents

08-44-11: Potency of several **alcohols** for inhibiting NMDA-activated currents are correlated with their **intoxicating potency**†, suggesting that responses to NMDA receptor activation may contribute to the neural and cognitive impairments associated with intoxication[294]. NMDA-activated currents are inhibited by **ethanol** over a concentration range that produces intoxication[295, 296]. Ethanol may inhibit the NMDA-activated ion current by interaction with a hydrophobic site on the NMDA channel protein[294]. *Comparative note:* Ethanol effects a number of receptor-stimulated and voltage-dependent ion fluxes with differing sensitivity. For an illustration of the relative potentiating effects on $GABA_A$-mediated Cl^- flux and the relative depressive effects on NMDA-, kainate- and voltage-gated Ca^{2+}-flux, *see Fig. 5 under ELG Cl $GABA_A$, entry 10.*

Non-selective inhibition of NMDAR-mediated calcium influx by dihydropyridines

08-44-12: The dihydropyridine **nitrendipine** suppresses NMDA/glycine-mediated calcium influx by a rapid and direct interaction with the NMDA receptor–channel complex[297]. Thus nitrendipine may exhibit **anticonvulsant** and **neuroprotectant** activity via a combined ability to modulate *both* NMDA-associated ion channels and L-type voltage-sensitive calcium channels[297] *(see Blockers under VLG Ca, 42-43).*

Table 9. *Some features of polyamine modulation of NMDA receptors (From 08-44-09)*

Feature	Description	Refs
'Concentration-dependent' effects of spermine modulation	Spermine can produce both potentiation *or* block of NMDA responses by binding to multiple sites in a potential-sensitive manner. Spermine potentiates the action of NMDA at micromolar concentrations but is less effective at millimolar concentrations[288]. At low concentrations (1–10 μM) spermine enhances NMDA receptor current in cultured cortical neurones by increasing channel opening frequency. At higher concentrations (> 10 μM), it produces additional voltage-dependent decreases in channel amplitude and average open time which limit its 'enhancing' action	289, 288
Voltage-dependence of spermine potentiation	Concentration-jump responses to 100 μM NMDA in the presence of 10 μm glycine reveals *potentiation* by 3 mM spermine at a membrane potential of +60 mV, but *depression* at −120 mV	290
Separate mechanisms of spermine potentiation or block	Potentiation results from both an increase in apparent affinity for glycine as well as an increase in maximum amplitude of responses to NMDA recorded in the presence of a saturating concentration of glycine. Spermine produces a voltage-dependent open-channel block of NMDA currents at high concentrations (> 100 μM) but has no effect on the block by the putative polyamine site competitive antagonist arcaine *(see Blockers, 08-43)*	290
Examples of block by other polyamines	The polyamines diethylenetriamine and 1,10-diaminodecane produce voltage-dependent block of responses to NMDA, with apparent equilibrium dissociation constants at 0 mV of 0.75 and 2.93 respectively. Block is voltage-dependent, and most likely due to binding of polyamines to sites within the ion channel *(see Voltage sensitivity, 08-42)*	291
Spermine modulation of co-agonist affinity	Spermine increases the affinity of the NMDAR for glycine	290

Comparative note: Polyamine modulation is not unique to NMDAR channels. As reviewed in ref.[287], polyamines modulate several Cl^- conductances, including Ca^{2+}-activated Cl^- channels *(see ILG Cl Ca, entry 25)* and $GABA_A$ receptor–channels *(see ELG Cl $GABA_A$, entry 10)*. Polyamine toxins have also been shown to modulate nicotinic acetylcholine receptor–channels *(see ELG CAT nAChR, entry 09)*, non-NMDA iGluRs *(see ELG CAT GLU AMPA/KAIN, entry 07)* and several subtypes of voltage-gated calcium channels *(see VLG Ca, entry 42)* and potassium channels *(see VLG K and INR K series entries)*. Important roles for polyamines have also been defined in the control of gene expression during cell differentiation and proliferation, and can mediate both increases and decreases of phosphorylation via polyamine-dependent protein kinases[292].

Modulation of NMDAR–channel fractional open time by neurosteroids
08-44-13: NMDAR currents in hippocampal neurones are approximately doubled in the presence of the **neurosteroid pregnenolone sulphate** (PS, 100 μM)[298]. The dose–response curve of PS action shows significant potentiation† at concentrations greater than 250 nM and shows a half-maximal effect at ~ 29 μM. Maximum potentiation† is reached within 25 s, with the potentiation† being fully reversed with 60 s of washout. At saturating concentrations of NMDA and glycine co-agonists, PS does *not* change the affinity between the co-agonists and the NMDA receptor. PS potentiates the **fractional open time**† of NMDA channels by increased frequency of openings of single NMDA-activated channels but does *not* affect single-channel conductances[298].

pH regulation of NMDA channel responses
08-44-14: Responses of mouse hippocampal neurones to NMDA or to low concentrations of glutamate (recorded in the absence of Mg^{2+} and with glycine in the extracellular superfusion solution) are antagonized by acidic pH and potentiated† by **alkaline extracellular solutions**[299,300]. Decreasing pH from 7.3 to 6.0 reduces NMDA responses to 33 ± 2% and an increase in pH from 7.3 to 8.0 potentiates it to 141 ± 6%[299].

Indirect modulation of hippocampal NMDAR by tolbutamide
08-44-15: The sulphonylurea **tolbutamide** (500 μM) has been shown to reversibly increase peak amplitudes and the steady-state† levels of NMDA- but not kainate-evoked iGluR currents in cultured hippocampal neurones[301]. The effect is observed at both low and saturated concentrations of glycine, and the affinity† of the NMDAR for glycine does not change in the presence of tolbutamide. Although tolbutamide has been generally characterized as a blocker of **K_{ATP}** *(see INR K ATP-i, entry 30)*, the action of tolbutamide on the NMDA-activated current is *not* mediated by K_{ATP} channels since the effect persists in the presence of intracellular CsCl at concentrations which induce total block† of all K^+ channels. Tolbutamide may therefore additionally modulate **intracellular messengers** influencing NMDAR–channel activity in this preparation *(see ref.[301])*.

Equilibrium dissociation constant

Similar [³H]-MK-801 binding in native cells and heterologous† NR1/NR2A combinations
08-45-01: Cells transiently co-expressing NR1 and NR2A yield a 10-fold increase in the number of **[³H]-MK-801-binding sites** compared to channels expressed from homomeric assemblies. Similar affinities for [³H]-MK-801 have been measured in HEK-293 cells transiently expressing NR1/2A combinations as in native adult rodent brains[12].

External Ca^{2+} contributes to 'unusually high' glycine affinities for NMDA receptors
08-45-02: The affinity of the NMDAR co-agonist glycine *(see Activation, 08-33)* is sensitive to the extracellular Ca^{2+} concentration in trigeminal neurones

(the apparent dissociation constant (EC_{50}) for glycine decreases with increasing external Ca^{2+} concentrations, increasing by about 3.7 times in Ca^{2+}-containing solutions)[236]. **Kinetic studies** of glycine binding to NMDA receptors indicate that external Ca^{2+} causes a decrease in the off rate† of the glycine binding, while having no effect on the on rate†[236] *(see also Activation, 08-33).*

Ligands

08-47-01: Commercially available radioligands† include **[³H]-AP5** ([³H]-D-2-amino-5-phosphopentanoic acid), **[³H]-CGS19755** ([³H]-*cis*-4-phosphonomethyl-2-piperidine carboxylic acid), **[³H]-MK-801** ([³H]-dizocilpine, 5-methyl-10,11-dihydro-5*H*-dibenzo[*a,d*]cyclohepten-5,10-imine), **[³H]-TCP** ([³H]-1-1-(2-thienyl)cyclohexylpiperidine) and **[³H]-CPP** ([³H]-3-(2-carboxypiperazin-4-yl)-propyl-1-phosphonic acid), which is a competitive† inhibitor of NMDA.

Radioligands for the glycine site
08-47-02: [³H]-5,7-Dichlorokynurenic acid (DCKA) and [³H]-L-689,560 are selective radioligands which have high affinity† for the glycine recognition domain on the NMDA receptor complex in brain synaptic membranes[302] *(see also Receptor antagonists, 08-51).*

Non-competitive antagonists for radioligand-binding studies
08-47-03: High-affinity [³H]-dextrorphan binding in rat brain is localized to a non-competitive† antagonist site of the activated NMDA receptor[303]. The iodinated monohydroxyl phenyl derivatives of **argiotoxin-636** *(see Blockers, 08-43)* retain NMDA-selective binding and can serve as non-competitive† antagonists for radioligand-binding assays[304].

Receptor/transducer interactions

Several other components and messenger molecules involved in coupling metabotropic† receptors to modulation of NMDAR-associated phenotypes are also described in Table 7.

Arachidonate as a diffusible 'retrograde messenger' for maintenance of glutamate release
08-49-01: Generation of post-synaptic **arachidonic acid** (AA) by phospholipase A_{2AA} (under conditions of repetitive NMDAR stimulation) has been suggested to act as a **'retrograde messenger'** (i.e. able to 'diffuse back') to pre-synaptic termini and potentiate† (maintain) glutamate release by a **positive-feedback** mechanism *(briefly reviewed in refs.*[106,305,306]). Arachidonic acid is able to sensitize **protein kinase C** (PKC) in pre-synaptic terminals, which can increase neuronal excitability (and thereby gluatamate release) by inhibiting the I_{KAA} channel in cell bodies of Purkinje cells[307]. *(See also Fig. 4 and Receptors/transducer interactions under ILG* KAA, 26-49.)

Pre-synaptic metabotropic glutamate receptors can also activate I_{KAA}
08-49-02: Pre-synaptic, **metabotropic**† **glutamate receptors** (mGluR or mGlu) have been shown to mediate activation of **protein kinase C**, with coupling of receptors to K^+ channel inhibition being dependent on low concentrations of **arachidonic acid**[308]. These '**positive feedback**' mechanisms have formed the basis of a model for *sustained release* of glutamate sufficient to establish **synaptic plasticity**† and inhibition of **phospholipase A_2** is known to block the induction of long-term potentiation *(see ILG K AA, entry 26 and* ref.[306]*). Note*: The involvement of mGluR in the induction of **short-term potentiation**† (STP) and mossy-fibre† LTP† processes are outlined under *Phenotypic expression, 08-14.*

Tachykinin-induced potentiation of NMDAR responses in dorsal horn neurones
08-49-03: Substance P and **neurokinin A** both potentiate NMDAR-induced currents in acutely isolated neurones from the dorsal horn of the rat[309]. However, substance P, but not neurokinin A, increases $[Ca^{2+}]_i$ in a sub-population of neurones, due to Ca^{2+}-influx through **voltage-sensitive Ca^{2+} channels**. Protein kinase C (PKC) may mediate NMDA-selective potentiation by **tachykinins** in these cells, since non-specific PKC activators (e.g. phorbol esters) enhance the effects of NMDA and **staurosporine** (a relatively non-specific protein kinase C inhibitor) inhibits the potentiation of NMDA currents[309].

Ca^{2+} signal amplification in 'retrograde signalling'
08-49-04: Arachidonic acid potentiation† induces a transient NMDA current[310] enabling **amplifications** in intracellular calcium concentrations to be induced by glutamate. Potentiation† of the NMDA receptor current is associated with increases in channel P_{open}†, with no apparent change in open-channel current.

Nitric oxide elevations and guanylate cyclase activation
08-49-05: Glutamate has been shown to induce the release of **nitric oxide** (**NO•**, **EDRF**, **endothelium-derived relaxing factor**)[113]. Nitric oxide is generated in a Ca^{2+}-influx-dependent manner, and its activity can account for observed **cGMP elevations** that take place in the CNS following NMDAR activation by glutamate *(see Protein interactions, 08-31).* For details of functional changes in the NMDAR induced by **nitroso compounds**†, *see Channel modulation, 08-44.*

Receptor agonists

Normal glutaminergic neurotransmission and glutamate-induced neuronal death
08-50-01: The excitatory amino acid† (EAA) **glutamate** is present in about 30% of central synapses and plays a central role in neuronal transmission under normal physiological conditions. Following pre-synaptic release following excitation† of nerve terminals†, **quanta** of glutamate diffuse across the **synaptic cleft**† and transiently occupy post-synaptic receptors (typically for 1–2 ms) before enzymatic degradation and/or re-uptake[61]. This

normal pattern of '**receptor use**' has been contrasted with '**receptor abuse**'[343] where accumulation of glutamate in the interstitial fluids during **cerebral oedema**† (e.g. associated with **trauma**† and **stroke**†) leads to persistent activation of its receptors[344, 345]. *Note:* In stroke†, neurones within the **ischaemic**† **area** (oxygen-deprived) die within a short time (several minutes) while in the **ischaemic penumbra** neurones are exposed to steadily increasing concentrations of glutamate and degenerate more slowly[343]. **Amplifications** of $[Ca^{2+}]_i$ and/or **sustained Ca^{2+}-influx**[105, 346–348] generally occur through receptor-coupled (as opposed to voltage-gated[349]) mechanisms. *Comparative note:* In general, Ca^{2+} channels play a significant role in *persistence* of **excitotoxic**† **phenotypes**[343, 3530] *(see Phenotypic expression, 08-14, and Protein interactions, 08-31).*

Endogenous metabolites as agonists for the NMDAR

08-50-02: **L-Homocysteic acid** (L-homocysteate) is presumed to be an **endogenous agonist** more selective for NMDA-sensitive receptors than the other endogenous ligands glutamate and aspartate[286]. Some endogenous metabolites of tryptophan (e.g. **quinolinic acid**) are active at similar potencies† to glutamate and NMDA[351, 352].

Amino acid co-agonism – the glycine site

08-50-03: By definition, all NMDA-type ionotropic† glutamate receptor subtypes are selectively 'agonized' by ***N*-methyl-D-aspartate**. Occupation of a **separate, allosteric 'glycine site'** (Gly_B, to distinguish it from the inhibitory glycine receptor described under *ELG Cl GLY, entry 11*) is also an **absolute requirement** for NMDA receptor activation. This requirement is observed experimentally as a dramatic potentiation† of NMDA responses by glycine[322], which therefore acts as a **co-agonist**. Glycine potentiation† in outside-out patches is characterized by increases in frequency of NMDAR channel opening[322].

Mechanism of glycine potentiation

08-50-04: Several experiments have shown that rapid application of glycine plus NMDA speeds the rate of receptor recovery from desensitization†[353, 354]. Patch-clamp experiments suggest the existence of two forms of desensitization† for the NMDA receptor in outside-out patches, however only one of these is dependent on the concentration of glycine. **Glycine-insensitive desensitization**† increases rapidly over the first few minutes of recording and largely occludes the **glycine-sensitive desensitization**† in outside-out patches. In patches that display no glycine-sensitive desensitization, the rate of glycine dissociation can be increased fourfold in the presence of glutamate, suggesting that the two binding sites are still allosterically† coupled. Thus, ligand binding at both types of sites can affect the affinity of the other type for its agonist[354]. In Mg^{2+}-free solutions, responses to glutamate application immediately *following* repetitive stimulation with glutamate plus glycine is increased by ~25–88%, returning to control levels over 10–15 min[355]. Enhancement of glutamate-induced currents is also seen following stimulation with solutions containing **aspartate plus glycine** or **NMDA plus glycine**. However, currents induced by aspartate alone are not potentiated[355]. A retrospective/review on the **glycine site** of the NMDAR has appeared[326].

Independence of NMDAR potentiation from glycine agonism at ionotropic glycine receptors
08-50-05: Glycine potentiation† of the NMDAR can be detected at a glycine concentration as low as 10 nM and is *not* mediated by the inhibitory strychnine-sensitive glycine receptor *(see ELG Cl GLY, entry 11)*. NMDAR potentiation is associated with increases in channel P_{open} activated by NMDA agonists[322]. Glycine can act as a **complete agonist**† at certain heteromultimers† expressed in *Xenopus* oocytes (e.g. epsilon 1/zeta 1 subunit combinations)[7]. *(See also Activation, 08-33, and Receptor antagonists, 08-51).*

Implications for glycine co-agonism from behavioural studies
08-50-06: Enhanced activation of the glycine co-agonist site on the NMDAR appears to facilitate one form of **associative learning** and may be used in other learning tasks. Peripheral injections of **D-cycloserine** (a partial agonist of the glycine site able to cross the **blood–brain barrier**†) doubles the learning rate in rabbit models[170] *(see also effects of monoclonal† antibodies which displace glycine from the NMDA receptor under Phenotypic expression, 08-14).*

Sensitivity of recombinant NMDARs to agonists – examples
08-50-07: Homomeric† NMDA receptors (NR1) expression-cloned in *Xenopus* oocytes from cDNA in plasmid vector pN60 *(see Isolation probe, 08-12)* display the following agonist sensitivities (measured as percentage steady-state current response compared to 100 μM NMDA in Mg^{2+}-free media supplemented with 10 μM glycine)[10]: 100 μM **NMDA** (control), 100%; 10 μM **L-glutamate**, 212 ± 15%; 100 μM **ibotenate**, 71 ± 20%; 200 μM **quisqualate**, 44 ± 10%; 100 μM **L-homocysteate**, 85 ± 19%; 100 μM NMDA without glycine (control), 35 ± 6%. Other agonists are ineffective in comparison (< 5%) and include 500 μM kainate, 50 μM AMPA (α-amino-3-hydroxy-5-methyl-4-isoxazolepropionate, *see ELG CAT GLU AMPA/KAIN, entry 07*), 100 μM 1S, 3R-ACPD (11-amino-cyclopentyl-1,3-dicarboxylate, an mGluR agonist) and 1 mM GABA.

Mixed agonism across iGluR molecular subtypes
08-50-08: In native tissues, the **dicarboxylic acids** L-glutamate and L-aspartate are **mixed agonists**† of the NMDA-, kainate- and AMPA-selective receptors and their effects are partially inhibited by all selective antagonists†. Definite increases in both the open time† and open-state probability† of NMDA-operated channels have been shown to be induced by *prolonged* application of glutamate to hippocampal slices *in situ*[356]. In spinal cord neurones, **L-proline** elicits inward current that can be *partially* antagonized by **D-AP5**[357].

Evidence for bicarbonate ion as a co-factor in agonist-dependent glutamate neurotoxicity
08-50-09: The ability of the (exogenous) neurotoxic glutamate agonist **BMAA** (β-*N*-methylamino-L-alanine) to open NMDA and AMPA channels in isolated membrane patches is strongly potentiated† by **bicarbonate**[358]. The **neurotoxic** and **neuroexcitatory** effects of two structural analogues of BMAA, **DL-2,4-diaminobutyrate** and **DL-2,3-diaminopropionate**, are also potentiated by bicarbonate. *Note*: BMAA is a neurotoxic glutamate agonist implicated in neuronal degeneration found in the Guam amyotrophic lateral sclerosis–Parkinsonism–dementia complex (**Guam disease**) – *see characteristics listed in ref.*[102].

Inhibition of pre-synaptic glutamate a[illegible]luzole

08-50-10: Sodium channel blockers, inclu[illegible]**lsant**† and **neuroprotective**† compound **riluzole**, block respo[illegible]A (IC_{50} = 18.2 μM) following expression of rat whole brain or cortex mRNA into *Xenopus* oocytes[359]. Riluzole also blocks responses to **kainic acid** (kainate, IC_{50} = 167 μM), **CNQX** (IC_{50} = 0.21 μM), **NBQX** (IC_{50} = 0.043 μM). **2-APV** *(see Receptor antagonists, 08-51)* yields an IC_{50} of 6.1 μM in this system[359]. The inhibition by both riluzole and 2-APV is reversible and does *not* appear to be use-dependent†, unlike that of the channel blocker MK-801 (dizocilpine). Riluzole acts in a direct but non-competitive† manner and does not interact with any of the known ligand-recognition sites on either the kainate or the NMDA receptor[359]. Characteristics of riluzole and other antagonists of glutamate release have have been reviewed[62].

Synaptic release of glutamate in hippocampal slices occurs in an 'all-or-none' manner

08-50-11: The substantial differences in reported sensitivities of the NMDA and non-NMDA receptors to glutamate agonist suggest that changes in **transmitter concentration** in the **synaptic cleft** can result in **differential modulation** of these two components of the EPSC†. However, pharmacological manipulation of *pre-synaptic* receptors affecting glutamate release in CA1 pyramidal cells of guinea-pig hippocampal slices (i.e. **baclofen** antagonism of **$GABA_B$ receptors** and non-selective agonism of **adenosine receptors** by **theophylline**) result in *parallel changes* of NMDA and non-NMDA receptor-mediated components of EPSCs over a 16-fold range[360]. Induction of long-term potentiation† (LTP†) in this preparation (by low-frequency synaptic stimulation in conjunction with depolarization to +30 mV, *see Fig. 1*) leads to *differential enhancement* of the *non-NMDA* receptor-mediated component of the EPSC†. Thus LTP† appears to occur through either modifications of post-synaptic receptors or through pre-synaptic changes involving increased transmitter concentration in the synaptic cleft[360].

Receptor antagonists (selective)

Pharmacologically distinct sites for antagonism of NMDAR responses

08-51-01: The multiple sites for antagonism on the NMDA receptor complex have been reviewed[311–314]. Conventionally, these have been separated into **pharmacologically distinct sites** based on the **characteristics of block** and the patterns of **additivity** or **overlap** in dose–response experiments. These sites include (i) the **transmitter-recognition site**, (ii) the **ion channel site**, (iii) the **glycine (co-agonist) site**, (iv) the **zinc (modulatory) site**, and (v) the **polyamine (modulatory) site**. The present 'classification' of sites is largely based on pharmacological criteria, and further discrete sites have been proposed[311]. Many studies provide support for antagonism at the sites represented schematically in Fig. 6. In due course it may be possible to resolve some of these discrete sites of antagonist action to defined amino

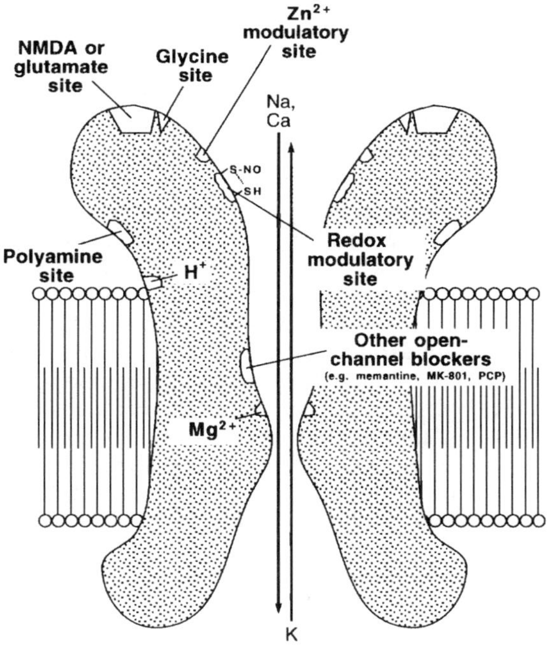

Figure 6. *Postulated sites of antagonist action at the NMDA receptor complex. Relative sizes and positions of sites on protein domains are shown on a hypothetical diagrammatic cross-section through a NMDAR. The site positions are not absolute and are for illustration only (see text). (From 08-51-01)*

acid loci (e.g. by means of site-directed mutagenesis†). The figure is a compilation of similar ones published in several reviews [129, 286, 287, 311].

General notes on modes of action for competitive versus non-competitive NMDAR antagonists

08-51-02: By definition, **competitive**† **antagonists** for receptor agonist sites can potentially inhibit normal physiological acivities in one brain region at concentrations which do not affect other regions. Glutamate accumulation to high levels within the **synaptic cleft**† in conditions such as **stroke**† may also 'out-compete' the effects of such competitive† antagonists[129]. By contrast, non-competitive† antagonists acting at 'modulatory' sites have a *potential* therapeutic advantage in that they are able to inhibit effects of excess glutamate in 'affected' areas of the brain, with relatively little direct influence on normal receptor function[270]. Examples of such '**modulatory sites**' include those regulated by **polyamines**†, **redox reagents**†, **channel blockade**, **zinc ions** and **pH**[129]. A summary of compounds shown to antagonize NMDA responses is listed in Table 10, categorized by their site of action. *For more comprehensive reviews, see refs.*[62, 311–314].

Table 10. *Known NMDA receptor antagonists and their features (From 08-51-02)*

Antagonist	Features	Refs
Competitive† *antagonists at glutamate-binding site* **D-AP5** **2-APV** **CPP** **CGS-19755** **CGP-37849** **LY-235959** **SDZEAA494** **MDL100453** **DAA**	D-2-Amino-5-phosphopentanoic acid (**D-AP5**) was the first NMDA receptor-specific antagonist (pA_2 5.2–5.9) *(see Protein interactions, 08-31)*. Similar compounds are *dl*-2-amino-5-phosphonovaleric acid (**2-APV**), D-2-amino-7-phosphonoheptanoic acid (AP7 or D-AP7) and 3-(2-carboxypiperazin-4-yl)-propyl-1-phosphonic acid (**CPP**). Other examples are **CGS-19755** (*cis*-4-phosphonomethyl-2-piperidine carboxylic acid, pA_2 6.0). **CGP-37849** (DL-(*E*)-2-amino-4-methyl-5-phosphono-3-pentanoic acid) and longer-chain glutamate analogues (e.g. D-α-aminoadipate, DAA). Generally, these compounds do *not* possess significant channel-blocking activity. Structure–activity relations of these and other compounds has been reviewed[311]. For molecular 'loci' of agonist-binding sites, *see Domain functions, 08-29*	***380***
Competitive antagonists at multiple sites **DDHB** and substituted derivatives	Substituted benzazepines are class of glutamate receptor antagonists that show competitive† action, significant potency at multiple sites, and a high degree of lipophilicity. 2,5-Dihydro-2,5-dioxo-3-hydroxy-1H-benzazepine (**DDHB**) and three substituted derivatives, 4-bromo-, 7-methyl-, and 8-methyl-DDHB, inhibit the activation of NMDA receptors at both the NMDA recognition site and the glycine allosteric site	***315***
Non-competitive† *(use-dependent, uncompetitive) open-channel antagonists* **Dizocilpine (MK-801)** **Ketamine** **Tiletamine** **Phencyclidine** **SKF10047** **Dextrorphan** **Dextro-methorphan** **Desipramine**	Dissociative anaesthetics including **dizocilpine** (**MK-801**), (+)-5-methyl-10,11-dihydro-5H-dibenzo-cyclohepten-5,10-imine maleate), **phencyclidine** (PCP, 'angel dust'), **ketamine**, **tiletamine** and **SKF10047**. High-affinity dextrorphan binding in rat brain has been localized to non-competitive antagonist sites also recognized by nanomolar concentrations of dizocilpine and TCP ([^{3}H]-1-(1-(2-thienyl)cyclohexyl)piperidine. **Dextromethorphan** and **remacemide** are well-tolerated clinically. Dizocilpine has been used to solubilize NMDA receptors from rat and porcine brain and in its tritiated form is a common radioligand for receptor binding and distribution studies *(see Protein distribution, 08-15)*. The *S*-enantiomer of ketamine is ~twice as potent as the *R*-enantiomer in exhibiting a voltage- and use-dependent blockade of NMDA receptor currents. Calculated relative	***303, 316–318***

Table 10. *C*

Antagonist		Refs
Memantine **CNS1102** **CNS1505** **Mg^{2+}** **Remacemide/ FPL12495** **Zn^{2+} (high μM)** **Mn^{2+}**	forward and backward rates suggest that enantiomer conformational differences influence the *dissociation* from the binding site more than the association with it. The tricyclic antidepressant **desipramine** (DMI, 20–50 μM) is a potent selective antagonist of responses to NMDA in mouse hippocampal neurones. The potency of DMI as an NMDA antagonist is highly voltage-dependent with the K_d increasing *e*-fold per 36 mV depolarization. At −60 mV, the K_d for DMI block of responses to NMDA is 10 μM and in the presence of Mg^{2+} block, channels do *not* bind DMI but do show a decrease in the open time and burst length distributions, consistent with binding of DMI to open channels. For details of block by ketamine and **memantine**, *see Blockers, 08-43*. For molecular loci of pore-lining domains, *see Domain conservation, 08-28, and Domain functions, 08-29*. For the dual blocking- and potentiating actions of Zn^{2+} ions, *see Blockers, 08-43, and Channel modulation, 08-44*	
Non-competitive antagonists, dose-dependent **Pyrazole**	Electrophysiological and biochemical studies have demonstrated that **pyrazole**, an inhibitor of alcohol dehydrogenase and a proposed therapeutic agent for treatment of alcoholic intoxication (i) activates NMDA receptors at low (~ 0.5 μM) concentrations but (ii) blocks NMDA receptors non-competitively at higher concentrations. *Comparative note:* Pyrazole does *not* interact significantly with the end-plate nicotinic acetylcholine receptor (AChR)	***319***
Competitive antagonists, glycine site **Indole-2-carboxylic acid** **Methoxyindole carboxylic acid**	**Indole-2-carboxylic acid** (I2CA) specifically and competitively inhibits glycine potentiation of NMDA-gated current. In solutions containing low levels of glycine, I2CA completely blocks the response to NMDA, suggesting that NMDA alone is not sufficient for channel activation. **Methoxyindole carboxylic acid** is a low-affinity competitive antagonist (K_d ~ 5 mM) at the glycine-binding site of the NMDAR and can be used to eliminate spurious gating due to glycine contamination from solutions, which is a common technical problem *(see Activation, 08-33)*. For molecular 'loci' of the glycine co-agonist site, *see Domain functions, 08-29*	***320***

Table 10. *Continued*

Antagonist	Features	Refs
Antagonists at the glycine (D-serine) site **Ha-966** **MNQX** **7-Cl-Kyn** **L-689,560** **(CNQX)** **L687414** **Felbamate** **ACPC**	Compounds include **HA-966** ([illegible] 1-hydroxy-pyrrolid-2-one), D-cycloserine, **MNQX** (5,7-dinitro-quinoxaline-2,3-dione), 7-dichlorokynureate (**7-Cl-Kyn**), 5,7-dichlorokynureate and **L-689,560** (+/−4-*trans*-2-carboxy-5,7-dichloro-4-phenylaminocarbonylamino-1,2,3,4-tetrahydroquinoline). *At high concentrations*, the *non*-NMDA receptor antagonist **CNQX** can reduce NMDA responses non-competitively by competing with and blocking the enhancing action of glycine on the NMDA receptor. For further details on the glycine co-agonist site, *see Receptor agonists, 08-50*. **Felbamate** is a tolerated, FDA-approved anti-convulsive agent which in addition to its antagonism of 'strychine-insensitive' glycine receptors (i.e. NMDAR) may also block voltage-gated sodium channels	*321–326, 327, 328*
Redox modulatory site **Nitroglycerin** **Nitroprusside** **Glutathione** **PQQ**	For mechanistic details of modulation at the redox site, *see Channel modulation, 08-44*. Compounds that can act as antagonists at this site include **nitroglycerin**, sodium **nitroprusside**, **glutathione** and pyrroloquinone (**PQQ**). Notably, agents affecting the redox site are well-tolerated in humans	*129*
H^+ ions	A fall in **pH** decreases channel activity non-competitively *(see Channel modulation, 08-44)*	*299, 300, 329*
Competitive antagonists, polyamine site (putative) **Arcaine** **Ifenprodil** **Eliprodil/ SL820715**	**Arcaine**, a putative competitive antagonist at the polyamine site on the NMDAR inhibits polyamine enhancement of NMDA-induced [^{3}H]-dizocilpine (MK-801) binding and also depresses binding in the absence of polyamines. In cultured hippocampal neurones, arcaine produces a concentration- and voltage-dependent block of NMDA-evoked inward currents ($K_d \sim 61$ μM at −60 mV). Increasing the dizocilpine concentration partially overcomes the arcaine effect, indicating a competitive interaction between arcaine and dizocilpine. Arcaine has also been shown to block the open NMDA receptor–channel, an action that is *independent* of the polyamine site. For further details of polyamine modulation, *see Channel modulation, 08-44.* **Ifenprodil** can discriminate between different combinations of heteromeric NMDAR: The affinity of recombinant NR1A/NR2A receptors expressed in oocytes for ifenprodil ($IC_{50} = 146$ μM) is ~ 400-fold lower than that of NR1A/NR2B receptors ($IC_{50} = 0.34$ μM). Part of the mechanism of action of ifenprodil at NR1A/NR2B receptors may involve non-competitive antagonism of the effects of glycine	*330, 331*

Table 10. *Continued*

Antagonist	Features	Refs
Competitive peptide antagonists, polyamine site **Conantokin-G** **Conantokin-T** **Argiotoxin**	A peptide from *Conus geographus* (cone snail) venom, **conantokin-G** (CntxG), competitively blocks with high affinity and specificity NMDAR-mediated currents in hippocampal neurones and has been reported as a 'reliable' probe for determination of NMDAR distribution *(see Protein distribution, 08-15)*. Conantokin-G peptide inhibits only ~ 70% of the elevation of intracellular free calcium produced by NMDA in cerebellar granule cells. The highly-related polypeptide **conantokin-T** also acts as a potent non-competitive inhibitor of polyamine responses of the NMDAR–channel. Chemical substitution of the highly conserved γ-carboxyglutamate residues as well as modification of the N- and C-termini of conantokin-G abolishes these responses. The derivative Tyr0-conantokin-G has been found to exhibit polyamine-like actions at ~ 7-fold greater potency than spermine. **Argiotoxin-636**, a component of the *Argiope* spider venom, has a higher affinity for the NMDAR than for kainate receptors, blocking the corresponding ion channels in a voltage-dependent manner. Modifications of the polyamine tail or the terminal arginine residue of synthetic argiotoxin-636 strongly reduces the blocking potency. The biology of *Conus* (cone snail) venom peptides which act on ion channels including the NMDAR has been reviewed[332]	*182*, *333*, *334*, *304*
Other antagonists at multiple sites	Bioxanthracenes of microbial origin (the ES-242s) represent novel NMDA receptor antagonists which interact with both the neurotransmitter recognition site and the ion channel domain	*335*

[a]For further details, see refs[62, 311–314].

The therapeutic significance of NMDA receptor antagonists

08-51-03: NMDAR antagonists have been examined as candidate **neuroprotective agents** for pathological conditions associated with acute and chronic '**overstimulation**' of NMDAR and non-NMDAR[40, 62, 92, 102, 121–125, 336] *(see also references under Phenotypic expression, 08-14)*. NMDA channel blockers can provide a 'therapeutic window' by preventing 'excitotoxic' calcium influx during **global**† or **focal**† **brain ischaemia**† elicited by conditions such as **stroke**†. The role of excitatory amino acid receptors in **epilepsy**, and the effectiveness of NMDAR antagonists in various *in vivo* and *in vitro* models of **epileptic seizure**† have been reviewed[127]. Effective therapeutic compounds may act to directly antagonize excitatory amino acid agonist binding to receptors, or may modulate uptake of endogenous agonist within the synaptic cleft†[129]. The **chemical structural requirements** for the development of potent† NMDA receptor antagonists have been discussed[337].

General note on the relationship of antagonist selectivity and their molecular 'targets'

08-51-04: The existence of **multiple subunit isoforms**† (and **splice variants**†) within functional extracellular ligand-gated channel complexes (coupled with known variabilities in expression patterns within the CNS) raises the possibility that selection of antagonists capable of modulating **subsets** of brain functions is a rational goal[129].

Antagonist sensitivities for recombinant NMDARs – examples

08-51-05: Homomeric† NMDA receptors (NR1) expression-cloned in *Xenopus* oocytes from cDNA in plasmid vector pN60 *(see Isolation probe, 08-12)* display the following **antagonist sensitivities** (measured as percentage reductions of the steady-state current response compared to a control of 100 μM NMDA in Mg^{2+}-free media[10]): 100 μM **NMDA** (control, without antagonists), 100%; 10 μM **D-APV**, $27 \pm 6\%$; 1 μM **CPP**, $47 \pm 7\%$; 1 μM **CGS-19755**, $43 \pm 5\%$; 50 μM **7-Cl-Kyn**, $2 \pm 1\%$; 1 μM (+) **MK-801**, $7 \pm 4\%$; 100 μM **GAMS**, $97 \pm 2\%$; 100 nM **Joro spider toxin**, $83 \pm 6\%$; 100 μM **CNQX**, $70 \pm 6\%$[10]. *Note:* **Differential sensitivity** to the open-channel blocker **dizocilpine** (**MK-801**) by certain NMDAR subunit combinations has been reported[267] *(for details, see Blockers, 08-43).*

Effects of NMDA antagonism in vivo *– behavioural studies*

08-51-06: A systematic study of relative potencies† for glutamate receptor antagonists administered *in vivo* reported **MK-801** and **CGP-37849** to have relatively high potency ($ED_{50} < 1$ mg/kg following three different administration procedures)[338]. A range of NMDA channel antagonists (including several listed above plus **ifenprodil**, **±-*N*-allylnormetazocine** and **dextromethorphan**) produce **impairment of learning** and **muscle relaxation** when administered i.c.v.† in rats[339]. In addition to 'temporary psychiatric disorders', systemic administration of allosteric† and isosteric† NMDAR antagonists has also been associated with disorders of **cardiorespiratory regulation**[340], **neuronal degeneration**[341] and reductions in **synaptic strength**†[48].

NMDA receptor antagonist neurotoxicity dependent on neuronal interconnections

08-51-07: Certain NMDA antagonists cause **neurotoxic side-effects** consisting of **pathomorphological** changes in neurones of the cingulate and retrosplenial cerebral cortices[341]. Following low doses these changes appear to be reversible, but higher doses can cause irreversible **neuronal necrosis**[342]. The neurotoxic action of MK-801 in adult rat cingulate cortex is potentiated by pre-treatment with the cholinergic† agonist **pilocarpine**, and can be abolished by co-administration of the cholinergic muscarinic antagonist **scopolamine** and agonists† which act at the $GABA_A$ receptor (e.g. **diazepam** and **barbiturates**)[212]. This *prevention of NMDA antagonist-induced neurotoxicity*

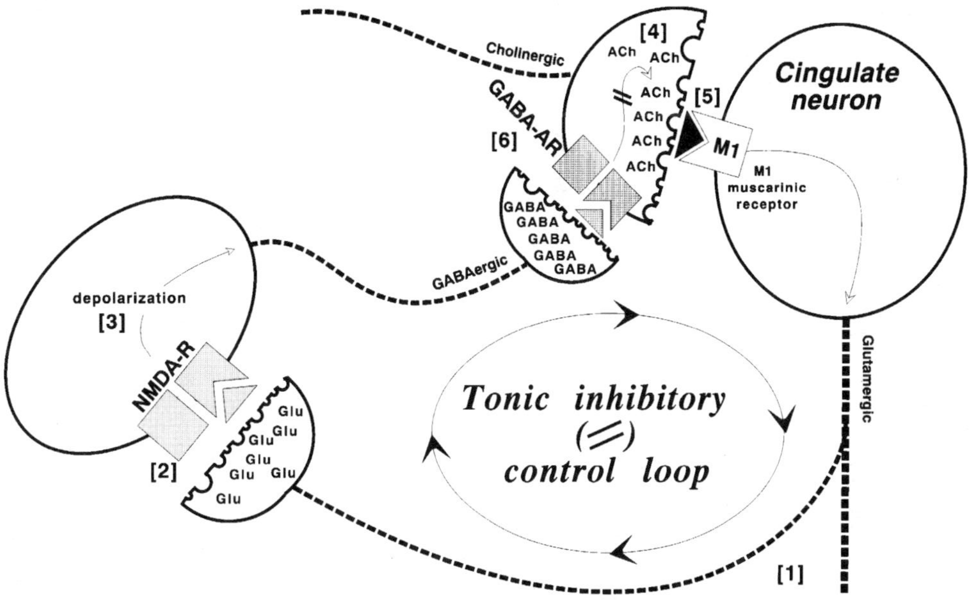

Figure 7. *NMDA receptor antagonist neurotoxicity dependent on neuronal interconnections. Glutaminergic (Glu) axon collaterals of* ***cingulate neurones*** *[1] feed back to* ***NMDA receptors*** *[2] expressed at GABAergic (GABA) neurones [3] to maintain* ***toxic inhibitory control*** *over the release of acetylcholine (ACh) from cholinergic neurones [4] and its actions at M1 muscarinic receptors [5] expressed on the cingulate neurone itself.* ***Pharmacological blockade*** *of the NMDA receptor [2] abolishes the inhibitory control over ACh release and subjects the cingulate neurone to a state of persistent cholinergic* ***hyperstimulation****. This hyperstimulation has been proposed as the main cause of the* ***pathomorphological effects*** *in the cingulate neurone following NMDA receptor–channel blockade (see Receptor antagonists, 08-51). Restoration of* ***GABAergic tone*** *(e.g. by administration of barbiturates opening the GABA receptor–channel at [6], even in the absence of GABA (see ELG CI GABA$_A$, entry 10) or by M1-selective antagonism at [5]) has been shown to* ***protect*** *cingulate neurones against the neurotoxic side-effects of NMDA antagonists. (Based on data from Olney* et al. *(1991)* Science **254***: 1515–18.) (From 08-51-07)*

by M1-selective anticholinergic and barbiturate drugs has been explained by considering the specific set of connections which modulate glutameric cingulate neuronal activity *(for a summary of these interactions, see Fig. 7)*[212].

Receptor inverse agonists

Putative inverse agonists at the polyamine site

08-52-01: Voltage-independent, dose-dependent, non-competitive† antagonism of NMDA-mediated currents by chloride transport blockers has been shown in cultured spinal cord neurones[361]. These agents include **furosemide** (a widely-used loop diuretic), and the related compounds **piretanide**, **bumetanide**, **niflumic acid** and **flufenamic acid**. The action of furosemide could arise from interaction with the Zn^{2+} inhibitory site (as blockade of NMDA-induced responses by furosemide and Zn^{2+} are additive). These agents may act as **inverse agonists** of the **polyamine site** and their action may explain the 'protective' effect that has been shown for some of these drugs in neuronal degeneration[361] *(see Phenotypic expression, 08-14).*

08-52-02: 1,10-Diaminodecane (**DA10**) acts as an inverse† agonist† at the **polyamine recognition site** of the NMDA receptor[362]. The effect of DA10 is not mediated by action of DA10 at the binding sites for glutamate, glycine, Mg^{2+} or Zn^{2+} but is attenuated by **diethylenetriamine** (**DET**). In hippocampal neurones, NMDA-elicited current is decreased by DA10, an effect opposite to that of **spermine**. The effects of spermine are also selectively blocked by DET[362] *(see also Channel modulation, 08-44).*

INFORMATION RETRIEVAL

Database listings/primary sequence discussion

08-53-01: *The relevant database is indicated by the lower case prefix (e.g. gb:) which should not be typed (see Introduction & layout of entries, entry 02). Database locus names and accession numbers immediately follow the colon. Note that a comprehensive listing of all available accession numbers is superfluous for location of relevant sequences in GenBank® resources, which are now available with powerful in-built* ***neighbouring*****†** ***analysis*** *routines (for description, see the Database listings field in the Introduction & layout of entries, entry 02). For example, sequences of cross-species variants or related gene family† members can be readily accessed by one or two rounds of neighbouring† analysis (which are based on pre-computed alignments performed using the BLAST† algorithm by the NCBI†). This feature is most useful for retrieval of sequence entries deposited in databases later than those listed below. Thus, representative members of known sequence homology groupings are listed to permit initial direct retrievals by accession number, author/reference or nomenclature. Following direct accession, however, neighbouring† analysis is strongly recommended to identify newly reported and related sequences.*

Nomenclature	Species, DNA source	Original isolate	Accession	Sequence/ discussion
Rat NMDA receptor NMDAR1=NR1	Rat. 'Fundamental subunit' expression-cloned from a rat forebrain cDNA library *(see Isolation probe, 08-12)*	Molecular mass ~ 105 kDa; 939 aa	gb: X63255	Moriyoshi, *Nature* (1991) **354**: 31–7.
NMDAR1 (rat) NMDAR1 gene exons 1 through 22	Sprague–Dawley genomic DNA, 4444 bp		gb: L08228	Hollmann, *Neuron* (1993) **10**: 943–54.
NMDAR1-LL	Alternative splice variant (rat)	960 aa; combinatorial RNA splicing altering the surface charge on the NMDAR	gb: X65227	Anantharam, *FEBS Lett* (1992) **305**: 27–30.
NMDAR1 (rat) NMDAR1-1a subunit	Sprague–Dawley strain; cDNA library	939 aa; zinc potentiates agonist-induced currents at certain splice variants	gb: U11418	Sullivan, J.M. and Boulter, unpublished *(at Jun 1994)*.
NMDAR1 (rat) NMDAR1-1a subunit	Sprague–Dawley strain; cDNA library	939 aa; zinc potentiates agonist-induced currents at certain splice variants	gb: U08261	Hollmann, *Neuron* (1993) **10**: 943–54.
NMDAR1 (rat) NMDAR1-1b subunit	Sprague–Dawley strain; cDNA library	959 aa; zinc potentiates agonist-induced currents at certain splice variants	gb: U08263	Hollmann, *Neuron* (1993) **10**: 943–54.
NMDAR1 (rat) NMDAR1-1b subunit	Rat brain cDNA library	93 aa; partial sequence	gb: S67814	Kusiak, *Brain Res. Mol. Brain Res.* (1993) **20**: 64–70.
NMDAR1 (rat) NMDAR1-2a subunit	Sprague–Dawley strain; cDNA library	901 aa; zinc potentiates agonist-induced currents at certain splice variants	gb: U08262	Hollmann, *Neuron* (1993) **10**: 943–54.

Nomenclature	Species, DNA source	Original isolate	Accession	Sequence/ discussion
NMDAR1 (rat) NMDAR1-2b subunit	Sprague–Dawley strain; cDNA library	923 aa	gb: U08264	Hollmann, *Neuron* (1993) **10**: 943–54.
NMDAR1 (rat) NMDAR1-3a subunit	Sprague–Dawley strain; cDNA library	923 aa; zinc potentiates agonist-induced currents at certain splice variants	gb: U08265	Hollmann, *Neuron* (1993) **10**: 943–54.
NMDAR1 (rat) NMDAR1-3b subunit	Sprague–Dawley strain; cDNA library	944 aa; zinc potentiates agonist-induced currents at certain splice variants	gb: U08266	Hollmann, *Neuron* (1993) **10**: 943–54.
NMDAR1 (rat) NMDAR1-4a subunit	Sprague–Dawley strain; cDNA library	886 aa; zinc potentiates agonist-induced currents at certain splice variants	gb: U08267	Hollmann, *Neuron* (1993) **10**: 943–54.
NMDAR1 (rat) NMDAR1-4b subunit	Sprague–Dawley strain; cDNA library	907 aa; zinc potentiates agonist-induced currents at certain splice variants	gb: U08268	Hollmann, *Neuron* (1993) **10**: 943–54.
hNR1	Human brain cDNA library	886 aa plus 17 aa signal peptide	gb: LO5666	Planells-Cases, *Proc Natl Acad Sci USA* (1993) **90**: 5057–61.
NMDAR1 (human)	Human 'key' subunit	938 aa (99% amino acid homology to rat NR1)	gb: D13515	Karp, *J Biol Chem* (1993) **268**: 3728–33.
hNR1-1 (human)	*Homo sapiens* (cDNA library: Stratagene 936205) female infant hippocampus	886 aa	gb: L13266	Foldes, *Gene* (1993) **131**: 293–8.
hNR1-2 (human)	*Homo sapiens* (cDNA library: Stratagene 936205) female infant hippocampus	604 aa	gb: L13267	Foldes, *Gene* (1993) **131**: 293–8.
hNR1-3 (human)	*Homo sapiens* (cDNA library: Stratagene 936205) female infant hippocampus	929 aa (918 aa mature chain)	gb: L13268	Foldes, *Gene* (1993) **131**: 293–8.

Nomenclature	Species, DNA source	Original isolate	Accession	Sequence/ discussion
hNR1-4 (human)	*Homo sapiens* (cDNA library: Stratagene 936205) female infant hippocampus	592 aa partial sequence	gb: U08106	Foldes, *Gene* (1994) In press.
NMDAR (human)	human terato-carcinoma line NT2	101 aa partial sequence	gb: S57708	Younkin, *Proc Natl Acad Sci USA* (1993) **90**: 2174–8.
hNR1N (human) splice variant	GRIN1	196 aa partial sequence	gb: U08107	Foldes, *Gene* (1994) In press.
zeta 1 (= ζ1) **GluR subunit NMDAR1=NR1**	Mouse. NMDA 'fundamental' or 'key' subunit: isolated through cross-homology screening	938 aa, $M_r \sim 105\,500$	gb: D10028	Yamazaki, *FEBS Lett* (1992) **300**: 39–45.
NMDAR1 splice variants NR1a and NR1b	NR1 splice variants NR1a and NR1b *(for differences, see Gene organization, 08-20)*	*See also Protein phosphorylation, 08-32, and Channel modulation, 08-44*	gb: L01632	Durand, *Proc Natl Acad Sci USA* (1992) **89**: 9359–63.
NMDAR2A (=epsilon 1) **NMDAR2B** (=epsilon 2) **NMDAR2C** (=epsilon 3) **NMDAR2D**	Rat brain cDNA species which form heteromeric receptors with subunit NR1; homology-based PCR screen	NR2A – 943 aa NR2B – 1445 aa NR2C – 1456 aa	gb: M91561 gb: M91562 gb: M91563	Monyer, *Science* (1992) **256**: 1217–21.
NMDAR2A (=epsilon 1) **NMDAR2B** (=epsilon 2) **NMDAR2C** (=epsilon 3) **NMDAR2D**	NMDA 'potentiating subunits'[8]	Isolated by PCR-RT†/rat brain cDNA library	gb: D13211 gb: D13212 gb: D13213 gb: D13214	Ishii, *J Biol Chem* (1993) **268**: 2836–43.
Epsilon 1 subunit (=NMDAR2A)	Mouse. Homology-based PCR screen	1464 aa (1445 aa, mature)	gim: 405314	Meguro, *Nature* (1992) **357**: 70–4.
Epsilon 2 subunit (=NMDAR2B)	Mouse. Homology-based PCR screen	1482 aa	gim: 405323	Kutsuwada, *Nature* (1992) **358**: 36–41.
Epsilon 3 subunit (=NMDAR2C)	Mouse. Homology-based PCR screen	1239 aa	gim: 405325	Katsuwada, *Nature* (1992)

Nomenclature	Species, DNA source	Original isolate	Accession	Sequence/ discussion
Epsilon 4 subunit	Mouse. Homology-based PCR screen	1323 aa	not found	Ikeda, *FEBS Lett* (1992) **313**: 34–8.
Glutamate-binding protein (GBP) of NMDA receptor complex plus four NMDAR cDNAs	Ligand-binding properties not reported; no direct evidence of channel activity or NMDA pharmacological specificity	Molecular mass ~ 57 kDa (including signal peptide). No homology to any other GluR or kainate-binding protein	not found	Kumar, *Nature* (1991) **354**: 70–3.
Pre-synaptic glutamate receptor–channel			not found	Smirnova, *Science* (1993) **262**: 430–3.

Note: The above table lists genes encoding subunits **NMDA-selective ionotropic**† **glutamate receptors** (iGluR). In database searches, these should not be confused with the gene nomenclatures used to describe the G protein-linked (**metabotropic**†) glutamate receptors in the $mGlu_1$–$mGlu_7$ series *(see Appendix A – Index of G protein-linked receptors, entry 56, and the notes below the main table in Database listings under ELG CAT GLU AMPA/KAIN, 07-53).*

Related sources & reviews

08-56-01: *See also Table 4.* Major quoted sources for this entry[54, 64, 128, 363, 364]; NMDA receptor physiology reviews[5, 286, 353, 365–367]; functional distinctions of NMDA and non-NMDA receptor–channels[241, 257, 259, 262, 263, 363, 368–370]; physiological and **pathophysiological** roles of excitatory amino acids during development[40]; pharmacology of the glutamate receptor family[371]; NMDARs and **Hebb-type synaptic plasticity**[61]; excitatory amino acid receptors in epilepsy[127]; cloning of excitatory amino acid receptors[19, 372]; commentary on original descriptions of cloned NMDARs[373]; sites for **antagonism**† on the NMDA receptor complex[311]; glutamate neurotoxicity and neuronal degeneration[40, 62, 92, 102, 122–124, 374]; pharmacological strategies (including NMDAR antagonism) aimed at limiting CNS tissue damage following **trauma**†[123, 336]; commentaries on NMDARs in synaptic **excitation**†[375, 376]; physiological roles of second messengers in regulation of NMDAR function[5]; reviews on roles of GluRs in CNS function and **memory**[53, 54, 61, 74, 186, 365]; **phosphorylation** of iGluRs in **synaptic plasticity**†[65]; protein kinase C function in LTP induction[143]; the **glycine site** of the NMDAR[326, 377, 378]; **polyamine modulation** of the NMDAR[285, 287]; NMDAR as a pharmacological target for **ethanol**[296]; *in vitro* and *in vivo* neurochemical characterization of NMDAR[379]; NMDAR **'stimulus-transcription coupling'** in neurones (i.e. trans-synaptic control of gene expression)[28, 34]; **computational implications** of NMDA receptor–channels[128, 172]; **permeation pathways** of neurotransmitter-gated ion channels[199]. *See also reviews listed under ELG CAT GLU AMPA/KAIN, entry 07, and Resource E – Ion channel book references, entry 60.*

Book references

08-56-02:
Hall, Z.W., Sargent, P. B., Scheller, R. H., Ross, E. M., Kennedy, M. B., Vale, R. D., Banker, G, Lemke, G., Anderson, D. J., Patterson, P. H., Marder, E. and Breakefield, X. O. (1992) *An Introduction to Molecular Neurobiology*. Sinauer Associates, Sunderland, Massachusetts.
Kozikowski, A. P. and Barrionuevo, G. (eds) (1991) *Neurobiology of the NMDA Receptor: From Chemistry to the Clinic*. VCH Verlagsgesellschaft, Germany.
Krogsgaard-Larsen, P. and Hansen, J. J. (eds) (1992) *Excitatory Amino Acid Receptors*. Ellis Horwood.
Lodge, D. (ed.) (1988) *Excitatory Amino Acids in Health and Disease*. Biological Council Symposia on Drug Action. John Wiley & Sons.
Plaitakis, A. (1984) In *The Olivopontocerebellar Atrophies* (eds R.C. Duviosin and A. Plaitakis), pp. 225–43. Raven Press, New York.
Watkins, J.C. and Collingridge, G. L. (1989) *The NMDA Receptor*. IRL Press.

Feedback

Error-corrections, enhancement and extensions

08-57-01: Please notify specific errors, omissions, updates and comments on this entry by contributing to its **e-mail feedback file** (*for details, see Resource J, Search Criteria & CSN Development*). For this entry, send e-mail messages To: **CSN-08@le.ac.uk,** indicating the appropriate paragraph by entering its **six-figure index number** (xx-yy-zz or other identifier) into the **Subject**: field of the message (e.g. Subject: 08-50-07). Please feedback on only **one specified paragraph or figure per message,** normally by sending a **corrected replacement** according to the guidelines in *Feedback & CSN Access* . Enhancements and extensions can also be suggested by this route (*ibid.*). Notified changes will be indexed via 'hotlinks' from the CSN 'Home' page (http://www.le.ac.uk/csn/) from mid-1996.

Entry support groups and e-mail newsletters

08-57-02: Authors who have expertise in one or more fields of this entry (and are willing to provide editorial or other support for developing its contents) can join its support group: In this case, send a message To: **CSN-08@le.ac.uk,** (entering the words "support group" in the Subject: field). In the message, please indicate principal interests (see *fieldname criteria in the Introduction for coverage*) together with any relevant **http://www site links** (established or proposed) and details of any other possible contributions. In due course, support group members will (optionally) receive **e-mail newsletters** intended to **co-ordinate and develop** the present (text-based) entry/fieldname frameworks into a 'library' of interlinked resources covering ion channel signalling. Other (more general) information of interest to entry contributors may also be sent to the above address for group distribution and feedback.

REFERENCES

[1] Tanabe, *Neuron* (1992) **8**: 169–79.
[2] Sugiyama, *Neuron* (1989) **3**: 129–32.
[3] Schoepp, *Trends Pharmacol Sci* (1993) **14**: 13–20.
[4] Nakanishi, *Neuron* (1994) **13**: 1031–37.
[5] Mayer, *Trends Pharmacol Sci* (1990) **11**: 254–60.
[6] Monyer, *Science* (1992) **256**: 1217–21.
[7] Meguro, *Nature* (1992) **357**: 70–4.
[8] Ishii, *J Biol Chem* (1993) **268**: 2836–43.
[9] Stern, *Proc R Soc Lond (Biol)* (1992) **250**: 271–7.
[10] Moriyoshi, *Nature* (1991) **354**: 31–7.
[11] Karp, *J Biol Chem* (1993) **268**: 3728–33.
[12] Cik, *Biochem J* (1993) **296**: 877–83.
[13] Sakimura, *Neuron* (1992) **8**: 267–74.
[14] Honoré, *Eur J Pharmacol* (1989) **172**: 239–42.
[15] Yamazaki, *FEBS Lett* (1992) **300**: 39–45.
[16] Henley, *Proc Natl Acad Sci USA* (1992) **89**: 4806–10.
[17] Smirnova, *Science* (1993) **262**: 430–3.
[18] Mackler, *Mol Pharmacol* (1993) **44**: 308–15.
[19] Seeburg, *Trends Pharmacol Sci* (1993) **14**: 297–303.
[20] Teichberg, *FASEB J* (1991) **5**: 3086–91.
[21] Usowicz, *Nature* (1989) **339**: 380–3.
[22] Tremblay, *Brain Res* (1988) **461**: 393–6.
[23] Weiland, *Endocrinology* (1992) **131**: 662–8.
[24] Matute, *Proc Natl Acad Sci USA* (1992) **89**: 3399–403.
[25] Ray, *Biochem Biophys Res Commun* (1993) **197**: 1475–82.
[26] Pleasure, *J Neurosci* (1992) **12**: 1802–15.
[27] Bading, *Science* (1993) **260**: 181–6.
[28] Armstrong, *Annu Rev Neurosci* (1993) **16**: 17–29.
[29] Komuro, *Science* (1993) **260**: 95–7.
[30] Hack, *Neuroscience* (1993) **57**: 9–20.
[31] Pearson, *Neurosci Lett* (1992) **142**: 27–30.
[32] Morgan, *Cell Calcium* (1988) **9**: 303–11.
[33] Lerea, *J Neurosci* (1992) **12**: 2973–81.
[34] Morgan, *Trends Neurosci* (1989) **12**: 459–62.
[35] Favaron, *Proc Natl Acad Sci USA* (1988) **85**: 7351–55.
[36] Spitzer, *J Neurobiol* (1991) **22**: 659–73.
[37] Blanton, *Proc Natl Acad Sci USA* (1990) **87**: 8027–30.
[38] Fox, *Nature* (1991) **350**: 342–4.
[39] Neve, *Proc Natl Acad Sci USA* (1989) **86**: 4781–4.
[40] McDonald, *Brain Res Rev* (1990) **15**: 41–70.
[41] Constantine-Paton, *Annu Rev Neurosci* (1990) **13**: 129–54.
[42] Kato, *Proc Natl Acad Sci USA* (1993) **90**: 7114–18.
[43] Ben-Ari, *Neurosci Lett* (1988) **94**: 88–92.
[44] Bowe, *Dev Brain Res* (1990) **56**: 55–61.
[45] McDonald, *Exp Neurol* (1990) **110**: 237–47.
[46] Hestrin, *Nature* (1992) **357**: 686–9.

[47] Cline, *J Neurosci* (1990) **10**: 1197–216.
[48] Cline, *Neuron* (1989) **3**: 413–26.
[49] Cline, *Neuron* (1991) **6**: 259–67.
[50] Kleinschmidt, *Science* (1987) **238**: 355–8.
[51] Udin, *Science* (1990) **249**: 669–72.
[52] Lipton, *Trends Neurosci* (1989) **12**: 265–70.
[53] Collingridge, *Trends Pharmacol Sci* (1990) **11**: 290–6.
[54] Bliss, *Nature* (1993) **361**: 31–9.
[55] Carmignoto, *Science* (1992) **258**: 1007–11.
[56] Agmon, *J Neurophysiol* (1992) **68**: 345–9.
[57] Masu, *Nature* (1991) **349**: 760–5.
[58] Brose, *J Biol Chem* (1993) **268**: 22663–71.
[59] Kutsuwada, *Nature* (1992) **358**: 36–41.
[60] Cotman, *Trends Neurosci* (1987) **10**: 263–5.
[61] Cotman, *Annu Rev Neurosci* (1988) **11**: 61–80.
[62] Lipton, *N Engl J Med* (1994) **330**: 613–22.
[63] Tsumoto, *Nature* (1987) **327**: 513–14.
[64] Malenka, *Trends Neurosci* (1993) **12**: 521–7.
[65] Raymond, *Trends Pharmacol Sci* (1993) **147**: 147–53.
[66] Collingridge, *Trends Neurosci* (1987) **10**: 288–93.
[67] Scheetz, *FASEB J* (1994) **8**: 745–52.
[68] Rauschecker, *Physiol Rev* (1991) **71**: 587–615.
[69] Tsumoto, *Jpn J Physiol* (1990) **40**: 573–93.
[70] Rauschecker, *Nature* (1987) **326**: 183–5.
[71] Cline, *Proc Natl Acad Sci USA* (1987) **84**: 4342–5.
[72] Gu, *Dev Brain Res* (1989) **47**: 281–8.
[73] Morris, *Nature* (1986) **319**: 774–6
[74] Eichenbaum, *Trends Neurosci* (1993) **16**: 163–4.
[75] Daw, *Annu Rev Neurosci* (1993) **16**: 207–22.
[76] Fox, *Neural Comput* (1992) **4**: 59–83.
[77] Fox, *J Neurosci* (1989) **9**: 2443–54.
[78] Fox, *J Neurophysiol* (1990) **64**: 1413–28.
[79] Kemp, *J Physiol* (1982) **323**: 377–91.
[80] Spierra, *Brain Res* (1988) **45**: 130–6.
[81] Salt, *Nature* (1986) **322**: 263–6.
[82] Suga, *Cold Spring Harbor Symp Quant Biol* (1992) **55**: 585–97.
[83] Stephens, *Psychopharmacology* (1986) **90**: 166–9.
[84] Croucher, *Science* (1982) **216**: 899–901.
[85] Meldrum, in Watkins and Collingridge (1989) *(see book refs).*
[86] Plaitakis, (1984) *(see book refs).*
[87] DeBelleroche, *Neurochem Pathol* (1984) **2**: 1–6.
[88] Spencer, *Lancet* (1986) **1**: 965.
[89] Weiloch, *Science* (1985) **230**: 681–3.
[90] Choi, *Annu Rev Neurosci* (1990) **13**: 171–82.
[91] Meldrum, *Clin Sci* (1985) **68**: 113–21.
[92] Rothman, *Trends Neurosci* (1987) **10**: 299–302.
[93] Herrling, in Watkins and Collingridge (1989) *(see book refs).*
[94] Bazil, *Brain Res* (1991) **555**: 149–52.
[95] Kubo, *Neurosci Lett* (1988) **87**: 69–74.

96 Hong, *Brain Res* (1992) **569**: 38–45.
97 Chen, *Nature* (1992) **356**: 521–3.
98 Dickenson, *Neurosci Lett* (1991) **121**: 263–6.
99 Dickenson, *Trends Pharmacol Sci* (1990) **11**: 307–9.
100 Foutz, *Neurosci Lett* (1988) **87**: 221–6.
101 Hartley, *J Neurosci* (1993) **13**: 1993–2000.
102 Meldrum, *Trends Pharmacol Sci* (1990) **11**: 379–87.
103 Malenka, *Trends Neurosci* (1989) **12**: 444–50.
104 Favaron, *Proc Natl Acad Sci USA* (1990) **87**: 1983–7.
105 Connor, *Science* (1988) **240**: 649–53.
106 Williams, *Nature* (1989) **341**: 739–42.
107 Kontos, *Science* (1980) **209**: 1242–5.
108 Barbour, *Nature* (1989) **342**: 918–20.
109 O'Brian, *Biochem Biophys Res Commun* (1988) **155**: 1374–80.
110 Lynch, *Neurosci Lett* (1986) **65**: 171–6.
111 Orrenius, *Trends Pharmacol Sci* (1989) **10**: 281–5.
112 Nichols, *Nature* (1990) **343**: 647–51.
113 Garthwaite, *Nature* (1988) **336**: 385–8.
114 Hibbs, *Biochem Biophys Res Commun* (1988) **157**: 87–94.
115 Beckman, *Proc Natl Acad Sci USA* (1990) **87**: 1620–4.
116 Zaidi, *J Membr Biol* (1989) **110**: 209–16.
117 Siman, *J Neurosci* (1989) **9**: 1579–90.
118 McCord, *Fed Proc* (1987) **46**: 2402–6.
119 Melloni, *Trends Neurosci* (1989) **12**: 438–44.
120 Kishimoto, *J Biol Chem* (1989) **264**: 4088–92.
121 Olney, *Exp Brain Res* (1971) **14**: 61–76.
122 Choi, *Neuron* (1988) **1**: 623–34.
123 Faden, *Trends Pharmacol Sci* (1992) **13**: 29–35.
124 Lipton, *Trends Neurosci* (1992) **15**: 75–9.
125 Coulter, *J Neurophysiol* (1992) **68**: 362–73.
126 Fisher, *Neurotrans Epilepsy* (1991) **11**: 131–45.
127 Dingledine, *Trends Pharmacol Sci* (1990) **11**: 334–8.
128 Gasic, *Annu Rev Physiol* (1992) **54**: 507–36.
129 Lipton, *Trends Neurosci* (1993) **16**: 527–32.
130 Lobner, *J Neurosci* (1993) **13**: 4861–71.
131 Nowak, *Neuron* (1992) **8**: 181–7.
132 Zeevalk, *J Neurochem* (1992) **59**: 1211–20.
133 Weiss, *Neuron* (1993) **10**: 43–9.
134 Wahlestedt, *Nature* (1993) **363**: 260–3.
135 Kohr, *J Neurosci* (1993) **13**: 3612–27.
136 Cherubini, *J Physiol Lond* (1991) **436**: 531–47.
137 Battaglia, *J Comp Neurol* (1988) **277**: 302–12.
138 Adrian, *Proc R Soc Lond Ser B* (1931) **109**: 1–18.
139 Russo, *Neuroscience* (1994) **61**: 191–7.
140 Belhage, *Neuroscience* (1993) **54**: 1019–34.
141 Copenhagen, *Nature* (1989) **341**: 536–9.
142 Brown, *Science* (1988) **242**: 724–8.
143 Ben-Ari, *Trends Neurosci* (1992) **15**: 333–9.
144 Swope, *FASEB J* (1992) **6**: 2514–23.

145 Mcbain, *Physiol Rev* (1994) **74**: 723–60.
146 Verhage, *Prog Neurobiol* (1994) **42**: 539–74.
147 Nakanishi, *Annu Rev Biophys Biomol Struc* (1994) **23**: 319–48.
148 Thomson, *Trends Neurosci* (1994) **17**: 119–26.
149 Larson, *Brain Res* (1986) **368**: 347–50.
150 Rose, *Neurosci Lett* (1986) **69**: 244–8.
151 Otto, *Hippocampus* (1991) **1**: 181–92.
152 Artola, *Trends Neurosci* (1993) **16**: 480–7.
153 Kauer, *Nature* (1988) **334**: 250–2.
154 Perkel, *Neuron* (1993) **11**: 817–23.
155 Lynch, *Nature* (1983) **305**: 719–21.
156 Bortolotto, *Neuropharmacology* (1993) **32**: 1–9.
157 Obenaus, *Neurosci Lett* (1989) **98**: 172–8.
158 Harvey, *Neurosci Lett* (1992) **139**: 197–200.
159 Johnston, *Annu Rev Physiol* (1992) **54**: 489–505.
160 Komatsu, *J Neurophysiol* (1992) **67**: 401–10.
161 Kullmann, *Neuron* (1992) **9**: 1175–83.
162 Grover, *Nature* (1990) **347**: 477–9.
163 Miyakawa, *Neuron* (1992) **9**: 1163–73.
164 Keller, *EMBO J* (1992) **11**: 891–6.
165 Asztely, *Eur J Neurosci* (1992) **4**: 681–90.
166 Aniksztejn, *Nature* (1991) **349**: 67–9.
167 Malgaroli, *Nature* (1992) **357**: 134–9.
168 Calabresi, *Eur J Neurosci* (1992) **4**: 929–35.
169 Walsh, *Neuroscience* (1993) **57**: 241–8.
170 Thompson, *Nature* (1992) **359**: 638–41.
171 Perouansky, *J Physiol London* (1993) **465**: 223–44.
172 Bekkers, *Cold Spring Harbor Symp Quant* (1990) **55**: 131–5.
173 Mott, *Science* (1991) **252**: 1718–20.
174 Brocard, *Neuron* (1993) **11**: 751–7.
175 Izumi, *Neurosci Lett* (1991) **122**: 187–90.
176 Collingridge, *Int Acad Biomed Drug Res* (1991) **2**: 41–9.
177 Radpour, *Neurosci Lett* (1992) **138**: 119–22.
178 Xie, *J Neurophysiol* (1992) **67**: 1009–13.
179 Sakurai, *J Neurochem* (1993) **60**: 1344–53.
180 Harris, *Neurosci Lett* (1986) **70**: 132–7.
181 Guthrie, *Nature* (1991) **354**: 76–80.
182 Benke, *Proc Natl Acad Sci USA* (1993) **90**: 7819–23.
183 Forsythe, *J Physiol Lond* (1988) **396**: 515–33.
184 Bekkers, *Nature* (1989) **341**: 230–3.
185 Malenka, *Mol Neurobiol* (1991) **5**: 289–95.
186 Izquierdo, *Trends Pharmacol Sci* (1991) **12**: 128–9.
187 D'Angelo, *Nature* (1990) **346**: 467–70.
188 Bekkers, *Proc Natl Acad Sci USA* (1991) **88**: 7834–8.
189 Takano, *Biochem Biophys Res Commun* (1993) **197**: 922–6.
190 Nakanishi, *Proc Natl Acad Sci USA* (1992) **89**: 8552–6.
191 Anantharam, *FEBS Lett* (1992) **305**: 27–30.
192 Sugihara, *Biochem Biophys Res Commun* (1992) **185**: 826–32.
193 Hollmann, *Neuron* (1993) **10**: 943–54.

194 Bai, *Biochim Biophys Acta* (1993) **1152**: 197–200.
195 Planellscases, *Proc Natl Acad Sci USA* (1993) **90**: 5057–61.
196 Burnashev, *Science* (1992) **257**: 1415–19.
197 Lee, *FEBS Lett* (1992) **311**: 81–4.
198 Sakurada, *J Biol Chem* (1993) **268**: 410–15.
199 Lester, *Annu Rev Biophys Biomol Struc* (1992) **21**: 267–92.
200 Mori, *Nature* (1992) **358**: 673–75.
201 Durand, *Proc Natl Acad Sci USA* (1992) **89**: 9359–63.
202 Ikeda, *FEBS Lett* (1992) **313**: 34–8.
203 Hahn, *Proc Natl Acad Sci USA* (1988) **1988**: 6556–60.
204 Chetkovich, *Proc Natl Acad Sci USA* (1991) **88**: 6467–71.
205 Silva, *Science* (1992) **257**: 201–6.
206 Zeilhofer, *Neuron* (1993) **10**: 879–87.
207 Chernevskaya, *Nature* (1991) **349**: 418–20.
208 Simpson, *J Neurochem* (1993) **61**: 760–3.
209 Conquet, *Nature* (1994) **372**: 237–43.
210 Chen, *Neuron* (1991) **7**: 319–26.
211 Artola, *Nature* (1987) **330**: 649–52.
212 Olney, *Science* (1991) **254**: 1515–18.
213 Novelli, *J Neurosci* (1987) **7**: 40–7.
214 Garthwaite, *Neuroscience* (1988) **26**: 321–6.
215 Garthwaite, *Trends Neurosci* (1991) **14**: 60–7.
216 Madison, *Annu Rev Neurosci* (1991) **14**: 379–97.
217 Malinow, *Nature* (1990) **346**: 177–80.
218 Bekkers, *Nature* (1990) **346**: 724–9.
219 Kullmann, *Nature* (1992) **357**: 240–4.
220 Davies, *Nature* (1989) **338**: 500–3.
221 Halpain, *Neuron* (1990) **5**: 237–46.
222 Lieberman, *Nature* (1994) **369**: 235–9.
223 Rosenmund, *J Physiol Lond* (1993) **470**: 705–29.
224 Okada, *Neurosci Lett* (1989) **100**: 141–6.
225 Williams, *Neurosci Lett* (1989) **107**: 301–6.
226 Lynch, *J Neurochem* (1991) **56**: 113–18.
227 Oliver, *Brain Res* (1989) **505**: 233–8.
228 Huang, *Neuroscience* (1992) **49**: 819–27.
229 Kelso, *J Physiol Lond* (1992) **449**: 705–18.
230 Hu, *Nature* (1987) **328**: 426–9.
231 Malenka, *Nature* (1986) **321**: 175–7.
232 O'Dell, *Nature* (1991) **353**: 558–60.
233 Bading, *Science* (1991) **253**: 912–14.
234 Lester, *Nature* (1990) **346**: 565–7.
235 Hestrin, *Neuron* (1990) **5**: 247–53.
236 Gu, *J Neurosci* (1994) **14**: 4561–70.
237 Clements, *Neuron* (1991) **7**: 605-13.
238 Jahr, *Science* (1992) **255**: 470–2.
239 Gibb, *J Physiol Lond* (1992) **456**: 143–79.
240 Edmonds, *Proc R Soc Lond [Biol]* (1992) **250**: 279–86.
241 Mayer, *J Physiol Lond* (1987) **394**: 501–27.
242 Jahr, *Proc Natl Acad Sci USA* (1993) **90**: 11573–7.

243 Reichling, *J Physiol Lond* (1993) **469**: 67–88.
244 Schinder, *FEBS Lett* (1993) **332**: 44–8.
245 Vyklicky, *J Physiol Lond* (1993) **470**: 575–600.
246 Chizhmakov, *J Physiol Lond* (1992) **448**: 453–72.
247 Legendre, *J Neurosci* (1993) **13**: 674–84.
248 Hestrin, *Neuron* (1992) **9**: 991–9.
249 Leonard, *Neuron* (1990) **4**: 53–60.
250 Rosenmund, *Neuron* (1993) **10**: 805–14.
251 Schneggenburger, *Neuron* (1993) **11**: 133–43.
252 Howe, *J Physiol Lond* (1991) **432**: 143–202.
253 Cull-Candy, *J Physiol Lond* (1989) **415**: 555–82.
254 Jahr, *Nature* (1987) **325**: 522–5.
255 Cull-Candy, *Nature* (1987) **325**: 525–8.
256 Cull-Candy, *J Physiol Lond* (1988) **400**: 189–222.
257 Ascher, *J Physiol Lond* (1988) **399**: 207–26.
258 Smith, *J Neurophysiol* (1991) **66**: 369–78.
259 Ascher, *J Physiol Lond* (1988) **399**: 247–66.
260 Keller, *J Physiol Lond* (1991) **435**: 275–93.
261 Wang, *Can J Physiol Pharmacol* (1992) **70**: 283–8.
262 Mayer, *Nature* (1984) **309**: 261–3.
263 Nowak, *Nature* (1984) **307**: 462–5.
264 Legendre, *J Physiol Lond* (1990) **429**: 429–49.
265 Westbrook, *Nature* (1987) **328**: 640–3.
266 Gottesman, *J Neurophysiol* (1992) **68**: 596–604.
267 Yamakura, *Neuroreport* (1993) **4**: 687–90.
268 MacDonald, *J Physiol Lond* (1991) **432**: 483–508.
269 Lodge, *Trends Pharmacol Sci* (1990) **11**: 81–6.
270 Chen, *J Neurosci* (1992) **12**: 4427–36.
271 Lin, *FASEB J* (1993) **7**: 479–85.
272 Netzer, *Eur J Pharmacol* (1993) **238**: 209–16.
273 Wright, *J Physiol Lond* (1991) **439**: 579–604.
274 Malenka, *Nature* (1989) **340**: 554–7.
275 Aizenman, *Neuron* (1989) **2**: 1257–63.
276 Tang, *Mol Pharmacol* (1993) **44**: 473–8.
277 Lei, *Neuron* (1992) **8**: 1087–99.
278 Lipton, *Nature* (1993) **364**: 626–32.
279 Ginsberg, *Stroke* (1976) **7**: 125–31.
280 Reynolds, *Br J Pharmacol* (1990) **101**: 178–82.
281 Levy, *Neurosci Lett* (1990) **110**: 291–6.
282 Sucher, *Neuroreport* (1990) **1**: 29–32.
283 Tang, *J Physiol Lond* (1993) **465**: 303–23.
284 Wieraszko, *Neuron* (1993) **10**: 553–7.
285 Williams, *Life Sci* (1991) **48**: 469–98.
286 Yoneda, *Neurosci Res* (1991) **10**: 1–33.
287 Scott, *Trends Neurosci* (1993) **16**: 153–60.
288 Rock, *Mol Pharmacol* (1992) **41**: 83–8.
289 Rock, *Mol Pharmacol* (1992) **42**: 157–64.
290 Benveniste, *J Physiol London* (1993) **464**: 131–63.
291 Bird, *J Biol Chem* (1993) **268**: 21486–8.

292 Morgan, *Biochem Soc Trans* (1990) **18**: 1080–3.
293 Vyklicky, *J Physiol Lond* (1990) **428**: 313–31.
294 Weight, *Ann NY Acad Sci* (1991) **625**: 97–107.
295 Lovinger, *Science* (1989) **243**: 1721–4.
296 Gonzalez, *Trends Pharmacol Sci* (1990) **11**: 137–9.
297 Skeen, *Mol Pharmacol* (1993) **44**: 443–50.
298 Bowlby, *Mol Pharmacol* (1993) **43**: 813–19.
299 Vyklicky, *J Physiol Lond* (1990) **430**: 497–517.
300 Traynelis, *Nature* (1990) **345**: 347–50.
301 Kwiecien, *Eur J Pharmacol* (1993) **249**: 325–9.
302 Yoneda, *J Neurochem* (1993) **60**: 634–45.
303 Franklin, *Mol Pharmacol* (1992) **41**: 134–46.
304 Draguhn, *Neurosci Lett* (1991) **132**: 187–90.
305 Dumuis, *Nature* (1988) **336**: 68–70.
306 Nicholls, *Nature* (1992) **360**: 106–7.
307 Linden, *J Neurosci* (1992) **12**: 3601.
308 Herrero, *Nature* (1992) **360**: 163–6.
309 Rusin, *J Neurochem* (1993) **60**: 952–60.
310 Miller, *Nature* (1992) **355**: 722–5.
311 Wong, *Annu Rev Pharmacol Toxicol* (1991) **31**: 401–25.
312 Watkins, *Trends Neurosci* (1987) **10**: 265–72.
313 Mayer, *J Neurophysiol* (1988) **60**: 645–63.
314 Watkins, *Trends Pharmacol Sci* (1990) **11**: 25–33.
315 Swartz, *Mol Pharmacol* (1992) **41**: 1130–41.
316 McKernan, *J Neurochem* (1989) **52**: 777–85.
317 Zeilhofer, *Eur J Pharmacol* (1992) **213**: 155–8.
318 Sernagor, *Neuron* (1989) **2**: 1221–7.
319 Pereira, *J Pharmacol Exp Ther* (1992) **261**: 331–40.
320 Huettner, *Science* (1989) **243**: 1611–13.
321 Grimwood, *Mol Pharmacol* (1992) **41**: 923–30.
322 Johnson, *Nature* (1987) **325**: 529–31.
323 Kleckner, *Science* (1988) **241**: 835–7.
324 Kemp, *Proc Natl Acad Sci USA* (1988) **85**: 6547–50.
325 Moroni, *Eur J Pharmacol* (1992) **218**: 145–51.
326 Kemp, *Trends Pharmacol Sci* (1993) **14**: 20–5.
327 White, *Epilepsia* (1992) **33**: 564–72.
328 McCabe, *J Pharmacol Exp Ther* (1993) **264**: 1248–52.
329 Tang, *Proc Natl Acad Sci USA* (1990) **87**: 6445–9.
330 Donevan, *Mol Pharmacol* (1992) **41**: 727–35.
331 Williams, *Mol Pharmacol* (1993) **44**: 851–9.
332 Olivera, *Science* (1990) **249**: 257–63.
333 Haack, *Neurosci Lett* (1993) **163**: 63–6.
334 Chandler, *J Biol Chem* (1993) **268**: 17173–8.
335 Toki, *J Biol Chem* (1992) **267**: 14884–92.
336 Rogawski, *Trends Pharmacol Sci* (1993) **14**: 325–31.
337 Bigge, *Biochem Pharmacol* (1993) **45**: 1547–61.
338 Bisaga, *Eur J Pharmacol* (1993) **242**: 213–20.
339 Murata, *Eur J Pharmacol* (1993) **239**: 9–15.
340 Abrahams, *Soc Neurosci Abstr* (1988) **14**: 937.

341 Olney, *Science* (1989) **244**: 1360–3.
342 Allen, *Science* (1990) **244**: 1360–3.
343 Manev, *FASEB J* (1990) **4**: 2789–96.
344 Engelsen, *Acta Neurol Scand* (1982) **74**: 337–55.
345 Benveniste, *J Cereb Blood Flow Metab* (1989) **9**: 629–39.
346 Manev, *Mol Pharmacol* (1989) **36**: 106–12.
347 Choi, *J Neurosci* (1987) **7**: 369–79.
348 Ogura, *Exp Brain Res* (1988) **73**: 447–58.
349 Holopainen, *Neurosci Lett* (1989) **98**: 57–62.
350 Pauwels, *Mol Pharmacol* (1989) **36**: 525–31.
351 Perkins, *Brain Res* (1982) **247**: 184–7.
352 Perkins, *J Pharmacol Exp Ther* (1983) **226**: 551–7.
353 Mayer, *Nature* (1989) **338**: 425–7.
354 Lester, *J Neurosci* (1993) **13**: 1088–96.
355 Zilberter, *Neuron* (1990) **5**: 597–602.
356 Kovalchuk, *Neuroscience* (1993) **54**: 557–9.
357 Henzi, *Mol Pharmacol* (1992) **41**: 793–801.
358 Weiss, *Neuron* (1989) **3**: 321–6.
359 Debono, *Eur J Pharmacol* (1993) **235**: 283–9.
360 Perkel, *J Physiol Lond* (1993) **471**: 481–500.
361 Lerma, *Mol Pharmacol* (1992) **41**: 217–22.
362 Williams, *Neuron* (1990) **5**: 199–208.
363 Nicoll, *Physiol Rev* (1990) **70**: 513–65.
364 Wisden, *Curr Opin Neurobiol* (1993) **3**: 291–8.
365 Collingridge, *Physiol Rev* (1989) **40**: 145–210.
366 Mayer, *Prog Neurobiol* (1987) **28**: 197–276.
367 Monaghan, *Annu Rev Pharmacol Toxicol* (1989) **29**: 365–402.
368 Ascher, *J Physiol Lond* (1988) **399**: 227–45.
369 Ozawa, *Jpn J Physiol* (1993) **43**: 141–59.
370 Peters, *Science* (1987) **236**: 589–93.
371 Shinozaki, *Prog Neurobiol* (1988) **30**: 399–435.
372 Henneberry, *Bioessays* (1992) **14**: 465–71.
373 Barnard, *Trends Pharmacol Sci* (1992) **13**: 12–13.
374 Siesjö, *Ann NY Acad Sci* (1988) **522**: 638–61.
375 Barnes, *Science* (1988) **239**: 254–6.
376 Foster, *Nature* (1987) **329**: 395–6.
377 Thomson, *Trends Neurosci* (1989) **12**: 349–53.
378 Blatz, *Trends Neurosci* (1987) **10**: 463–7.
379 Wood, *Neurochem Res* (1990) **15**: 217–30.
380 Murphy, *Br J Pharmacol* (1988) **95**: 932–8.

William J. Brammar **Entry 09**

NOMENCLATURES

Abstract/general description

09-01-01: The nicotinic acetylcholine receptor (nAChR) is a ligand-gated, non-selective, cation channel that is activated by binding of the chemical neurotransmitter acetylcholine. There are two major subtypes of nAChR: the muscle type and the neuronal type. In vertebrate skeletal muscle, the nAChR is concentrated at the **endplate** of **neuromuscular junctions**†, where it binds the neurotransmitter acetylcholine (ACh) and mediates excitatory† transmission and initiation of the **action potential**†. In neural tissues, the nAChRs are present on **autonomic neurones** and **adrenal chromaffin cells** in the peripheral nervous system and on many neurones in the **central nervous system**. Neuronal nAChRs are probably involved in the control of electrical excitation and release of neurotransmitters.

09-01-02: The nAChRs are targets for several animal and plant **toxins**, **local anaesthetics**, **sedatives**† and **hallucinogenic drugs**, all acting as antagonists and inhibiting channel function. **Nicotine** itself is a receptor agonist and probably has its addicting psychoactive effects through actions on nAChRs in the central nervous system.

09-01-03: In nearly all cases, both the neuronal and the muscle nAChRs mediate 'fast' inward currents in response to activation by acetylcholine. The neuronal channels, unlike the muscle isoforms, are modulated to pass increased currents by **external Ca^{2+}**, and can play a role in activating Ca^{2+}-dependent pathways via Ca^{2+}-influx.

09-01-04: The nAChRs from *Torpedo* electric organ and skeletal muscle are composed of five subunits, $\alpha_2\beta\gamma\delta$; in adult muscles of some species an ε subunit displaces the γ subunit.

09-01-05: The neuronal nAChRs show considerable **diversity** in their **subunit composition** and function. Two main types of subunit have been identified, comprising eight distinct α and four β subunits. The muscle α-subunit is referred to as $\alpha1$ and the β-subunit as $\beta1$. The neuronal subunits are designated $\alpha2$–$\alpha9$ and $\beta2$–$\beta5$. The β-subunits from avian systems tend to be called 'non-α' (nα).

09-01-06: The general structure of the *Torpedo* nAChR has been determined to 9 Å resolution by electron microscopy and image-reconstruction† techniques. The nAChR is composed of a ring of five membrane-spanning subunits delineating a central cation-conducting pore†. The channel opens transiently when acetylcholine is released from nerve terminals into the **synaptic cleft**†, and depolarizes the post-synaptic membrane.

Category (sortcode)

09-02-01: ELG CAT nAChR, i.e. extracellular ligand-gated cation channels integral to the nicotinic acetylcholine receptor. The suggested **electronic retrieval code** (unique embedded identifier or **UEI**) for 'tagging' of new articles of relevance to the contents of this entry is UEI: **NACHR-NAT** (for reports on native† channel properties) and UEI: **NACHR-HET** (for reports or reviews on channel properties applicable to heterologously† expressed recombinant† subunits encoded by cDNAs† or genes†). *For a discussion of the advantages of UEIs and guidelines on their implementation, see the section on Resource J under Introduction and layout, entry 02, and for further details, see Resource J – Search criteria & CSN development, entry 65.*

Channel designation

Channel designations presently in use

09-03-01: The ionotropic† acetylcholine-gated cation channels are generally designated as nicotinic acetylcholine receptors (nAChRs). The muscle form of the receptor closely resembles that found in the electric organ of the electric ray, *Torpedo californica*, while the many isoforms found in the nervous system, collectively known as 'neuronal channels', are less closely related and are electrophysiologically and pharmacologically distinguishable.

Current designation

09-04-01: Generally of the form $I_{agonist}$, i.e. I_{ACh}.

Gene family

09-05-02: Sequences encoding isoforms of the subunits of muscle and neuronal nAChRs have been determined for several species, as shown in Table 1.

Xenopus muscle subunits

09-05-02: Two distinct muscle α subunit cDNAs†, $\alpha 1_a$ and $\alpha 1_b$, derived from mRNAs encoded by different genes, have been cloned from ***Xenopus* embryos**. Both $\alpha 1_a$ and $\alpha 1_b$ encode a protein with a predicted signal peptide and a mature protein of 437 amino acids[23].

Subtype classifications

09-06-02: A change in the electrophysiological properties of the nAChR channels in vertebrate muscle is associated with the switch from the gamma (γ) subunit in embryonic muscle to the epsilon (ε) subunit in the adult channel[24] *(For details, see Developmental regulation, 09-11.)*

09-06-02: The neuronal channels can be constructed from various combinations drawn from eight α variants and four different β subunits. Each known subunit can affect receptor kinetic parameters, receptor pharmacology and/or channel conductance[24–26]. *(For details, see ELECTROPHYSIOLOGY* and *PHARMACOLOGY sections.)*

Table 1. *nAChR subunits specified by cloned sequences from various species (From 09-05-01)*

Species	Subunit	Transcript size (kb)	Encoding[a]		Mature peptide, M_r	Refs
			Amino acids (aa)	Signal peptide (aa)		
Torpedo	α	2.3	437	24	50 150	***1, 2***
	β	2.0	469	24	53 681	***3***
	γ	2.1	489	17		***36***
	δ	6.0	501	21	57 565	***3***
Rat	$\alpha2$		484	27	55 500	***4***
	$\alpha3$		474	25	54 800	***5***
	$\alpha4$		600		67 100	***6***
	$\alpha5$		424		48 800	***7***
	$\alpha6$		463	30	53 300	***8***
	$\alpha7$		480	22	54 200	***9***
	$\alpha9$		457	22		***218***
	$\beta2$		475		54 300	***10***
	$\beta3$		434		50 200	***11***
	$\beta4$		475		53 300	***12***
	n$\alpha2$		499		56 100	***13***
	ε		473	20		***220***

Chicken	α	2.8	437	19	50 150	*14*
	α2		505	23	58 100	*14*
	α3		474	22	54 800	*14*
	α4		599		68 400	*14*
	α5		425		49 000	*15*
	α6		463		54 100	*8*
	α7	7, 3	480	22	54 600	*16, 17*
	α8		481		55 200	*16*
	β2 (nα, nα1)		473		54 000	*14, 16*
	β3 (nα2)		435		50 100	*8*
	β4 (nα3)		467		53 500	*15*
	γ		492	22	56 484	*18*
	δ		495	18	57 215	*18*
Human	α		437	20	49 694	*19*
	α3		474	29	54 100	*20*
	α5	2.7, 2.1	446		51 100	*21*
	α7	5.9, 2.6, 1.3	502			*219*
	β2		477	25	54 700	*22*
	β4		456		51 400	*8*

[a]Encoding data show the number of amino acid residues in the specified channel subunit, determined from the longest open reading frame[†] and the predicted position of signal peptide cleavage, with signal peptide residues and the predicted molecular weight listed separately. In many cases the position of signal cleavage has not been determined or predicted, and the number of amino acids encoded is then determined directly from the open reading frame. Sequences encoding four different *Drosophila* subunits, one cockroach subunit, and four goldfish subunits, all of the neuronal type, have been determined and are detailed in ref.[8].

Trivial names

09-07-01: Endplate channels (nAChRs at the motor endplate); nicotinic receptor.

EXPRESSION

Cell-type expression index

09-08-01: *See mRNA distribution, 09-13, and Protein distribution, 09-15.*

Channel density

nAChR channel density at the vertebrate neuromuscular junction

09-09-01: Endplate† **channels** are 'by definition' expressed at the vertebrate **neuromuscular junction**† in high density. Electron microscope autoradiography has been used to show a nAChR density of 10 000–20 000 channels per μm^2 at the neuromuscular junction[27].

nAChR channel density at neuronal sites

09-09-02: The density of nAChR channels in the synaptic membrane of the chicken **ciliary ganglion** has been measured by quantitative electron microscopic autoradiography and [^{125}I]-**neuronal-bungarotoxin (n-Bgt)**, which blocks nAChR function in this tissue. The n-Bgt binds to two sites, one of which is shared with α-Bgt. When the α-Bgt-binding sites are blocked, [^{125}I]-n-Bgt binds selectively to synaptic sites at a density of approximately $600/\mu m^2$ [28].

09-09-03: The total number of ACh-activatable nicotinic channels in bovine **adrenal chromaffin cells** has been estimated at 2600[29].

α7 channels in neuroblastoma cell line

09-09-04: The human **neuroblastoma cell line SH-SY5Y** expresses an α-Bgt-sensitive neuronal nAChR with the properties of the α7 homomeric channel. The channel density measured by [^{125}I]-α-Bgt binding is 2400 per cell[219].

nAChR channel densities obtained by heterologous expression

09-09-05: A high density of nAChR channels can be obtained by **transient expression**† of cRNAs† injected into *Xenopus* oocytes. Estimates of 2×10^9 molecules of nAChR per cell (3.2 fmol per oocyte) were obtained by α-Bgt binding to oocytes 2 days after microinjection of cRNAs encoding all four mouse nAChR subunits[30]. Expression of the human α7 cRNA in *Xenopus* oocytes produces 7.2×10^8 α-Bgt-binding sites per oocyte, with 63% of the expressed monomers assembled into pentamers at the cell surface[219].

nAChR channels are associated with cytoskeletal components

09-09-06: Nicotinic receptors are known to associate with components of the **cytoskeleton**†[31–33], and such interactions may control channel mobility, subcellular distribution and local channel density *(for details, see Protein interactions, 09-31).*

Special note on determination of channel density by immunoassay

09-09-07: There is an important cautionary note for the interpretation of data relating to channel density determined by ligand-binding or quantitative immunoassay. It is clear that antisera and ligand-binding assays can detect both active and inactive channels at the cell surface, and that the quantitative relationship between the two forms can vary. For example, analysis of chicken **ciliary ganglion neurones** grown in normal and high K^+ media shows that the ACh sensitivity is reduced several-fold by elevated K^+, but the number of nAChRs detected by antibody-binding remains unchanged. Electrophysiological estimations of the number of functional channels are 2000 per cell under normal growth conditions, and 700 per cell in high K^+ conditions. The number of antibody-binding sites is 100 000 per cell under both conditions. The ciliary ganglion neurones have two classes of nAChR on their surface: a large excess of **'non-activatable channels'** and a small fraction of ACh-activatable channels, the size of the latter pool being sensitive to growth conditions[34].

Cloning resource

Original isolates

09-10-01: Sequences encoding subunits of the nAChR were first obtained from cDNA libraries prepared from poly(A)$^+$ RNA from the electric organ of the electric ray, *Torpedo californica*[35] or *T. marmorata*[2]. The candidate clones were recognized by their hybridization† to mixed synthetic oligonucleotide probes† designed on the basis of primary amino acid sequence from the *Torpedo* α[1], β, δ[3] and γ[36] subunits, then verified by DNA sequencing.

Cloning sequences encoding muscle and neuronal nAChR subunits

09-10-02: Clones encoding the subunits of muscle or neuronal nAChR channels were isolated from cDNA libraries prepared from the RNA of appropriate tissues or from genomic DNA libraries by their cross-hybridization† to probes derived from cloned *Torpedo* sequences. The use of cloned sequences as probes to screen libraries at reduced stringency† allowed the isolation of homologous† sequences.

Cultured cell lines as a source of nAChR mRNA

09-10-03: A number of cell lines in culture express the genes encoding forms of the nAChR and can be used as a source of mRNA from which to generate cDNAs corresponding to channel subunits. The **mouse myogenic cell line, Sol 8**, undergoes a developmental change in culture and switches from expressing the foetal form of muscle nAChR, $\alpha_2\beta\gamma\delta$, to the adult form, $\alpha_2\beta\varepsilon\delta$[37]. The human **rhabdomyosarcoma cell line TE671** is a rich source of human muscle nAChR mRNA[21]. The human **neuroblastoma line IMR 32** is the source from which the human $\alpha5$ subunit cDNA was cloned[21] and the human **small cell lung carcinoma cell line NCI-N-592** also produces high levels of $\alpha5$ subunit mRNA[21]. The cDNA for the human $\alpha7$ subunit was generated from mRNA isolated from the **SH-SY5Y cell line**, derived from a neuroblastoma of a sympathetic adrenergic ganglion[219].

Developmental regulation

Developmental gene-switching in skeletal muscle

09-11-01: The γ subunit of the embryonic or foetal muscle nAChR is replaced by the homologous† ε subunit in adult soleus muscle nAChR[38]. Direct comparison of the nAChR channels of foetal and adult bovine skeletal muscle by single-channel recording† with outside-out† patches shows that the channels are distinguishable by their differences in single-channel current, slope conductance† and their duration after activation by ACh *(see Table 2)*. Single channels obtained by expression of the cRNAs encoding bovine α, β, γ and δ subunits in *Xenopus* oocytes closely resemble the channels from foetal muscle in all respects, while those derived by expression of cRNAs encoding α, β, δ and ε subunits are indistinguishable from the channels from adult bovine muscle[38] (see Fig. 1).

Table 2. *Single-channel properties of two forms of bovine nAChRs expressed in muscle and* Xenopus *oocytes (From 09-11-01)*

nAChR channels from	Conductance[a] γ (pS)	Average duration[b]	
		τ (−100) (ms)	H (mV)
Foetal muscle	40 (4)	11.0 (4)	108
Adult muscle	59 (2)	5.6 (6)	93
Oocytes with AChRγ	39 (7)	10.4 (7)	106
Oocytes with AChRϵ	59 (6)	5.3 (6)	83

[a] γ represents the slope of the linear I–V relationships determined with outside-out† patches.
[b] τ (−100) represents the calculated duration of elementary currents at −100 mV membrane potential. H is the shift in membrane potential† that causes an e-fold change in the average duration of elementary currents.
Data taken from ref.[38].

Changes in γ and ε mRNAs during muscle development

09-11-02: Northern blot† analyses of RNA from bovine diaphragm and leg muscles[38] and from rat muscle show reciprocal changes in the level of γ and ε subunit-specific mRNAs during post-natal development.

09-11-03: The embryonic form of the channel permits more current flow (smaller conductance but a significantly longer open time), so **miniature endplate currents**† can promote spontaneous muscle contractions in developing muscle[39]. Spontaneous muscle activity induces differentiation of the neuromuscular junction.

Developmental regulation by electrical activity in primary cultures

09-11-04: In primary muscle cultures, spontaneous or stimulated electrical activity of the myotubes represses nAChR biosynthesis, whereas treatment with pharmacological agents that inhibit electrical activity increases nAChR levels. The Ca^{2+} channel blocker **verapamil** (10 mM) causes a 2.5-fold increase in surface AChR, measured by [^{125}I]-α-Bgt binding, and a 7.9-

fold increase in α subunit mRNA in cultured chicken myotubes[40]. The Na^+ channel blocker **TTX** (0.5 mM) generates 2.9-fold and 14.9-fold increases in these respective levels. These effects are almost eliminated by the **Ca^{2+} ionophore A23187**.

09-11-05: The effects of TTX in elevating nAChR and α subunit mRNA are enhanced by 50 mM **dantrolene**, a blocker of Ca^{2+} efflux from the sarcoplasmic reticulum[40].

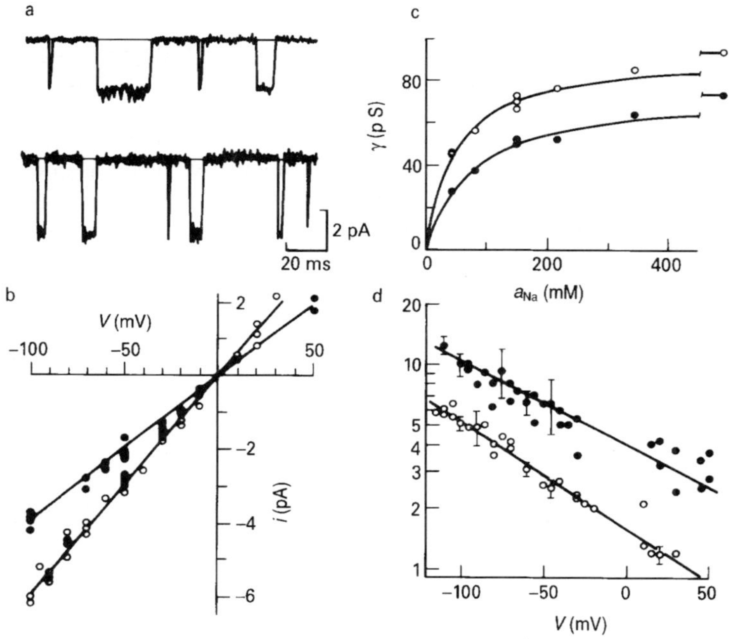

Figure 1. *Properties of ACh-activated single-channel currents in* Xenopus *oocytes injected with cRNAs encoding bovine nAChR α, β, γ and δ subunits ($AChR_\gamma$) or with cRNAs encoding the α, β, δ and ε subunits ($AChR_\epsilon$). (a) Single-channel currents activated by 0.5 μM ACh, recorded using outside-out patches at a membrane potential of −60 mV. Upper trace, recording from an oocyte injected with $AChR_\gamma$ RNAs: mean current amplitude 2.4 pA. Lower trace, recording from an oocyte injected with $AChR_\epsilon$ RNAs: mean current amplitude 3.6 pA. (b) Single-channel I–V relations of the $AChR_\gamma$ (filled symbols) and the $AChR_\epsilon$ (open symbols). The slope conductances are 39 pS and 59 pS for the $AChR_\gamma$ and $AChR_\epsilon$ channels, respectively. (c) Conductance (γ) versus Na^+ activity (a_{Na}) for* $AChR_\gamma$ (filled symbols) and $AChR_\epsilon$ (open symbols). Half-saturating Na^+ activities are 73 mM for $AChR_\gamma$ and 46 mM for $AChR_\epsilon$. (d) Average duration (*t*) of elementary currents versus membrane potential on semilogarithmic coordinates. Filled symbols, $AChR_\gamma$ channel; open symbols, $AChR_\epsilon$ channel. Figure reproduced from Mishina *et al. Nature* (1986) **321:** 406–411, with kind permission. (*From 09-11-01*)

The region of the δ subunit gene promoter responsible for response to electrical activity

09-11-06: Using a reporter† gene construction in which the promoter† of the rat δ subunit gene controls the expression of a luciferase-coding sequence, a region of the promoter that confers **regulation by electrical activity** has been defined[41]. The electrical stimulation depresses expression of the luciferase reporter gene during transient transfection of primary rat myotubes. Deletion-mapping† experiments identify a minimal sequence of 102 bp that confers regulation in response to electrical stimulation.

Involvement of PKC in the control of nAChR synthesis

09-11-07: The **protein kinase C** inhibitor **staurosporine** causes a concentration-dependent increase in surface AChR and α subunit mRNA levels. The maximal effect, a five-fold increase in surface receptor concentration and a 10-fold increase in α mRNA level, is obtained at staurosporine concentrations of 10–30 ng/ml, with a half-maximal response at approximately 1 ng/ml. (Staurosporine inhibits protein kinase C activity half-maximally at 1.2 ng/ml[42].) Prolonged treatment with the **phorbol ester TPA**, which results in down-regulation of PKC levels, also increases surface nAChR and α subunit mRNA levels.

Developmental regulation of nAChR synthesis in muscle cell-lines

09-11-08: The levels of the nAChR increase dramatically in terminally differentiating muscle. In the skeletal muscle cell line C2 the levels of the nAChR as measured by α-Bgt binding increase 10- to 100-fold during differentiation. Northern blot† determinations show that the level of α and δ subunit mRNAs increase 10-fold and 15-fold, respectively, during differentiation. Nuclear run-on† experiments show that these increases are accounted for by changes in the rates of transcription of the α and δ genes[37].

The α subunit mRNA is affected by electrical activity

09-11-09: The mature α mRNA and its precursor forms are found to vary in parallel throughout all the treatments that modulate the electrical activity of the cultured myotubes†, supporting the idea that the regulation is occurring at the level of transcription[40]. The mechanism by which PKC and Ca^{2+} are involved in the regulation of ACh biosynthesis by electrical activity in developing muscle is not understood.

Developmental gene-switching in cultured cell lines

09-11-10: Cells from the mouse **myogenic cell line Sol 8** can be cultured with a feeder layer of cells from the embryonic mesenchymal cell line 10T1/2 and a serum-free medium, to form **contracting myotubes**† for 2 weeks. Under these conditions, Sol 8 myotubes undergo a maturation process characterized by sequential expression of two phenotypes: an early 'embryonic' phenotype typified by the expression of the nAChR γ subunit transcripts and low-conductance ACh-activated channels, and a late 'adult' phenotype characterized by the expression of nAChR ε subunit transcripts, the decreased accumulation of γ subunit transcripts and the appearance of high-conductance ACh-activated channels. These observations indicate that expression of functional adult-type AChR is an intrinsic feature of the Sol 8 muscle cells and does not require the presence of the motor nerve[43].

Influence of synaptic connections on the developmental regulation of neuronal nAChR

09-11-11: The detection of specific nAChR mRNAs in the developing brain by hybridization† techniques has suggested that the expression of nAChR genes can be increased, sometimes transiently, when neurones make **synaptic connections**.

09-11-12: Neurones in the chicken **lateral spiriform nucleus** do not express detectable levels of $\alpha 2$ mRNA until embryonic day 11 (E11), when fibres containing choline acetyltransferase first enter the nucleus[44].

09-11-13: In the chicken **optic lobe**, expression of the $\beta 2$ gene is increased $>$ 10-fold between E6 and E12, after which it rapidly declines[45]. *In situ* hybridization† shows that $\beta 2$ expression occurs in a **rostrocaudal gradient** and is coincident with the invasion of the **tectum** by **retinal afferents**. Removal of the eye cup at E2 results in $>$ 10-fold reduction of expression of $\beta 2$ compared with controls. This suggests that $\beta 2$ expression in optic lobe neurones is transiently stimulated by arriving retinal afferents[8].

Developmental regulation of $\alpha 7$ subunit gene expression in chick optic tectum

09-11-14: Transcripts from the gene encoding the $\alpha 7$ isoform of chick nAChR transiently accumulate in the developing **optic tectum** between E5 and E16. They are present in both the deep and the superficial layers of E12 tectum[17]. There is excellent correlation between the levels of $\alpha 7$ mRNA and α-Bgt-binding activity in the developing optic tectum of the chick embryo. Toxin-binding activity increases sharply between E10 and hatching, decreasing to a low plateau in the neonate and adult[46]. The decrease in toxin binding may coincide with the maturation of **cholinergic† synapses**[46].

Regeneration of optic nerve stimulates nAChR subunit gene expression

09-11-15: In adult goldfish retina, expression of $\alpha 3$ and three β genes is elevated 15 days after crushing of the **optic nerve**, when **retinal ganglion** cell axons are reforming connections in the optic tectum. Surgical sectioning of the optic nerve to prevent the reforming of connections from the retinal ganglia blocks the increase in nAChR gene expression. It appears that retinal ganglion cells are stimulated to express their nAChR genes when their terminals reach their target[47].

mRNA distribution

Distribution of muscle subunit mRNAs

09-13-01: The mRNA for the nAChR α subunit in chick **latissimus dorsi** is located in discrete regions that co-localize with neuromuscular junctions identified by histochemical staining for acetylcholinesterase[48].

mRNAs encoding muscle subunits in Xenopus

09-13-02: The genes encoding the two *Xenopus* muscle α subunits show different patterns of expression. The $\alpha 1_a$ mRNA is found in skeletal muscle at all developmental stages and in oocytes: the $\alpha 1_b$ mRNA is also expressed throughout muscle development, but is not detectable in oocytes[23].

Distribution of neuronal subunit mRNAs

09-13-03: Neuronal nAChR genes show differences in their patterns of expression among brain regions, suggesting that different genes are expressed by unique subsets of neurones. The sites of expression of the genes encoding the various subunits of neuronal nAChRs are shown in Table 3.

Phenotypic expression

Function of nAChR in the electric organ of marine rays

09-14-01: The richest naturally occurring sources of the nACh receptor are the electric organs of the marine electric rays, such as *Torpedo californica* or *T. marmorata*, or the electric eel, *Electrophorus*. The tissue of the electric organ consists of parallel stacks of flat cells (electrocytes or **electroplax**†), each cell being innervated on one face by a cholinergic† nerve. The stacks of cells are such that current cannot flow from one side of a cell to the other. Simultaneous stimulation of the nerves causes depolarization of the innervated faces of each electroplax, producing a potential difference between the extracellular fluid of each cell in the stack. The transcellular potentials across each cell in the stack add to produce a large electrical discharge, which in the electric eel can be hundreds of volts.

Functional expression of nAChR at the neuromuscular junction

09-14-02: At the neuromuscular junction†, the nAChRs in the plasma membrane of the skeletal muscle cells are activated by the ACh released from the **motor neurones**. The resulting transient opening of the nAChR cation channels produces an influx of Na^+ ions, causing localized depolarization† of the muscle cell membrane. This depolarization opens voltage-gated Na^+ channels, resulting in further Na^+-influx and *enhanced* depolarization. The consequent opening of neighbouring voltage-activated Na^+ channels leads to the self-propagating depolarization of the plasma membrane characteristic of the **action potential**†.

Functional expression of nAChR in neurones

09-14-03: Numerous neuronal nAChRs have been recognized by electrophysiological methods, immunological cross-reaction and biochemical purification. The neuronal channels differ from the *Torpedo* and muscle nAChRs by their containing only two types of subunit, α and β, and most of them are *not* sensitive to α-Bgt. Sequences encoding eight neuronal α subunits and four neuronal β subunits have been identified by molecular cloning. *In situ* hybridization† of brain sections with probes for different subunit mRNAs shows that each α and β subunit gene is expressed in a different, regionally specific manner. The functions of the different molecular species of nAChR in the different areas of the brain are poorly understood.

Table 3. *Sites of expression of the genes encoding various subunits of neuronal nAChRs (From 09-13-03)*

Subunit	Cell or tissue type	Refs
$\alpha 2$	**Lateral spiriform nucleus** of chick diencephalon (E11 to neonate); Neurones do not express detectable levels of $\alpha 2$ mRNA until embryonic day 11 (E11), when fibres containing choline acetyltransferase first enter the nucleus. The rat $\alpha 2$ probe highlights a small number of **Purkinje cells** in the cerebellar cortex. It is not known whether this represents a distinct sub-population of cells, or whether all Purkinje cells can transiently express the $\alpha 2$ gene	*44, 49*
$\alpha 3$	Chicken ciliary ganglion (E18: 900 copies mRNA per neurone); superior cervical ganglion (E10); goldfish retinal ganglia	*50, 15, 47*
$\alpha 4$	Adult chicken cerebrum, cerebellum, optic lobe	*14*
$\alpha 5$	Chicken **ciliary ganglion** (E18: 200–300 copies mRNA per neurone); human **neuroblastoma cell line**, IMR 32; human **small cell lung carcinoma cell line**, NCI-N-592	*50, 21*
$\alpha 7$	*In situ* hybridization† shows that the rat $\alpha 7$ gene is highly expressed in **olfactory regions**, the **hippocampus**, the **hypothalamus**, the **amygdala** and the **cerebral cortex**. Chicken E18 parasympathetic **ciliary ganglion** (1800 copies mRNA per neurone); amacrine and ganglion cells in chick retina express the $\alpha 7$ subunit, detected by immunohistochemistry	*9, 50, 62*
$\alpha 8$	The $\alpha 8$ subunit is the major subtype in chick **retina**, where its distribution has been studied by immunohistochemistry *(see Protein distribution, 09-15)*	*62*
$\alpha 9$	**Hypophyseal gland** of the rat embryo at stage E16: the mRNA is restricted to the **pars tuberalis** of the adenohypophysis: also found in the adult rat pars tuberalis, at the ventral surface of the median eminence. The $\alpha 9$ subunit mRNA is also detectable in the E16 rat **nasal epithelium** of the olfactory turbinates and in the skeletal muscle of the **tongue** of the developing rat. *In situ* hybridization of serial sections of the adult brain failed to detect $\alpha 9$ mRNA. The $\alpha 9$ subunit mRNA can be detected in rat **cochlea** by RT-PCR†, using $\alpha 9$-specific primers† designed to span an intron† of the $\alpha 9$ gene and thus distinguish cDNA from genomic DNA. *In situ* hybridization on sections of rat cochlea shows that the $\alpha 9$ gene is expressed in both the **inner and outer hair cells** of all cochlear turns	*218*
$\beta 2$	Chick ciliary ganglion (E18: 200–300 copies per neurone): In the chicken **optic lobe**, expression of the $\beta 2$ gene is increased >10-fold between E6 and E12, after which it rapidly declines. The $\beta 2$ expression occurs in a **rostrocaudal gradient** and is coincident with invasion of	*50, 8*

Table 3. *Continued*

Subunit	Cell or tissue type	Refs
	the **tectum** by **retinal afferents**. Removal of the eye cup at E2 results in >10-fold reduction of expression of β2 compared with controls. This suggests that β2 expression in optic lobe neurones is transiently stimulated by arriving retinal afferents	
β4	Chick ciliary ganglion (E18: 200–300 copies mRNA per neurone); superior cervical ganglion (E10); Purkinje cells of rat cerebellar cortex	*50, 15, 49*

An autoimmune disease involving antibodies against nAChR

09-14-04: **Myasthenia gravis**, an autoimmune† disease characterized by muscular weakness and fatiguability, results from a breakdown in **immune tolerance** of the nAChR. **Autoantibodies**† to the nAChR are found, and immune complexes (IgG† and complement†) are deposited at the post-synaptic membranes, causing interference with and subsequent destruction of the nAChR (reviewed by [51]). A number of infectious agents, including HSV† 1[52] and several Gram-negative† bacteria[53], encode molecules that immunologically cross-react with the nAChR α-chain sequence and might initiate the breakdown of self-tolerance†.

09-14-05: Immunization of mice with nAChR purified from *Torpedo* electric organ causes a disease similar to human myasthenia gravis, termed **experimental autoimmune myasthenia gravis (EAMG)**, susceptibility to which correlates with the **H-2 haplotype**†. Peptides derived from the murine muscle nAChR α subunit strongly stimulate T helper† cells from immunized H-2d mice[54].

Protein distribution

Distribution of nAChRs in adult skeletal muscles

09-15-01: The nAChR channels at the neuromuscular junction†, localized by immunogold electron microscopy†, are concentrated at the **crests of the post-synaptic folds** and immediately surrounding membrane foldings[55].

Distribution of nAChRs in foetal muscle or denervated adult muscle

09-15-02: In foetal skeletal muscle, the nAChR protein is found throughout the muscle cell membrane. If adult skeletal muscle is **denervated**, nAChRs are synthesized and appear throughout the entire surface of the muscle. This increase in cell surface receptors is preceded by an accumulation of nAChR mRNAs in the muscle[56].

Distribution of neuronal nAChRs

09-15-03: Many of the nAChRs in brain appear to be located on nerve terminals, where their role is presumed to be the modulation of transmitter release. Radioligand-binding studies in the cat **visual cortex** demonstrate the presence of **pre-synaptic nAChRs**[57]. Immunochemical studies with monoclonal antibodies against electric organ nAChRs show labelling of the **lateral spiriform nucleus (SpL)** and specific layers of the chicken **optic tectum**, which is the principal site of termination of SpL neurones and contains axon† terminals with nAChR immunoreactivity[58].

nAChRs on the non-synaptic surfaces in the optic tectum

09-15-04: Electron microscopic examination of immunolabelled nAChRs in the **optic tectum** of the frog shows that the receptors are present on the non-synaptic surfaces of vesicle-bearing profiles[59]. It is suggested that ACh released from cholinergic† terminals in the **nucleus isthmi of the optic tectum** binds nAChRs on retinal afferents† and modifies their release properties[8].

Subunits combinations in ciliary ganglia

09-15-05: At least five neuronal nAChR genes, α3, α5, α7, β2 and β4, are expressed in chick **ciliary ganglia**. Immunopurified† nAChR from embryonic chick ciliary ganglia has been shown by Western blotting† with subunit-specific monoclonal antibodies to contain α3, α5 and β4 subunits[60]. Antibody specific for the α3 subunit removes 80% of the β4 subunit and 73% of the α5 subunit from a membrane extract; similarly, anti-β4 removes 38% of the α3 subunit and 56% of the α5 subunit. Sequential immunoaffinity† purification of the receptors using anti-α3 followed by anti-β4 antibodies yields receptors that contain substantial amounts of the α5 gene product. These findings support the conclusion that a significant proportion of the receptors from synaptic sites in chick ciliary ganglia contain the co-assembled α3, α5 and β4 subunits. The same nACh receptors lack the α7 subunit, but this is present in the distinguishable, α-Bgt-binding nAChR from non-synaptic sites in chick ciliary ganglia[60].

Neuronal α7 subunits

09-15-06: The screening of a chick brain cDNA library with synthetic oligonucleotide† probes† based on N-terminal peptide sequences of an α-bungarotoxin-binding protein subunit led to the isolation of a cDNA clone encoding a novel α subunit, termed α7[16].

Prevalence of α7 subunits in nAChR from cerebellum

09-15-07: The receptor affinity purified from chick **cerebellum** by binding to α-bungarotoxin contains at least three subunits of apparent mol. wt 52 000, 57 000 and 67 000. The use of monoclonal antibodies specific for the α7 subunit demonstrated that 75% of the molecules present in the purified preparation are of the α7 subtype and that this antibody labels the 57 000 band in a Western blot†. Reconstruction experiments in planar lipid bilayers show that this α-Bgt-binding protein forms a carbachol-gated cation-selective channel whose opening is blocked by **(+)-tubocurarine**[61].

Distribution of neuronal α8 subunits

09-15-08: Low-stringency screening of a chick brain cDNA library with a cDNA probe for the α7 subunit revealed a second cDNA clone encoding a distinguishable α-related subunit, now termed α8[16]. Immunoprecipitation† and immunohistochemistry† have identified a nAChR subtype that contains α8 subunits, but not α7 subunits, as the major subtype in chick **retina**. This subtype has a lower affinity for α-Bgt than does the subtype containing only α7 subunits. The subtype containing only α7 subunits comprises 14% of the α-Bgt-sensitive nAChRs in hatchling chick retina. The subtype containing α8 subunits (but no α7 subunits) accounted for 69%, and the α7α8 subtype accounted for 17%[62].

09-15-09: Amacrine, bipolar, and ganglion cells display α8 subunit immunoreactivity, and a complex pattern of labelling is evident in both the **inner and outer plexiform layers**. In contrast, only amacrine and ganglion cells exhibit α7 subunit immunoreactivity, and the pattern of α7 detection in the inner plexiform layer differs from that of α8 subunit labelling. These disparities suggest that the α-Bgt-sensitive nAChR subunits are differentially expressed by different populations of retinal neurones. In addition, the distribution of α-Bgt-sensitive nAChR subunit immunoreactivity differs from that of α-Bgt-insensitive nAChR subunits[62].

Subcellular locations

Subcellular location of nAChR in ciliary ganglion neurones

09-16-01: Two classes of nAChR have been identified on chick **ciliary ganglion neurones**, where they occupy different subcellular locations. One class is concentrated in **post-synaptic membrane** and is responsible for mediating synaptic transmission through the ganglion. The other, which binds α-bungarotoxin, is located predominantly in **non-synaptic membrane**[63].

Transcript size

09-17-01: The sizes of the mRNAs encoding subunits of the nAChR are as follows:

α subunit	(*T. californica*)	2.3 kb[1]
	(mouse)	1.8 kb[37]
	(chicken)	2.8 kb[14]
β subunit	(*T. californica*)	2.0 kb[3]
γ subunit	(*T. californica*)	2.1 kb[3]
δ subunit	(*T. californica*)	6.0 kb[3]
	(mouse)	3.3 kb[37]
α5 subunit	(human)	2.7 kb and 2.1 kb[21]
α7 subunit	(chicken)	7 kb (major) and 3 kb (minor)[17]
	(human cell line)	5.9 kb, 2.6 kb and 1.3 kb[219]

SEQUENCE ANALYSES

Chromosomal location

Location of genes encoding muscle subunits in the mouse

09-18-01: The chromosomal locations of the genes encoding four muscle subunits of the nAChR in the mouse have been determined by RFLP† analysis of DNA from crosses between *Mus musculis domesticus* (DBA/2) and *Mus spretus* (SPE). The α gene maps to chromosome 17, the β-gene to chromosome 11 and the γ and δ genes are closely linked on mouse chromosome 1[64].

Encoding

09-19-01: *See Gene family, 09-05, for a list of the proteins encoded by the genes of the nAChR family in various species.*

Classification of subunits as α or β

09-19-02: Assignment of a neuronal subunit to the 'α' class is based on the conservation of the adjacent Cys residues at the positions homologous to Cys192 and Cys193 of the *Torpedo* α subunit[1]. Subunits lacking the two adjacent Cys residues are generally designated β, but investigators working with sequences from chick and goldfish prefer the term 'non-α'

Muscle subunits required for function

09-19-03: Fully functional nAChR channels are expressed by the co-injection of the four cRNAs encoding the α, β, γ and δ subunits of the muscle channel into *Xenopus* oocytes[65]. When combinations of RNAs encoding only a subset of the subunits are injected, the α subunit is essential for activity, together with either the γ or δ subunit: the β subunits are dispensable[66,67].

09-19-04: Six different combinations of three or more subunit RNAs produce significant numbers of functional channels. The order of combinations yielding the greatest amount of current is $\alpha\beta\gamma > \alpha\beta\delta = \alpha\delta\varepsilon > \alpha\delta\gamma > \alpha\delta > \alpha\gamma$. The extent to which a channel type with three different subunits is expressed is highly dependent upon the ratios of RNAs coding for the different subunits and is critically dependent upon the order of injection of the RNAs[68].

Neuronal subunits required for channel function

09-19-05: The RNAs encoding only two neural subunits, α and non-α (also called β), are sufficient to encode functional channels in *Xenopus* oocytes[5]. Eight distinct α and four different β subunits have been identified in neuronal tissues. Neural β subunits can substitute for muscle β subunits in forming functional channels in *Xenopus* oocytes[10].

09-19-06: At least five neuronal nAChR genes, α3, α5, α7, β2 and β4, are expressed in chick **ciliary ganglia**. Immunopurified† nAChR from embryonic chick ciliary ganglia has been shown by Western blotting† with subunit-specific monoclonal antibodies to contain α3, α5 and β4 subunits[60] *(see Protein distribution, 09-15).*

The α7, α8 and α9 subunits form homo-oligomeric channels

09-19-07: The chick α7 and α8 and the rat α9 nAChR subunits assemble into functional homo-oligomeric† channels, responding to acetylcholine, when the corresponding cRNAs are singly injected into *Xenopus* oocytes[17,217,218].

Most α-Bgt-binding proteins in brain contain the α7 subunit
09-19-08: The mature α7 protein (479 residues) has moderate homology with all other α and non-α nAChR subunits and probably assumes the same transmembrane topography. A bacterial fusion protein containing residues 124–239 of α7 binds labelled α-Bgt[17]. *In situ* hybridization† maps of α7 mRNA closely resemble the pattern of [^{125}I]-α-Bgt binding in rat brain, suggesting that most α-Bgt-binding proteins in the tissue contain α7 subunits. The α8 subunits occur less commonly, representing only 15% of the α-Bgt-binding complexes and tending to be associated with the more abundant α7 subunits[16].

nAChR from insect CNS
09-19-09: A nAChR purified from the **cockroach CNS** by a-Bgt-binding has an overall size of about 300 kDa, but produces a single band of 65 kDa on denaturing gels[69]. Reconstitution into lipid bilayers produces channels that are gateable by nicotinic agonists and blocked by the antagonist, **(+)-tubocurarine**[70]. Sequences encoding a cockroach subunit, αL1, have been cloned and expressed in *Xenopus* oocytes, where they specify functional channels gated by nicotine and blocked by **α-Bgt** and **n-Bgt**.[71]

Subunit composition of functional neuronal nAChR channels
09-19-10: The single-channel conductance† and current amplitude† of neuronal nAChR can be manipulated by changing the charged residues immediately downstream of the M2 region. Change of E266 to K in the α4 subunit reduces the single-channel conductance of α4/β1 channels, while change of the analogous residue (K260) in β1 to E increases the single-channel conductance. When a combination of cDNAs encoding α4E266 and α4K266 is co-injected with cDNA encoding β1E260 into *Xenopus* oocyte nuclei, channels with three different amplitudes are detected in inside-out patches[72]. This finding is the prediction if the functional nAChR contains two α subunits.

Evidence for the pentameric nature of neuronal nAChR
09-19-11: Similar experiments co-injecting cDNAs specifying two distinguishable β subunits plus one α subunit (β1, β1E260 plus α4) result in four distinguishable current amplitudes, as predicted if there were three β subunits per functional channel. Thus the functional neuronal nAChR is a pentamer, of composition α2β3[72].

09-19-12: When the nAChRs synthesized following injection of chicken α4 and β2 mRNAs into *Xenopus* oocytes are labelled with [^{35}S]-methionine, 1.46 times more label is found in β subunits than in α subunits, after correction for their methionine content[73]. This ratio is very close to the value of 1.5 expected for a stoichiometry† of α2β3.

Gene organization

Introns and exons in the α subunit genes
09-20-01: The chicken[14] and human[19] α genes have nine coding **exons**†. In the human α gene, the lengths of the eight **introns**† are ~ 4.9 kb, 111 bp, ~ 1.7 kb, ~ 3.1 kb, ~ 0.4 kb, ~ 3.4 kb, ~ 1.2 kb and 324 bp in the 5′ to 3′ direction[19].

Intron–exon structure of the α2, α3 and α4 subunit genes
09-20-02: The chicken α2, α3, α4 and non-α genes and the rat α2 and α3 genes all have six protein-encoding exons†, the fifth of which is large and encodes protein sequences homologous to those specified by exons 5 through 8 of the α gene[14]. The positions of the **exon–intron boundaries** in α2–α5 and nα1–nα3 are exactly conserved[17].

Intron–exon structure of the α7 subunit gene
09-20-03: The chick α7 gene contains 10 exons†, the first four of which exactly match the corresponding exons in other nAChR subunit genes[17]. The other six exons do not correspond to any of the exons in the other muscle or neuronal nAChR genes.

Intron–exon structure of the α9 subunit gene
09-20-04: The mouse gene encoding the α9 subunit of the nAChR has five exons and an intron–exon structure that differs from that of all other known nAChR genes. In contrast with other nAChR genes in which the intron–exon boundaries of the first four exons are conserved *(see above)*, exons III and IV in the α9 gene are fused[218]. (*Note:* The intron–exon structure of the α9 gene was determined by comparing rat cDNA sequence with *mouse* genomic sequence[218], so that the information presently available strictly applies only to the *mouse* α9 gene.)

Intron–exon structure of the δ and γ subunit genes
09-20-05: The chick genes encoding the δ and γ subunits both contain 12 **exons**†. The homologous exons of the two genes are very similar in size and the **splice sites**† are exactly conserved. The corresponding introns† of the two genes differ sharply in length and sequence[18]. The two genes are very closely linked in the chicken genome, with only 740 bp between the last codon of the δ gene and the translation-initiation codon of the γ gene. The **intergenic region** contains a single canonical **polyadenylation† site**, 77 bp downstream of the δ gene translation-termination codon[18]. The human δ gene also contains 12 exons, and the sizes and positions of its **introns** have been determined[74].

The α subunit gene promoter confers tissue specificity
09-20-06: A region of upstream† sequence of the chicken α subunit gene lying between −110 and −45, was shown to confer **tissue- and stage-specific gene expression** on a reporter gene† following transfection into chicken primary myotubes or the mouse C2.7 myogenic cell line[75]. This region of DNA interacts with several nuclear proteins from muscle cells and differentiated myotubes, including an **Sp1†-like factor** and a **G stretch-binding protein**, which bind to overlapping sites immediately upstream of the TATA† box. Several proteins interacting with a sequence similar to the SV40 core enhancer† appear during *in vitro* differentiation of myoblasts into myotubes, and the concentration of some of these increases after **denervation** of leg muscle in newborn chicks[75].

The δ subunit gene promoter

09-20-07: The promoter region of the chicken nAChR δ subunit gene lacks typical TATA† and CCAAT† boxes, and **transcription starts** at six major and seven minor sites between −110 and −30 with respect to the translation-initiation site. Two sites, at positions −77 and −66, give rise to about 50% of all transcripts. A transcription **enhancer**† was located by deletion mapping† to the region from −207 to −146. The enhancer is active in fibroblasts and differentiated muscle cells, but not in myoblasts[76].

The transcription-start site for the δ gene

09-20-08: The **transcription-start site** for the mouse δ gene has been mapped 55 bp upstream of the translation-initiation codon. A sequence TAAACCA at positions −33 to −27 relative to the transcription-start site is presumed to serve as the TATA box†, and a GATTG sequence, complementary to the CAATC† box, occupies −66 to −62. CG-rich sequences having homology with known **AP-2†-binding sites** are centred at positions −55, −180 and −210[77].

A sequence in the δ-gene promoter necessary for muscle-specfic expression

09-20-09: A 54 bp region of 5′-flanking DNA from the murine gene encoding the δ subunit, occupying −148 to −95, is necessary and sufficient for **muscle-specific gene-expression**. Deletion of this sequence results in a 50-fold reduction in expression in myotubes, while fusion of the sequence upstream of the c-*fos* basal promoter confers myotube-specific gene expression on an otherwise weak, non-tissue-specific promoter. The muscle-specific transcription factor **MyoD1** does not bind to the 54 bp region, but other nuclear proteins from myotubes are able to bind this element[78].

Consensus sequences for muscle-specific regulatory motifs in the promoters of genes encoding nAChR subunits

09-20-10: The consensus sequences for nAChR gene enhancers contain several motifs that have previously been implicated in tissue-specific gene expression. There are two copies of the CANNTG motif present in the **MyoD1** target sequence characteristic of many muscle-specific regulatory regions: these flank a '**M-CAT**' motif and an overlapping TGCCTGG sequence, both of which have been proposed as **muscle-specific regulatory motifs**[78, 79].

A sequence common to several nAChR subunit gene promoters

09-20-11: A sequence common to the chicken α gene and the mouse β, γ and δ genes has been termed the subunit homologous upstream element (**SHUE**) **box**. The 13 bp SHUE box sequence, all four copies of which contain CCCTGG/C, is located at −155 in the mouse δ gene and −75 in the chicken α gene[77]. The function of the SHUE box has not been determined.

Close linkage of the δ and γ subunit genes

09-20-12: The genes encoding the δ and γ subunits are very closely linked in the chicken, mouse and human genomes. The γ gene lies 740 bp downstream of the δ gene in chicken[18] and 5000 bp downstream in mouse[77]. This, together with the high degree of sequence homology and identical **exon–intron structure**, supports the idea that the two genes have undergone a relatively recent **tandem duplication**[18]. The human γ and δ genes can be isolated on a single genomic DNA fragment of 15.7 kb, with δ upstream† of the γ gene[74].

Clustering of the nα3, α3 and α5 genes in chicken
09-20-13: The three genes encoding the nα3, α3 and α5 subunits are clustered in the chicken genome. Gene nα3 lies 5′ of α3 and is transcribed in the same direction, whereas α5 is located on the 3′ side of α3 and is transcribed from the opposite DNA strand. The nα3 and α3 genes are separated by about 5 kb, and α3 and α5 by less than 1 kb[15]. There is also an **α3 gene cluster**, containing genes encoding β4, α3 and α5 subunits, in the rat genome[7].

Homologous isoforms

Homology amongst muscle subunits
09-21-01: The muscle α, β, γ and δ subunits show sequence similarities with each other and with subunits from other channel types, including the mouse serotonin-gated ion channel and several glycine and GABA receptors. The β subunits show approximately 42–45% sequence identity, the δ subunits have about 40% identity and the γ subunits 37–40% identity with the α subunits.

Homology between all nAChR subunits
09-21-02: There is strong conservation of amino acid sequence amongst all known subunits of the nAChR. This conservation is not evenly distributed throughout the polypeptide chains, being strongest in the region that includes the transmembrane domains M1, M2 and M3 *(see Domain conservation, 09-28)*. The percentages of amino acid identity between the subunits of the muscle and neuronal nAChRs are shown in Table 4.

Xenopus *muscle α subunits*
09-21-03: The two *Xenopus* muscle α isoforms are 89% similar to each other, and they show similar levels of similarity (85–89%) to the mammalian or *Torpedo* α subunit[23].

Protein molecular weight (purified)

The nAChR purified from Torpedo *electric organ*
09-22-01: The nAChR purified from *Torpedo* electric organ is a 250 kDa pentamer, formed by four different types of subunits (α, β, γ, δ) with $\alpha2\beta\gamma\delta$ stoichiometry[80]. The individual subunits, each of which is glycosylated†, have apparent molecular masses of 40 (α), 50 (β), 60 (γ) and 65 (δ) kDa[81].

The chicken brain nAChR
09-22-02: An nAChR immunopurified† from chicken brain is constituted of subunits of 49 kDa and 59 kDa measured by electrophoretic mobility on

Table 4. *Amino acid identity (%) between nAChR subunits (From 09-21-02)*

	α	α2	α3	α4	α5	α6	α7	α8	β	β2	β3	β4	γ	δ
α	—	48	53*	51*	45*		38*		41^					36^
α2	50*	—					40*							
α3	52	58					39*							
α4	53	68	59	—			41*							
α5	44			55			38*				67*			
α6	41	49	59	49	45		38*			43		47		
α7	38*	40*	39*	41*	38*	38*		82*	35*	35*	38*			32*
α8							82*							
α9	37						38	38						
β	41^						35*							41^
β2							35*		45					
β3	45				68		38*		40	44				
β4	43					48*			43	64	44			
γ														49*
δ	36^						32*		41^					

Footnotes quoted are the percentage amino acid identities between different subunits of rat, chicken (*) or *Torpedo* (^) nAChR subunits.

denaturing gels[82]. Antibodies raised against this material are able to immunoprecipitate† receptors containing 49 kDa and 75 kDa components. In each case, the larger subunits are affinity labelled† by **[^{3}H]-MBTA** and have N-terminal† sequences homologous to those of α subunits[83]. Subsequent studies have revealed that the N-terminal sequence of the 75 kDa band corresponds to that of the α4 subunit[84] and the N-terminal sequence of the 49 kDa band to that of the β2 subunit[85]. The 59 kDa band probably includes both α2 and α3 subunits[86].

09-22-03: The α7 subunit immunopurified† from chicken brain has an apparent M_r of 57 kDa on denaturing gels[87].

Purified nAChR from chicken ciliary ganglia

09-22-04: Purification of nAChRs from extracts of chicken **ciliary ganglia** yields a fraction showing three components on denaturing gels, at 49, 52 and 60 kDa[88]. These have been identified, using subunit-specific antibodies, as the α5, β4 and α3 subunits, respectively. It has not been established that these components assemble into a single nAChR.

The nAChR purified from rat brain

09-22-05: An nAChR immunopurified from rat brain with anti-chicken nAChR mAb 270 contained subunits of apparent M_r 52 and 80 kDa[89]. The N-terminal sequence of the 80 kDa subunit corresponds to that of α4[90] and that of the 52 kDa subunit to the N-terminal sequence of the β2 subunit[85]. Antibodies to these two components remove >90% of the high-affinity [^{3}H]-nicotine- or [^{3}H]-cytisine-binding sites from detergent-solubilized rat brain extracts, suggesting that nAChRs comprised of α4 and β2 subunits are the majority species in rat brain[91].

Protein molecular weight (calc.)

09-23-01: *See data in Table 1 under Gene family, 09-05.*

Southerns

Southern blots of chicken genomic DNA

09-25-01: Blots of chicken genomic DNA digested with *Eco*RI, *Bam*HI and *Hin*dIII showed single bands hybridizing to δ- and γ-specific probes. The two probes reveal the same-sized *Eco*RI band, but distinct bands on an *Eco*RI–*Hin*dIII double-digest, consistent with the two genes being very closely linked on a single *Eco*RI fragment *(see Gene organization, 09-20)* and with their being unique in the chicken genome[18].

Southern analysis of Xenopus *genomic DNA*

09-25-02: Southern blots of *Xenopus* genomic DNA probed with $\alpha 1_a$ and $\alpha 1_b$ probes indicate the presence of two different genes, each present in a single copy per genome[23].

STRUCTURE & FUNCTIONS

Domain arrangement

Transmembrane domains

09-27-01: Each nAChR subunit contains a large, hydrophilic†, extracellular N-terminal domain, four hydrophobic† transmembrane domains and a short, extracellular C-terminal region. The sequences of the four strongly hydrophobic segments (M1–M4) and that of a region that suggests an amphipathic† helix (MA) are strongly conserved amongst different subunits and across species. Covalent labelling of all four segments with **photo-reactive† phospholipids** supports the contention that the M1–M4 regions are membrane-spanning[92, 93].

Secondary structure of transmembrane domains

09-27-02: There is no direct evidence that M1–M4 are α-helical and they are at least candidates for β-sheet-formers according to secondary structure prediction algorithms[80].

The M1 domain is accessible to open-channel blockers

09-27-03: The proposed open-channel structure of the nAChR suggests that the putative α-helix M1 is exposed at the interstices (clefts) between the M2 helices. This interpretation is supported by labelling of M1 helices with open-channel blockers such as **quinacrine** *(see refs[96, 97])*.

Domain conservation

Conserved Cys residues in all subunits

09-28-01: *All* the subunit genes encode proteins with two cysteines separated by 13 residues that align with C128 and C142 of the muscle α subunit.

Invariant Cys192 and 193 in α subunits

09-28-02: All α subunits from muscle and electric organ have adjacent cysteines at positions corresponding to muscle α192 and 193. Neuronal subunits with this feature are designated as α subunits; those without it are designated non-α (avian species) or β (mammalian species). *(Note that the designation 'β' does not mean that such a subunit most closely resembles the muscle β subunit. The neuronal β subunits represent members of a heterogeneous group united by their common lack of the two adjacent cysteines!)*

Similarities between muscle and neuronal subunits

09-28-03: There is approximately 60% identity between the muscle and neural α subunits over the first 320 residues and in the MA and M4 regions near the C-termini. The neural sequences contain an insertion of 60–160 amino acid residues in the cytoplasmic region between M3 and MA. The similarity between muscle and neural non-α sequences is slightly less (about 44% in the first 350 residues)[80].

Overall sequence conservation
09-28-04: There is striking conservation of sequence amongst *all* the subunits of nAChR channels. Amongst the first 320 amino acids, one-third are conserved. The strongest conservation is between residues 224 and 320, which includes the M1, M2 and M3 regions, where there is about 50% identity. There is also about 25% similarity in the first 223 residues that constitute the N-terminal† extracellular domain. The conservation is much less for the large cytoplasmic domain between M3 and M4[80].

An invariant Pro–Cys in M1
09-28-05: There is an invariant Pro–Cys (221 and 222) in the centre of the M1 helices of the α- and non-α-subunits of the nAChR. The Pro will introduce a bend of about 20° into an α-helix†, disrupting local hydrogen-bonding† and leaving amide and carbonyl groups free to interact with water or a permeating ion[98]. It has been suggested that the Pro may be a focus for **conformational changes** involving *cis–trans* isomerization† of the peptide bond[80].

Similarity between nAChR and ryanodine receptor
09-28-06: The region encompassing segments M2 and M3 of the nAChR show some sequence similarity with the M2–M3 region of the **ryanodine receptor** subunits *(see ILG Ca Ca RyR-Caf, entry 17).*

Domain functions (predicted)

Acetylcholine-binding sites on the α subunits
09-29-01: The **acetylcholine-binding sites** are primarily located on the α subunits, though since the two sites are not equivalent they may also involve residues on other subunits[99].

09-29-02: Amino acid residues involved in forming the binding site have been identified by covalent affinity labelling†[100,101] and by site-specific mutagenesis†[65] to be Tyr190, Cys192, Cys193 and Tyr198 on the *Torpedo* α subunit. These residues, which are conserved amongst all known α subunits, lie within 10 Å of a carboxylate group that is involved in binding the quaternary ammonium group in acetylcholine.

Aromatic residues implicated in ACh binding
09-29-03: *In vitro* mutagenesis† experiments have implicated aromatic† residues in the vicinity of the two Cys residues (192 and 193) as being part of the ACh-binding site[102], or of being involved in the **coupling of agonist binding to channel activation**[103].

09-29-04: Two other peptide loops, one including Trp86 and Tyr93 and the other Trp149 and Tyr151, are also accessible to bound affinity ligands[104,105]. Mutations substituting Phe residues for the amino acids homologous to Tyr93, Trp149 or Tyr190 in the neuronal homo-oligomeric α7 receptor decreased the apparent affinity for acetylcholine 10–100-fold, as well as affecting the binding of α-Bgt[106].

Acid residues on the δ and ε subunits are important for ACh binding

09-29-05: ACh and all other potent agonists and **competitive antagonists** contain a positively charged **quaternary ammonium**† **group**. There are negatively charged residues in the δ subunit sufficiently close to αCys192/Cys193 to contribute to the binding of ACh. These were first located within the *Torpedo* δ subunit between residues 164 and 224 by use of a cross-linking† agent that reacts with sulphydryls at one end and with carboxyls at the other[107]. Each of the 12 acidic residues in this region of the mouse δ subunit was changed to the corresponding amide. The mutation δD180N led to a 100-fold decrease, and the change εE189Q to a 10-fold decrease, in the apparent affinity of the receptor for ACh[108].

ACh-binding sites are at the interfaces between subunits

09-29-06: Taken together, the data suggest that the two non-identical **ACh-binding sites** are formed at the interfaces between the α and γ and the α and δ subunits. Residues identical to δAsp180 and δGlu189 are present in all δ, γ and ε subunits, and this conservation is consistent with the location of these residues at the ACh-binding sites. In contrast, in all muscle-type β subunits, His or Asn corresponds to δAsp180 and Gln aligns with δGlu189. This may account for the inability of β subunits to form an ACh-binding site with α[68, 99, 109].

αTyr190, αTyr198 and αAsp200 are involved in the coupling of ACh binding to channel opening

09-29-07: Wild-type *Torpedo* nAChRs expressed in *Xenopus* oocytes are half-maximally activated ($K_{1/2}$) by 20 μM acetylcholine with a Hill coefficient of 1.9. Substitution of αY190 and αY198 with Phe residues (αY190F, αY198F) or αD200 with Asn (αD200N) altered the $K_{1/2}$ to 408, 117 and 75 μM, respectively, with no effect on the Hill coefficient†.

Mutant receptors with altered response to partial agonists

09-29-08: The above mutant receptors show altered responses to the partial agonists **phenyltrimethylammonium (PTMA)** and **tetramethylammonium (TMA)**. While wild-type receptors are half-maximally activated by 73 μM PTMA and 2 mM TMA, αY190F, αY198F, and αD200N receptors are not activated by PTMA and TMA by concentrations of up to 500 μM or 5 mM, respectively. However, PTMA and TMA bind to the mutant receptors with the same affinity as to the wild-type, acting as competitive antagonists. The αY190F, αY198F and αD200N mutations thus have their major effect on the **coupling of ligand binding to opening** of the channel[110].

α-Bgt-binding site: αCys128 and αCys142 are essential for α-Bgt binding

09-29-09: In addition to the two Cys residues (192 and 193) conserved in the α subunit sequences, two other cysteines (C128 and C142 in α subunits) are found at homologous positions in *all* AChR subunits, as well as in GABA and glycine receptors. Mutant forms of the *T. californica* nAChR with substitutions of Ser for Cys128 or Cys142 in either the α or β subunits are able to associate with other normal subunits, although the efficiency of association of the mutant α subunits with the δ subunit is reduced[111]. The mutations in the α subunit abolish detectable α-Bgt binding in whole oocytes, whereas the mutations in the β subunit result in decreased total binding of α-Bgt and no detectable surface binding.

09-29-10: All the subunits mutated at residues 128 or 142, when co-expressed with the other normal subunits in *Xenopus* oocytes, produce small acetylcholine-activated currents. (This observation is at variance with earlier studies[112], where mutation of C128 or C142 abolished the response of the receptor to ACh.) The functional ACh-gated channel formed with mutant α subunits, but not mutant β subunits, is *not* blocked by α-Bgt. Thus, a disulphide bond† between Cys128 and Cys142 of the nAChR α or β subunits is not essential for acetylcholine binding, but this disulphide bond on the α subunit is important for formation of the **α-Bgt-binding site**[111].

Determination of the location of the α-Bgt-binding site

09-29-11: The location of the **α-Bgt-binding site** in the *Torpedo* α subunit has been determined by assaying the binding of the labelled toxin to overlapping synthetic peptides each of 20 amino acids. The peptide α181–200 is by far the most efficient in binding the toxin, with α55–74 also showing significant binding[113].

The influence of the oxidation state of αCys192/Cys193 on α-Bgt binding

09-29-12: The importance of the potential disulphide link between the adjacent Cys residues α192 and 193 for **α-Bgt binding** can be assessed from the effects of chemical oxidation, reduction and carboxymethylation† of the α181–200 peptide on the efficiency of binding. Neither oxidation nor reduction of the Cys residues affects α-Bgt binding, but binding is drastically reduced by carboxymethylation of the reduced -SH groups[114]. Clearly the **Cys192–Cys193 disulphide bridge** is not important for the α-Bgt-binding reaction, but binding is affected by modification of these two Cys residues. These findings suggest that the binding sites for acetylcholine and α-Bgt are very close to each other and must overlap.

Affinity-labelling of residues in the pore

09-29-13: Affinity-labelling techniques† together with those of site-directed mutagenesis† and heterologous expression have helped define the **pore† structure** of the nAChR[115]. The transmembrane **domain M2** has been shown to label with compounds known to bind within the pore[116, 117]. **Non-competitive inhibitors** bind with high specificity to a defined region of M2 of several subunits, particularly Ser, Thr and Leu residues within the consensus MTLSINVLL sequence at the beginning of M2[118].

Evidence for the α-helical nature of M2

09-29-14: The cholinergic† non-competitive† antagonist† **3-(trifluoromethyl)-3-(*m*-[^{125}I]-iodophenyl)diazirine ([^{125}I]-TID)** binds to the nAChR with similar affinity in the presence and absence of 50 μM carbamylcholine, and acts as a photoaffinity† label. In the absence of agonist, [^{125}I]-TID specifically labels homologous aliphatic† residues (βL257, δL265, βV261 and δV269) in the M2 region. In the presence of agonist, labelling of these residues is reduced approximately 90%, and the distribution of labelled residues is broadened to include a homologous set of serine residues at the N-terminus† of M2. In the β subunit, residues βS250, βS254, βL257 and βV261 are all labelled in the presence of carbamylcholine. This pattern of labelling supports an α-helical model for M2, with the labelled face forming the **lumen**† of the ion channel.

Agonist causes rearrangement of M2

09-29-15: The redistribution of label in the resting and desensitized† states provides direct evidence for **agonist-dependent rearrangement** of the M2 helices. The efficient labelling of the resting state channel in a region capable of structural change also suggests that the aliphatic† residues labelled by [^{125}I]-TID form a permeability barrier to the passage of ions that is removed on gating† the channel[119].

Negatively charged rings flanking M2 enhance permeability to cations

09-29-16: Mutations of amino acids including those polar† groups within[120] and negatively charged rings bracketing[121] segment M2 affect ion permeation at the single-channel level. A change of Ser248 (within M2) to Ala decreased the outward single-channel currents and the residence† time of the **open-channel blocker QX-222**[120]. The negative charges at αAsp238, αGlu241 and αGlu262 were also implicated as determinants of the channel's permeability[121]. These negatively charged residues flanking M2, which are largely conserved among the various subunits, are believed to confer a net negative charge to the channel entrance and enhance the permeability to cations[121–123].

Residue αThr244 is involved in ion selectivity

09-29-17: Mutations of αThr244, within M2, affect the channel's ability to discriminate amongst monovalent cations[124–126].

09-29-18: The T244D mutation in the M2 segment of the neuronal α7 nAChR subunit changes the **selectivity** of the homo-oligomeric channel expressed in *Xenopus* oocytes. The α7D244 channel exhibits larger currents than the wild-type α7 channel and is activated at lower ACh concentrations. The relative ionic permeability of wild-type AChRα7 to K^+ is $P_K/P_{Na} = 1.2$, and to Ba^{2+}, $P'_{Ba}/P_{Na} = 1.4$. The α7D244 channel is less selective in discriminating between K^+ and Na^+, $P_K/P_{Na} = 0.95$, but exhibits a marked increase in permeability to Ba^{2+}, $P'_{Ba}/P_{Na} = 3.7$. In addition, only the mutant receptors are permeable to Mg^{2+}[127].

αLeu247 lies in the lumen of the channel

09-29-19: In the homo-oligomeric nAChR composed of the chick brain α7 subunit, mutations of the highly conserved Leu247 residue in the M2 segment suppress inhibition by the channel **blocker QX-222**, indicating that this residue, like others from M2, faces the lumen† of the channel[128].

09-29-20: The same L247T mutations also decrease the **rate of desensiti-**

zation† of the response, increase the apparent affinity for acetylcholine and abolish current rectification†. Moreover, unlike wild-type α7, which forms channels with a single conductance level (46 pS), the Thr247 mutant has an additional conducting state (80 pS) active at low acetylcholine concentrations. In addition, antagonists of the wild-type receptor, **dihydro-β-erythroidine**, **hexamethonium** and **(+)-tubocurarine**, act as agonists with the L247T mutant and activate the novel conducting state[128]. It is suggested that the L247T mutation makes one of the high-affinity desensitized states of the nAChR conductive[129].

αThr264 is located in the narrow region of the pore

09-29-21: Substitution of the Thr264 in the transmembrane segment M2 of the α subunit of the rat nAChR affects channel conductance†. Mutation of the residues at homologous positions in the β, γ, and δ subunits shows the conductance to be inversely related to the volume of the amino acid residue, suggesting that residues at this position form part of the **channel narrow region**. Exchanges of residues between subunits does not change the conductance, suggesting a ring-like structure formed by homologous amino acids in the subunits of the pentameric channel. Channels in which the narrow region is formed by four serines and one valine have the same conductance if the valine is located in the α, β, or γ subunits, but it is smaller if the valine is located in the δ subunit. These results suggest a **structural asymmetry** of the AChR channel in its narrow region formed by the hydroxylated† amino acids of α, γ and δ subunits, where the δ subunit serine is a main determinant of the channel conductance[130]. In addition, for a given size of side-chain, the conductance is consistently higher with a polar† rather than a hydrophobic† side-chain[131], suggesting a 'catalytic' role for the polar rings in the translocation of cations through the channel pore[132].

Mutations causing changes from cationic to anionic selectivity

09-29-22: The M2 region of nAChR has strong homology to the analogous region of the anion-selective glycine and $GABA_A$ receptors. Substitution of amino acids within or near the M2 region of the α7 nAChR subunit by different residues from the GlyR α1 subunit drastically changed the properties of the channel, converting its selectivity from cationic to anionic[133]. Three amino acid differences, changes of Glu237 to Ala, Val251 to Thr and the addition of a Pro after residue 236, are sufficient to produce channels that were 500-fold more sensitive to ACh, are activated by the competitive antagonist **dihydro-β-erythroidine (DHβE)**, no longer show inward rectification† of whole cell currents and are ***anion-selective.***

The M1 to M2 spacing affects ion selectivity

09-29-23: Deletion of the extra Pro residue from this mutant α7 channel gives functional channels that show the enhanced sensitivity to ACh, do not desensitize rapidly, are cation selective but do not pass Ca^{2+} currents. **Inversion of ion selectivity** is also achieved by the addition of either an Ala or a Pro residue, following position 236, to a cation-selective α7 mutant. These data point to the importance of the length of the segment spacing M1 and M2 in determination of the ion selectivity[133].

The α7 T251 channel shows additional conducting states

09-29-24: The single substitutions, Glu237 to Ala or Val251 to Thr, do not invert the ion selectivity, but the Thr251 mutation does change the apparent affinity for ACh, response to DHβE and rectification† properties of the channel. Single-channel recordings from outside-out† patches containing the Thr251 α7 nAChR show **multiple conducting states**, including one of 54 pS, similar to the wild-type A state, and one of 86.3 pS, corresponding to a **desensitized D* state** of the nAChR seen with the Thr247 mutant[129].

The Glu237Ala change in α7 affects Ca^{2+}-selectivity

09-29-25: The alteration of Glu237 to Ala in the α7 nAChR results in a receptor that responds to ACh normally and shows rapid desensitization†. Although this mutant channel is permeable to cations, it has lost the ability to conduct Ca^{2+} [133].

Conclusions from observations with mutant channels

09-29-26: Structural interpretations of the data obtained from studies with nAChR channels altered in the pore region indicate that the wide **entrance vestibules** of the nAChR pore contain net negative charges which can attract cations and are particularly important in attracting divalent cations[123]. Upon channel opening, the permeant cations rapidly pass through the uncharged, tapering region of the pore†, lined by the M2 sequences from each of the subunits. The geometry of the M2 segments, influenced by the M1–M2 spacer region, is crucial in determining ion selectivity of the channel. The narrowest region of the open pore, in the region of Thr244α, is very short (estimated to contain as few as six water molecules) and is likely to be the only region of the channel where strong interactions between permeant ions occurs[134].

Closed structure of the nAChR channel

09-29-27: The closed structure of the channel is more difficult to investigate, for obvious reasons, but experiments with M2 peptides have suggested that an association to block the pore†, with the appropriate stability, could be achieved by interactions between the M2 α-helices of the subunits[135]. The electron microscopic images of the *Torpedo* AChR suggest that amino acid residues come closest to the axis of the pore just below the middle of the bilayer when the channel is closed. This coincides with the position within M2 of the highly conserved Leu residues that are known to be in the narrow region and to face the lumen† of the channel *(see above)*. It has been suggested that the bulky, hydrophobic† Leu side-chains of the five α-helices associate to create a barrier of limited stability that forms the **gate**[136].

09-29-28: The MA segment, located between M3 and M4, is part of the cytoplasmic domain and does not form part of the pore†. Deletion of sequences encoding part of MA in the α subunit eliminates channel function in the *Xenopus* oocyte system. When MA and the 20 preceding amino acids are removed, 3% of the native activity is obtained[137].

Structural changes on desensitization

09-29-29: Alterations in protein structure produced by binding of cholinergic agonists to purified nAChR reconstituted into lipid vesicles can be detected by Fourier-transform infrared spectroscopy† and differential scanning calorimetry†. Spectral changes indicate that the exposure of the nAChR to the agonist carbamylcholine, under conditions which drive the AChR into the **desensitized state**, produces alterations in the protein secondary structure. Quantitative estimation of these agonist-induced alterations reveals no significant changes in the percentage of α-helix†, but a decrease in β-sheet† structure, concomitant with an increase in less-ordered structures. Agonist binding also results in a concentration-dependent increase in the protein's thermal stability, as indicated by the temperature dependence of the infrared spectrum and by calorimetric analysis, further suggesting that nAChR **desensitization**† induced by the cholinergic agonist involves significant rearrangements in the protein structure[138].

Predicted protein topography

Labelling studies

09-30-01: Given four membrane-spanning peptides (M1–M4) in each subunit *(see Domain arrangement, 09-27)*, the N- and C-terminal regions must be on the same side of the membrane. Antibody and chemical labelling[118, 139] specify the N-terminal†, C-terminal† and M2–M3 linker regions as extracellular, with the small M1–M2 and larger M3–M4 linkers intracellular. This determination suggests the MA region to be intracellular, at or near the membrane–cytoplasm interface.

Electron microscopic analysis of channel structure at 17 Å resolution

09-30-02: Electron microscopy and 3-dimensional image-reconstruction† has provided the structure of the post-synaptic *Torpedo marmorata* nAChR channel to a resolution of 17 Å[94]. Post-synaptic membranes formed into tubular vesicles suspended in thin films of ice permitted the receptors (organized into helical arrays) to be seen from all angles, revealing the relation of the lipid bilayer and the peripheral protein *(Fig. 2)*.

A pentameric, barrel-stave arrangement of subunits

09-30-03: The nAChR is a pseudosymmetric pentamer, with the subunits in a barrel-stave arrangement *(see Fig. 2)*. The channel complex is about 120 Å long, projecting about 60 Å into the extracellular space and about 20 Å into the intracellular solution. The diameter at the extracellular end of the channel is about 80 Å, narrowing to about 50 Å in the membrane-spanning region. The **conduction pathway** can be seen to consist of a narrow central pore† across the bilayer, terminating in entrances 20–25 Å wide *(for review see ref.[80])*.

Structure of the mouse muscle nAChR

09-30-04: The structure of the channel obtained by expressing cRNAs encoding the four mouse muscle subunits in *Xenopus* oocytes can be

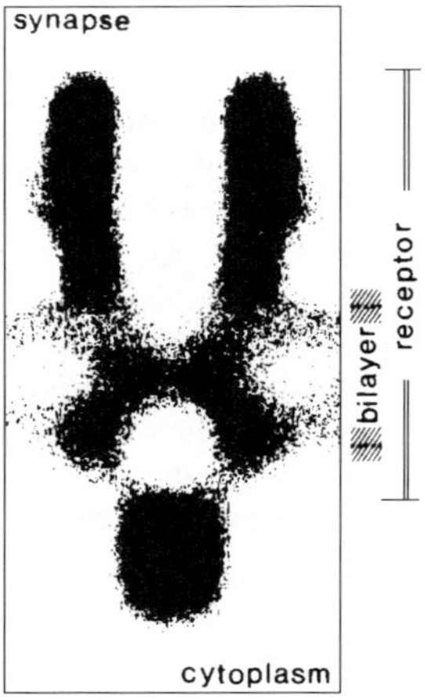

Figure 2. *Axial section through the cylindrically averaged structure of the* Torpedo marmorata *nAChR, showing details of the channel in relation to the lipid bilayer and the peripheral 43 kDa protein, at the bottom of the figure. (Reproduced with permission from Toyoshima (1988)* Nature ***336:*** *247–50.) (From 09-30-02)*

visualized by atomic force microscopy†. The pentameric structure with a central pore is observed on the extracellular face of the membrane. The angle between the two α subunits was 128° and the unit cell about 10 nm diameter[30].

Analysis of the Torpedo *channel at 9 Å resolution*

09-30-05: Analysis at higher resolution (9 Å) has been obtained by recording images at different levels of defocus and averaging data using helical diffraction† methods[140]. This method allows some identification of secondary structure, particularly the α-helical† rods within each subunit. In the synaptic part of each subunit there are three rods, oriented perpendicular to the plane of the bilayer. Two of the rods line the entrance to the channel, with the third on the outside. In the region of the receptor that spans the bilayer, each subunit has only one visible rod: since this forms the lining of the pore it is assumed to be the M2 transmembrane helix. This rod kinks near its midpoint, where it is closest to the **axis of the pore**†, and tilts outwards on either side. It is flanked on the lipid-facing sides by a continuous rim of density that is interpreted to be β-sheet†.

Position of the M2 domain in the structure

09-30-06: Alignment between the three-dimensional densities and the sequence of M2 places the charged residues at the ends of M2 symmetrically bestride the bilayer, and a highly conserved Leu residue (Leu251 of the α subunit) at the position of the kink. A model is suggested in which the side-chains of the Leu residues at the kink project into the pore† to form a **hydrophobic ring**, closing the channel by making a barrier that hydrated ions cannot cross[140] *(Fig. 3).*

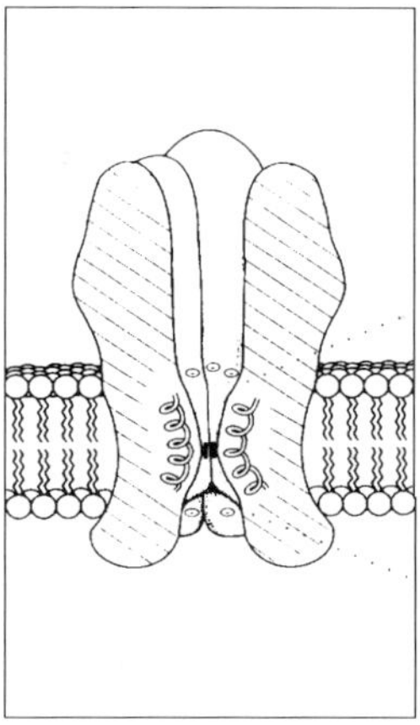

Figure 3. *Cut-away model to show the basic features of the nAChR–channel. The channel is built from a pentameric association of similar subunits, arranged around a central pore† that forms the pathway for the cations. Most of the protein mass of the channel is on the synaptic side of the bilayer, and this region includes the binding sites for acetylcholine and α-bungarotoxin. The pore is narrow across the lipid bilayer and is closed by the juxtaposition of hydrophobic amino acid side-chains in the region of the gate. The wider entrance and exit from the pore contain charged groups to screen out anions. The size of the pore at different levels is determined by the shape and orientation of the M2 helix and the nature of the amino acid side-chains projecting from it. (Reproduced with permission from Unwin (1993)* Cell **72** *(Suppl.): 31–41.) (From 09-30-06)*

Chemical modification of residues in M2

09-30-07: A combination of *in vitro* mutagenesis† to generate cysteine residues in M2 of the mouse muscle α subunit, followed by covalent chemical modification of the cysteines with small, charged, sulphydryl-specific reagents suggests that residues Ser248 to Thr254 of the M2 domain constitute a β-strand[141]. The side-chains of Cys residues substituting for Ser248, Leu250, Ser252 and Thr254 are all exposed to the reagent in the closed nAChR channel, and that of Leu251 is exposed on channel opening. These results are inconsistent with an α-helical structure, and strongly suggest a β-strand† conformation for this part of the M2 transmembrane domain.

A synthetic peptide model for the pore region of the nAChR channel

09-30-08: A **synthetic channel protein**, T5M2d, that emulates the presumed pore-forming structure of the nAChR has been generated by assembling five helix-forming peptide modules at the lysine ε-amino groups† of the 11-residue template {K*AK*KK*PGK*EK*G}, where * indicates the attachment sites. Helical modules represent the sequence of the M2 segment of the *Torpedo californica* nAChR δ subunit. Purified T5M2δ migrates in SDS–PAGE with an apparent M_r of approximately 14 000, consistent with a protein of 126 residues. When reconstituted in planar lipid bilayers, the T5M2δ polypeptide complex forms cation-selective channels with a single-channel conductance in symmetric 0.5 M KCl of 40 pS, a value close to the 45 pS characteristic of authentic purified *Torpedo* nAChR, recorded under similar conditions. These results support the contention that a bundle of five amphipathic† α-helices is a plausible structural motif for the **inner bundle that forms the pore**† of the pentameric nAChR channel[142].

Influence of lipids on the structure of the nAChR
09-30-09: The gross secondary structure of the purified *Torpedo californica* nAChR has been examined by Fourier-transform infrared resonance speç-troscopy† in reconstituted dioleoylphosphatidylcholine membranes in H_2O and D_2O. The secondary structure of nAChR in H_2O was calculated to contain about 19% α-helix, 42% β-structure†, 24% turns and 15% unordered. The secondary structure content in D_2O was estimated to be 14% α-helix, 37% β-structure, 29% turns and 20% unordered. In the presence of phosphatidic acid the β-structure content in D_2O increased significantly from 37% to 42%. This suggests that an ionic interaction between negatively charged lipid head-groups and positively charged peptide side-chains may stabilize a β-structure conformation. The inclusion of **cholesterol** in the reconstituted membranes significantly increased the α-helix† content from 14% to 17%. These results support the hypothesis that cholesterol may induce a transmembrane region to undergo an **unordered-to-helix transition** which is necessary to maintain the integrity of the ion channel[143].

Protein interactions

Synchronous initiation of subunit interactions in assembly
09-31-01: The assembly of *Torpedo* nAChR subunits is temperature-sensitive *in vivo* , a phenomenon that can be used to allow synchronous initiation of the **assembly of subunits** synthesized at the non-permissive temperature[144].

Order of subunit interactions in receptor assembly
09-31-02: In the mouse fibroblast cell line, All-11, that expresses the *Torpedo* nAChR subunits, the earliest identifiable complexes are $\alpha\beta\gamma$ trimers, which begin to be detectable within minutes of the temperature-shift. The δ subunit is added next, $\alpha\beta\gamma\delta$ tetramers beginning to appear at about 1 h, followed by the addition of the second α subunit from about 12 h onwards to form the $\alpha_2\beta\gamma\delta$ pentamer[145]. Studies with cell lines expressing restricted pairs of subunits show that the **assembly pathway** is kinetically determined, with $\alpha\delta$ complexes being formed much more slowly than $\alpha\beta$, $\beta\gamma$ or $\alpha\gamma$ complexes[145].

Subunit folding during the association process
09-31-03: In addition to subunit assembly, there are discernible **subunit folding** events occurring during the association process. Thus the **formation**

of the α-Bgt-binding site occurs on the α subunit hours after it has assembled into the trimer. The formation of an epitope† recognized by a specific monoclonal antibody, MAb14, is even further delayed, occurring just before the δ subunits are added to the trimers[145].

The extracellular N-terminal domain contains the information for subunit association

09-31-04: Co-expression of truncated, N-terminal† fragments consisting solely of the extracellular N-terminal domain of the α, δ or γ subunits of the mouse muscle nAChR with the four wild-type subunits in transfected COS cells blocks surface expression of the AChR. The formation of αδ heterodimers, an early step in the assembly pathway of the AChR, is inhibited. Immunoprecipitation and sucrose gradient sedimentation experiments show that the N-terminal fragment of the α subunit forms a specific complex with the intact δ subunit. Thus the extracellular N-terminal domains of the α, δ, and γ subunits contain the **information necessary for specific association** of the nAChR subunits[146].

A 43 kDa membrane protein is essential for nAChR aggregation in post-synaptic membranes

09-31-05: A **43 kDa peripheral membrane protein** is present in equimolar amounts with AChR in AChR-rich membranes, in close association with the β subunit of the AChR. This 43 kDa protein is essential for the characteristic **aggregation of the AChR** in the post-synaptic† membrane. Co-expression of cDNAs for each of the four subunits of both foetal (α, β, γ, δ) and adult (α, β, ε, δ) AChR with expression constructs encoding the 43 kDa protein in fibroblast cell lines leads to spontaneous clustering of AChR patches, compared with the uniform distribution obtained in the absence of the 43 kDa protein. The 43 kDa protein co-localises with the AChRs, and is able to aggregate into characteristic clusters when expressed alone. The 43 kDa protein has been proposed to form the key link between the AChR and the **cytoskeleton**[147].

58 and 87 kDa peripheral membrane proteins

09-31-06: Other **post-synaptic peripheral membrane proteins**, with molecular masses of 58 and 87 kDa, have been identified in *Torpedo* electric organ and mammalian neuromuscular junctions†. The **58 kDa protein** is concentrated at the synapse but has a broader distribution than the nAChR, including tissues such as kidney from which the nAChR is absent[148]. The **87 kDa protein** is more restricted to the synaptic region than the 58 kDa protein, but is also present in the sarcolemma† extrasynaptically. It has a restricted **tissue distribution**, being expressed in electric organ, muscle and brain, all tissues that express nAChR.

Dystrophin associates with the 58 kDa and 87 kDa proteins

09-31-07: Co-immunoprecipitation with anti-87 kDa protein antibodies shows that **dystrophin** (a membrane-associated cytoskeletal protein that is the product of the Duchenne muscular dystrophy [DMD] locus) and the

58 kDa protein associate with the 87 kDa protein in *Torpedo* post-synaptic membranes[148, 149]. (The 87 kDa protein, the 58 kDa protein and dystrophin have also been identified in vertebrate skeletal muscle sarcolemma, where dystrophin is proposed to link the extracellular matrix with the cytoskeleton[150].) The 87 kDa protein is a substrate for kinases, and it is suggested that its association with other synaptic proteins may be regulated by phosphorylation[149].

Protein phosphorylation

Multiple kinases can act on nAChR

09-32-01: The *T. californica* nAChR is phosphorylated by at least five different protein kinases; **PKA** that preferentially phosphorylates the γ and δ subunits[151]; **cyclic AMP-dependent protein kinase** that phosphorylates mainly γ and δ subunits[152]; **PKC** that modifies δ subunits; a **Ca^{2+}/calmodulin PKII** that phosphorylates β and δ subunits; and a **tyrosine-specific PK** that phosphorylates β, δ and γ subunits[153]. The nAChRs in skeletal muscle are also phosphorylated by similar protein kinases[154, 155].

09-32-02: Many of the phosphorylation sites have been identified: they are located on the major intracellular loop of each subunit, between the third and fourth transmembrane helices. These phosphorylation sites are conserved in the sequences of most nAChR subunits from a wide range of species[156].

Consensus sites for phosphorylation on the α7 subunit

09-32-03: Consensus sites exist for the phosphorylation of rat α7 nAChR subunits at S365 by **cAMP-dependent protein kinase**, of Thr415 and Ser427 by casein kinase II and of Tyr442 by **tyrosine kinase**[9].

Desensitization and phosphorylation

09-32-04: Nicotinic receptor–channels can be desensitized† more rapidly following cAMP-dependent phosphorylation[152, 156-158] *(see Receptor/transducer interactions, 09-49)*. cAMP-dependent processes may increase insertion of pre-existing nAChRs into the plasma membrane in the absence of new protein synthesis[159].

In vitro *phosphorylation by PKA*

09-32-05: The single-channel properties of purified AChRs from *T. californica* reconstituted in lipid bilayers are altered by *in vitro* phosphorylation with **PKA**. Notably, the spontaneous **open-channel probability** of phosphorylated AChRs (1.0 mol of phosphate incorporated per mol of AChR) is increased 40-fold over that of unphosphorylated receptors. Channel activation by PKA is correlated with AChR phosphorylation and is abolished by α-Bgt (200 nM). Like the unphosphorylated AChR, the phosphorylated channel has two distinct open states, short- and long-lived. The relative frequency of the long openings and the magnitude of both time constants increase 4–5-fold after phosphorylation, as they do with agonist-mediated activation[151].

PKC and aggregation of nAChR
09-32-06: Agrin, a protein isolated from the electric organ of *T. californica*, induces the formation of specializations† on chick myotubes in culture, at which several components of the post-synaptic apparatus, including nAChRs, are concentrated. The process of accumulation of the nAChRs into agrin-induced specializations† involves lateral migration of receptors already on the surface of the myotube, a process that does not require protein synthesis[160]. The formation of **agrin-induced nAChR aggregates** is blocked by the **phorbol ester TPA**, an activator of PKC, which can also disperse pre-existing aggregates[160].

Agrin stimulates nAChR phosphorylation
09-32-07: The addition of **agrin** to overnight cultures of chick myotube cultures, in the presence of $[^{32}P]$-H_3PO_4, causes a three-fold increase in phosphorylation of the nAChR β subunit, and a 20–30% increase in the phosphorylation of the γ and δ subunits. An inhibitor of protein serine kinases, **H-7**, blocks agrin-induced phosphorylation of the γ and δ subunits, but fails to inhibit induced phosphorylation of the β subunit or nAChR aggregation[161].

Tyrosine phosphorylation precedes aggregation
09-32-08: Immunoblotting experiments with affinity-purified antiphosphotyrosine antibodies demonstrate that **tyrosine phosphorylation** of the β subunit is induced by agrin-treatment of myotubes. Three treatments that block agrin-induced receptor aggregation, low pH (pH 6.5), **TPA** (50 nM) and addition of polyanions (0.05 mg/ml dextran sulphate), also block the agrin-induced increase in tyrosine phosphorylation of nAChRs[161]. Tyrosine phosphorylation of nAChRs precedes the aggregation of nAChRs by a few hours. The agrin-induced nAChR aggregates stain with anti-phosphotyrosine antibodies and correspond in location with the sites of labelling by rhodamine-conjugated α-Bgt[161].

Basic fibroblast growth factor induces nAChR aggregation
09-32-09: Beads coated with **basic fibroblast growth factor (bFGF)** induce the nAChRs of cultured *Xenopus* myocytes to aggregate at the site of the myocyte–bead contact. The **bFGF receptor** is a protein tyrosine kinase, and tyrosine kinase inhibitors stop the induction of clustering by the bFGF-coated beads[162].

ELECTROPHYSIOLOGY

Activation

Activation by acetylcholine
09-33-01: The *Torpedo* nAChR expressed in *Xenopus* oocytes is half-maximally activated by 20 mM **acetylcholine**[163].

Activation of nAChR by extracellular ATP
09-33-02: Extracellular ATP (10 μM) increases the spontaneous opening frequency of nAChR of rat skeletal muscle cells in the cell-attached†

configuration from 0.3 to 4.7 s^{-1}, without changing mean opening times (0.6 ms). When delivered through a separate drug pipette after first forming a gigaseal†, ATP increases ACh-activated single-channel open probability in a dose-dependent fashion. (100 μM and 500 μM ATP increases ACh channel activity induced by 0.1 μM ACh by 46% and 63%, respectively.) These concentrations of ATP are believed to be in the range found in the synaptic cleft immediately after release from the cholinergic motor nerve terminals[164].

09-33-03: The non-hydrolysable ATP analogue, **ATP-γ-S**, also enhances ACh-activated channel opening, but ADP, AMP and adenosine, up to 1 mM, are without effect. Because the facilitating effect is observed when the ATP is administered through a separate pipette, but cannot be seen with outside-out† patches, it is likely that the coupling of the ATP to the AChR occurs through an intracellular mediator[165].

Activation of neuronal nAChR by external Ca^{2+}

09-33-04: Neuronal nAChRs (but not muscle nAChRs) are strongly modulated by physiological concentrations of **external Ca^{2+}**. For neuronal nAChR species $\alpha2\beta2$, $\alpha3\beta2$, $\alpha3\beta4$ and $\alpha4\beta4$ expressed in *Xenopus* oocytes, and for native nAChRs in bovine **chromaffin cells**, ACh-induced currents increase as the $[Ca^{2+}]_o$ increases in the range 0.1–30 mM. The effect is Ca^{2+}-specific, is not due to Ca^{2+} carrying the additional current and occurs despite a reduction in the single-channel current amplitude[166].

Current–voltage relation

Muscle nAChR channels show a linear I–V relationship

09-35-01: The native nAChR channels of adult bovine skeletal muscle show a linear I–V relationship, with a conductance of 59 pS and an average open time of 5.6 ms[38]. The channels of foetal bovine muscle have a conductance of 40 pS and an average open time of 11.0 ms. These values are accurately reflected by those obtained from the expression of cRNAs encoding the α, β, δ and ε subunits and the α, β, γ and δ subunits respectively injected into *Xenopus* oocytes[38].

Neuronal nAChR channels show inward rectification

09-35-02: The currents through nAChR channels of central† and peripheral† neurones contrast with those passed by the muscle nAChR in displaying pronounced **inward rectification**†. In rat sympathetic† neurones, rectification of whole-cell currents results primarily from a reduced probability of channel opening at depolarizing potentials[167]. In rat **PC12 cells**, channels close more rapidly at positive potentials to cause the rectification[168].

09-35-03: The whole-cell nACh currents in cultured post-natal rat **hippocampal neurones** are of two distinct types. One class exhibits rapid and profound desensitization† and is sensitive to inhibition by α-Bgt. The second class activates slowly and exhibits no desensitization during prolonged agonist applications. This slow current is insensitive to α-Bgt. Both the fast and slow responses exhibit **inwardly rectifying**† current–voltage relationships and pass little current at positive membrane potentials[169].

Homomeric α7 and α8 channels show inward rectification

09-35-04: The ACh-induced currents through chick homomeric α7 and α8 channels expressed in *Xenopus* oocytes show very strong inward rectification† at potentials above −20 mV[17,217].

Current–voltage relationship of homomeric α9 channels

09-35-05: The I–V curve obtained from *Xenopus* oocytes injected with rat α9 cRNA is non-linear, with a maximal, ACh-induced inward current at −50 mV. Currents are strongly reduced at potentials negative to −50 mV, and at more positive holding potentials up to −25 mV. Strong rectification† then occurs up to a holding potential of +20 mV[218].

Inactivation

Neuronal nAChR channels show varying inactivation kinetics

09-37-01: The α-Bgt-sensitive nACh currents in rat hippocampal neurones decay rapidly (few milliseconds) in the continuing presence of agonist[169], and the chick α7 and α8 homomultimers† expressed in *Xenopus* oocytes behave similarly[17,217]. This behaviour contrasts with that of the α-Bgt-insensitive nicotinic responses of rat **hippocampal neurones**[169] and the **cerebellum** of adult rats[170], where nicotine-induced excitations do not show rapid desensitization.

The nAChR channel currents of embryonic muscle rapidly desensitize

09-37-02: Pulses of acetylcholine (ACh) applied to outside-out† patches of embryonic-like mouse muscle membrane elicit channel currents which decline rapidly (τ_d = 10–60 ms) due to desensitization†. About half of the channels recover from desensitization in 300 ms[171].

Kinetic model

Classical kinetic model

09-38-01: The classical model for activation of the neuromuscular nAChR can be represented by:

$$A + R \underset{k_{-1}}{\overset{k_{+1}}{\rightleftharpoons}} AR + A \underset{k_{-2}}{\overset{k_{+2}}{\rightleftharpoons}} A_2R \underset{\alpha}{\overset{\beta}{\rightleftharpoons}} A_2R^*$$

In this model, the closed receptor (R) undergoes sequential binding of two molecules of agonist (A), with association constants k_{+1} and k_{+2} and dissociation rate constants k_{-1} and k_{-2} , followed by a conformational change to the open state (R*), governed by opening rate constant β and closing rate constant α[172].

Model allowing for opening of singly ligated receptor

09-38-02: The above model is almost certainly too simplistic: there is evidence that short-duration openings can arise from singly ligated receptors, for example. This would require that the above scheme be extended to include the activation of AR to an AR* state, as well as the direct conversion of AR* to A_2R^* by binding of a second molecule of agonist. For a discussion of kinetic models for nAChR activation *see ref.*[173].

Selectivity

Non-selective cation currents

09-40-01: The nAChRs are generally cation specific, but relatively non-selective amongst cations. Every monovalent or divalent cation that can fit through a pore† of 0.65 nm diameter is permeant in endplate† channels. Permeant ions include not only the alkali metal† and alkaline earth† cations but also **organic cations** such as triaminoguanidinium, choline and histidine[174].

External Ca^{2+} modulates neuronal nAChRs

09-40-02: Note that physiological concentrations of extracellular Ca^{2+} can bind to neuronal nAChRs (but not the muscular form) and directly *modulate* channel activity, independently of permeation or second messenger-based processes[166] *(see Activation, 09-33).*

Ca^{2+} conductance measurements

09-40-03: The elementary slope conductance† of the nAChR channel from rat **medial habenula neurones** is 11 pS in pure external Ca^{2+} (100 mM) and 42 pS in standard solution. The Ca^{2+} influx through nAChRs results in the rise of $[Ca^{2+}]_i$ to the micromolar range. This increase is maximal below −50 mV, when Ca^{2+} influx through voltage-activated Ca^{2+} channels is minimal. The Ca^{2+} influx via the nAChRs activates a Ca^{2+}-dependent Cl^- conductance and causes a decrease in the $GABA_A$ response that outlasts the rise in $[Ca^{2+}]_i$[175].

09-40-04: The α-Bgt-sensitive nAChRs of chick **ciliary neurones**, activated by 1 μM nicotine, pass inward Ca^{2+} currents, detectable with Ca^{2+}-sensitive fluorescent dyes but not by whole-cell recording. The effect is sensitive to 20 μM **(+)-tubocurarine** and to nM α-Bgt[63]. When the neurones are treated with 10 μM nicotine, the α-Bgt-insensitive nAChR channels are activated and lead to increased $[Ca^{2+}]_i$[63].

Ca^{2+} selectivity of the α7 channel

09-40-05: The homo-oligomeric† α7 neuronal nAChR is exceptionally permeable to Ca^{2+}. Determinations based on the reversal potential shifts in 1 mM and 10 mM Ca^{2+} (+29 mV) suggest a permeability ratio (P_{Ca}/P_{Na}) of about 20 for the rat α7 channel[9]. This contrasts with a permeability ratio (P_{Ca}/P_{Na}) of 0.2 for the muscle nAChR[176], 0.7 for rat **parasympathetic cardiac neurones**[8] and about 1.5 for the nAChRs present in **chromaffin cells**[176] or in **PC12** cells[177].

Ca^{2+} influx through α7 channels activates a Cl^- channel

09-40-06: When the homo-oligomeric α7 channel of the rat is expressed in *Xenopus* oocytes, a significant proportion of the current through the channels is carried by Ca^{2+}: this Ca^{2+} influx activates a Ca^{2+}-sensitive Cl^- channel. High levels of α7 transcripts are detected by *in situ* hybridization† in the olfactory areas, the **hippocampus**, the **hypothalamus**, the **amygdala**, and the **cerebral cortex**. These results imply that α7-containing receptors may play a role in activating calcium-dependent mechanisms in specific neuronal populations of the adult rat limbic system[9].

Ba^{2+} selectivity of the rat α7 channel

09-40-07: The nicotine-activated rat α7 neuronal nAChR, expressed in *Xenopus* oocytes, shows an inwardly rectifying† current in external barium: in calcium the response is larger and has a linear I–V relation. The permeability ratio P_{Ba}/P_{Na} of the rat α7 receptor has been estimated at about 17[178].

The α8 and α9 homomeric channels are permeable to Ca^{2+}

09-40-08: The chick α8 subunit, which has 62% amino acid identity with the α7 subunit and an identical M2 domain, also forms homomeric channels permeable to Ca^{2+} when produced by expression of α8 cRNA in *Xenopus* oocytes[217]. The more distantly related α9 subunit from rat, which has 38% identity to the α7 and α8 subunits *(see Homologous isoforms, 09-21)*, also forms a homomeric channel that is permeable to Ca^{2+} when expressed in *Xenopus* oocytes[218]. In both these cases, influx of Ca^{2+} triggers a Ca^{2+}-sensitive Cl^- current.

Mutant channels with increased selectivity for divalent cations

09-40-09: The replacement of Thr244 by Asp in the putative channel-forming M2 segment of the chick neuronal α7 nAChR subunit produces a marked change in the selectivity of the homo-oligomeric ion channel produced by expression in *Xenopus* oocytes. The relative ionic permeability of wild-type AChRα7 to K^+ is $P_K/P_{Na} = 1.2$, and to Ba^{2+}, $P'_{Ba}/P_{Na} = 1.4$. In contrast, AChRα7D244 is less selective in discriminating between K^+ and Na^+, $P_K/P_{Na} = 0.95$, but exhibits a remarkable increase in permeability to Ba^{2+}, with $P'_{Ba}/P_{Na} = 3.7$. Furthermore, only the mutant channels are permeable to Mg^{2+}. The ring of negatively charged residues in the putative pore-forming segment of the nAChR increases the permeability to divalent cations[127].

Single-channel data

Single channels show inward rectification

09-41-01: In both neurones and muscle, ion permeation through single channels shows **inward rectification**†, behaviour that is dependent on internal Mg^{2+}. In the absence of Mg^{2+} or Ca^{2+} on either side of the membrane, single-channel I–V plots are linear, though whole-cell currents continue to show inward rectification[179].

Single-channel conductances of neuronal channels

09-41-02: The unitary conductance of nAChR channels in rat **sympathetic neurones** varies from 26 to 48 pS, with a mean of 36.8 pS, in 1 mM Ca^{2+}. Removal of divalent cations from the external solution increases the unitary channel conductance. Altering the main permeant ion in divalent-free solutions gives the following conductance sequence: K^+ (93 pS) > Cs^+ (61 pS) > Na^+ (51 pS) > Li^+(23 pS). Replacement of Na^+ by Cs^+ in the external solution considerably reduces the current evoked by ACh in whole-cell recordings and the channel-opening frequency in outside-out† patches[180].

Single-channel conductances of different subunit combinations

09-41-03: Single-channel conductances of avian and rat subunit combinations produced in *Xenopus* oocytes are as follows: α4nα1, 20 pS; α2β2, 33.6 pS and 15.5 pS; α3β2, 15.4 pS and 5.1 pS; α4β2, 13.3 pS. The α4β2 combination also has a secondary conductance state but this has not been characterized[181].

Multiple conductances in a single cell

09-41-04: There is strong electrophysiological evidence for functional heterogeneity of neuronal nAChRs, based on several open-channel conductances being observed under identical conditions in a single cell. Three or more populations of conductances are observable in rat **PC12 cells**[182], bovine **chromaffin cells**[183] and chicken **sympathetic neurones**[184].

Bursting behaviour of nAChR

09-41-05: Both neuronal and muscle nAChR channels display **bursting behaviour**. Burst durations vary amongst different channel types: several nAChR channels on **autonomic neurones** show burst lengths with time constants in the range of 5–10 ms[184, 185].

PHARMACOLOGY

Blockers

Non-competitive blockers

09-43-01: A heterogeneous group of pharmacological agents, collectively known as **non-competitive**† **blockers**, can inhibit the ionic permeability of muscle and neuronal AChR without affecting the agonist site. Such blockers include aminated **local anaesthetics**† (e.g. procaine, lidocaine, dibucaine, proadifen), **sedatives**†, such as chlorpromazine, and various **toxins**. Studies with ^{3}H-labelled non-competitive blockers, including **histrionicotoxin** (a frog toxin), **phencyclidine** (a halucinogen) and **chlorpromazine**, define high-affinity binding sites, sensitive to histrionicotoxin (K_d 0.16 μM), and low-affinity sites insensitive to histrionicotoxin (K_d > 100 μM). The high-affinity sites are present in one copy per receptor pentamer, with the low-affinity sites being 10–30-fold more common[186].

Non-competitive blockers interact with the open channel

09-43-02: Labelling of the nAChR by photoaffinity† derivatives of non-competitive blockers is enhanced by carbamylcholine and inhibited by histrionicotoxin. The rate of covalent association of chlorpromazine with the high-affinity site increases 100–1000-fold ($k_{on} = 10^7$ $M^{-1}s^{-1}$) when acetylcholine is added in the concentration range effective for activation of the channel *in vitro*. Competitive antagonists block this effect. The rate of [^{3}H]-chlorpromazine incorporation declines on prolonged exposure of the channel to acetylcholine, with a time course and concentration dependence similar to those of the rapid desensitization† of the ion-flux responses of the native membranes of *T. californica*. These observations suggest that the **non-competitive blockers** freely diffuse to their high-affinity binding sites within the pore† of the open channel[187].

Specific amino acids contacted by non-competitive blockers
09-43-03: The amino acids photolabelled† by [^{3}H]-**chlorpromazine** have been identified by peptide-mapping† and sequencing experiments. The residues αSer248, βSer254, βLeu257, γThr253, γSer257, γLeu260 and δSer262 are all labelled by [^{3}H]-chlorpromazine and protected by **phencyclidine**. The labelled serines on all subunits occupy homologous positions within the putative M2 helix that lines the pore of the nAChR[188] *(Fig. 4)*. Experiments with alternative non-competitive blockers reach very similar conclusions[118, 189].

Progesterone as a blocker of neuronal nAChR
09-43-04: The major brain nAChR is assembled from two subunits termed α4

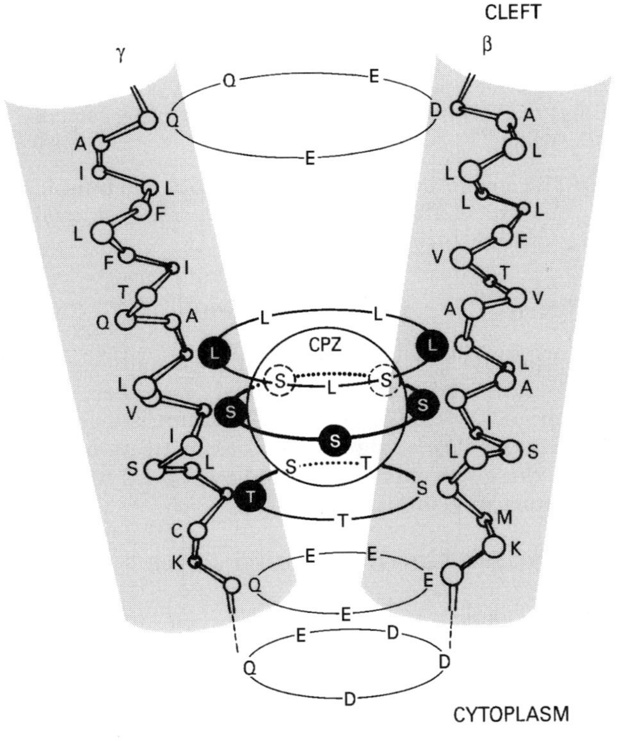

Figure 4. *Model of the high-affinity site for chlorpromazine within the nAChR channel. The M2 domains are shown as transmembrane helices, quasi-symmetrically arranged around the central axis of the molecule. Only the M2 helices of the β and γ subunits are shown in detail. The α-carbons of the amino acids are shown, with the residues designated by the standard single-letter code. The central sphere represents the space occupied by a chlorpromazine molecule in all possible orientations at its binding site. (Reproduced with permission from Revah* et al. *(1990)* Proc Natl Acad Sci USA ***87****: 4675–9.) (From 09-43-03)*

and nα1 (Note nα1 is equivalent to β2, see table 1, p235. When produced in *Xenopus* oocytes, these subunits reconstitute a functional receptor that is inhibited by **progesterone** concentrations similar to those found in serum. The steroid interacts with an extracellular site on the channel protein, and the inhibition does not require agonist, is voltage independent and does not affect receptor desensitization†. The inhibition by progesterone is not competitive, though it may involve interaction with the ACh-binding site, and is independent of the ionic permeability of the receptor[190].

Toxins as blockers of nAChR

09-43-05: The ganglionic antagonist **neosurugatoxin** (2 nM) causes almost complete blockade of ACh responses in *Xenopus* oocytes expressing α2β2, α3β2 or α4β2 combinations, but is ineffective against the muscle receptor formed by microinjection of α1, β1, γ and δ subunit RNAs[191].

09-43-06: Lophotoxin (10 μM), an inhibitor of neurotransmission at neuromuscular junctions† and autonomic† ganglia, completely blocks the muscle α1β1γδ and the neuronal α4β2 nAChRs, but only partially blocks the α2β2 and α3β2 species[191].

09-43-07: The **α-conotoxins G1A and M1** block the muscle α1β1γδ receptor at 100 nM, but are ineffective against α2β2, α3β2 or α4β2 combinations at concentrations up to 10 μM[191].

α-Bgt-sensitive and α-Bgt-resistant channels in hippocampal neurones

09-43-08: Two types of whole-cell nACh currents can be distinguished in cultured post-natal rat **hippocampal neurones**: one showing rapid desensitization† and sensitivity to α-bungarotoxin (IC_{50} of 1–2 nM[63]); the second activating slowly, showing no desensitization and insensitive to the toxin. Both these currents are blocked by 0.1–1.0 mM **(+)-tubocurarine** and 0.1–1.0 mM **mecamylamine**[169].

A plant toxin inhibits ACh-induced currents in hippocampal neurones

09-43-09: Picomolar **methyllycaconitine**, a toxin from the seeds of *Delphinium brownii*, inhibits acetylcholine- and anatoxin-induced whole-cell currents in cultured foetal rat **hippocampal neurones**. This antagonism was specific, concentration-dependent, reversible and voltage-independent. The toxin also inhibits [^{125}I]-α-Bgt binding to adult rat hippocampal membranes, protects against the α-Bgt-induced blockade of nicotinic currents, and shifts the concentration–response curve of acetylcholine to the right, suggesting a competitive mode of action. Low concentrations of methyllycaconitine (1–1000 fM) decrease the frequency of anatoxin-induced single-channel openings, with no detectable decrease in the mean channel open time[192].

Alcohols as blockers of nAChR

09-43-10: The ***n*-alcohols** block the activity of the nAChR from cultured rat r[illegible]hain alcohols (pentanol to octanol) cause currents t[illegible] fully open and closed state level, the number of g[illegible]reasing with alcohol concentration. Nonanol and d[illegible]duration of bursts of openings but do not cause an i[illegible]of short closed intervals within a burst. Beyond decanol there is a decline in the ability of the *n*-alcohols to affect nAChR function, and a saturated solution of dodecanol has no significant effect. The IC_{50} values determined from the effect on the total charge carried per burst are as follows: **hexanol**, 0.53 ± 0.14 mM; **heptanol**, 0.097 ± 0.02 mM; **octanol**, 0.04 mM and **nonanol**, 0.16 ± 0.035 mM. The K_d values, determined from the ratios of the blocking and unblocking rate constants†, decrease with increasing chain length from 8 mM for pentanol to 0.15 mM for octanol[193].

Anaesthetics as blockers of nAChR

09-43-11: The volatile anaesthetics **halothane and isoflurane** block the activated nAChR channels of cultured BC3H1 mouse tumour cells in cell-attached patches. Both halothane and isoflurane shorten the duration of individual opening events and cause openings to group in bursts†. The slower time constant of channel open-time distributions is decreased 50% by approximately 0.25% isoflurane (0.12 mM) or 0.30% halothane (0.15 mM) at room temperature. Total open time per burst is also decreased by each agent[194].

Other blockers of nAChR channels

09-43-12: The alkaloid **strychnine**, a classical blocker of glycine-gated Cl^- channels, blocks the rat α7 nAChR expressed in *Xenopus* oocytes with an IC_{50} of 0.35 μM[9]. The homomeric α9 channel from rat, expressed in *Xenopus* oocytes, is strikingly sensitive to blockage by strychnine, with an IC_{50} of 0.02 μM[218] *(see Receptor antagonists, 09-51).*

09-43-13: Hexamethonium (2.5 μM) and **decamethonium** (10 μM) block the function of the avian α4nα1 nAChR in a voltage-dependent manner, suggesting that they enter the channel pore[195].

Channel modulation

09-44-01: *See Protein phosphorylation, 09-32.*

Potentiation of nicotinic response by external Ca^{2+}

09-44-02: The nicotinic response of the nAChRs of neurones from the rat **medial habenular nucleus** (MHb) is strongly potentiated by extracellular Ca^{2+}. The amplitude of whole-cell currents evoked by acetylcholine is increased up to 3.5-fold by 4 mM external Ca^{2+}. The potentiation†, which is rapidly reversed on removal of Ca^{2+}, is due to a change in conductance, and is observed at both negative and positive potentials. The effect of external Ca^{2+} is similar when internal Ca^{2+} is buffered by 10 mM BAPTA or 0.5 mM EGTA, suggesting that the Ca^{2+} acts at an external site and *not* at an internal site after entry through the nAChR channel.

Ca^{2+} increases the frequency of chan[illegible]

09-44-03: Single-channel recordings with ou[illegible]atches show that the frequency of opening of ACh-activated channels is increased by a factor of 3.1 after addition of 4 mM Ca^{2+}, the magnitude of the **potentiation**† increasing linearly with external [Ca^{2+}] up to 4 mM. The potentiation is also achieved with Sr^{2+} (4.5-fold at 4 mM) and Ba^{2+} (3.2-fold at 4 mM), but *not* with 4 mM Mg^{2+}. These data are interpreted as indicating allosteric† modulation of nAChR by external Ca^{2+} to favour a transition to a distinct, ACh-activatable state[196].

Negative effect of divalent cations on single-channel conductance

09-44-04: Note that the divalent cations Ca^{2+}, Sr^{2+}, Ba^{2+} and Mg^{2+} at 4 mM all decrease the single-channel conductance of nAChR at negative (but not positive) potentials by about 50%, a finding that has been interpreted as a **screening effect** of the divalent cations at the entrance to the channel. This inhibitory effect contrasts with the potentiation† that is seen with Ca^{2+}, Ba^{2+} and Sr^{2+}, but not with Mg^{2+}, occurring at both negative and positive potentials[196] *(see above).*

Equilibrium dissociation constant

Dissociation constant for α-bungarotoxin

09-45-01: The dose–response curve for the blocking of the nAChR of chick **ciliary ganglion** neurones by α-Bgt yields an IC_{50} of 1–2 nM. This is in good agreement with the K_d of 1.4 nM determined from Scatchard† analysis of [^{125}I]-α-Bgt binding to the target on neurones[63] *(Table 5).*

Table 5. *Affinities of cholinergic agents for neuronal nAChRs (From 09-45-01)*

	K_i values (nM)	
Ligand	α-Bgt-sensitive	α-Bgt-insensitive
α-Bgt	0.7	>10 000
Nicotine	18	3100
(+)-Tubocurarine	600	8300
α-Cobratoxin	1.7	113
Methyllycaconitine	2.8	138

K_i values were determined by equilibrium binding experiments in which the ligands competed against [^{125}I]-α-Bgt in the case of the α-Bgt-sensitive channels or against [^{125}I]-n-Bgt in the case of the α-Bgt-insensitive channels. Data taken from ref.[63]

α-Bgt interaction with the α7 channel

09-45-02: The IC_{50} for the blocking of the chick α7 nAChR expressed in *Xenopus* oocytes by α-Bgt is 0.73 nM[17], in close agreement with the EC_{50} (0.35–0.61 nM) of the complex between the toxin and its binding proteins in CNS membranes[46], suggesting that most of the high affinity α-Bgt-binding proteins in chick brain may contain the α7 subunit.

Dissociation constants of various nAChR species for acetylcholine
09-45-03: The apparent dissociation constants of acetylcholine for the nAChR vary with the species of receptor. The K_d values for α4nα1, α4nα3, α3nα1, α3nα3 and α7 channels are 0.77, 4.8, 5.6, 158 and 115 μM, respectively[197], while that for the homopentameric α8 receptor is 1.9 μM[217].

Binding of agonists and antagonists to homomeric channels
09-45-04: The binding characteristics of a range of agonists and antagonists to homomeric α7 (chicken and human), α8 (chicken) and α9 (rat) channels expressed in *Xenopus* oocytes are summarized in Table 6.

Table 6. *Binding of agonists and antagonists to homomeric nAChR channels (From 09-45-04)*

	Homomeric channel			
	α7 (chick)[221]	α7 (human)[219]	α8 (chick)[217]	α9 (rat)[218]
	EC50 or IC50 (μM)			
Agonist				
Acetylcholine	110	79.2	1.9	10
L-Nicotine	7.8	40.2	1.0	30[a]
Cytisine	18	71.4	1.0	[a]
DMPP	30[b]	25.5	6.5	ca 50[b]
TMA	800	101	10	
Antagonist				
α-Bgt	0.003	0.0022	0.0014	[c]
Atropine	7.1	125	0.4	1.3
Curare	0.14	0.7	0.6	0.3
Strychnine	0.52	7.5	0.8	0.02

Nicotine (0.1 μM to 1 mM) and cytisine do *not* act as agonists for the α9 channel. Nicotine reduces the currents induced by ACh, with an IC_{50} of 30 μM[218].
[b]DMPP (1,1-dimethyl-4-phenylpiperazonium) is a weak, partial agonist with chicken α7 and rat α9 channels, but a full agonist for human α7 channels.[218,219]
[c]The IC_{50} for α-Bgt on α9 channels has not been determined, but the currents induced by 100 mM ACh are completely eliminated by 100 nM α-Bgt[218].

Hill coefficient

Hill coefficient for acetylcholine
09-46-01: The nAChR channels are predominantly activated by the binding of two molecules of acetylcholine, with a Hill coefficient† of 1.7–2.0[173]. Several types of channel opening can be discerned electrophysiologically and it is likely that a class of very brief openings arises when only a single molecule of acetylcholine is bound[173, 198-200] *(see Kinetic model, 09-38).*

Hill coefficient for α-Bgt binding
09-46-02: The chick α7 channel expressed in *Xenopus* oocytes is sensitive to α-Bgt (IC_{50} = 0.3 nM[217]; 0.73 nM[17]), with a Hill coefficient quoted as 1.3[217] or at least 6[17].

Ligands

Snake venom neurotoxins

09-47-01: Two snake venom neurotoxins, **α-bungarotoxin (α-Bgt)** and **neuronal bungarotoxin (n-Bgt**; also known as κ-bungarotoxin, bungarotoxin 3.1 and toxin F) are frequently used to characterize neuronal nAChR subtypes. The α-Bgt blocks most muscle nicotinic receptors and some, but not all, cloned or biochemically purified neuronal receptors from the chick, rat, and insect nervous systems.

Amino acid sequence motif common to α-Bgt-sensitive subunits

09-47-02: The sequence motif Cys–Cys–X–X–Pro–Tyr is common to all known α-subunits in neuronal receptors blocked by α-Bgt. In α-Bgt-insensitive neuronal receptors, the Pro is replaced by Ile: other amino acid substitutions may also be important. **Neuronal bungarotoxin** blocks some, but not all, α-Bgt-sensitive and -insensitive receptor subtypes expressed in oocytes, and in both vertebrate and invertebrate nervous systems. Non-α subunits help determine the ability of neuronal bungarotoxin to block functional neuronal receptors[201].

The α7 channel is very sensitive to α-Bgt

09-47-03: The α7 channel is distinguished amongst neuronal channels by its sensitivity to α-Bgt. The chick α7 channel expressed in *Xenopus* oocytes has a K_i of 0.73 nM for α-Bgt[17], while the rat α7 isoform is completely blocked by nanomolar amounts of toxin[9].

Sensitivities of different subunit combinations

09-47-04: None of the rat or avian receptors formed in *Xenopus* oocytes by expression of RNAs encoding α2, α3, α4, β2 and β4 subunits is functionally blocked by nanomolar α-Bgt *(see Table 7)*. The homomeric chick α7 and α8 channels produced in *Xenopus* oocytes are blocked by α-Bgt with EC_{50} values of 0.0003 μM and 0.0014 μM, respectively[217].

Table 7. *Toxin sensitivities of neuronal nAChR subunit combinations expressed in* Xenopus *oocytes (From 09-47-04)*

nAChR composition	α-Bgt (0.1 μM)	n-Bgt (0.1 μM)	NSTX (2 nM)	LTX-1 (10 μM)
α2β2	0	0	+++	++
α3β2	0	+++(0.01 μM)	+++	++
α4β2	0	+	+++	+++
α4nα1[a]	0 (0.3 μM)	0 (0.5 μM)		
α3β4[a]	0	0		
α7[a]	+++			
α8[a]	+++			

[a]Avian subunit combinations.
0, no blockade at indicated concentration; +, ++, partial blockade at indicated concentration; +++, nearly complete blockade at indicated concentration.
NSTX, neosurugatoxin; LTX-1, lophotoxin.

09-47-05: Neuronal Bgt strongly blocks the rat α3β2 combination at 0.01 μM, and also antagonises the α4β2 combination, though with 10–100-fold less sensitivity.

α-Bgt-sensitive central nicotinic responses

09-47-06: There are a few cases in which α-Bgt blocks central nicotinic responses. In the rat **hippocampus,** for example, nicotinic responses are blocked by 20 nM n-Bgt or 300 nM α-Bgt[202]. Such nAChRs, strongly blocked by relatively low concentrations of both α-Bgt and n-Bgt, have been characterized in chicken **cochlear hair cells**[203] and in insects[204].

Kappa-flavotoxin antagonizes α3 subtypes of nAChR

09-47-07: Kappa-flavotoxin (κ-FTX), a snake neurotoxin that is a selective antagonist of certain neuronal nAChRs, is related to but distinct from n-Bgt. The κ-FTX binds with high affinity to α3 subtypes of neuronal AChRs. The region of the neuronal AChR α3 subunit forming the binding site for κ-FTX has been identified using overlapping synthetic peptides covering the α3 sequence. The 'prototope' for κ-FTX-binding was identified within amino acid residues 51–70 of the α3 subunit[205].

Other toxins

09-47-08: Neosurugatoxin specifically blocks neuronal nicotinic receptors in mouse neuromuscular junctions† (when partially blocked with tubocurarine or with low Ca^{2+}/high Mg^{2+} tyrode solution). Neosurugatoxin (3–10 μM) depresses indirect twitches and produces waning of indirectly elicited tetanic contractions under these conditions[206].

09-47-09: The coral neurotoxin **lophotoxin** irreversibly inhibits the *Torpedo* AChR by binding covalently to αY190[207].

An immunoactive polypeptide blocks nicotinic responses

09-47-10: Thymopoietin, a thymus-derived polypeptide involved in the modulation of immune function, inhibits the binding of $[^{125}I]$-α-Bgt to membranes throughout the brain. Thymopoietin potently blocks nicotinic receptor-mediated responses in muscle cells, interacts with $[^{125}I]$-α-Bgt-binding sites on rat **PC12 cells** and blocks nicotine-induced inhibition of **neurite outgrowth**[208]. It has been suggested that extrasynaptic α-Bgt-binding proteins serve to monitor extrasynaptic ACh levels and hence regulate neurite outgrowth by raising free intracellular $[Ca^{2+}]$[63].

Chemical ligands: cyclic compounds that activate

09-47-11: The cyclic compound **1,1-dimethyl-4-acetylpiperazinium iodide** and its trifluoromethyl analogue (**F3-PIP**) interact with nAChRs from both *Torpedo* electroplaque† and **BC3H-1 cells** at lower concentrations than the acyclic derivatives, *N,N,N,N'*-tetramethyl-*N'*-acetylethylenediamine iodide and its fluorinated analogue (F3-TED). In measurements of the initial interaction with the nAChR, the PIP compounds have an affinity approximately one order of magnitude higher than that of the TED compounds. Longer incubations indicate that the PIP compounds are able to induce a time-dependent shift in receptor affinity consistent with desensitization†, whereas the TED compounds are unable to induce such a shift. The activation of single-channel currents by the cyclic compounds occurs at concentrations in the micromolar range, approximately two orders of magnitude lower than for the acyclic compounds, but the TED compounds exhibit a larger degree of channel blockade than the PIP compounds[209].

Chemical ligands: tricyclic antidepressants inhibit neuronal nAChRs

09-47-12: Two structurally related **tricyclic antidepressants** inhibit neuronal nAChR currents in **human neuroblastoma (SY-SY5Y) cells**. Both **desipramine** and **imipramine** reversibly inhibit inward currents evoked by application of the nAChR agonist dimethylphenylpiperazinium iodide (30–300 μM) with IC_{50} values of 0.17 μM and 1.0 μM respectively (holding potential −70 mV). The degree of current inhibition is unaffected by agonist concentration. The effects of desipramine are voltage-independent over the range −40 to −100 mV, and inhibition caused by imipramine only increases very slightly with membrane hyperpolarization† over the same range[210].

Receptor/transducer interactions

09-49-01: At the neuromuscular junction†, motoneurones release **calcitonin gene-related peptide (CGRP)**† *in addition to* acetylcholine. Post-synaptic receptors for CGRP coupled to G_s and adenylyl cyclase can elevate the level of cAMP required for desensitization via phosphorylation *(see Protein phosphorylation, 09-32).*

Receptor agonists (selective)

The different molecular species of nAChR show distinct agonist potencies

09-50-01: The rat α7 nAChR expressed in *Xenopus* oocytes displays the following order of agonist sensitivity: **nicotine** > **cytisine** > **DMPP** > ACh[9] *(see Equilibrium dissociation constant, 09-45).* The receptors from neurones of the rat **interpeduncular nucleus** show the order of potency cytisine > ACh > nicotine; those on neurones from the **medial habenula** have nicotine > cytisine > ACh[213].

The α9 homomeric channel has mixed nicotinic–muscarinic pharmacology

09-50-02: The homomeric nAChR channel obtained by expression of cRNAs encoding the rat α9 subunit in *Xenopus* oocytes shows an unusual pharmacological profile, with characteristics of both a nicotinic and a muscarinic acetylcholine receptor *(see Receptor antagonists (selective), 09-51).* The channels are activated by ACh (EC_{50} 10 μM), but not by nicotine or the nicotinic agonist, cytisine. The muscarinic agonists, bethanecol and pilocarpine are also ineffective. Surprisingly, both the nicotinic agonist **1,1-**

dimethyl- [illegible] **DMPP)** and the muscarinic agonist **oxotremo**[illegible] as partial agonists, eliciting maximum response[illegible]erved with ACh[218]. The unusual pharmacology [illegible]ingly similar to that of the cholinergic† receptor [illegible]**lear hair cells**[218].

Receptor antagonists (selective)

Forskolin antagonizes by interaction with the β subunit of nAChR
09-51-01: Forskolin at micromolar concentrations acts as a non-competitive inhibitor of the *Torpedo* electroplax (K_i 6.5 μM) and mouse muscle (K_i 22 μM) nAChRs expressed in *Xenopus* oocytes. The antagonist reduces the number of channel openings per unit time by interaction with the γ subunit of the nAChR[211].

Inhibitors of phosphate-transfer reactions as antagonists of nAChR
09-51-02: The single-channel and macroscopic ACh-induced currents produced by expresssion of the muscle channel in *Xenopus* oocytes are modified by **IBMX (3-isobutyl-1-methylxanthine)**, a phosphodiesterase inhibitor, and by **H-7 [1-(5-isoquinolinylsulphonyl)-2-methylpiperazine]**, a non-specific inhibitor of protein kinase activity. Both IBMX (IC_{50} 475 μM at −30 mV) and H-7 (IC_{50} 160 μM) directly inhibit ACh-induced currents *independently of their action on phosphorylation.* H-7 preferentially inhibits the open nAChR channel, but there is also some inhibition of the closed conformation. This inhibition is voltage-dependent, decreasing *e*-fold per 34 mV depolarization. A similar inhibition is also produced by 30 μM **HA-1004**, an analogue of H-7 that does *not* inhibit protein kinase activity[212].

A potent antagonist from Delphinium *seeds*
09-51-03: Methyllycaconitine, a toxin from the seeds of *Delphinium brownii*, inhibits acetylcholine- and anatoxin-induced whole-cell currents in cultured foetal rat **hippocampal neurones**, at picomolar concentrations. This antagonism is specific, concentration-dependent, reversible, and voltage-independent. Methyllycaconitine also inhibits α-Bgt binding to adult rat hippocampal membranes. Concentrations of methyllycaconitine in the 1–1000 fM range decreased the frequency of anatoxin-induced single-channel openings, with no detectable decrease in the mean channel open time[192].

The α9 receptor–channel shows an unusual pharmacological profile
09-51-04: The homomeric channel produced by expression of the cRNA encoding the rat α9 subunit in *Xenopus* oocytes displays novel patterns of response to classical nicotinic and muscarinic agonists and antagonists[218]. The classical cholinergic agonists **nicotine** and **muscarine** reduce the currents induced by ACh with IC_{50} values of 30 μM and 75 μM respectively. The α9 channel is also blocked by the nicotinic antagonist, **(+)-tubocurarine** (IC_{50} = 0.3 μM) and by the muscarinic antagonist, **atropine** (IC_{50} = 1.3 μM). (The homomeric α8 channel is also sensitive to atropine, with an IC_{50} of 0.4 μM[217] – see *Equilibrium dissociation constant, 09-45.)* The alkaloid **strychnine**, a blocker of glycine-gated chloride channels, is a potent

...t of α9 channels, with an IC_{50} of 0.02 μM. The α9 channels are ... to both α-Bgt and κ-Bgt. Although IC_{50} values for these toxins ... been determined, ACh-induced currents through α9 channels are ...y, but reversibly, blocked by 100 nM toxin[218].

INFORMATION RETRIEVAL

Database listings/primary sequence discussion

09-53-01: *The relevant database is indicated by the lower case prefix (e.g. gb:), which should not be typed (see Introduction & layout of entries, entry 02). Database locus names and accession numbers immediately follow the colon. Note that a comprehensive listing of all available accession numbers is superfluous for location of relevant sequences in GenBank® resources, which are now available with powerful in-built* ***neighbouring*** † ***analysis*** *routines (for description, see the Database listings field in the Introduction & layout of entries, entry 02). For example, sequences of cross-species variants or related gene family*† *members can be readily accessed by one or two rounds of neighbouring*† *analysis (which are based on pre-computed alignments performed using the BLAST*† *algorithm by the NCBI*†*). This feature is most useful for retrieval of sequence entries deposited in databases later than those listed below. Thus, representative members of known sequence homology groupings are listed to permit initial direct retrievals by accession number, author/reference or nomenclature. Following direct accession, however, neighbouring*† *analysis is strongly recommended to identify newly reported and related sequences.*

Nomenclature	Species, DNA source	Original isolate	Accession	Sequence/ discussion
nAChR α	Calf skeletal muscle cDNA library	437 + 20 aa	em: X02509	Noda, *Nature* (1983) **305**: 818–23.
	Mouse muscle cDNA library	457 aa	em: X03986	Isenberg, *Nucleic Acids Res* (1986) **14**: 5111.
	T. californica electric organ cDNA library	437 + 24 aa	gb: J00963	Noda, *Nature* (1982) **299**: 793–7.
	T. marmorata electric organ cDNA library	437 + 24 aa	gb: M25893	Devillers-Thiery, *Adv Exp Med Biol* (1984) **181**: 17–29.
α-1a	*Xenopus laevis* developmental stage 17 cDNA library	457 aa	em: X17244	Hartman, *Nature* (1989) **343**: 372–5.
nAChR β	Mouse myoblast cDNA library	478 + 23 aa	gb: M14537	Buonanno, *J Biol Chem* (1986) **261**: 16451–8.

Nomenclature	Species, DNA source	Original isolate	Accession	Sequence/ discussion
	Mouse genomic DNA library	entire β gene, including introns	gb: J04699	Buonanno, *J Biol Chem* (1989) **264**: 7611–16.
	T. californica electric organ cDNA library	469 + 24 aa	gb: J00964	Noda, *Nature* (1983) **301**: 251–5.
nAChR δ	Chicken genomic DNA library	497 + 18 aa		Nef, *Proc Natl Acad Sci USA* (1984) **81**: 7975–9.
	Rat genomic DNA library	1145 bp of DNA sequence upstream of translation-start	em: X66531	Chahine, *Development* (1992) **115**: 213–19.
	T. californica electric organ cDNA library	501 + 21 aa	gb: J00965	Noda, *Nature* (1983) **301**: 251–5.
nAChR γ	Chicken genomic DNA library	492 + 22 aa		Nef, *Proc Natl Acad Sci USA* (1984) **81**: 7975–9.
	Mouse myogenic cell line cDNA library	497 + 22 aa	em: X03818	Yu, *Nucleic Acids Res* (1986) **14**: 3539–55.
	Mouse muscle cell line cDNA library		gb: M30514	Boulter, *J Neurosci* (1986) **16**: 37–49.
	T. californica electric organ cDNA library	490 + 16 aa	gb: J00966	Ballivet, *Proc Natl Acad Sci USA* (1982) **79**: 4466–70.
nAChR ϵ	Mouse muscle cDNA library	493 aa	em: X55718	Gardner, *Nucleic Acids Res* (1990) **18**: 6714.
	Mouse genomiç DNA library	493 aa	gb: J04698	Buonanno, *J Biol Chem* (1989) **264**: 7611–16.
	Rat cDNA	493 aa	em: X13252	Criado, *Nucleic Acids Res* (1988) **16**: 10920.
	Rat genomic DNA	Promoter region and exon 1	gb: L19594	Goldman, unpublished (1993).
nAChR α2	Goldfish (*Carassius auratus*) retinal cDNA library	462 aa	em: X14786	Cauley, *J Cell Biol* (1989) **108**: 637–45.
	Drosophila melanogaster head cDNA library	535 + 41 aa (*sad* gene-product)	em: X52274	Jones, *FEBS Lett* (1990) **269**: 264–8.
			em: X53583	Sawruk, *EMBO J* (1990) **9**: 2671–7.

Nomenclature	Species, DNA source	Original isolate	Accession	Sequence/ discussion
	Rat genomic DNA library	exons 1-6, encoding 511 aa	gb: M20292-M20297 gb: L10077	Wada, *Science* (1988) **240**: 330–4.
nAChR α3	Chicken genomic DNA library	496 aa	gb: M37336	Couturier, *J Biol Chem* (1990) **265**: 17560–7.
	Goldfish (*C. auratus*) retinal cDNA library	512 aa	em: X54051	Hieber, *Nucleic Acids Res* (1990) **18**: 5293.
	Human T-cell cDNA library	503 aa	gb: M37981	Mihovilovic, *J Exp Neurol* (1990) **111**: 175–80.
	Human neuro-blastoma cell line cDNA library	502 aa	gb: M86383	Fornasari, *Neurosci Lett* (1990) **111**: 351–6.
		474 aa (mature)	em: X03440	Boulter, *Nature* (1986) **319**: 368–74.
	Rat PC12 cell line cDNA library	499 aa	gb: L31621	Boulter, *Proc Natl Acad Sci USA* (1987) **84**: 7763–7.
nAChR α6	Rat brain cDNA library	463 + 30 aa	gb: L08227	Boulter, unpublished (1993).
nAChR α7	Chicken brain cDNA library	479 + 23 aa[a]		Couturier, *Neuron* (1990) **5**: 847–56.
	Human neuro-blastoma cell line cDNA library	502 aa	em: X70297	Peng, *Mol Pharmacol* (1994) **45**: 546–54.
	Rat brain cDNA library	480 + 22 aa	gb: M85273	Séguéla, *J Neurosci* (1993) **13**: 596–604.
nAChR α8	Chicken brain cDNA library	481 aa (mature)		Schoepfer, *J Neurosci* (1990) **5**: 35–48.
nAChR α9	Rat olfactory epithelium cDNA library	457 + 22 aa		Elgoyhen, *Cell* (1994) **79**: 705–15.
nAChR β2	Goldfish (*C. auratus*) retinal cDNA library	460 aa (C-terminal fragment)	em: X54052	Hieber, *Nucleic Acids Res* (1990) **18**: 5307.
	Human foetal brain cDNa library	502 aa	em: X53179	Anand, *Nucleic Acids Res* (1990) **18**: 4272.
nAChR β3	Goldfish (*C. auratus*) retinal cDNA library	438 + 28 aa	gb: M29529	Cauley, *J Neurosci* (1990) **10**: 670–83.

Nomenclature	Species, DNA source	Original isolate	Accession	Sequence/ discussion
	Human adult brainstem cDNA library	423 aa[b]	em: X67513	Willoughby, *Neurosci Lett* (1993) **155**: 136–9.
	Rat brain cDNA library	434 + 30 aa	gb: J04636	Deneris, *J Biol Chem* **264**: 6268–72.
nAChR β4	Rat genomic DNA	Promoter region and 5′ end	gb: L22646	Hu, *J Neurochem* **62**: 392–5.

[a]The position of cleavage of the leader sequence to generate the N-terminus of the mature α7 polypeptide cannot be unambiguously determined. The numbering shown is according to the placement of Couturier *et al.*
[b]May be incomplete at the 5′ end.

Related sources & reviews

09-56-01: Major quoted sources[8, 80, 108, 136, 214]; diversity of muscle and neuronal nAChR channels[24, 25, 95, 215]; architecture of nAChR[98, 108, 118, 136, 214]; activation of the nAChR channels[173]; site-directed mutagenesis of nAChR coding sequences[115, 132]; surface charges and channel function[122]; phosphorylation of channel subunits[156, 216]; snake venom toxins and nAChR[201]; α-Bgt-binding proteins[208]. *See also Resource E – Ion channel book references, entry 60.*

Feedback

Error-corrections, enhancement and extensions

09-57-01: Please notify specific errors, omissions, updates and comments on this entry by contributing to its **e-mail feedback file** (*for details, see Resource J, Search Criteria & CSN Development*). For this entry, send e-mail messages To: **CSN-09@le.ac.uk,** indicating the appropriate paragraph by entering its **six-figure index number** (xx-yy-zz or other identifier) into the **Subject**: field of the message (e.g. Subject: 08-50-07). Please feedback on only **one specified paragraph or figure per message,** normally by sending a **corrected replacement** according to the guidelines in *Feedback & CSN Access* . Enhancements and extensions can also be suggested by this route (*ibid.*). Notified changes will be indexed via 'hotlinks' from the CSN 'Home' page (http://www.le.ac.uk/csn/) from mid-1996.

Entry support groups and e-mail newsletters

09-57-02: Authors who have expertise in one or more fields of this entry (and are willing to provide editorial or other support for developing its contents) can join its support group: In this case, send a message To: **CSN-09@le.ac.uk,** (entering the words "support group" in the Subject: field). In the message, please indicate principal interests (see *fieldname criteria in* *the Introduction for coverage*) together with any relevant **http://www site links** (established or proposed) and details of any other possible contributions. In due course, support group members will (optionally) receive **e-mail newsletters** intended to **co-ordinate and develop** the present (text-based) entry/fieldname frameworks into a 'library' of interlinked resources covering ion channel signalling. Other (more general) information of interest to entry contributors may also be sent to the above address for group distribution and feedback.

REFERENCES

[1] Noda, *Nature* (1982) **299**: 793–7.
[2] Devillers-Thiery, *Proc Natl Acad Sci USA* (1983) **80**: 2067–71.
[3] Noda, *Nature* (1983) **301**: 251–5.
[4] Wada, *Science* (1988) **240**: 330–4.
[5] Boulter, *Nature* (1986) **319**: 368-74.
[6] Goldman, *Cell* (1987) **48**: 965–73.
[7] Boulter, *J Biol Chem* (1990) **265**: 4472–82.
[8] Sargent, *Annu Rev Neurosci* (1993) **16**: 403–43.
[9] Séguéla, *J Neurosci* (1993) **13**: 596–604.
[10] Deneris, *Neuron* (1988) **1**: 45–54.
[11] Deneris, *J Biol Chem* (1989) **264**: 6268–72.
[12] Duvoisin, *Neuron* (1989) **3**: 487–96.
[13] Isenberg, *J Neurochem* (1989) **52**: 988–91.
[14] Nef, *EMBO J* (1988) **7**: 595–601.
[15] Couturier, *J Biol Chem* (1990) **265**: 17560–7.
[16] Schoepfer, *Neuron* (1990) **5**: 35–48.
[17] Couturier, *Neuron* (1990) **5**: 847–56.
[18] Nef, *Proc Natl Acad Sci USA* (1984) **81**: 7975–9.
[19] Noda, *Nature* (1983) **305**: 818–23.
[20] Fornasari, *Neurosci Lett* (1990) **111**: 351–6.
[21] Chini, *Proc Natl Acad Sci USA* (1992) **89**: 1572–6.
[22] Anand, *Nucleic Acids Res* (1990) **18**: 4272.
[23] Hartman, *Nature* (1990) **343**: 372–5.
[24] Steinbach, *Annu Rev Physiol* (1989) **51**: 353–65.
[25] Steinbach, *Trends Neurosci* (1989) **12**: 3–6.
[26] Steinbach, *Ciba Found Symp* (1990) **152**: 53–61.
[27] Fertuck, *J Cell Biol* (1976) **69**: 144–58.
[28] Loring, *J Neurosci* (1987) **7**: 2153–62.
[29] Maconochie, *J Physiol* (1992) **454**: 129–53.
[30] Lal, *Proc Natl Acad Sci USA* (1993) **90**: 7280–84.
[31] Jasmin, *Nature* (1990) **344**: 673–5.
[32] Froehner, *J Cell Biol* (1991) **114**: 1–7.
[33] Froehner, *Neuron* (1990) **5**: 403–10.
[34] Margiotta, *J Neurosci* (1987) **7**: 3612–22.
[35] Noda, *Nature* (1983) **302**: 528–32.
[36] Ballivet, *Proc Natl Acad Sci USA* (1982) **79**: 4466–70.

[37] Buonanno, *J Biol Chem* (1986) **261**: 11452–5.
[38] Mishina, *Nature* (1986) **321**: 406–11.
[39] Jaramillo, *Nature* (1988) **335**: 66–8.
[40] Klarsfeld, *Neuron* (1989) **2**: 1229–36.
[41] Chahine, *Development* (1992) **115**: 213–19.
[42] Tamaoki, *Biochem Biophys Res Commun* (1986) **135**: 397–402.
[43] Pinset, *EMBO J* (1991) **10**: 2411–18.
[44] Daubas, *Neuron* (1990) **5**: 49–60.
[45] Matter, *EMBO J* (1990) **9**: 1021–6.
[46] Wang, *Brain Res* (1976) **114**: 524–9.
[47] Hieber, *J Neurochem* (1992) **58**: 1009–15.
[48] Fontaine, *EMBO J* (1988) **7**: 603–9.
[49] Wada, *J Comp Neurol* (1989) **284**: 314–15.
[50] Corriveau, *J Neurosci* (1993) **13**: 2662–71.
[51] Steinman, *FASEB J* (1990) **4**: 2726–31.
[52] Schwimbeck, *J Clin Invest* (1989) **84**: 1174–80.
[53] Stefannson, *New Engl J Med* (1986) **312**: 221–5.
[54] Bellone, *Eur J Immunol* (1991) **21**: 2303–10.
[55] Flucher, *Neuron* (1989) **3**: 163–75.
[56] Merlie, *J Cell Biol* (1984) **99**: 332–5.
[57] Parkinson, *Exp Brain Res* (1988) **73**: 553–68.
[58] Britto, *J Comp Neurol* (1992) **317**: 325–40.
[59] Sargent, *J Neurosci* (1989) **9**: 563–73.
[60] Vernallis, *Neuron* (1993) **10**: 451–64.
[61] Gotti, *Neuroscience* (1992) **50**: 117–27.
[62] Keyser, *J Neurosci* (1993) **13**: 442–54.
[63] Vijayaraghavan, *Neuron* (1992) **8**: 353–62.
[64] Heidmann, *Science* (1986) **234**: 866–8.
[65] Mishina, *Nature* (1984) **307**: 604–8.
[66] Kullberg, *Proc Natl Acad Sci USA* (1990) **87**: 2067–71.
[67] Kurosaki, *FEBS Lett* (1987) **214:** 253–8.
[68] Liu, *J Physiol* (1993) **470**: 349-63.
[69] Breer, *J Neurosci* (1985) **5**: 3386–92.
[70] Hanke, *Nature* (1986) **321**: 171–4.
[71] Marshall, *EMBO J* (1990) **9**: 4391–8.
[72] Cooper, *Nature* (1991) **350**: 235–8.
[73] Anand, *J Biol Chem* (1991) **266**: 11192–8.
[74] Shibahara, *Eur J Biochem* (1985) **146**: 15–22.
[75] Piette, *EMBO J* (1989) **8**: 687–94.
[76] Wang, *EMBO J* (1990) **9**: 783–90.
[77] Crowder, *Mol Cell Biol* (1988) **8**: 5257–67.
[78] Baldwin, *Nature* (1989) **341**: 716–20.
[79] Mar, *Proc Natl Acad Sci USA* (1989) **85**: 6404–8.
[80] Andersen, *Physiol Rev* (1992) **72**: S89-S158.
[81] Raftery, *Science* (1980) **208**: 1454–7.
[82] Whiting, *Biochemistry* (1986) **25**: 2082–93.
[83] Whiting, *J Neurosci* (1988) **7**: 4005–16.
[84] Whiting, *Mol Brain Res* (1991) **10**: 61–70.
[85] Schoepfer, *Neuron* (1988) **1**: 241–8.

[86] Lindstrom, *CIBA Found Symp* (1990) **152**: 23–52.
[87] Schoepfer, *Neuron* (1990) **5**: 35–48.
[88] Halvorsen, *Neuroscience* (1990) **10**: 1711–18.
[89] Whiting, *Proc Natl Acad Sci USA* (1987) **84**: 895–9.
[90] Whiting, *FEBS Lett* (1987) **213**: 55–60.
[91] Flores, *Mol Pharmacol* (1992) **41**: 31–7.
[92] Blanton, *Biochemistry* (1992) **31**: 3738–50.
[93] Giraudet, *Biochemistry* (1985) **24**: 3121–7.
[94] Toyoshima, *Nature* (1988) **336**: 247–50.
[95] Dani, *Curr Opin Cell Biol* (1989) **1**: 753–64.
[96] Karlin, *Trends Pharmacol Sci* (1986) **7**: 304–8.
[97] Karlin, *Harvey Lect* (1989) **85**: 71–107.
[98] Eisenman, *Annu Rev Biophys Biophys Chem* (1987) **16**: 205–26.
[99] Blount, *Neuron* (1989) **3**: 349–57.
[100] Kao, *J Biol Chem* (1984) **259**: 11662–5.
[101] Middleton, *Biochemistry* (1991) **30**: 6987–97.
[102] Tomaselli, *Biophys J* (1991) **60**: 721–7.
[103] Oiki, *Biophys J* (1992) **62**: 28–30.
[104] Dennis, *Biochemistry* (1988) **27**: 2346–57.
[105] Galzi, *J Biol Chem* (1990) **265**: 10430–7.
[106] Galzi, *FEBS Lett* (1991) **294**: 198–202.
[107] Czakowski, *J Biol Chem* (1991) **266**: 22603–12.
[108] Karlin, *Current Opin Neurobiol* (1993) **3**: 299–309.
[109] Golind, *J Membr Biol* (1992) **129**: 297–309.
[110] Oleary, *J Biol Chem* (1992) **267**: 8360–5.
[111] Sumikawa, *J Biol Chem* (1992) **267**: 6286–90.
[112] Mishina, *Nature* (1985) **313**: 364–9.
[113] Conti-Tronconi, *Biochemistry* (1990) **29**: 6221–30.
[114] McLane, *Biochemistry* (1991) **30**: 4925–34.
[115] Dani, *Trends Neurosci* (1989) **12**: 125–8.
[116] Giraudat, *Proc Natl Acad Sci USA* (1986) **83**: 2719–23.
[117] Hucho, *FEBS Lett* (1986) **205**: 137–42.
[118] Galzi, *Annu Rev Pharmacol* (1991) **31**: 37–72.
[119] White, *J Biol Chem* (1992) **267**: 15770–83.
[120] Leonard, *Science* (1988) **242**: 1578–81.
[121] Imoto, *Nature* (1988) **335**: 645–8.
[122] Green, *Annu Rev Physiol* (1991) **53**: 341–59.
[123] Dani, *J Gen Physiol* (1987) **89**: 959–83.
[124] Charnet, *Neuron* (1990) **4**: 87–95.
[125] Villarroel, *Proc R Soc Lond B Biol Sci* (1991) **243**: 69–74.
[126] Villarroel, *Biophys J* (1992) **62**: 196–208.
[127] Ferrermontiel, *FEBS Lett* (1993) **324**: 185–90.
[128] Bertrand, *Proc Natl Acad Sci USA* (1993) **90**: 6971–5.
[129] Revah, *Nature* (1991) **353**: 846–9.
[130] Villarroel, *Proc R Soc Lond [Biol]* (1992) **249**: 317–24.
[131] Imoto, *FEBS Lett* (1991) **289**: 193–200.
[132] Changeux, *Trends Pharmacol Sci* (1992) **13**: 299–301.
[133] Galzi, *Nature* (1992) **359**: 500–5.
[134] Dani, *J Neurosci* (1989) **9**: 884–92.

[135] Langosch, *Biochim Biophys Acta* (1991) **1063**: 36–44.
[136] Unwin, *Cell* (1993) **72**: 31–41.
[137] Mishina, *Nature* (1985) **313**: 364–9.
[138] Castresana, *FEBS Lett* (1992) **314**: 171–5.
[139] Karlin, *Harvey Lect* (1991) **85**: 71–107.
[140] Unwin, *J Mol Biol* (1993) **229**: 1101–24.
[141] Akabas, *Science* (1992) **258**: 307–10.
[142] Montal, *FEBS Lett* (1993) **320**: 261–6.
[143] Butler, *Biochim Biophys Acta* (1993) **1150**: 17–24.
[144] Ross, *J Cell Biol* (1991) **113**: 623–6.
[145] Green, *Cell* (1993) **74**: 57–69.
[146] Verrall, *Cell* (1992) **68**: 23–31.
[147] Phillips, *Science* (1991) **251**: 568–70.
[148] Butler, *J Biol Chem* (1992) **267**: 6213–18.
[149] Wagner, *Neuron* (1993) **10**: 511–22.
[150] Ibraghimov-Beskrovnaya, *Nature* (1992) **355**: 696–702.
[151] Ferrer, *Proc Natl Acad Sci USA* (1991) **88**: 10213-17.
[152] Huganir, *Nature* (1986) **321**: 774–6.
[153] Hopfield, *Nature* (1988) **336**: 677–80.
[154] Miles, *Mol Neurobiol* (1988) **2**: 91–124.
[155] Ross, *J Biol Chem* (1987) **262**: 14640–7.
[156] Huganir, *Crit Rev Biochem Mol Biol* (1989) **24**: 183–215.
[157] Huganir, *Neuron* (1990) **5**: 555–67.
[158] Mulle, *Proc Natl Acad Sci USA* (1988) **85**: 5728–32.
[159] Higgins, *J Cell Biol* (1988) **107**: 1157–65.
[160] Wallace, *J Cell Biol* (1988) **107**: 267–8.
[161] Wallace, *Neuron* (1991) **6**: 869–78.
[162] Peng, *Neuron* (1991) **6**: 237–46.
[163] O'Leary, *J Biol Chem* (1992) **267**: 8360–5.
[164] Smith, *J Physiol* (1990) **432**: 343–54.
[165] Lu, *J Physiol* (1991) **436**: 45–56.
[166] Vernino, *Neuron* (1992) **8**: 127–34.
[167] Mathie, *J Physiol* (1990) **427**: 625–55.
[168] Ifune, *J Physiol* (1992) **457**: 143–65.
[169] Zorumski, *Mol Pharmacol* (1992) **41**: 931–6.
[170] de la Garza, *Neuroscience* (1987) **23**: 887–91.
[171] Franke, *Neurosci Lett* (1992) **140**: 169–72.
[172] Katz, *J Physiol* (1957) **138**: 63–80.
[173] Lingle, *J Membr Biol* (1992) **126**: 195–217.
[174] Hille, *Ionic channels of excitable membranes*, 2nd edn. (1992) Sinauer Associates, Sunderland, Mass.
[175] Mulle, *Neuron* (1992) **8**: 135–43.
[176] Vernino, *Neuron* (1992) **8**: 127–34.
[177] Sands, *Brain Res* (1990) **560**: 38–42.
[178] Sands, *Biophys J* (1993) **65**: 2614–21.
[179] Ifune, *Proc Natl Acad Sci USA* (1990) **87**: 4794–8.
[180] Mathie, *J Physiol* (1991) **439**: 717–50.
[181] Papke, *Neuron* (1989) **3**: 589–96.
[182] Ifune, *Proc Natl Acad Sci USA* (1990) **87**: 4794–8.

[183] Cull-Candy, *J Physiol* (1988) **402**: 255–78.
[184] Moss, *Neuron* (1989) **3**: 597–607.
[185] Kuba, *Pflugers Arch* (1989) **414**: 105–12.
[186] Heidman, *Biochemistry* (1983) **22**: 3112–27.
[187] Heidman, *Proc Natl Acad Sci USA* (1994) **81**: 1897–1901.
[188] Revah, *Proc Natl Acad Sci USA* (1990) **87**: 4675–9.
[189] Pedersen, *J Biol Chem* (1992) **267**: 10489–99.
[190] Valera, *Proc Natl Acad Sci USA* (1992) **89**: 9949–53.
[191] Luetje, *J Neurochem* (1990) **55**: 632–40.
[192] Alkondon, *Mol Pharmacol* (1992) **41**: 802–8.
[193] Murrell, *J Physiol* (1991) **437**: 431–48.
[194] Wachtel, *Ann NY Acad Sci* (1991) **625**: 116–28.
[195] Bertrand, *Proc Natl Acad Sci USA* (1990) **87**: 1993–7.
[196] Mulle, *Neuron* (1992) **8**: 937–45.
[197] Couturier, *J Biol Chem* (1990) **265**: 17560–7.
[198] Colquhoun, *J Physiol* (1985) **369**: 501–57.
[199] Jackson, *J Physiol* (1988) **397**: 555–83.
[200] Sine, *J Gen Physiol* (1990) **96**: 395–437.
[201] Loring, *J Toxicol-Toxin Rev* (1993) **12**: 105–53.
[202] Alkondon, *J Recept Res* (1991) **11**: 1001–21.
[203] Fuchs, *Proc R Soc (Lond) B* (1992) **248**: 35–40.
[204] Pinnock, *Brain Res* (1988) **458**: 45–52.
[205] Mclane, *Biochemistry* (1993) **32**: 6988–94.
[206] Hong, *Neuroscience* (1992) **48**: 727–35.
[207] Abramson, *J Biol Chem* (1989) **264**: 12666–72.
[208] Clarke, *Trends Pharmacol Sci* (1992) **13**: 407–13.
[209] Mcgroddy, *Biophys J* (1993) **64**: 325–38.
[210] Rana, *Eur J Pharmacol* (1993) **250**: 247–51.
[211] Aylwin, *Mol Pharmacol* (1992) **41**: 908–13.
[212] Reuhl, *J Neurophysiol* (1992) **68**: 407–16.
[213] Mulle, *J Neurosci* (1991) **11**: 2588–97.
[214] Changeux, *Q Rev Biophys* (1992) **25**: 395–432.
[215] Role, *Curr Opin Neurosci* (1992) **2**: 254–62.
[216] Leidenheimer, *Trends Pharmacol Sci* (1991) **12**: 84–7.
[217] Gerzanich, *Mol Pharmacol* (1994) **45**: 212–20.
[218] Elgoyhen, *Cell* (1994) **79**: 705–15.
[219] Peng, *Mol Pharmacol* (1994) **45**: 546–54.
[220] Criado, *Nucleic Acids Res* (1988) **16**: 10920.
[221] Akondon, *J Pharmacol Exp Ther* (1993) **265**: 1455–73.

ELG Cl GABA$_A$

Inhibitory receptor-channels gated by extracellular gamma-aminobutyric acid

Edward C. Conley

Entry 10

NOMENCLATURES

Abstract/general description

10-01-01: GABA$_A$ receptor–channels (**GABA$_A$R**) are mediators of post-synaptic inhibition† in the brain responding to the major neurotransmitter **γ-aminobutyric acid** (GABA). GABA is a neutral amino acid released from GABAergic† neurones, which form approximately 30% of all central† synapses†. GABA also acts at G protein-linked receptors (GABA$_B$ subtypes) which do not contain integral ion channels but are coupled to separate calcium and potassium channel proteins *(see Receptor/transducer interactions, 10-49).*

10-01-02: Activation of neuronal GABA$_A$ receptors increases inward chloride current through an integral channel which **hyperpolarizes** the post-synaptic cell and inhibits synaptic activity *(but see next paragraph).* Generally, GABA$_A$ receptors underlie **fast IPSP**† (inhibitory post-synaptic potentials†) components following stimulation of excitatory afferents† (with **slow IPSP** components often being mediated by GABA$_B$, adrenoceptor and 5-HT receptor responses). GABA$_A$-mediated conductances influence **impulse initiation** and the amplitude† and duration of excitatory post-synaptic potentials† (EPSPs†), especially the expression of the NMDA-mediated component *(see ELG CAT GLU NMDA, entry 08).*

10-01-03: All CNS neurones, as well as glial† cells, exhibit GABA$_A$ responses, although in the latter are characterized by an 'excitatory, depolarizing' outward current response to GABA *(see Cell-type expression index, 10-08).* GABA$_A$ receptor–channels have also been characterized in the **peripheral nervous system** (e.g. uterus, ileum, lactotrophs†, retinal cells, pancreas, pituitary intermediate lobe and adrenal chromaffin cells). A class of GABA$_A$ receptors that modulate the release of GABA itself has been studied in **pre-synaptic nerve fibres**. These **GABA$_A$ autoreceptors**† have a novel agonist† and antagonist† profile.

10-01-04: Both the GABA$_A$ receptor and the glycine receptor *(described in the entry ELG Cl GLY, entry 11)* are hetero-oligomeric† (**pentameric**) protein complexes in their native† state, composed of subunits with molecular weights of ~50–60 kDa. **Multiple combinations** of a large set of **distinct channel structural gene products** underlies a **large functional heterogeneity** of GABA$_A$ channel complexes. The genes encoding GABA$_A$ receptor–channels form part of the extracellular ligand-gated channel gene superfamily†. At least 16 different receptor subunit genes have been separated into the sequence homology† subfamilies† **alpha** (α_1–α_6); **beta** (β_1–β_4); **gamma** (γ_1–γ_3); **delta** (δ), and **rho** (ρ1, ρ2) *(see Table 1).* This large number of **distinct isoforms**† revealed through molecular cloning† has enabled detailed **subunit co-expression studies** to begin.

10-01-05: Generally, GABA$_A$ subunit genes display **differential**† **spatial**†

expression and co-expression of multiple subunit mRNAs within single cells is commonplace. 'Dynamic control' of $GABA_A$ receptor gene expression can potentially lead to diversity in **subcellular locations**, **ligand affinities**† and **ion channel gating**† properties. Phosphorylation† of *in trans* gene regulatory factors† (coupled to extracellular messenger† molecule reception) may also regulate $GABA_A$ subunit gene expression. In common with other ion channels, **post-transcriptional modifications**† of $GABA_A$ protein subunits (principally by **phosphorylation**†) are also important modulators of physiological function.

10-01-06: The mechanism for **assembling** and **'targeting'** of protein complexes with different **isoform composition** may be central to modulation of $GABA_A$ receptor distribution in neurones. Assembly of $GABA_A$Rs from constituent subunits does *not* proceed randomly to form all possible combinations, but certain subunit combinations are **'preferred intermediates'** during the assembly process – e.g. transiently† expressed $GABA_A$R subunits in mouse L929 fibroblast cells, an 'ordered assembly process' appears to produce a **'preferred final form'** of the receptor–channel ($\alpha_1\beta_1\gamma_2$S) *(see Domain arrangement, 10-27).*

10-01-07: In general, there is a high degree of sequence conservation† within the putative **transmembrane segments M1–M3** between the $GABA_A$R and the **glycine receptor–channels** (GlyR, *see ELG Cl GLY, entry 11*), indicating their likely importance in **chloride channel** formation. $GABA_A$R and GlyR also display several invariant residues at homologous positions and relatively **high positive charge density** within eight residues of the ends of their transmembrane domains on the extracellular sides. The existence of **channel vestibules**† containing these **net positive charges** are consistent with the **anion-selective permeability** properties of $GABA_A$ receptor–channels and they may have the property of attracting anions. This arrangement is in contrast to vestibules of cation-selective receptor–channels which possess 'rings' of *negatively* charged residues *(see ELG CAT nAChR, entry 09).*

10-01-08: $GABA_A$ receptor–channels are 'positively modulated' by several classes of clinically important drugs (such as the **barbiturate** CNS depressants, **benzodiazepine** anxiolytic/anticonvulsant/sedative-hypnotics, and several anaesthetic compounds). $GABA_A$ responses also partially mediate sensitivity to **alcohol** (ethanol) and a range of **convulsant**† **compounds**. The **binding affinity**† as well as the **modulatory efficacy**† of these and other drugs change with **receptor subunit composition** and this factor offers clear potential for the development of more selective therapeutic compounds. A large number of studies have indicated that in *most* cases, modulatory drugs do not interact directly with GABA-binding sites, but influence receptor–channel function by binding to additional **allosteric**†**-modulatory sites** on the $GABA_A$ complex.

10-01-09: The availability of a rich diversity of cloned $GABA_A$ subunit genes and powerful techniques such as site-directed mutagenesis†, domain swopping†, photoaffinity labelling† and microsequencing† has enabled a process of **'mapping' key amino acid loci** implicated in modulator binding and function to begin. In addition, mechanisms underlying many different

$GABA_A$ functions and properties are yielding to such analyses, including those involving or determining ligand binding/recognition, co-operativity† for agonist binding, channel conductance characteristics, modulator efficacy, mechanisms of channel block, rectification properties, allosteric modulation by drugs, domain topography, subunit arrangements, assembly patterns, dose–response relationships, phosphomodulation properties, desensitization kinetics, etc. Such approaches will define in detail the functional criteria for the operation of the $GABA_A$ receptor family *in vivo*.

Category (sortcode)

10-02-01: ELG Cl $GABA_A$, i.e. extracellular ligand-gated chloride channels activated by gamma-aminobutyric acid. The suggested **electronic retrieval code** (unique embedded identifier or **UEI**) for 'tagging' of new articles of relevance to the contents of this entry is UEI: GABAA-NAT (for reports or reviews on native† channel properties) and UEI: GABAA-HET (for reports or reviews on channel properties applicable to heterologously† expressed recombinant† subunits encoded by cDNAs† or genes†). *For a discussion of the advantages of UEIs and guidelines on their implementation, see the section on Resource J under Introduction and layout, entry 02, and for further details, see Resource J – Search criteria & CSN development, entry 65.*

Channel designation

10-03-01: $Cl_{GABA\text{-}A}$; $GABA_AR$ (~$GABA_A$ receptor); ionotropic GABA receptors.

Current designation

10-04-01: Usually designated as $I_{GABA\text{-}A}$ or $I_{Cl\ (GABA\text{-}A)}$; other designations used for $GABA_A$ currents include $I_{G\text{-}Actx}$ (typical, expressed from cerebral cortex mRNA in oocytes)[1] as distinct from $I_{G\text{-}Aret}$ (expressed from retinal mRNA in oocytes)[1] *(see Cell-type expression index, 10-08).*

Gene family

10-05-01: The genes encoding $GABA_A$ receptor–channels form part of the extracellular ligand-gated channel gene superfamily† *(described under ELG Key facts, entry 04).* For updates on the agreed nomenclatures, refer to the latest IUPHAR Nomenclature Committee recommendations via the CSN *(see Feedback & CSN access, entry 12).*

$GABA_A$ subunit families and subfamilies

10-05-02: Multiple types of **subunit associations** underlie the great **pharmacological and biochemical diversity** of the $GABA_A$ receptor–channel family[2] *(see Predicted protein topography, 10-30, and Protein interactions, 10-31).* At least 16 different receptor subunit genes have been separated into **sequence homology subfamilies**†. There is typically ~70–80% amino acid sequence identity between members of a subfamily and ~30–40% between members of different subfamilies. Basic distinguishing features of the genes encoding members of the subunit families **alpha** (α_1–α_6); **beta** (β_1–β_4); **gamma** (γ_1–γ_3); **delta** (δ), and **rho** (ρ_1 and ρ_2) are listed in Table 1.

Table 1. *Distinguishing characteristics of genes encoding subunits of GABA$_A$ receptor–channels (From 10-05-02)*

Subunit	Transcript size (kb)[a]	Encoding	
		ORF[b]	Signal peptide[c]
α_1	Rat/mouse: 3.8 + 4.2[d]	Bovine: 456 aa; Rat: 455 aa; Mouse: 455 aa; Chicken: 455 aa; Human: 456 aa[e]	Rat: 27 aa M_r rat: 48.8 kDa[f] 51 kDa (recombinant† by Western† blotting, WB)
α_2	Rat: 3.0	Bovine: 451 aa; Rat: 451 aa; Mouse: 451 aa; Human: 451 aa[e]	Rat: 28 aa 53 kDa by WB
α_3	Rat: 4.2	Rat: 493 aa; Mouse 492 aa; Bovine: 492 aa[e]	Rat: 28 aa 59 kDa by WB
α_4	Bovine: 4.0	Rat: 552 aa; Bovine: 556 aa[e,f]	Rat: 35 aa Rat: 59 kDa by WB Bovine: 59 kDa by WB
α_5	Rat: 2.8	Rat: 464 aa[e]	Rat: 31 aa
α_6	Rat: 3.2[g]	Rat: 443 aa[e]	Rat: 19 aa 57 kDa by photoaffinity† labelling with [^{3}H]-Ro154513
β_1	Rat: 12.0[h]	Human: 474 aa: Bovine: 474 aa; Rat: 474 aa	Rat: 25 aa M_r bovine: 51.4 kDa
β_2	Rat: 8.0[d, i]	Human: 474 aa; Rat: 474 aa	Rat: 24 aa 57 kDa by WB
β_3	Rat/mouse: 6.0 + 2.5	Human: 473 aa Rat: 473 aa; Chicken: 476 aa	Rat: 25 aa 54–59 kDa by WB
β_4	[j]	Bovine β_4: 459 aa Chicken: 488 aa (AS)	Bovine: β_4: 25 aa
β_4'		Chicken β_4': 463 aa (AS)	Chicken β_4': 27 aa
δ	Rat: 2.0 + 3.0	Rat: 449 aa Mouse: 449 aa	Rat: 16 aa 54 kDa by WB
γ_1	Rat: 3.8	Rat: 465 aa	Rat: 35 aa

Table 1. *Continued*

Subunit	Transcript size (kb)[a]	Encoding	
		ORF[b]	Signal peptide[c]
γ_{2S}	Rat/mouse γ_{2S}: 4.2 + 2.8[d, k]	Rat γ_{2S}: 428 aa (mature) *See also Database listings, 10-53*	Rat γ_{2S}: 38 aa 43–47 kDa by WB
γ_{2L}	[d, k]	Bovine γ_2 442 aa; rat γ_{2L}: 436 aa (mature, AS[k]) *See also Database listings, 10-53*	Rat γ_{2S}: 38 aa M_r bovine γ_2: ~50.4 kDa
γ_3		Rat: 467 aa; Mouse: 467 aa	Mouse: 17 aa
ρ_1	Bovine: 3.9 + 4.8 + 3.1	Human: 473 aa	Human: 15 aa
ρ_2		Human: 465 aa	

For further subunit-specific characteristics, see *mRNA distribution, 10-13, Phenotypic expression, 10-14, Protein distribution, 10-15, Subcellular locations, 10-16, Protein molecular weight (purified), 10-22, Domain arrangement, 10-27, Domain conservation, 10-28, Domain functions, 10-29, Predicted protein topography, 10-30, Protein interactions, 10-31, Protein phosphorylation, 10-32, Activation, 10-33, Dose–response, 10-36, Single-channel data, 10-41, Blockers, 10-43, Channel modulation, 10-44, Equilibrium dissociation constant, 10-45, Receptor antagonists, 10-51, Database listings, 10-53 and Related sources and reviews, 10-56.*

[a]Shows transcript size as detected by Northern† analysis of total mRNA from the species shown.
[b]Shows the number of amino acid residues in the specified channel subunit. For a more detailed breakdown, *see Database listings, 10-53*. (AS) indicates a product of alternative splicing†.
[c]Shows typical contributory length to total ORF by cleaved signal peptide plus experimentally determined molecular masses for recombinant† subunits by SDS–PAGE† and Western blotting† (WB) procedures (predicted mol. wt values can be derived from sequence analysis following retrieval using accession numbers† in *Database listings, 10-53*). Note that native† cellular preparations may reveal 'microheterogeneity' of protein bands due to cross-reactivity† of epitopes† within the protein family†.
[d]α_1 mRNA is often co-localized with mRNA encoding β_2 and γ_2 subunits. *For significance, see Domain arrangement, 10-27.*
[e]The α_6 subunit domain structure is illustrated in the [PDTM], Fig. 1.
[f]Largest β subunit, very low expression in rat and mouse.
[g]Also reported as 2.7 kb.
[h]Low-abundance transcript.
[i]Highest abundance of β mRNAs.
[j]β4 is alternatively spliced† in the M3–M4 loop, producing an extra 12 ribonucleotides (encoding the 4 amino acids VREQ).
[k] γ_{2S} and γ_{2L} are highly expressed in neurones (cf. γ_1 subunit mRNA, which is highly expressed in astrocytes). For further information on alternatively spliced forms γ_{2S} and γ_{2L}, *see Protein phosphorylation, 10-32, and Domain functions, 10-29.*

$GABA_A$ receptor gene nomenclatures

10-05-03: A series of **gene names** for those encoding $GABA_A$ receptor subunits are also in use – ***gabra1–gabra6***, ***gabrb1–gabrb4***, ***gabrd***, ***gabrg1–gabrg3*** and, by convention ***gabrr***). These names also appear in un-italicized uppercase within sequence database entries.

Functional co-expression studies of recombinant subunits

10-05-04: The large number of **distinct subunit isoforms**† revealed through molecular cloning† has enabled detailed **co-expression studies** to begin. These studies ultimately seek to relate electrophysiological and pharmacological parameters measured *in situ* on native† cells to contributions of individual protein subunits. Electrophysiological and pharmacological experiments have suggested a remarkable **heterogeneity** of $GABA_A$ receptor complexes within individual cells *(reviewed in ref.[3])*.

Theoretical and actual numbers of subunit combinations

10-05-05: From the large number of **combinatorial possibilities** (the 15 *known* subunits would give ~151 887 subunit combinations with 1–5 subunits co-expressed) the total number of combinations *actually used* in native cells is unknown. By means of RT-PCR† techniques, *single* neuronal cells have been reported to co-express up to 14 distinct $GABA_A$ channel transcripts *(see mRNA distribution, 10-13). Note:* **'Preferential' arrangements** of subunit co-assemblies have been established *(see Domain arrangement, 10-27)*.

Subtype classifications

Rationalization of subtype classifications incorporating both molecular and pharmacological characteristics

10-06-01: As described in the PHARMACOLOGY section of this entry, multiple distinct **drug-binding** and other **modulatory sites** of $GABA_A$ receptors have been characterized by pharmacologists. The multiplicity of these sites, and the **large diversity of genes** encoding $GABA_A$ receptors *(see Table 1 under Gene family, 10-05)* have resulted in significant problems for those involved in proposing agreed schemes of **receptor nomenclature**. At the time of going to press, there is no 'universally accepted' nomenclature for $GABA_AR$. For updates on the agreed nomenclatures, refer to the latest **IUPHAR Nomenclature Committee recommendations** via the CSN *(see Feedback & CSN access, entry 12)*.

Existing classifications based on functional properties

10-06-02: Features of a provisional **$GABA_A$ receptor subtype classification** ($GABA_{A1}$, $GABA_{A2}$, $GABA_{A3}$) based on the *functional properties* of their **allosteric modulatory centres**† have been summarized[4–6] *(see also Blockers, 10-43, Channel modulation, 10-44, Receptor agonists, 10-50, and Receptor antagonists, 10-51)*. Distinctions between the $GABA_A$-related **'central'-type** benzodiazepine receptors (as described in this entry) and the $GABA_A$-unrelated **'peripheral'-type** benzodiazepine receptors are outlined in the footnote to Table 6 under *Ligands, 10-47*.

Novel GABA receptor subtypes mediating 'excitatory, depolarizing' responses
10-06-03: Distinct receptor subtypes may be responsible for '**excitatory**†, **depolarizing**†' responses of some $GABA_A$ receptors[7]. These novel subtypes may be associated with **voltage-dependent anion channels** which are capable of regulating chloride flux through the $GABA_A$ integral channel (and vice versa). Excitatory†, depolarizing† $GABA_A$ subtypes may depend on co-expression of specific types of ion transporter proteins also be involved in mediation of **neurotrophic**† **responses** during development *(see Cell-type expression index, 10-08, Developmental regulation, 10-11, and Phenotypic expression, 10-14).*

Trivial names

10-07-01: The $GABA_A$ receptor; the GABA receptor–channel; the A receptor; the bicuculline-blocked GABA receptor.

EXPRESSION

Cell-type expression index

GABA-inhibitory responses are ubiquitous in nervous tissue
10-08-01: GABA is considered to be the major neurotransmitter mediating **inhibitory**† **neurotransmission**† in **higher brain functions**, although the wide distribution of GlyR subtypes in the CNS *(see ELG Cl GLY, entry 11)* may also suggest parallel roles for **glycinergic**† **transmission**. All CNS neurones, as well as glial† cells, exhibit $GABA_A$ responses[8,9]. The established subunit diversity *(see Table 1)* and the multiplicity of sites for pharmacological modulation demonstrate the considerable **heterogeneity** of $GABA_A$ complexes throughout the nervous system. Patterns of $GABA_A$ expression in brain have been reviewed[10,11].

10-08-02: The *well-characterized* neuronal preparations expressing $GABA_A$ subtypes quoted within this entry are as follows:

cultured rat cerebellar neurones
dissociated rat sympathetic neurones
chick cerebral cortical neurones
porcine pituitary intermediate lobe
mouse spinal cord neurones
cerebellar Purkinje cells
cultured hippocampal pyramidal cells
cultured spinal cord neurones
adult/foetal dorsal root ganglion neurones
synaptoneurosomal preparation from rat cerebral cortex
embryonic kidney line (heterologous)
superior cervical ganglion neurones
turtle cone photoreceptors
rat/cat dorsal root ganglion neurones
bovine adrenal medulla chromaffin cells
pre-synaptic peptidergic nerve terminals *(see Subcellular locations, 10-16)*

Retinal GABA$_A$ channels
10-08-03: RNA extracted from **bovine retina** expresses GABA responses composed of two pharmacologically distinct Cl^- currents, one mediated by GABA$_A$ receptors ($I_{G\text{-}Aret}$) and the other by 'atypical' GABA receptors[1] that are resistant to bicuculline and are not activated by baclofen ($I_{G\text{-}BR}$) *(see Blockers, 10-43).*

Excitatory, depolarizing responses of GABA$_A$-like channels in vertebrate glial cells
10-08-04: Astrocytes and oligodendrocytes are **depolarized** by GABA, and pharmacological experiments indicate that the channels underlying these responses are related to neuronal GABA$_A$ receptor–channels *(for review, see ref.[12])*. The **'glial'** and **'neuronal' subtypes** are distinguishable by the actions of the benzodiazepine **inverse agonist**† **DMCM** *(see the field Receptor inverse agonists, 10-52).* Inverse agonists† such as DMCM augment the GABA response in astrocytes, whereas in neurones (and cells of the oligodendrocyte lineage) the response is decreased *(for discussion of possible underlying mechanisms, see Protein interactions, 10-31 and refs[12–15]).* *Notes*: 1. As judged by *in situ* hybridization†, the α_2 subunit has been reported[16] to be highly expressed in the **Bergmann glial cell layer**. 2. The property of chloride *efflux from glial GABA-activated channels* may serve to maintain or 'buffer' extracellular chloride for *influx through neuronal GABA$_A$ channels* at the synaptic cleft† *(for discussion, see ref.[12])*. 3. **Glial–neuronal interactions** involving GABA$_A$ and glutamate receptor–channel activities may mediate **trophic**† **responses** and/or **synaptic plasticity**† phenotypes both during development and in the adult *(see ref.[12] and ELG CAT GLU NMDA, entry 08).*

GABA$_A$ channels in the PNS
10-08-05: GABA$_A$ receptor–channels are also expressed in the **peripheral nervous system**† (e.g. in the uterus, ileum, lactotrophs†, retinal cells, pituitary intermediate lobe and adrenal chromaffin cells). Cultured melanotrophs† from frog pituitary can be maintained as a 'pure' population of **endocrine cells** enriched with GABA$_A$ receptors[17].

Pancreatic GABA$_A$ responses
10-08-06: GABA has a distinct role in **pancreas** (where it is co-secreted with **insulin** from **beta cells**) mediating part of the inhibitory† action of glucose on **glucagon secretion** (i.e. by activation of GABA$_A$ receptor–channels[18]).

Other novel functional classes of GABA$_A$ receptor
10-08-07: A class of GABA$_A$ receptors that modulate the release of GABA itself has been studied in **pre-synaptic nerve fibres**. These **GABA$_A$ autoreceptors** have a novel agonist† and antagonist† profile *(see Receptor agonists, 10-50, and Receptor antagonists, 10-51).*

Specification of subtype co-expression
10-08-08: The relationship between **heterogeneity** of native† receptor properties and specification of subtype co-expression is largely unknown and the subject of ongoing investigation.

Channel density

$GABA_A$ channel loss with neurodegeneration

10-09-01: The density of **benzodiazepine type I† receptors** in the substantia nigra pars reticulata is reduced by 34% following 6-hydroxydopamine-induced degeneration of dopaminergic pars compacta neurones *(see also Protein distribution, 10-15, and Subcellular locations, 10-16). Note:* Significant increases in $GABA_A$ receptor expression in the lower CNS may serve a **'compensatory' function** for losses of glycinergic† receptor function in the mouse mutant *spastic (see Phenotypic expression under ELG Cl GLY, 11-14).*

Cloning resource

Cell-line models for developmental expression of $GABA_A$ receptors

10-10-01: Immortalized† hybrid clones that express characteristics of differentiated† neurones derived from the cerebellar and brainstem regions have been established by **somatic cell fusion** between a hypoxanthine phosphoribosyltransferase† (HPRT)-deficient neuroblastoma, N18TG2, and newborn mouse cerebellar/brainstem neurones. Immortalized† hypothalamic (GT1-7) neurones expressing functional $GABA_A$ receptors have also been established in culture[19]. *Note:* Such **immortalized cells**† express a range of **developmentally specific** neuronal proteins and are likely to prove important for studies of the molecular basis of **neuronal differentiation**† and **degeneration**.

Developmental regulation

Control of $GABA_A$ receptor subunit expression at the transcriptional and post-transcriptional levels

10-11-01: In common with other ion channel gene families, **cell type-selective developmental expression** of specific $GABA_A$ gene† isoforms† is regulated by tissue-specific and ontogenetic† factors (e.g. **transcription factors/morphogens**) that are likely to act directly on sequences *in cis*† to the $GABA_A$ structural† genes. 'Dynamic control' of gene expression can potentially lead to diversity in **subcellular locations, ligand affinities**† and **ion channel gating**† properties. Phosphorylation† of *in trans* gene regulatory factors† (coupled to extracellular messenger† molecule reception) may also regulate $GABA_A$ subunit gene expression *(in addition to the well-characterized mechanisms for post-transcriptional modulation of the protein structures themselves – see Protein phosphorylation, 10-32, and ELG Key facts, entry 04).* Notably, certain $GABA_A$ receptors (excitatory† depolarizing† subtypes, *see Subtype classifications, 10-06, Cell-type expression index, 10-08, and Protein interactions, 10-31)* appear to mediate **trophic**† **responses** to GABA during development[20]. GABA agonists† can regulate **transcriptional activation** of $GABA_A$ subunit genes† in both cell culture and animal models[21,22].

Developmental 'routing' of multiple subunit expression

10-11-02: The mechanism for **assembling** and **'targeting'** protein complexes of different **isoform composition** may be central to modulation of $GABA_A$

receptor distribution in neurones[23] *(see Subcellular locations, 10-16).* Placement of receptors at *specific locations* is predicted to depend (at least in part) on the subunit composition of the complex if different subunits have **specific routing properties**[23]. Neuronal phenotype (i.e. **plasticity**†) may be directed by expression-control of specific isoforms†, which in turn could be influenced by developmental differentiation†[24,25] and the '**excitatory drive**' that increases expression of transcription factors†[26].

Stereotypical order of ion channel gene expression in developing spinal cord

10-11-03: In developing populations of cells from several regions of embryonic rat spinal cord, functional **sodium channels** appear *prior* to **GABA$_A$ receptors**, which in turn emerge prior to **kainate-activated glutamate receptors**. This **stereotypical pattern** of sequential channel development occurs individually on most cells in each region[27].

Zinc ion sensitivity dependent on developmental stage

10-11-04: GABA$_A$ receptors of the rat superior cervical ganglion exhibit a **developmental increase** in resistance to blockade by **zinc ions**[28]. Zn^{2+}-antagonism of GABA-elicited membrane currents in cultured rat cerebellar neurones also declines by 50% ~48 days after birth *(for further details on Zn^{2+} blockade, see Blockers, 10-43).*

Developmental delay symptoms in genetic disorders linked to GABA$_A$ β_3 subunit gene deletion

10-11-05: Many patients with the maternally derived genetic disorder **Angelman syndrome**† (AS) carry chromosomal deletions† mapping to the same *region* as the gene encoding the **GABA$_A$ β_3 subunit** *(for details, see Phenotypic expression, 10-14).* AS patients are characterized by **developmental delay**†, **seizures**†, **inappropriate laughter** and **ataxic**† **movements**. Deletions in the same chromosomal region are common in subjects affected by **Prader–Willi syndrome**† (PWS) who display **infantile hypotonia**†, **childhood hyperphagia**†**/obesity**, **developmental delay** and **hypergonadism**†.

mRNA distribution

Co-expression of GABA$_A$ subunit genes

10-13-01: Generally, GABA$_A$ subunit genes display **differential**† **spatial**† **expression**[29–31] and expression of multiple subunit mRNAs within single cells is commonplace *(see examples in Table 2).* The mRNAs encoding the α_1, β_2 and γ_2 subunits are **co-localized** in many areas of the CNS – e.g. cerebellar Purkinje cells, hippocampal pyramidal cells and mitral cells of the olfactory bulb *(see also Predicted protein topography, 10-30).*

Phenotypic expression

GABA$_A$ receptor-associated behavioural phenotypes

10-14-01: Inbred strains of mice differ markedly in behaviour thought to be related to GABA$_A$ receptors, including sensitivity to **convulsants**† and

Table 2. *Relative distributions of $GABA_A$ receptor mRNAs (From 10-13-01)*

Feature/subunit	Characteristics	Refs
Co-expression of multiple subunit types	Co-expression of multiple subunit types is common (e.g. rat cerebrellar granule cells which express α_1, α_4, α_6, β_1, β_2, β_3, γ_2 and δ; olfactory bulb mitral cells express α_1, β_1, β_2, β_3 and γ_2 mRNAs; hippocampal dentate granule cells express all subunit mRNAs with the possible exception of α_6). In general, the expression level of an mRNA encoding a given subunit is variable between cell types	*10*
α subunit data	α_1 mRNA is abundantly expressed in cerebellar granule cells, but is also detected throughout the forebrain, including hippocampal dentate granule cells. α_1 but *not* α_2 mRNA is expressed in the pars reticulata, but *not* in the pars compacta of the rat substantia nigra *(see also Channel density, 10-09, Protein distribution, 10-15, and Subcellular locations, 10-16).*	*33*
	α_2 mRNA is of high abundance in the hippocampus, cerebellum and Bergman glia. α_3 mRNA is mostly restricted to the forebrain, associated with dentate granule cells but absent from cerebellum. α_4 mRNA is present in high concentrations within the hippocampus. α_6 mRNA is uniquely confined to the cerebellar granule cells of the cerebellum. *(See also a review of α_1, α_3 and α_6 distribution in ref.*[32]*)*	*32*
δ subunit data	δ subunit mRNA in the brain has a similar distribution pattern to the areas of GABA radioligand binding which is *not* associated with benzodiazepine binding. δ subunit mRNA is particularly prominent in rat **cerebellar granule cells**. It has also been suggested that δ subunits are associated with high-affinity $GABA_A$ receptors as they follow the distribution of $[^3H]$-muscimol binding. Generally, the location of δ and γ_2 subunit message appears distinct	*34*
γ subunit data	γ_1 mRNA is predominant in neuronal cell populations within the **limbic system** (amygdala, septum) and in the hypothalamus. γ_2 mRNA distribution resembles that of $GABA_A$/benzodiazepine receptors labelled with $[^3H]$-flunitrazepam. γ_3 mRNA distribution is similar to γ_2 but is less abundant. Generally, the location of γ_2 and δ subunit message appears distinct	*35*

Table 2. Continued

Feature/subunit	Characteristics	Refs
ρ subunit data	Human $\rho1$ and $\rho2$ mRNA is expressed (and probably restricted to) retinal tissue *(see references under rho subunits in Database listings, 10-53)*	36, 37
Absolute mRNA transcript numbers	Determination of absolute amounts of mRNA (using competitive PCR† templates to native transcripts† encoding 14 distinct subunits of the $GABA_AR$) in primary cultures of rat cerebellar granule neurones and cerebellar astrocytes show the latter to contain two orders of magnitude lower mRNA than the former. Granule cells express 14 different subunit mRNAs, while astroglial cultures do *not* express detectable amounts of α_6 and the γ_{2L} message. In direct comparison by competitive cDNA amplification[38], α_1, α_5 and α_6 mRNAs are prominent in cultured granule cells, the α_1 and α_2 mRNAs are abundant in cultured astrocytes, while β_1 and β_3 mRNAs are abundantly expressed in both cultures. The γ_{2S} and γ_{2L} mRNAs constitute the great majority of γ subunit mRNAs in neurones, while the γ_1 subunit mRNA is the most abundant γ subunit mRNA in astrocytes *(see also Channel modulation, 10-44)*	38

anticonvulsants†, **alcohol** and **diazepam**. Dysfunctional $GABA_A$ receptor proteins (or the expression-control† of their genes) may play a role in CNS 'hyperexcitability' states such as **epilepsy**† and **anxiety disorders**, although *direct* evidence of their involvement is lacking.

10-14-02: $GABA_A$ receptor dysfunction may be related to some forms of **myoclonus**† (quick involuntary shock-like movement disorders; reviewed in ref.[39]). Blockade of $GABA_A$ receptors or GABA synthesis regularly evokes **convulsive seizures**†, but administration of many GABA agonists† and some **GABA uptake blockers** *paradoxically* may evoke myoclonus (i.e. effects are receptor type-dependent). Enhancement of GABA function is effective as an anticonvulsant† strategy in several animal seizure† models[40] and **receptor subtype-selective drugs** may aid the development of efficacious compounds for the various types of epilepsy†.

Increased expression of $GABA_A$ receptors in the mouse spa *mutants*
10-14-03: The mutant mouse **spastic** (**spa**) acquires a severe motor disease about 2 weeks after birth[41]. *Spastic* mice display a fundamental disturbance in the 'balance' of excitatory† and inhibitory† impulses, which is likely to be due to aberrant regulation of normal **glycine receptor–channel gene switching** from 'foetal/neonatal' to 'adult' isoforms *(for details, see next paragraph and Phenotypic expression under ELG Cl GLY, 11-14)*. Notably,

facilitation† of $GABA_A$ responses also alleviates symptoms of affected mice *(see next paragraph).*

Alleviation of spa *symptoms is specifically mediated by $GABA_A$ receptors*

10-14-04: The **developmental transition**† from a 'foetal/neonatal' glycine receptor ($GlyR_N$) to the 'adult' GlyR complex ($GlyR_A$) is perturbed in the **spastic** mice *(see previous paragraph and ref.[41])*. Changes in function or structure of the GlyR proteins appear unaffected in *spastic* mice, factors which provide evidence for a regulatory rather than a structural effect of the *spastic* mutation[42]. A significant increase in $GABA_A$ receptors in the lower CNS may serve as a **compensatory function** in *spastic* mice, counteracting (in part) losses in glycinergic† function. Pharmacological facilitation† of $GABA_A$ responses, for example by administration of **benzodiazepine agonists**† or **aminooxyacetic acid** (which reduces degradation of endogenous GABA by inhibiting **GABA transaminase**) also alleviates symptoms of affected mice *(see Phenotypic expression under ELG Cl GLY, 11-14).*

$GABA_A$ channel activation by ethanol

10-14-05: Application of **ethanol** *facilitates* GABA responses in oocytes injected with mRNA from '**long-sleep**†' mice but *antagonizes* responses in oocytes injected with mRNA from '**short-sleep**' mice[43]. Ethanol inhibits **baroreflex bradycardia**† through $GABA_A$ and $GABA_B$ receptor-dependent potentiation† of the actions of endogenous GABA in the dorsal vagal complex[44]. For an illustration showing the **relative potentiating**† **effects** on $GABA_A$-mediated Cl^- flux and the *depression* of NMDA-, kainate- and voltage-gated Ca^{2+} flux, *see Fig. 5 under Channel modulation, 10-44).* A review on alcohol intoxication† and $GABA_A$ receptor activation is given in ref.[45].

$GABA_A$ channel activation by volatile anaesthetics

10-14-06: $GABA_A$ receptor complexes serve as common 'molecular target' sites for a variety of structurally diverse **anaesthetic molecules**[46]. The major volatile anaesthetics **isoflurane**, **halothane** and **enflurane** open anion-selective channels in rat hippocampal neurones which are sensitive to $GABA_A$ antagonists[46,47]. Like several other molecules with general anaesthetic properties (such as **barbiturates**, **steroids** and **etomidate**), these volatile anaesthetics have a **GABA-mimetic**† effect on vertebrate central neurones in culture. *Note:* Drugs which direct 'GABA-agonist'† and 'GABA-modulatory' properties will tend to produce overall *depression* of the CNS by increasing inhibitory† synaptic influence and by direct hyperpolarization† of neurones[47].

Mediation of excitatory responses by $GABA_AR$

10-14-07: In addition to a major inhibitory† role of post-synaptic cell **hyperpolarization**†, $GABA_A$ receptors have been shown to mediate *excitatory* responses in morphologically identified **interneurones**† in the presence of **4-aminopyridine** or following the elevation of **extracellular K^+** concentrations. In these cells, GABA can function to **synchronize firing** of inhibitory interneurones. The 'enhanced output' of the interneurone† population gives rise to **giant IPSPs**† in the **principal cells**[7], consistent with the role of GABA acting as an inhibitory† neurotransmitter.

Transformation of inhibitory to excitatory GABA responses

10-14-08: GABAergic† **inhibitory**† **responses** have also been shown to undergo **'long-term transformation'** into excitatory† **responses** following **pairing** of exogenous GABA with post-synaptic depolarization† or following pairing of pre- and post-synaptic stimulation. This **synaptic transformation** is due to a shift from a net increase of conductance to a net decrease of conductance in response to GABA[48]. Compare with responses of the NMDA receptor–channel with concomitant depolarizing stimuli – *see Phenotypic expression under ELG CAT GLU NMDA, 08-14.*

Excitatory transients in astrocytes

10-14-09: Stimulation of both $GABA_A$ and **$GABA_B$**† **receptors** evokes Ca^{2+} transients within **type 1 astrocytes** in rat cortical astroglial primary cultures[49]. Ca^{2+} responses are (most commonly) in a single-phase curve or are biphasic (i.e. have an initial rise that persists at the maximal or submaximal level). Both types of **Ca^{2+} responses** appear with some latency† following activation of Ca^{2+} channels and/or via release from internal **Ca^{2+}-mobilizing sites** (stores). GABA-evoked responses are reduced following incubation with GABA uptake blockers[49].

Phenotypic studies relating to specific $GABA_A$ receptor subunits (summary)

10-14-10: α subunit data**:** Cerebellar **motor control**† and drug-induced **behavioural impairment** has been linked to point mutations in α_6 subunit-containing $GABA_A$ receptor subtypes[50]. 'Alcohol-non-tolerant' rats are highly susceptible to impairment of **postural reflexes**† induced by benzodiazepine-site agonists† such as diazepam[50]. It has been found that the α_6 gene of 'alcohol non-tolerant' rats is expressed at wild-type† levels but carries a point mutation generating an **arginine (R)-to-glutamine (Q) substitution** at position 100. In consequence, $\alpha_6(Q100)\beta_2\gamma_2$ receptors show diazepam-mediated potentiation† of GABA-activated currents and diazepam-sensitive binding of $[^3H]$-Ro154513.

10-14-11: β subunit data**:** Deletions of the proximal† long arm† of human chromosome 15 (bands 15q11q13) are found in the majority of patients with two distinct genetic disorders – **Angelman syndrome**† (AS, maternally derived deletions) and **Prader-Willi**† **syndrome** (PWS, paternally-derived deletions). Deletion of the gene for $GABA_A$ receptor subunit β_3 (GABRB3), which maps† to the AS/PWS region, was found in AS and PWS subjects with interstitial† cytogenetic† deletions†[51]. *(For descriptions of typical symptoms of the Angelman and Prader–Willi syndromes, see Phenotypic expression, 10-14.)*

Contribution of $GABA_A$ receptors in pain reception

10-14-12: Antinociception† (i.e. inhibition of **pain receptors**) produced by microinjection of **L-glutamate** into the ventromedial medulla of the rat (nucleus reticularis gigantocellularis pars alpha) is antagonized† by **bicuculline** and enhanced by **diazepam**. These observations support the hypothesis that the response is mediated *in part* by an action of GABA at GABA$_A$ receptors in the spinal cord[52].

Protein distribution

Systematic studies of GABA$_A$ distribution in substantia nigra
10-15-01: The regional, cellular and subcellular distribution of GABA$_A$/benzodiazepine receptors has been systematically investigated by light and electron microscopy in the rat substantia nigra[33]. These studies employed a combination of quantitative **radioligand autoradiography**†, **immunohistochemical**† **localization** and ***in situ* hybridization** using antisense† probes. Moderate to high densities of GABA$_A$/benzodiazepine receptors are present throughout the full extent of the **substantia nigra pars reticulata**† with a very low density of receptors in the **substantia nigra pars compacta**†. The pars reticulata displays mainly central benzodiazepine type I† receptors with the highest density of receptors being in the caudal pars reticulata with lower densities of receptors in the middle and rostral levels of the pars reticulata. GABA$_A$ β_2/β_3 immunoreactivity is observed as a **punctate**† **distribution** on dendrites and neuronal cell bodies in the pars reticulata[33] *(see also Channel density, 10-09, and Subcellular locations, 10-16).*

Benzodiazepine immunoreactive sites in brain
10-15-02: Monoclonal† antibodies against benzodiazepines *(see Channel modulation, 10-44)* show endogenous 'immunoreactivity' in rat brain in an order compatible with the regional distribution of GABA$_A$ receptors in the brain (medial septum > amygdala > hippocampus > cerebral cortex > cerebellum).

10-15-03: α subunit data: The α-variant GABA$_A$ proteins are **differentially expressed**† in the CNS and can be photoaffinity†-labelled with benzodiazepines†.

Subcellular locations

Post-synaptic GABA$_A$ receptor–channels
10-16-01: The majority of GABA$_A$ receptors (located **post-synaptically**) are apposed to transmitter-release sites and effect post-synaptic cell hyperpolarization† (cf. sympathetic ganglion neurones, which *depolarize* in response to GABA – *see also glial cell responses under Cell-type expression index, 10-08).*

Different GABA$_A$ responses due to receptor heterogeneity and subcellular locations
10-16-02: Multiple types of GABA$_A$ responses recorded in rat hippocampal pyramidal cells are consistent with heterogeneous receptor **subunit composition** and locations of expression. At the **cell soma**†, responses are

hyperpolarizing† and enhanced by diazepam, whereas responses on **dendrites** are depolarizing†, pentobarbital-enhanced, and show higher sensitivity to picrotoxinin and bicuculline.

Secretory control functions by pre-synaptic GABA$_A$ receptor–channels

10-16-03: Activation of GABA$_A$ receptor–channels in the membranes of small pre-synaptic **peptidergic nerve terminals** of the posterior pituitary weakly depolarizes† the nerve terminal membrane and blocks action potentials. In this way, GABA limits **secretion** by retarding the spread of excitation into the terminal arborization†[53]. GABA$_A$ **autoreceptors** *(see Receptor agonists, 10-50, and Receptor antagonists, 10-51)* are located **pre-synaptically**.

Immunocytochemical distribution studies

10-16-04: Using immunocytochemical detection, a differential† **subcellular distribution** has been determined for the α_6 versus the α_1 and β_2/β_3 GABA$_A$ subunits in granule cells of the rat cerebellar cortex: α_6 subunit are detectable only at **synaptic sites** while α_1 and the β_2/β_3 subunits are located at both synaptic and **extrasynaptic** sites[54].

10-16-05: GABA$_A$ β_2/β_3 immunoreactivity is observed associated with the pre- and post-synaptic membranes of **axodendritic synaptic complexes** along the length of small- to large-sized smooth **dendrites** in the pars reticulata. Approximately 80% of 'immunopositive synapses' show *equal* staining of **pre- and post-synaptic membranes** and are associated with **small axon terminals** ($< 1.0\,\mu$m). Approximately 20% of 'immunopositive synapses' display immunoreactive thickening of the **post-synaptic membrane** and are associated with **large axon terminals** ($> 1.0\,\mu$m) which are in synaptic contact with large mainstem **dendrites**†[33] *(see also Channel density, 10-09, and Protein distribution, 10-15).*

Non-uniform (clustered) distribution of GABA$_A$ channels

10-16-06: Based on the distribution of single channels in outside-out† membrane patches of cultured rat cortical neurones, GABA$_A$-activated channels exist almost exclusively in **clusters**. This **non-uniform distribution** is present both in innervated† and un-innervated† neurones[55].

Sorting and subcellular targeting of GABA$_A$ channels

10-16-07: Assembly patterns and specifications of GABA$_A$ receptor subunits determine **sorting and localization** of protein complexes in **polarized† cells**[23]. Confocal† microscopy and subunit-specific immunoblot† analyses of epithelial cells (transfected with the cDNAs encoding the α_1 and β_1 GABA$_A$ subunits) show that the α_1 subunit is targeted to the **basolateral†** surface, and that the β_1 subunit is sorted to the **apical†** membrane. In cells where α_1 and β_1 isoforms are co-expressed, assembly of the β_1 with the α_1 subunit **'re-routes'** the α_1 subunit to the **apical surface**[23]. For implications of 'subunit routing' in control of neuronal phenotype, *see Developmental regulation, 10-11.*

Transcript size

10-17-01: *See Table 1 under Gene family, 10-05.*

SEQUENCE ANALYSES

For likely functional contributions of single subunits in homomeric† receptors and those in co-expressed heteromeric† complexes, see α, β, δ, γ and ρ subunit-specific data. The symbol [PDTM] denotes an illustrated feature on the channel protein domain topography model. (Fig.1)

Chromosomal location

Related $GABA_A$ genes are located on different chromosomes

10-18-01: Three isoforms of the $GABA_A$ α subunit are located on **different chromosomes**[56]. *Comparative note: Diverse* chromosomal locations of related channel gene family† members appears to be common *(for example, see Chromosomal location under VLG K Kv1-Shak, 48-18)* . In general, **gene separation** can occur by gene duplication† and recombinational transfer† or via retroviral or retrotransposon† excision and insertion events within the eukaryotic genome[57].

Chromosome linkages† to segregated† disease markers†

10-18-02: For possible linkage of $GABA_A$ gene function to chromosome deletions associated with the **Angelman syndrome** and **Prader–Willi syndrome**, *see Phenotypic expression, 10-14.*

Encoding

10-19-01: *Supplementary to Table 1 under Gene family, 10-05.* Open reading frames† of cDNA sequences encoding $GABA_A$ subunits predict **polypeptide lengths** between 423 and 521 amino acids with an average of ~450 amino acids.

Gene organization

Variable untranslated regions

10-20-01: Independently isolated $GABA_A$ receptor cDNAs can exhibit variable lengths of **untranslated** (**non-translated**) **sequences**, suggesting expression control† can operate at the level of translation†.

Homologous isoforms

10-21-01: Amino acid sequences of $GABA_A$ subunits are generally **well-conserved**† across vertebrate species. For detailed analyses, compare sequences retrievable by the accession numbers listed under *Database listings, 10-53.*

Protein molecular weight (purified)

10-22-01: α subunit data: The $GABA_A$ '*α* **subunit**' was originally affinity†-purified as a **50–53 kDa protein** *(see also Table 6 under Ligands, 10-47)*. *β* subunit data: The 'original' affinity-purified *β* **subunit** (**55–57 kDa**) migrates at a similar position to the **recombinant**† α_3 **subunit** on SDS–PAGE gels†[58]. Estimates of the native† molecular mass of the solubilized $GABA_A$ receptor–channel range between **220 kDa**[59] and **355 kDa**[60]. *Note:* The molecular weight of native† (functional) receptor–channels depends upon the precise **subunit stoichiometry**† *(see individual subunit data)*. *Supplementary note:* For the predicted **pentameric arrangement** of subunits *(see Predicted protein topography, 10-30)*, the molecular weight of native channels would be of the order ~250 kDa.

Protein molecular weight (calc.)

10-23-01: Calculated molecular weights derived from summated amino acids in $GABA_A$ subunits can be derived from analysis of sequences via the references/accession numbers given under *Database listings, 10-53.*

Sequence motifs

Invariant features of the extracellular ligand-gated ion channel superfamily

10-24-01: Inhibitory† receptor–channel types (e.g. those gated† by glycine and $GABA_A$) and excitatory† receptor–channel types (e.g. those gated† by nicotinic acetylcholine and 5-HT_3) of the extracellular ligand-gated ion channel superfamily† are related by both structural similarities and primary sequence identity *(see ELG Key facts, entry 04)*. An **invariant feature** of all members of this receptor superfamily is the presence of an extracellular **disulphide loop**† **motif**†, suggested to form part of antagonist†-binding sites[61] *(see Sequence motifs under ELG Cl GLY, 11-24)*. Descriptions of other sequence motifs† with functional significance are described under *Domain arrangement, 10-27, Domain conservation, 10-28, Domain functions, 10-29, and Protein phosphorylation, 10-32).*

STRUCTURE & FUNCTIONS

For likely functional contributions of single subunits in homomeric† receptors and those in co-expressed heteromeric† complexes, see ***α, β, δ, γ and ρ subunit-specific data****. The symbol [PDTM] denotes an illustrated feature on the channel protein domain topography model. (Fig.1)*

Domain arrangement

Common domain structures

10-27-01: All of the $GABA_A$ subunit subtypes are predicted to have a similar overall **domain structure** *(see [PDTM], Fig. 1)*. This includes a **signal**

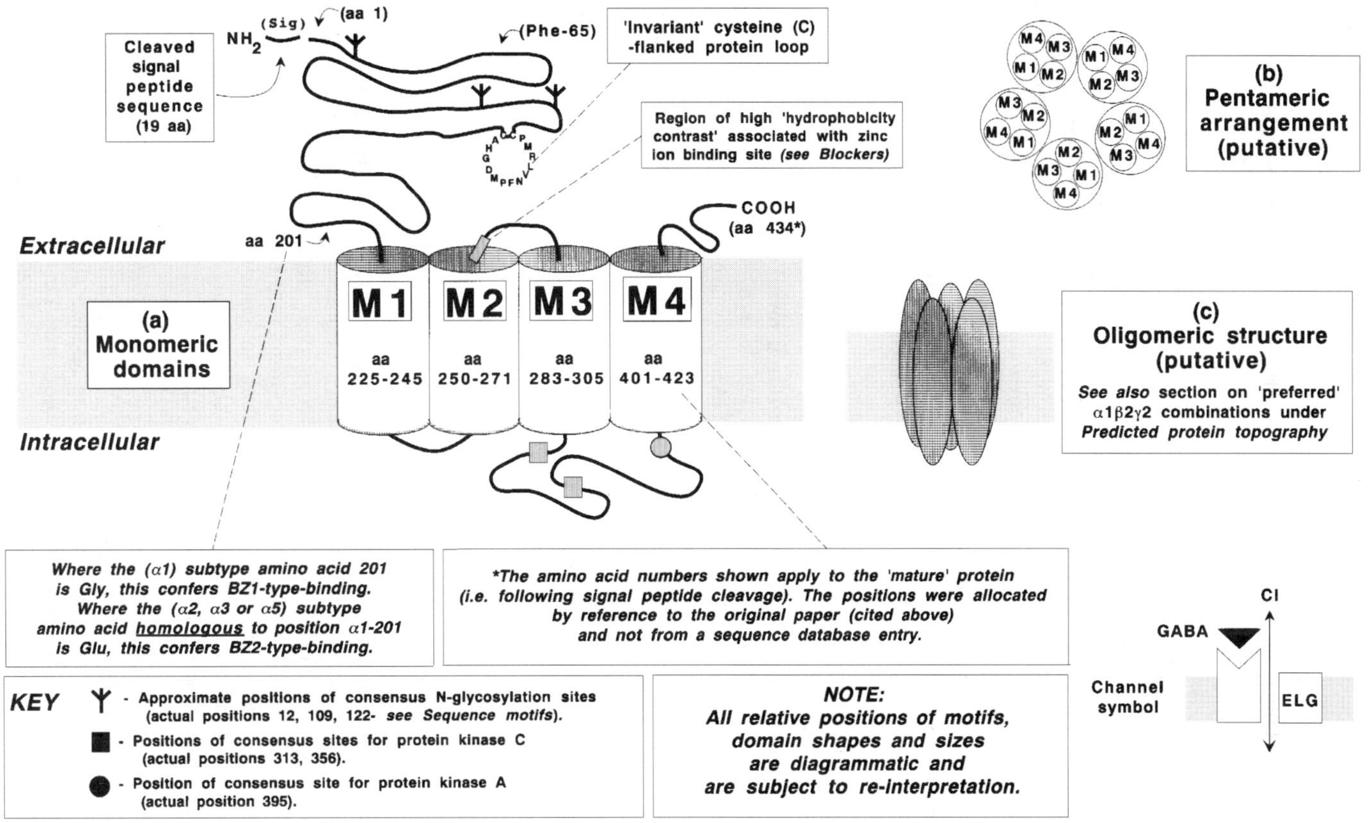

Figure 1. *Generalized monomeric protein domain topography model [PDTM] for the GABA$_A$ receptor–channel subunit series α, β, δ and γ. Residue numbers and motif positions apply to the* **rat GABA$_A$ α$_6$ subunit** (see ref.[147]). *(From 10-27-01)*

sequence† *(see [PDTM], Fig. 1)*, a large **N-terminal** (putatively extracellular) domain *(see [PDTM], Fig. 1)*, three putative **transmembrane domains** (M1–M3) *(see [PDTM], Fig. 1)*, a large variable sequence (putatively cytoplasmic) **M3–M4 loop** *(see [PDTM], Fig. 1)* and a **fourth transmembrane domain** (M4) near the **C-terminus** *(see [PDTM], Fig. 1) (see also Domain functions, 10-29).*

Preferential arrangements of subunit co-assembly

10-27-02: Assembly of $GABA_A$ receptors from constituent subunits does *not* proceed randomly to form all possible combinations, but certain subunit combinations are **preferred intermediates** during the **assembly process**[62]: For example, when varying combinations of α_1, β_1, and γ_{2S} subunits are transiently† expressed in mouse L929 fibroblast cells, **predominant assembly** of functional $\boldsymbol{\alpha_1\beta_1}$ and $\boldsymbol{\alpha_1\beta_1\gamma_{2S}}$ $GABA_AR$ forms occur. Notably, functional $\alpha_1\gamma_{2S}$ and $\beta_1\gamma_{2S}$ $GABA_ARs$ are *not* detected[62]. $GABA_ARs$ co-expressed with and without the γ_{2S} **subunit** differ in their **GABA** and **diazepam pharmacology**, and single-channel recordings suggest that the two predominant types of $GABA_ARs$ ($\alpha_1\beta_1$ and $\alpha_1\beta_1\gamma_{2S}$) have different conductance† and gating† properties.

10-27-03: Further co-expression studies to those described above have also shown that assembly of subunits into mature $GABA_ARs$ can arise from an '**ordered process**' to produce a '**preferred form**' of the receptor–channel ($\boldsymbol{\alpha_1\beta_1\gamma_{2S}}$)[63]. Intrinsic† kinetic properties of $\boldsymbol{\alpha_1\beta_1}$ and $\boldsymbol{\alpha_1\beta_1\gamma_{2S}}$ assemblies provide further evidence that $\alpha_1\beta_1$ $GABA_ARs$ are rarely (if ever) formed *upon co-expression of all three subunits.* This feature is apparently conserved across several channel types which are assembled from many possible discrete subunits[63] *(see also Activation, 10-33, and Single-channel data, 10-41). Comparative note:* An 'ordered' assembly process has also been observed for nicotinic ACh receptors *(see ELG CAT nAChR, entry 09).*

Domain conservation

Anionic selectivity determinants in the M2 domain

10-28-01: In general, there is a high degree of sequence conservation† of **transmembrane segments M1–M3** between the $GABA_AR$ and the **glycine receptor–channels** (GlyR), indicating their likely importance in **chloride channel** formation[64,65]. Introduction of three conserved amino acids present in the M2 segment of $GABA_A$ and glycine receptors into the M2 segment of an $\alpha7$ nicotinic receptor subunit is sufficient to convert this **cation-selective** channel type into an **anion-selective** channel gated by acetylcholine[66] *(for further details, see Domain functions under ELG Cl GLY, 11-29).*

Conserved anionic receptor–channel amino acid sequence motifs

10-28-02: Both $GABA_A$ (subunits α, β, γ, δ, and ρ) and glycine receptor–channels display: (i) an **invariant proline residue** at mid-position in the **M1 domain**; (ii) a common **hydroxy-rich sequence** Thr–Thr–Val–Leu–Thr–Met–Thr (Ser) and a total of eight Ser or Thr in each **M2 domain**; (iii) a **proline residue** at the fourth position preceded by a phenylalanine residue in the

M4 domain; (iv) relatively **high positive charge density** within eight residues of the ends of the transmembrane domains on the *extracellular* sides (by contrast, cation-selective ELG channels display negatively charged residues in addition to positive charges bordering M2[67,68]); (v) potential to form a *β*-**loop**† between Cys139 and Cys153, with 8 of 15 positions being identical or highly conserved in all subunits[69]. *Note:* A **conserved G-TTVLTM-T motif**† has aided further isolation of multiple subunit types in cross-hybridization† screens.

Conservation with invertebrate proteins

10-28-03: The **zeta polypeptide** from the freshwater mollusc *Lymnaea stagnalis* exhibits between 30 and 40% identity to vertebrate $GABA_A$ and glycine receptor subunit sequences[70]. Furthermore, locations of six out of seven introns† occur at similar relative positions as those found in vertebrate $GABA_A$ receptor genes. The zeta polypeptide mRNA is expressed in the adult nervous system but it can also be detected in peripheral tissues[70].

Domain functions (predicted)

N-*Glycosylated GABA-binding extracellular domains*

10-29-01: The N-terminal (putatively extracellular) domain of $GABA_AR$ contains three *N*-glycosylation† sites *(see [PDTM], Fig. 1)* and two Cys residues which help form a protein loop† *(see [PDTM], Fig. 1)* may function in binding GABA and other ligands[71].

Pore-lining domain M2

10-29-02: By comparison with nicotinic receptor–channels *(see ELG CAT nAChR, entry 09)*, the **M2 domain** *(see [PDTM], Fig. 1)* is thought to line the **channel pore** (see also paragraph on conversion of cation- to anion-selective channels[66] in *Domain conservation, 10-28)*.

Point mutations within Drosophila *$GABA_A$A M2 domains are associated with insecticide resistance*

10-29-03: The **cyclodiene insecticides** and **picrotoxinin** (PTX) are $GABA_A$ receptor antagonists† which competitively displace each other from the same binding site. The field-isolated *Drosophila* mutant Rdl ('Resistant to dieldrin') is insensitive to both PTX and cyclodienes and carries a **resistance-associated point mutation** (alanine to serine) within the M2 pore-lining domain. The widespread occurrence of this mutation in *Drosophila* populations promotes resistance to PTX/cyclodienes and could underlie > 60% of reported cases of **insecticide resistance**[72] *(see also the Database listings table for $GABA_A$ β subunits)*.

'Clustering' of phosphorylation sites in the M3–M4 loop

10-29-04: The large (putatively) **cytoplasmic loop** between M3 and M4 frequently contains **consensus**† **phosphorylation sites** *(see [PDTM], Fig. 1 and Protein phosphorylation, 10-32)*.

Positive charges in the channel vestibule

10-29-05: The existence of **channel vestibules**† containing **net positive charges** are consistent with **anion-selective permeability** properties of $GABA_A$ receptor–channels and they may have the property of attracting anions[73] *(compare with vestibules of the cation-selective nAChR – Domain functions, under ELG CAT nAChR, 09-29).*

Evidence for benzodiazepine antagonist sites on $GABA_A$ α-subunits

10-29-06: α subunit data**:** The **α variant** in a GlyR complex appears to determine the affinity for **ligand binding/recognition** at the **benzodiazepine (BZ) site**. Pharmacologically distinct **BZ-binding sites** can be 're-created' by co-expression of *any one* of six recombinant† α subunits, a β subunit variant, and the γ_2 subunit *(see Gene family, 10-05). Comparative note:* Recombinant† receptors containing $GABA_A$ **α4** and **α6** subunits do *not* recognize benzodiazepines but bind the β-carboline **Ro174513**.

Subunit contribution to benzodiazepine type I† and type II† pharmacology

10-29-07: The subtype of α subunit expressed in combination with other subunits affects **GABA affinity** and other pharmacological properties: $GABA_A$ 'α subtype functions' include (i) an association of **α_1 subunits** with **benzodiazepine type I**† receptor pharmacology with (ii) an association of **α_5** subunits contributing to **positive co-operativity**† of agonist binding and **benzodiazepine type II**† receptor pharmacology with low affinity for zolpidem *(see also following paragraphs).*

Relationships of recombinant† subunit combination and display of BZ_1 or BZ_2-type pharmacology

10-29-08: Benzodiazepine type I† (BZ_1) pharmacology is associated with high-affinity† binding of ethyl-β-carboline-3-carboxylate methyl ester (**β-CCM**, *see Table 5 under Channel modulation, 10-44*) and **CL218872**. In heterologous† expression systems, $GABA_A$ **$\alpha_1\beta_1\gamma_2$** subunit combinations exhibit **BZ_1-type** pharmacology (i.e. producing $GABA_A$ receptors displaying high-affinity binding for central benzodiazepine receptor ligands[29]). The subunit combinations **$\alpha_2\beta_1\gamma_2$** and **$\alpha_3\beta_1\gamma_2$** exhibit binding characteristics of benzodiazepine type II† (BZ_2) receptors. A summary of key amino acid loci implicated in modulator binding (as determined by chimaeric† cDNA constructions, site-directed mutagenesis† and photoaffinity-labelling†/microsequencing† procedures) are shown in Table 3 *(see also footnote to Table 6 under Ligands, 10-47).*

'Domain swop†' experiments determining loci of benzodiazepine binding

10-29-09: Exchange of nucleotide† sequences between the $GABA_AR$ variants **α_1** and **α_6** shows that a portion of the large extracellular domain determines sensitivity toward **benzodiazepine (BZ) ligands**[75]. A **single histidine** in the α_1 variant is essential for **BZ agonist**† **binding** of $GABA_A$ receptors. This residue also plays a role in determining high-affinity† binding for BZ antagonists and is present in α_1, α_2, α_3, and α_5 but is represented by an arginine in α_6[75] *(for clarification, see Table 3).*

Large $GABA_A$ current responses require α subunit expression
10-29-10: From injection of recombinant† $GABA_A$ subunit mRNAs in *Xenopus* oocytes, the presence of *at least one* **α subunit** in the combination is required for induction of **'large' currents** (I_{max} > 3000 nA). However,

Table 3. *Amino acid loci implicated in modulator binding within $GABA_A$ receptor–channel complexes (From 10-29-08)*

Construction/site-directed mutant/photoaffinity label	Functional implication	Refs
Chimaeras† with **mixed domains** of α_1 and α_3 subtypes	Where the (α_1) subtype amino acid 201 is Gly, this confers BZ_1-type-binding *([PDTM], Fig. 1).* Where the (α_2, α_3 or α_5) subtype amino acid *homologous to* position α_1-201 is Glu, this confers BZ_2-type binding	*74*
α_1 subunit cDNAs co-expressed with α_6 and γ subunits – (**'Domain-swops**†' – *see also 10-29-09*)	Mutation of the α_6 subunit cDNA (Arg100His) confers ability to bind [^{3}H]-flunitrazepam. (*Note:* Native α_6 receptors do not bind the ligand, and His100 is found in homologous positions in α_1, α_2, α_3 and α_5)	*75*
Microsequencing of cyanogen bromide†-cleaved peptide fragments following labelling with **[^{3}H]-flunitrazepam**	In separate studies, sites of attachment of the [^{3}H]-flunitrazepam photolabel† has been localized to residues 59–148 of the bovine α_1 subtype[76] and to residues 8–297 (with *S. aureus* V8† digestion)[77]; microsequencing† has revealed photolabel† attachment at the aromatic residues Phe226 and Tyr231, close to the beginning of the first transmembrane domain region[76] *([PDTM], Fig. 1)*	*76, 77*
Photoaffinity†-labelling of **Phe65** in affinity†-purified bovine α_1 with **[^{3}H]-muscimol**	Labelling was first localized to a chymotryptic substrate (beginning at Thr61) conserved in bovine and rat α_1–α_3 and α_5 of rat, but not in any β subunit. The residue carrying the label was localized by microsequencing† to Phe65 on α_1 which is conserved among all $GABA_A$ receptor α, γ_2 and δ subunits *([PDTM], Fig. 1)*	*78*
Site-directed mutations† of **Phe64Leu** in the rat α_1 subunit	Phe64Leu mutants exhibit marked decreases in agonist and antagonist affinity when co-expressed with β_2 and γ_2 subunits in *Xenopus* oocytes, consistent with the bovine Phe64 homologue† identified by photoaffinity labelling *(see this table, row immediately above) ([PDTM], Fig. 1)*	*79*

oocyte expression in the *absence* of α subunit mRNAs confirms that they are not *absolutely* required for channel formation or GABA-dependent gating *in vitro*.

Role of $GABA_A$ β subunits in receptor assembly and channel electrophysiological properties

10-29-11: β subunit data: Generally, **$GABA_A$ β subunits** are required for **efficient assembly** of complexes, and the presence of different β subunit types in recombinant complexes *influences* **GABA affinity**, **channel conductance**[80] and **modulator efficacy**† (e.g. to benzodiazepines, barbiturates and steroids[81, 82]). Although photo-labelling† studies implicate β subunits in GABA binding, channels composed *exclusively* of α, β or δ subunits can be gated by GABA, implying that the agonist-binding site is *not* unique to the β subunit. GABA-activated currents in cells heterologously† expressing GABA receptors incorporating the **β_2 subunit** show **faster desensitization**† and greater **outward rectification**† *(see Current–voltage relation, 10-35)*. **β_2-containing complexes** also exhibit a **shorter mean open time**† than receptors composed of $\alpha_1\gamma_2$ subunits[80].

$GABA_A$ γ subunit functions

10-29-12: γ subunit data: $GABA_A$ γ subunits *influence* **benzodiazepine binding**, although the α *subunits* determine whether the pharmacology matches 'classical' benzodiazepine type I† or type II† pharmacology – *see above*. γ subunits appear absolutely required for **allosteric modulation** of $GABA_A$ receptor–channels: **γ_1 subunits** contribute '**atypical**†' benzodiazepine responses; **γ_{2S} subunits** contribute **typical**† benzodiazepine sensitivity and **Zn^{2+}-insensitivity**. When the γ subunit is present in recombinant $GABA_A$ complexes expressed in heterologous† cells, the GABA-activated channels generally have a **larger conductance**†[80] *(see also Channel modulation, 10-44)*.

$GABA_A$ ρ subunit expression associated with non-$GABA_A$/$GABA_B$ responses in retinal bipolar cells

10-29-13: ρ subunit data: The cloned **$GABA_A$ ρ_1 (rho-1) subunit** is highly expressed in the **retina**[36] and, with the exception of its high picrotoxinin sensitivity (IC_{50} < 200 nM) ρ_1 homomultimer†, channels display an essentially similar pharmacological profile to that of the novel GABA receptor–channels expressed in retinal bipolar cells (i.e. non-$GABA_A$/$GABA_B$, insensitive to bicuculline, barbiturates, benzodiazepines and baclofen – *see Blockers, 10-43*)[83]. Expression of the GABA ρ1 receptor subunit cDNA in *Xenopus* oocytes yields a pharmacologic profile charaç-teristic of the **'$GABA_C$' responses** *(for discussion, see ref.*[84]).

Predicted protein topography

$GABA_A$ receptor subunits for functional hetero-oligomeric complexes

10-30-01: Functional diversity of native $GABA_A$ receptors is likely to arise through formation of different **hetero-oligomeric**† (**pentameric**) arrangements from subsets of the 16 (presently known) $GABA_A$ gene isoforms† (α_1–α_6; β_1–β_4,

γ_1–γ_3, δ, and ρ_1, ρ_2). Although there is little *direct* information on the subunit composition and/or topography of *native* receptor–channels, a major (preferred) subtype may be $\alpha_1\beta_2\gamma_2$ since the mRNAs encoding these subunits are often co-localized in the CNS[34] *(see below, but also see ref.[81])*. The slope of the **concentration–response curve** is steeper for the 'preferred' $\alpha_1\beta_2\gamma_2$ combination compared with the $\alpha_1\beta_2$ or $\alpha_1\gamma_2$ subunit combinations[80]. *(For further details on experiments establishing 'preferential' arrangements of subtypes, see Domain arrangement, 10-27.)*

Recombinant subunit assemblies resembling native $GABA_A$ receptors

10-30-02: Recombinant† $GABA_A$ $\alpha_1\beta_2\gamma_2$ **subunit complexes** exhibit pharmacological and electrophysiological properties resembling those of native† $GABA_A$ receptors. The **combination** $\boldsymbol{\alpha_5\beta_2\gamma_2}$ also produces 'consensus' properties of the vertebrate $GABA_A$ receptor–channel, including those for **co-operativity of gating**, **drug modulation** and $\mathbf{K_a}$[81]. *Note:* Substitution of α_1 or α_3 in $GABA_A$ complexes appears to have little effect on the channel properties.

Properties of recombinant homomultimeric $GABA_A$ receptor–channels

10-30-03: Single-subunit channels (homomultimers† of any one of the subtypes) have been reported to display **GABA-dependent gating**†, **barbiturate potentiation**†, **bicuculline** and **picrotoxinin blockade**, **multiple conductance states** and **desensitization**† *(but see ref.[81])*. *Note:* Because of their relative instability, homomultimeric† $GABA_A$ receptor–channels are *unlikely* to form native† receptor complexes *in vivo*.

Invertebrate $GABA_A$ protein topography and domain arrangement

10-30-04: β subunit data: Molluscan β subunits are capable of *replacing* vertebrate β subunits in heterologous† cell co-expression experiments with the bovine $GABA_A$ receptor α_1 subunit, suggesting that *invertebrate* $GABA_A$ receptors also exist as hetero-oligomeric† complexes *in vivo*[85].

Protein interactions

Subunit protein interactions displaying selective pharmacology

10-31-01: Co-expression of certain recombinant† **'$\boldsymbol{\alpha}$ with $\boldsymbol{\beta}$'** and **'$\boldsymbol{\alpha}$ with $\boldsymbol{\gamma}$'** subunits can yield complexes with the pharmacological properties of **$GABA_A$ benzodiazepine type I**† or **type II**† **receptors** *(see refs[86,87] and Channel modulation, 10-44)*. *Comparative note:* Receptors expressed in heterologous† cells which include the **$GABA_A$ α_1 variant** generally display the known properties of **benzodiazepine type I**† receptors. Receptors containing the **$GABA_A$ α_2, α_3 or α_1 variants** *generally* have much in common with the **$GABA_A$ benzodiazepine type II**† receptors.

Homogeneous electrophysiological properties from multiple co-expressed receptor proteins

10-31-02: Co-injection of *six different* $GABA_A$ subunit isoforms have been shown to result in an apparently '**homogeneous population**' of $GABA_AR$ ion channels, indicating that the protein assembly process selects '**appropriate subunits**' to build functional channels[81].

A requirement for α_1/β_1 subunit interactions for functional protein assembly in insect cells

10-31-03: In Sf9 (insect) cell hosts for recombinant† baculoviruses† driving† expression of human $GABA_A$ α_1 and β_1 subunit cDNAs, fluorescently labelled monoclonal† antibodies to α_1 subunits show fluorescence of the plasma membrane *only* in cells **co-infected** with both α_1 and β_1 recombinant virus constructs[88].

Relative stability of co-expressed $GABA_A$ subunit complexes in oocytes

10-31-04: α subunit data: Co-injection of bovine α_1 **and** β_1 **subunit** mRNAs produce $GABA_A$ channels that are **more stable** than subunit mRNAs supporting expression of homomultimers†. Together, α_1 and β_1 replicate many properties of native† $GABA_A$ receptors *(see above)*. Quantitative immunoprecipitation† studies in a range of cell types suggest that native† receptors can contain either a single α variant or multiple α variants.

10-31-05: γ subunit data: 'Typical' **benzodiazepine potentiation**† following co-expression† of **$GABA_A$ α_1 and β_1 subunits** requires the *additional* presence of one or more γ **subunits** (e.g. α_1 β_1 γ_2). Substitution of δ for γ has been shown to *abolish* benzodiazepine binding.

Complex interactions of excitatory and inhibitory receptor–channel proteins – an example

10-31-06: The post-synaptic current† (PSC) in the outer molecular layer of the adult rat **dentate gyrus** produces a depolarizing post-synaptic potential† (DPSP†) in **granule cells** and is *part-mediated* by $GABA_A$, NMDA and AMPA receptor proteins in the ratio 1 : 0.1 : 0.2 (as estimated by peak† currents)[89]. The peak current ratio† is essentially constant over a range of stimulus intensities that produce compound PSC amplitudes of 80–400 pA[89]. Simultaneous stimulation of pre-synaptic fibres from both the **perforant path** and **interneurones** results in a large depolarizing $G_{GABA\text{-}A}$ component that inhibits the granule cell by '**shunting**' the excitatory PSCs. The $G_{GABA\text{-}A}$ component decreases the peak DPSP amplitude† by ~35%, shunts 50% of the charge transferred to the soma by the excitatory PSC, and completely inhibits the NMDA receptor-mediated component of the DPSP. Similar kinetics of the $GABA_A$ and NMDA PSCs effectively inhibit the NMDA PSC. The (more rapid) AMPA PSC is less affected by the $G_{GABA\text{-}A}$, so that granule cell excitation *under simultaneous stimulation* is primarily due to AMPA receptor activation[89].

Binding of voltage-dependent anion channel (VDAC) protein by benzodiazepine ligands

10-31-07: A 36 kDa polypeptide, originally purified on a *benzodiazepine affinity column*† has shown channel-forming activity in lipid bilayer membranes that is 'virtually identical' to the **VDAC** (**voltage-dependent**

anion channel, isolated from mitochondria of various sources[90] – *see MIT (mitochondria), entry 37).* Notably, this distribution is consistent with the VDAC being equivalent to the **'peripheral-type' benzodiazepine receptor**, described in the footnote under Table 6 *under Ligands, 10-47.* Rat hippocampal cDNA† clones isolated using probes synthesized following microsequencing† of the 36 kDa polypeptide show ~24% amino acid identity to yeast VDAC, and over ~70% identity to the human B lymphocyte VDAC. Antisera to the 36 kDa polypeptide is able to precipitate **[^{3}H]-muscimol**-binding activity, indicating a **'tight' association** with the $GABA_A$ receptor protein *in vitro*[90].

Depolarizing actions of GABA in glial cells explained by differential expression of transporters

10-31-08: In contrast to the hyperpolarizing actions of GABA in most neurones, **astrocytes** and **oligodendrocytes** are **depolarized** by GABA (although the underlying receptor–channel appears to be related to 'neuronal' $GABA_A$ receptors) *(see Cell-type expression index, 10-08).* In oligodendrocytes, there is a marked increase in intracellular Cl^- concentration owing to the functional expression of a **bumetanide-sensitive $Na^+/K^+/Cl^-$ transport system**. In astrocytes, increased intracellular Cl^- involves co-expression of a **$Na^+/K^+/2Cl^-$-co-transporter** and **Cl^-/HCO_3^--exchanger**[91]. These differences in **transporter protein expression** can account for GABA-gated Cl^- channels giving rise to **Cl^- efflux** from glial cells. For the dependence of current direction on **electrochemical driving force**†, *see Current type, 10-34, and Voltage sensitivity, 10-42.*

Protein phosphorylation

$GABA_A$ protein phosphorylation in vitro

10-32-01: The $GABA_A$ receptor is phosphorylated *in vitro* by **protein kinase A** (PKA), **protein kinase C** (PKC) and an unidentified **receptor-associated kinase** *(reviewed in ref.*[92]*).* PKC generally '**down-modulates**' different recombinant $GABA_A$ channels[93] *(see below). Comparative note:* Like the nAChR *(entry 09),* the $GABA_A$ receptor does *not* appear to be phosphorylated by calmodulin kinase II on any of its constituent subunits.

Functional consequences of $GABA_A$R phosphorylation – protein kinase A

10-32-02: In general, phosphorylation of $GABA_A$ subunits will have differential effects depending on the receptor complex **subunit composition** and the **number and loci of phosphorylation sites** within these subunits *(see subunit-specific data below). In vivo,* **cAMP-dependent phosphorylation** may increase the rate of $GABA_A$ **receptor desensitization**† (in chick cerebral cortical neurones) or may enhance receptor function by phosphorylation following β_1-adrenoceptor stimulation (in cerebellar Purkinje cells). PKA inhibits GABA-evoked Cl^- currents (in mouse spinal cord neurones) both in whole-cell and single-channel patch preparations. PKA decreases frequency† of channel opening without affecting either channel conductance† or mean channel open time† *(cf. Protein phosphorylation under ELG CAT nAChR, 09-32).*

Amino acid residues sensitive to phosphorylation by the endogenous protein kinase C of oocytes

10-32-03: The functional effects of **protein kinase C**-mediated '**down-regulation**' of $GABA_A$ currents have been studied by mutations altering serine or threonine residues of consensus phosphorylation sites for PKC in the large intracellular (M3–M4) domain of subunits α_1, β_2, and γ_{2S}[94]: Following co-expression with wild-type† subunits (to yield α_1 β_2 γ_{2S} combinations) 14 individual mutations did *not* affect the level of expression of GABA current. However two mutations, $\boldsymbol{\beta_2}$ **(Ser410)** and $\boldsymbol{\gamma_{2S}}$ **(Ser327)** result in significant reductions of the 'down-regulatory' effect of the PKC activator 4 β-phorbol 12-myristate 13-acetate (10 nM). Other combinations of co-expressed subunits suggest that phosphorylation of *both* sites is required for a 'full, PKC-mediated down-regulation' of $GABA_A$ currents[94]. *Note:* In these experiments, PKC-dependent phosphorylation appeared to 'preferentially inactivate' a non-desensitized form or state of the $GABA_A$ receptor[95].

Presence of phosphorylation consensus sites in $GABA_A$ subunit sequences (summary)

10-32-04: α subunit data: Phos/PKC: $\boldsymbol{\alpha_5}$ – 3 protein kinase C sites. $\boldsymbol{\alpha_6}$ – 2 protein kinase C sites. Phos/PKA: 1 protein kinase A site. β subunit data: Phos/PKA: 1 protein kinase A site. The intracellular domain of the $\boldsymbol{\beta_1}$ subunit is phosphorylated on Ser409 by both PKA and PKC[96]. *In vitro* phosphorylation of β subunit isolated from porcine and rat brain has been demonstrated. *Note:* The loci of these consensus† sites (and those described below) can be derived from primary† sequence data retrievable by the accession numbers listed under *Database listings, 10-53.*

Extra phosphorylation sites present in alternatively spliced variants of $GABA_A$ receptor proteins

10-32-05: γ subunit data: There are two **alternatively spliced**† forms of the γ_2 subunit (γ_{2S}/γ_{2L}). The 'long form' (γ_{2L}) incorporates an **additional exon**† of eight amino acids (**LLRMFSFK**) inserted in the first portion of the **M3–M4 intracellular loop** *(see [PDTM], Fig. 1)*. This exon contains a **'new' phosphorylation site** for protein kinase C (Ser343) in addition to the PKC site at Ser327 (on γ_{2S} and γ_{2L})[96,97]. *Note:* Phosphopeptide fragments of γ_{2L} can be phosphorylated *in vitro* by protein kinase C. Regulation of this splicing† event may be under developmental control and could account for cell type-selective patterns of phosphorylation. *Note:* Both the γ_1 and γ_2 subunit sequences have a consensus **tyrosine kinase** phosphorylation site[29,98].

10-32-06: ρ subunit data: ρ subunit sequences possess 3 'consensus' protein kinase C sites.

Links between $GABA_A$ phosphorylated splice variants and ethanol sensitivity

10-32-07: Alternative splicing† of the $\mathbf{GABA_A}$ **subunit** $\boldsymbol{\gamma_{2L}}$ can result in '**ethanol-sensitive**' or **'ethanol-insensitive' forms** of $GABA_A$ complexes (when combined with α and β subunits)[99]. Site-directed mutagenesis† of the γ_{2L} protein kinase C consensus† phosphorylation site *(see above)* has shown that it is critical for modulation by ethanol but *not* benzodiazepines. Inhibition of phosphorylating enzyme activity in oocytes expressing a $\alpha_1\beta_1\gamma_{2L}$ can also prevent ethanol enhancement. These data have been taken to infer that **phosphorylation/dephosphorylation status** of specific sites on the $GABA_AR$ can act as a 'control mechanism' for neuronal responses to alcohol exposure[99] *(see also Channel modulation, 10-44).*

ELECTROPHYSIOLOGY

Activation

$GABA_ARs$ display multiple open state, closed state, and 'bursting' behaviour

10-33-01: Main conductance states† of the $GABA_ARs$ in cultured mouse spinal cord neurones open singly and in **bursts**† of openings. The channel opens into at least three **open states** in this preparation: **O1** ($\tau \sim 1.0 \pm 0.2$ ms), **O2** ($\tau \sim 3.7 \pm 0.4$ ms), **O3** ($\tau \sim 11.3 \pm 0.5$ ms), which do not vary over the range 0.5–5 μM GABA[100]. The **transition O1 (shortest) → O2/O3** shifts in relative frequency† to give increases in long open times† with increasing agonist concentration. The channel has several **closed states**, the two shortest time constants being concentration-independent, with three longer time constants, each *decreasing* with *increasing* agonist concentration[100] *(see also ref.[101])*. The **bi-liganded receptor state (A_2R)** is then converted to an **activated state**† **(A_2R^*)** leading to **bursts**† (~15–40 ms, with 2–3 interruptions per burst). Short-lived bursts (~1–2 ms) are also present, which are consistent with the openings, possibly due to a **mono-liganded (AR) state** of the receptor.

Kinetic properties of $GABA_A$ channel activation depend on subunit composition

10-33-02: Recombinant $\boldsymbol{\alpha_1\beta_1}$ and $\boldsymbol{\alpha_1\beta_{11}\gamma_{2S}}$ heteromultimers† transiently† expressed in mouse L929 fibroblast† cells differ in their open state† and burst† properties. On average, $\boldsymbol{\alpha_1\beta_{11}\gamma_{2S}}$ $GABA_ARs$ open for almost three times the duration of the $\boldsymbol{\alpha_1\beta_1}$ combination, $GABA_ARs$ (6.0 versus 2.3 ms, respectively) and have three openings per burst† *(see also previous paragraph).* $\alpha_1\beta_1$ $GABA_ARs$ open predominantly as single opening **bursts**†[63]. These kinetic properties have been used to determine **assembly patterns** of heteromultimeric† $GABA_ARs$ *(see Domain arrangement, 10-27).*

Typical conductance values of multiple substates

10-33-03: The **main conductance state**† observed in mammalian neurones varies between tissues – e.g. ~ ≤20 pS in **hippocampal neurones** and ~30 pS for **spinal cord neurones** (the former is similar to that observed for γ_2-containing recombinant complexes[80]). Whole-cell current† properties, such as **desensitization**† and the slope of the concentration-response curve for γ_2-containing complexes are similar to the native† $GABA_A$ receptors of bovine adrenal chromaffin cells and pancreatic islet cells. α subunit data: Multiple **subconductance states**† exist for homomeric† **α subunits**, principally 19 pS and 28 pS. β subunit data: Multiple subconductance states† have been shown for homomeric† **β subunits**, mainly at 18 pS and 27 pS.

Dose-dependent activation and desensitization properties

10-33-04: Typically, $GABA_A$ receptor–channel (in mouse spinal cord neurones) openings are **agonist†-concentration dependent** in the range 0.5–5 μM GABA. However, application of GABA to cultured neocortical neurones from rat produces a **desensitizing† response**, i.e. one that declines over several seconds, even in the continued presence of agonist[102]. There are *two* activation and **desensitization phases** to the GABA-induced chloride current of frog sensory neurones at high agonist concentrations (~30 μM) compared to only one at lower agonist concentrations. *Note:* This behaviour may reflect receptor heterogeneity at the single-cell level *(see Subcellular locations, 10-16, and Inactivation, 10-37).*

Pharmacological modulation of open duration and frequency

10-33-05: $GABA_A$ receptor 'positive modulators' such as **pentobarbital** (a benzodiazepine-site agonist†) increase average channel open duration† without increasing opening frequency, whereas **picrotoxinin** *(see Blockers, 10-43)* slightly reduce average channel open duration† and reduce opening frequency†.

Activation by single molecules of GABA

10-33-06: The benzodiazepine† **chlordiazepoxide** (CDPX) modulates $GABA_AR$ so that *one molecule* of bound GABA becomes sufficient to open the channel[103]. This effect may be explained by a property of CDPX yielding a closed state† which can result in channel opening mediated by a *single* GABA-binding site *(see also Dose–response, 10-36)*. Alternatively, CDPX may act at one of the channel opening binding sites without a postulated, second closed conformational state†[103].

Vesicular and non-vesicular GABA agonist release

10-33-07: The majority of endogenous GABA release occurs from **vesicles†** (whose release can be evoked experimentally by **K^+-induced depolarization†**). A second type of release (endogenous glutamate-dependent release) is **non-vesicular**†[104]. The glutamate-sensitive release is mainly mediated by **NMDA receptor–channels** and consists of a single, sustained phase. This phase is insensitive to **nocodazole**, partly inhibited by **verapamil** and can be blocked by **SKF 89976A**. The blocking action of **Co^{2+} ions** has also been attributed to a block of NMDA-associated ion channels[104] *(see ELG CAT GLU NMDA, entry 08).*

Current type

General dependence of $GABA_A$ current direction and magnitude on driving force

10-34-01: Activation of $GABA_A$ receptors leads to either inwardly *or* outwardly directed Cl^- ion movement *(see Channel modulation, 10-44)*, depending on the **electrochemical driving force†** (i.e. determined by the resting potential and the Cl^- gradient – *see Current–voltage relation, 10-35)*. During the activation of excitatory† receptors, $GABA_A$ channels act to stabilize the cell resting potential†. A striking example of current 'direction' being influenced by driving force is the excitatory†, depolarizing† responses to GABA seen in **glial cells** *(described under Cell-type expression index, 10-08). Note:* The *magnitude* of the current also depends on the electrochemical driving force† *(see Fig. 3 under Voltage sensitivity, 10-42).*

Current–voltage relation

Rectification properties of native GABA$_A$R channels

10-35-01: Current–voltage (I–V) relations† of GABA-activated currents obtained from whole-cell measurements in some preparations (e.g. mouse cultured spinal neurones, 145 mM Cl^- intracellularly and extracellularly) display **outward rectification**†. In voltage-jump† experiments, the **'instantaneous'**† **I–V relations**† are linear†, while the **steady-state**† **I–V relations**† are outwardly rectifying† indicating that the gating† of GABA$_A$R channels is voltage-sensitive in this preparation[73] *(see Voltage sensitivity, 10-42).*

Dose–response

GABA$_A$ channel open time depends on agonist concentration

10-36-01: In general, open-time† frequency histograms† are shifted to longer times as GABA concentration is increased from 0.5 to 5 μM. GABA concentration–response curves are sigmoidal and have Hill coefficients† of about 2 in several preparations *(see Hill coefficient, 10-46).*

Low-concentration GABA responses are enhanced by chlordiazepoxide

10-36-02: The benzodiazepine **chlordiazepoxide** (CDPX) shows an enhancement of GABA$_A$ channel opening (10-fold at ~0.3 μM GABA) which decreases with increasing GABA concentration. At maximal agonist† responses (GABA concentrations > 1000 μM), *no* enhancement by CDPX occurs[103]. Typically, the half-response concentration is reduced from 80 to 50 μM with CDPX. In the presence of CDPX, channel opening occurs with only **one bound GABA molecule**, whereas in its absence, channel opening with two bound GABA molecules appears much more favourable *(see also Activation, 10-33).*

Subunit dependence of agonist affinity† and co-operative gating†

10-36-03: α subunit data: The GABA$_A$ **α_5 subunit** appears to be important for high GABA affinity† and co-operativity† of GABA gating[81]. β subunit data: In mRNA co-injection studies in oocytes, the presence of **β_1 or β_2 subunits** are *not* required for expression of GABA-gated ion currents displaying agonist co-operativity†[81]. γ subunit data: The GABA$_A$ **γ_2 subunit** affects GABA gating† of channels expressed in oocytes from cloned RNAs: γ_2 subunit added to subunit combinations $\alpha_1\beta_1$ or $\alpha_1\beta_2$ result in a 7-fold decrease in $K_{a(GABA)}$[81].

Dose–response of $GABA_A$ receptor inactivation properties
10-36-04: Desensitization† of native† $GABA_A$ receptors expressed in cultured rat hippocampal neurones is agonist dose-dependent (higher concentrations of GABA induces both larger and faster desensitization). Desensitization phenotypes may vary with the 'passage number' of the cultured cells. *Note:* $GABA_A$ receptors are *not* desensitized without first being activated[105].

Inactivation

Modulation of desensitization kinetics
10-37-01: In common with other extracellular ligand-gated receptor–channels, $GABA_A$ receptors **desensitize**† in the continued presence of agonist. GABA-binding sites specific for the initiation of desensitization† are distinct from those mediating the opening of the channel *(see Channel modulation, 10-44).* $GABA_A$ channel modulators such as **bicuculline** slow desensitization† while agents such as **diazepam** enhance the rate of desensitization†. In cultured cortical neurones, single-channel responses often '**fail to recover**' after only a few exposures to agonist†. *Functional note:* Desensitization† of $GABA_A$ responses may help regulate **cortical inhibition**†, especially under conditions of intense excitatory† and inhibitory† synaptic activation[102].

Kinetic models

Modelling of multiple open and closed states – an example
10-38-01: For *recombinant*† $\alpha_1\beta_1$ $GABA_A$ channels in chinese hamster ovary (CHO) cells[106], **open duration**† **frequency distributions**† are fitted best with the sum of *two* exponential functions†, suggesting that the $\alpha_1\beta_1$ $GABA_A$ receptor–channel has at least two open states†. Distributions of closed durations† between main conductance level† openings are fitted best with *multiple* exponential functions†, suggesting that the $\alpha_1\beta_1$ $GABA_A$ receptor–channel also has several closed states†. *Comparative note:* Native† $GABA_A$ main conductance† and subconductance† levels are characterised by longer openings and display at least three open states *(see Single-channel data,* 10-41). Kinetic models for $GABA_AR$ equilibrium single-channel gating† properties have been reviewed[6].

Rundown

Phosphorylation of $GABA_A$ receptor may be required to maintain responsiveness (examples)
10-39-01: The $GABA_A$ receptor has been shown to undergo '**rundown**'† in cultured hippocampal pyramidal cells and cultured spinal cord neurones *(see also paragraphs below).* Unlike desensitization†, rundown is *not* dependent upon agonist† application. $GABA_A$ channel rundown is sometimes accelerated by application of **alkaline phosphatase** and retarded by **ATPγS, low intracellular Ca^{2+},** Mg-ATP (possibly in 'complexed' form) and the **calmodulin inhibitor W7** (the latter two factors indicate that the 'endogenous phosphatase' may be the Ca^{2+}-dependent phosphatase **calcineurin**). In cultured rat cerebellar granule cells, the peak Cl^- current

elicited by GABA decreases (runs down) with time of **cell registration**†, with a time constant of 7.3 min[107], while 'residual' responsiveness is maintained thereafter. The kinase (or another associated protein) involved in 'maintaining' the GABA$_A$ channel activity has not been characterized.

An ATP-receptor site associated with GABA$_A$R rundown phenomena
10-39-02: Based on the action of ATP and specific ATP analogues, it has been shown that the GABA$_A$Rs native† to dissociated **nucleus tractus solitarii (NTS) neurones** possess an **intracellular ATP-sensitive binding site** (i.e. an ATP receptor) which can modulate the channel[108]: Following fast application of 2 mM ATP, the amplitude of I_{Cl} elicited by 10^{-5} M GABA does *not* show any time-dependent decrease (i.e. apparent rundown) over 60 min. In the *absence* of intracellular ATP, the amplitude of GABA-induced I_{Cl} decreases with time. *Note:* Removal of **intracellular Mg^{2+}** induces rundown, even in the presence of ATP[108] *(see also next paragraph).*

Effective and non-effective agents in prevention of rundown
10-39-03: In neurones of the nucleus tractus solitarii, activation or inhibition of **dephosphorylation processes** (by alkaline phosphatase and the phosphatase inhibitor okadaic acid respectively) does *not* affect the GABA response in the presence of 2 mM ATP and 3 mM Mg^{2+}. Rundown can also be prevented by **2 mM ADP plus Mg^{2+}** or **Mg^{2+} plus ATPγS** (adenosine-5′-*O*-3-thiotriphosphate), however application of Mg^{2+} plus 2 mM adenosine, AMP, cyclic AMP, AMP-PNP (adenylimido-diphosphate) or ADPβS (adenosine-5′-*O*-2-thiodiphosphate) can *not* prevent rundown in this preparation[108].

Selectivity

GABA$_A$ receptor–channels are chloride-selective
10-40-01: E_{rev}† in response to GABA is given by the **chloride equilibrium potential**†: When the external Cl$^-$ concentration is reduced, E_{GABA}† is shifted in the depolarizing† direction by ~51.5 mV per 10-fold change in external Cl$^-$, which is close to the shift predicted by the Nernst equation† for a selective increase in Cl$^-$ conductance[109]. Determination of E_{rev}† in native preparations of mouse spinal cord neurones with different anions has derived the permeability ratios† for a variety of anions with respect to chloride *(see Fig. 2)*. Permeability† falls with increasing **ionic diameter** (i.e. the **hydrated anion size**† or **Stokes diameter**†), reaching zero at about 5.6 Å *(see Fig. 2)*. *Comparative note:* These data indicate the internal diameter of the GABA$_A$ channel to be slightly smaller than that of the nAChR (estimated at ~ 7.4 Å, *see ELG CAT nAChR, entry 09*).

Conductance versus permeability sequences for GABA$_A$ receptors – evidence for anionic binding sites
10-40-02: The **conductance sequence**† obtained from single-channel current measurements for the GABA$_A$R in mouse spinal cord neurones is Cl$^-$ > Br$^-$ > I$^-$ > SCN$^-$ > F$^-$ which is almost in *reverse order* to the **selectivity**† **(permeability**†**) sequence** SCN$^-$ > I$^-$ > Br$^-$ > Cl$^-$ > F$^-$ *(see Fig. 2)*. This 'reversal' indicates that the rate of ion transport is limited by **anion binding**

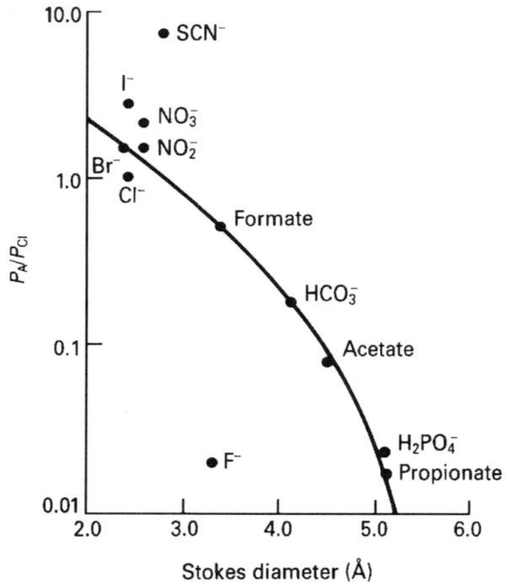

Figure 2. *Ionic permeability† through native GABA$_A$ receptor–channels. Permeability falls with increasing* ***ionic diameter;*** *the* ***hydrated anion size† or Stokes diameter†****, reaching zero at about 5·6Å. (Reproduced with permission from Bormann (1987)* J Physiol Lond ***385****: 243–86.) (From 10-40-01)*

inside the channel. This, and other experiments indicate that GABA$_A$ channels contain at least **two anionic binding sites per pore**[73]. *Comparative note:* These properties are shared with **glycine receptor–channels** *(for a direct comparison of the permeability properties of the GABA$_A$ and glycine receptor–channels, see Selectivity under ELG Cl GLY, 11-40).*

Permeability sequence for large polyatomic anions

10-40-03: The permeability sequence† for **large polyatomic anions** through GABA$_A$ channels in mouse spinal cord neurones has been determined as formate > bicarbonate > acetate > phosphate > propionate. *Comparative note:* Glycine receptor–channels in the same preparation *(see ELG Cl GLY, entry 11)* are *not* measurably permeant to phosphate and propionate[73].

Similarity of GABA$_A$R and GlyR ionic selectivity filters

10-40-04: *Note:* GABA and glycine agonists activate channels with almost identical permeation properties in cultured hippocampal neurones[110]. Comparative features of the permeation pathway for a range of neurotransmitter-gated ion channels have been reviewed[111]. For the **molecular determinants** of anionic selectivity, *see Domain conservation, 10-28, and Domain functions, 10-29.*

Single-channel data

Multiple conductance levels of native GABA$_A$ receptor–channels – examples

10-41-01: Patch-clamp† studies of single channels with permeant† chloride ions show prominent **multiple conductance levels**: The typical *range* of GABA$_A$ single-channel conductances (145 mM symmetrical† Cl^-) in mouse spinal cord neurones have been quoted[6] as **27–30 pS** (main† conductance level, responsible for > 95% of the current through the channel) with a subconductance† level range of **17–19 pS** and an infrequent **11–12 pS** level. With **thiocyanate** as the permeant ion†, the main conductance level† reduces to 22.5 pS. *Note:* With *mixtures* of chloride (84%) and thiocyanate ions (16%), main conductance states reduce to ~12 pS. This **anomalous mole fraction**† effect has been explained by invoking two *interacting* binding sites in each channel.

Single-channel conductance studies of native channels – examples

10-41-02: GABA-activated ion channels determined by **power spectra**† measurements in cultured rat cerebellar neurones display single-channel conductances† of ~22.7 pS (which are Zn^{2+}-insensitive) and ~ 14 ± 2 pS (which are picrotoxinin-insensitive). With **high GABA concentrations** (~10–2000 μM), the main state conductance† of native† $GABA_AR$ in superior cervical ganglion neurones is ~30 pS with other (less frequently observed) levels at 15–18 pS, 22–23 pS, together with (less well-defined) levels of 33–36 pS and 7–9 pS[112].

'Diagnostic' single-channel conductance levels of recombinant channels

10-41-03: The main conductance level of $GABA_AR$ channels has been shown to vary with subunit type composition. For example, $\alpha_1\beta_1$ and $\alpha_1\beta_1\gamma_{2S}$ heteromultimers† transiently† expressed in mouse L929 fibroblast† cells display single-channel openings to both main conductance† (15 and 29 pS respectively) and subconductance† levels (10 and 21 pS respectively), with greater than 90% of the total current through the main conductance level openings[63]. These kinetic† properties have been used to determine **assembly patterns** of heteromultimeric† $GABA_ARs$ *(see Domain arrangement, 10-27)*. Single-channel analyses for recombinant $\alpha_1\beta_1$ $GABA_A$ channels in chinese hamster ovary (CHO) cells[106] have shown the kinetic† properties of the $\alpha_1\beta_1$ main conductance level to *differ* from those of the native† spinal cord neurone 27 pS main conductance† level and the 19 pS subconductance† level[106] *(see also Kinetic model, 10-38)*.

Steroid modulation of single $GABA_A$ channel properties

10-41-04: Neurosteroid regulation of $GABA_A$ receptor **single-channel** kinetic properties have been systematically studied *(see Channel modulation, 10-44)*.

Comparison with 'retinal-type' GABA-activated channels

10-41-05: The novel GABA receptor–channels expressed in **retinal bipolar cells**[83] (**$GABA_C$** or '**non-$GABA_A$/$GABA_B$**'- *see Blockers, 10-43*) display a single-channel conductance† of ~7.4 pS with an open time† of ~150 ms (range 120–180 ms) in symmetrical† Cl^- concentrations of 145 mM.

Voltage sensitivity

Sensitivity to ionic driving force

10-42-01: $GABA_A$ currents depend mainly on the **electrochemical driving force**†, i.e. on the resting potential† and the Cl^- gradient† *(see Fig. 3 and*

Current type, 10-34). In rat locus coeruleus (LC) neurones, the time constant† of GABA current decay is decreased by membrane hyperpolarization†, possibly due to a voltage-dependent† change in receptor or channel

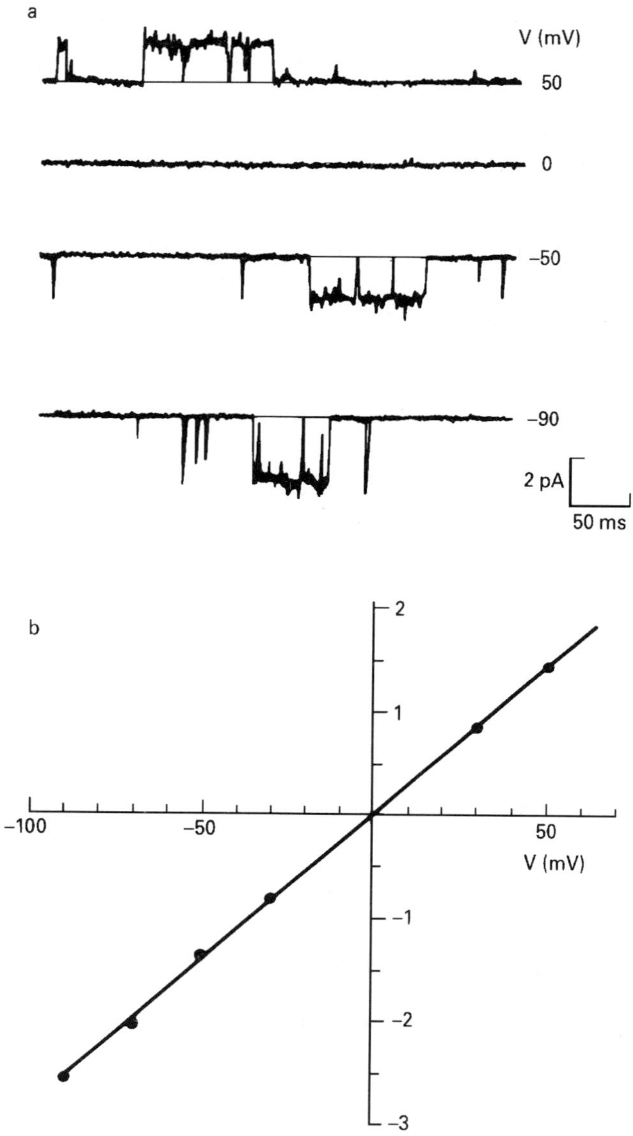

Figure 3. *Dependence of GABA$_A$ receptor–channel current on electrochemical driving force. (Reproduced with permission from Bormann (1987)* J Physiol Lond ***385****: 243–86.) (From 10-42-01)*

kinetics[109]. GABA exerts a stronger inhibitory effect on LC neurones at depolarized† than at hyperpolarized† membrane potentials, which could serve as a **negative feedback** mechanism to control **neuronal excitability**†[109]. An example of current 'direction' being influenced by driving force is the excitatory†, depolarizing† responses to GABA seen in **glial cells** *(described under Cell-type expression index, 10-08).*

PHARMACOLOGY

*Note: A comprehensive treatment of $GABA_A$ receptor pharmacology is difficult for several reasons. The large volume of available information cannot be adequately covered in a reasonable space, leading to problems of choice over what to include and omit. This section therefore reports an overview, with citation of specialist reviews. More emphasis has been placed on the **molecular aspects** of $GABA_A$ and less on **behavioural effects**, although these are outlined. Compound names are also listed in Resource C – Compounds and proteins, entry 58. For a diagrammatic summary of the important pharmacological modulatory 'sites' including those acting as **ionic pore blockers**, see Channel modulation, 10-44.*

Blockers

10-43-01: For a summary of $GABA_A$ receptor–channel blockers, *see Table 4.*

Presence of a γ subunit determines Zn^{2+}-insensitivity of recombinant $GABA_A$ complexes

10-43-02: γ subunit data**:** The presence of a **γ subunit** appears to confer **insensitivity to Zn^{2+} ion block**: Receptors formed from α_1 and β_2 subunits are *sensitive* to Zn^{2+} (half-blocking concentration ~0.56 μM) whereas those formed from α_1, β_2 *and* γ_2 subunits are *insensitive* to Zn^{2+} blockade (10 μM Zn^{2+} for 2 min, plus 10 μM GABA in the presence of 10 μM Zn^{2+})[117]. This **subunit-dependent pattern** of Zn^{2+} blockade is illustrated in Fig. 4.

Novel characteristics of retinal bipolar cell GABA receptor–channels

10-43-03: GABA receptor–channels expressed in rat **retinal bipolar cells** have been characterized as *insensitive* to $GABA_A$ antagonists (e.g. bicuculline) and $GABA_B$ agonists (e.g. baclofen) but can be selectively activated by the GABA analogue ***cis*-4-aminocrotonic acid (CACA)**. These novel channels have a single-channel conductance† of 7 pS and an open time† of 150 ms. *Note:* '$GABA_C$'-type receptors have also been characterized as **bicuculline- and baclofen-insensitive** GABA receptors[123]. The retinal bipolar cell channels[83] are also *not* sensitive to modulation by flunitrazepam, pentobarbital and alphaxalone and are only 'slightly blocked' by picrotoxinin *(for the association of these 'non-$GABA_A$/$GABA_B$' with **ρ_1 subunit expression**, see also Domain functions, 10-29).*

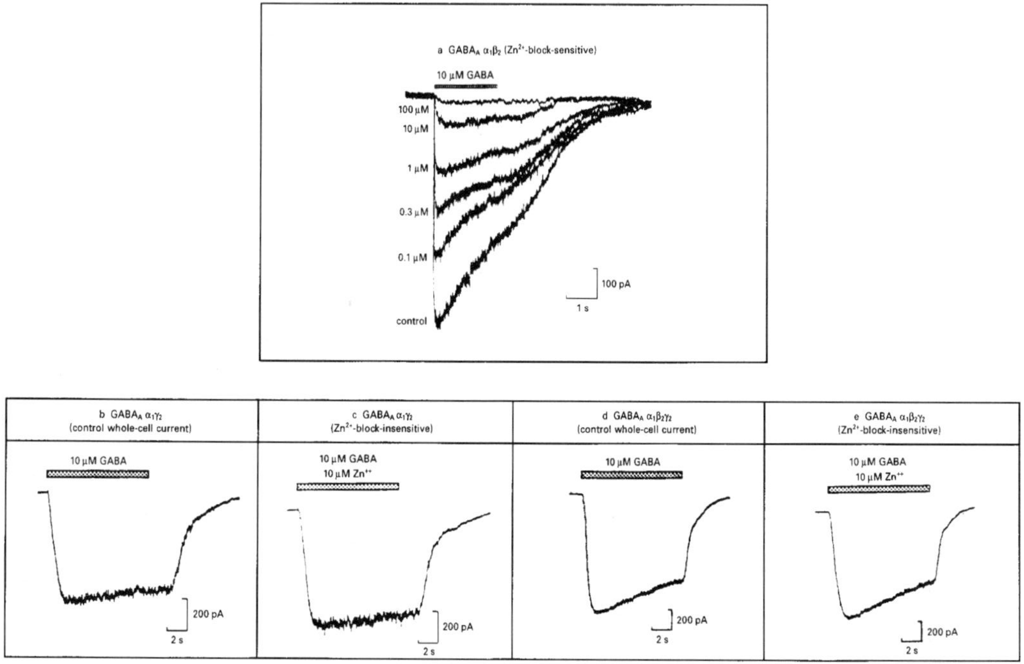

Figure 4. *Presence of a γ subunit determines Zn^{2+}-insensitivity of recombinant $GABA_A$ complexes. (a) Whole-cell currents elicited by GABA (10 μM, indicated by bar) in fibroblast cells expressing only **$GABA_A$ α_1 and β_2 subunits** display **sensitivity** to increasing micromolar concentrations of Zn^{2+}. (b–e) $GABA_A$ receptors reconstituted from subunit combinations which **include a γ subunit become insensitive to Zn^{2+}-block**. For example, $\alpha_1\gamma_2$ (compare panels b and c) or $\alpha_1\beta_2\gamma_2$ (compare panels d and e).* Note: *The control traces in b and d show whole-cell inward currents recorded following application of 10 μM GABA. The traces in c and e show currents recorded in the same cells following equilibration of the bathing solution with 10 μM Zn^{2+} for 2 min and application of 10 μM GABA in the presence of 10 μM Zn^{2+}. (Traces reproduced with permission from Draguhn (1990)* Neuron, **5**: *781–8.) (From 10-43-02)*

Table 4. *Summary of GABA$_A$ receptor–channel blockers (From 10-43-01)*

Blocker	Description and examples	Refs
Bicuculline block	**Bicuculline** is a blocker of GABA$_A$ channels which reduces GABA IPSPs† and in behavioural studies, can cause **convulsions**† in a dose-dependent manner	*See Receptor antagonists, 10-51*
Picrotoxinin block	**Picrotoxinin** can *both* block the chloride receptor–channel and elicit outward currents *(see also Domain functions, 10-29, and Channel modulation, 10-44)*. In behavioural studies, 'post-training blockade'† of the GABA$_A$-related chloride channel has been reported to enhance **memory**	***113***
Divalent cation block by Zn^{2+}	Antagonism† of GABA-elicited chloride currents in clutured rat cerebellar neurones to Zn^{2+} ions declines by 50% ~48 days after birth. Zn^{2+} at 25–100 μM does *not* affect the single-channel conductance, nor does it induce any rapid closure/blocking events. A novel GABA$_A$ **Zn^{2+}-binding site** has been proposed for some (but not all) GABA$_A$R *(see [PDTM], Fig. 1)*. The binding site is predicted to be within an area of 'high hydrophobicity contrast' (probably at a site located on the **extracellular part** of the GABA$_A$ receptor–channel complex). Zn^{2+} ions appear to bind to an 'un-liganded' or 'mono-liganded' state of the GABA$_A$R, which stabilizes the channel in the **closed conformation**†. *Functional notes:* 1. Zn^{2+} ions are good candidates for modulators of neuronal excitability† – they are present at high concentrations in brain (especially the synaptic vesicles of mossy fibres† in the hippocampus) and are released during neuronal activity[114, 115]. 2. Both **pro-convulsant**† and CNS-depressant† actions of Zn^{2+} have been reported: Zinc is a potent non-competitive† antagonist of NMDA responses on cultured hippocampal neurones[115]. Thus, Zn^{2+} antagonism of GABA$_A$ responses may underlie observed pro-convulsant activity while Zn^{2+} antagonism of NMDA responses may underlie their CNS-depressant properties. *See also ELG CAT GLU NMDA, entry 08, and paragraph 10-43-02*	***116***

Table 4. *Continued*

Blocker	Description and examples	Refs
Divalent cation block by Co^{2+}, Cd^{2+}, Mn^{2+} and Ni^{2+} (but *not* Ca^{2+}, Mg^{2+} and Ba^{2+})	Some native† cell preparations (e.g. turtle cone photoreceptors) exhibit GABA-activated currents which are blocked by **Co^{2+}**, **Cd^{2+}** and **Ni^{2+}** at micromolar concentrations. These ions also block GABA-activated currents from channels reconstituted from *purified* α_1 and β_2 subunit proteins, but in these cases **Zn^{2+}** is more effective *(see above and ref.[117])*. The divalent cations Ca^{2+}, Mg^{2+} and Ba^{2+} appear to have little or no blocking activity at similar concentrations to those described above	*118* *See also Receptor antagonists, 10-51.*
Penicillin block	There is some evidence for the antibiotic penicillin G acting as an **open-channel blocker**† at the $GABA_A$ receptor. Penicillin G reduces average channel open-duration† and increases average burst† duration without altering single-channel conductance. *Notes:* 1. Penicillin is **negatively charged** at physiological pH, and may therefore interact with positively charged amino acids within the channel vestibule *(see Domain functions, 10-29)*. 2. At high doses, penicillin G acts as a **convulsant** and can inhibit picrotoxinin binding. 3. Penicillin G can also induce potentiation and block of currents through the **glycine receptor-channel** *(see Channel modulation under ELG Cl GLY, 11-44)*	*119, 120*
Ro54864 block	$GABA_A$ channels are blocked at *micromolar* levels by the 4′-chloro-derivative of diazepam **Ro54864** (which is *also* an *agonist* at *nanomolar* concentrations for non-$GABA_A$ **'peripheral-type'**† benzodiazepine receptors – *see footnote to Table 6 under Ligands, 10-47*). On $GABA_A$ **'central-type'** receptors (this entry) Ro54864 occurs at a site close to (but *not* identical to) the picrotoxinin site *(see Channel modulation, 10-44)*. In behavioural studies, Ro54864 can cause convulsions† and reportedly, facilitate memory-formation. The effects of Ro54864 can be antagonized† by the isoquinoline carboxamide **PK11195**, which is generally used as a specific 'peripheral†-type' GABA receptor antagonist. In addition to PK11195, the phenylquinolines **PK8165** and **PK9084** appear to modulate the $GABA_AR$ by binding at the Ro54864 site	*121*
Anti-secretory factor (ASF) block	The **antisecretory factor** (ASF) has been reported to non-selectively block neuronal chloride channels including those activated by GABA	*122*

Channel modulation

A diagrammatic summary of characterized modulatory 'sites' on GABA$_A$ receptor–channels

10-44-01: Native† GABA$_A$Rs contain specific **binding sites** for multiple classes of pharmacological and (putatively) natural modulators of receptor–channel function. As noted in Fig. 5 and Table 5, a large number of studies have helped organize these 'sites' into **'discrete'** or **'overlapping' structural/functional categories** (tentatively assumed to be at different loci on the intact receptor–channel protein complex). Apart from **direct effects** on the channel molecular complex (e.g. by open-channel blockers), modulation may frequently occur via **allosteric† modulation†** of GABA-binding properties (conversely, binding of modulator molecules may be allosterically modulated by GABA itself). For details of the types of functional changes observed following application of various agonists†, antagonists†, channel blockers† and other modulators, *see below and the fields indicated in Fig. 5.*

GABA$_A$ channel modulation by widely used CNS-depressant drugs

10-44-02: Chloride flux† through the activated GABA$_A$ receptor–channel is subject to **allosteric modulation†** by several **CNS-depressant† drugs** including the **benzodiazepines**, including the anxiolytic† **diazepam (Valium)** and **flunitrazepam**. Fluctuation analysis† suggests these drugs increase the effect of GABA$_A$ agonists† (i.e. increase chloride ion flux†) by increasing the **frequency of channel openings**[124] without altering mean open time† or single-channel conductance†. It has also been proposed that benzodiazepines enhance GABAergic† function by increasing **receptor–ion channel coupling**, rather than by increasing GABA$_A$ receptor affinity† for its agonist†[125]. *Notes:* 1. It is possible (though unproven) that benzodiazepines reduce the rate of entry of a desensitized† state of the GABA$_A$R without altering gating†. 2. Benzodiazepines are also used clinically in **anti-epileptic†**, **muscle relaxant** and **hypnotic** regimes. 3. For a description of the distribution of benzodiazepine binding sites and benzodiazepine-like epitopes† in brain, *see Protein distribution, 10-15.* 4. Distinctions between the GABA$_A$-related **'central'-type** benzodiazepine receptors (as described in this entry) and the GABA$_A$-unrelated **'peripheral'-type** benzodiazepine receptors are outlined in the footnote to Table 6 under *Ligands, 10-47.*

Association of endogenous† benzodiazepines and memory processes

10-44-03: In behavioural studies, several modulators of GABA$_A$ receptor function have been reported to 'down-regulate' **memory†** and may reflect channel modulation by **endogenous benzodiazepines** *(for a review, see ref.*[121]*).* In humans, synthetic benzodiazepines in clinical use and **muscimol** (a structural analogue of GABA) have also been reported to depress memory functions.

Other GABA$_A$ modulatory sites

10-44-04: In addition to benzodiazepine sites, several distinct drug-binding sites[126] of GABA$_A$ receptors have been characterized, each with its own set of agonists† and antagonists†, as summarized in Table 5. Broadly, these sites can be classified as those for **barbiturates**, **GABA**, and the **chloride**

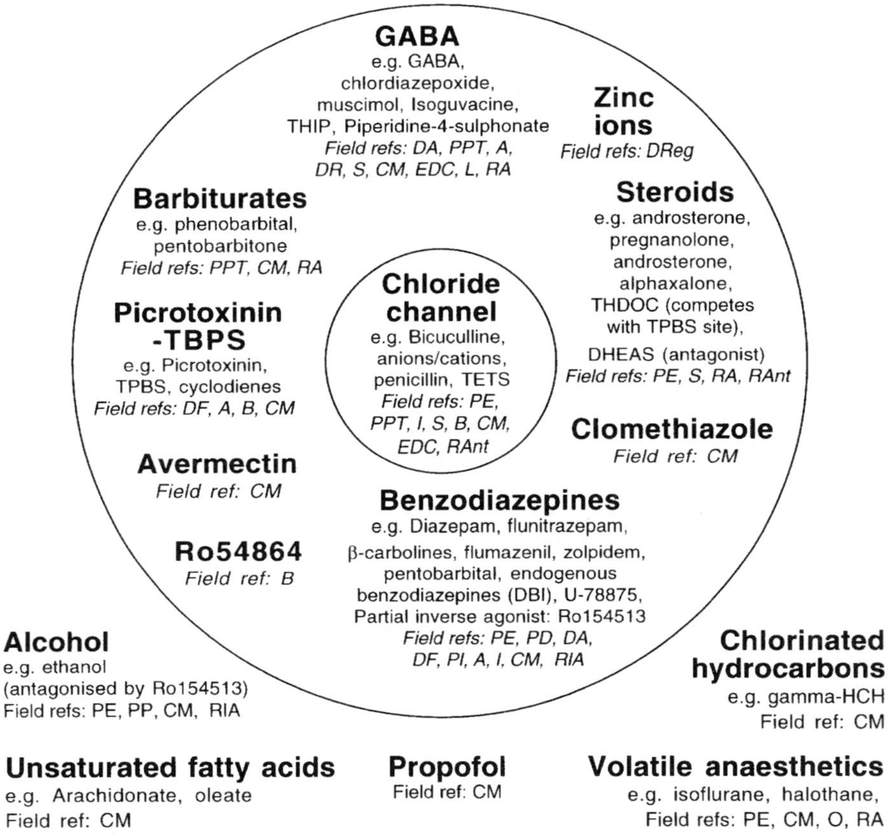

Figure 5. *Principal binding sites for endogenous and pharmacological modulators of $GABA_A$ receptor–channel function. Key to fieldname references: A, Activation, 10-33; B, Blockers, 10-43; CM, Channel modulation, 10-44; DA, Domain arrangement, 10-27; DF, Domain functions, 10-29; DR, Dose–response, 10-36; DReg, Developmental regulation, 10–11; EDC, Equilibrium dissociation constant, 10-45; I, Inactivation, 10-37; L, Ligands, 10-47; O, Openers, 10-48; PD, Protein distribution, 10-15; PE, Phenotypic expression, 10-14; PI, Protein interactions, 10-31; PP, Protein phosphorylation, 10-32; PPT, Predicted protein topography, 10-30; RA, Receptor agonists, 10-50; RAnt, Receptor antagonists, 10-51; RIA, Receptor inverse agonists, 10-52; S, Selectivity, 10-40. (From 10-44-01)*

Table 5. *$GABA_A$ receptor–channel modulation – general notes, examples and references (see also* Receptor agonists, 10-50 *and* Receptor antagonists, 10-51). *(From 10–44-01)*

Modulator class	
Barbiturate modulation (general notes)	The **barbiturate** drugs, e.g. **phenobarbital** which at ~50 μM potentiates† (enhances) the inhibitory action of GABA by increasing average channel open durations† for the subunit combination $\boldsymbol{\alpha_1, \beta_1, \gamma_2}$, heterologously† expressed in embryonic kidney cell lines. Barbiturates both allosterically† enhance **GABA binding** and can **mimic GABA** in its absence. Barbiturates in clinical use have **sedative-hypnotic**†, **anaesthetic**† and **anti-convulsant**† properties. **Pentobarbitone** can directly activate $GABA_A$ Cl^- channels at high concentrations (⩾50 μM). *Notes:* 1. Related *non*-barbiturates such as the anaesthetics **etazolate**, **etomidate** and **LY81067** also enhance $GABA_A$ currents as measured by radiotracer† ion flux or electrophysiological methods. 2. 'Active' barbiturates can allosterically† **inhibit the binding of antagonists** (picrotoxinin/benzodiazepine antagonists) and can also enhance binding of other agonists (benzo-diazepine inverse agonists/GABA-like agonists)
Benzodiazepine modulation	*See paragraphs 10-44-04 and 10-44-05 and Receptor agonists, 10-50*
'Neurosteroids' and **steroid anaesthetic modulation**	The **neurosteroids**, commonly 3α-hydroxy ring A-reduced metabolites of progesterone and deoxycorti-costerone (e.g. **3α-OH-DHP**, 3α-hydroxy-5α-pregnan-20-one or **THDOC**, 3α-,21-dihydroxy-5α-pregnan-20-one), compete with the high-affinity† binding site of the convulsant **TBPS** (*tert*-butylbicyclophosphorothionate) to potentiate† GABA-activated Cl^- currents in synapto-neurosomal preparation from rat cerebral cortex. Neurosteroid modulation occurs at nanomolar concen-trations and can directly elicit bicuculline-sensitive Cl^- currents at micromolar concentrations *(see also paragraph 10-44-08).* The steroid† anaesthetic† **alphaxalone** (3α-hydroxy-5-α-pregnane-11,20-dione) increases amplitude and duration of $GABA_A$ chloride current in cultured spinal cord neurones and can directly activate channels at high concentration. The mechanism of modulation by **pregnane steroids** appears distinct from that of **barbiturates**[127]. By entropy/enthalpy calculations, the anaesthetic $GABA_A$ channel modulators **alphaxalone**, **propofol** and **pentobarbitone** have been determined to interact with the $GABA_A$ receptor at *distinct* recognition sites[128]

Table 5. *Continued*

Modulator class	
Picrotoxinin modulation	The non-competitive†, non-selective $GABA_AR$ antagonist **picrotoxinin** (a non-nitrogenous convulsant† of plant origin) reduces GABA IPSPs†. Picrotoxinin binds preferentially to agonist-bound forms of the receptor at a site distinct from GABA and appears to stabilize an **agonist-bound shut state**, enhancing the occurrence of a desensitized† state or an allosterically† blocked state. At concentrations which significantly reduce the amplitude of whole-cell GABA currents, picrotoxinin does *not* alter spectral time constants† or single-channel conductance in dissociated rat sympathetic neurones (as estimated by current noise analysis†)[129]. In behavioural studies, picrotoxinin can induce convulsions†. Other drugs acting at the picrotoxinin site include **tutin** (natural), the synthetic bicyclophosphate 'cage' compounds (e.g. **TBPS**, *see above*), synthetic polycyclic convulsants (e.g. **pentylentetrazole, PTZ**) and 'convulsant benzodiazepines' (e.g. **Ro5-3663** and **Ro5-4864**, which are *not* active at the 'benzodiazepine site'). *Note:* Picrotoxinin is able to induce **complete block** of GABAergic† transmission, and can close $GABA_A$ channels that have been opened by steroids, barbiturates or avermectin. Picrotoxinin-binding appears to require a **multi-subunit complex** and picrotoxinin may act as a **slow channel blocker**† or else alter the **intrinsic gating**† of the channel once GABA is bound
β-Carboline modulation	The **β-carbolines**, e.g. methyl- (β-CCM), ethyl (β-CCE) and propyl (β-CCP) esters of β-carboline-3-carboxylate have **anxiogenic† effects** by reducing the action of GABA agonists. Several β-carbolines act as **full or partial inverse† agonists** at the benzodiazepine receptor site *(for further details see Receptor inverse agonists, 10-52)*. Methyl 6,7-dimethoxy-4-ethyl-β-carboline-3-carboxylate (**DMCM**) 'negatively modulates' chloride currents elicited by GABA in granule cells but 'positively modulates' $GABA_A$ currents in astrocytes. This **differential modulation** is also observed with recombinant $GABA_A$ receptors containing a γ_1 instead of a γ_2 subunit. Quantitative mRNA determinations in both cell types suggest key molecular determinant responsible for DMCM-positive modulatory effects in astroglial native $GABA_ARs$ is the presence of the **γ1 subunit** in the assembled receptor complex[38] *(for other comparative findings, see mRNA distribution, 10-13)*
Volatile gas anaesthetic modulation	$GABA_A$ receptor complexes serve as common but generally non-selective 'molecular target' sites for a variety of structurally diverse **volatile anaesthetic molecules**[46] (e.g. **isoflurane**, **halothane**, and **enflurane**). A brief description of $GABA_A$ current elicitation following administration of volatile anaesthetics appears under *Phenotypic expression, 10-14*

Table 5. *Continued*

Modulator class	
Intravenous anaesthetic modulation (propofol)	The intravenous general anaesthetic **propofol** (2,6-diisopropylphenol)[130, 131] may exert part of its effect through $GABA_AR$: At low doses, propofol dose-dependently potentiates† GABA-activated currents while at high doses it is capable of directly activating bicuculline-sensitive chloride flux. Propofol may act at a site distinct from other sedative-hypnotic sites (barbiturate/benzodiazepine/ GABA) *(for brief review, see ref.[5])*
Avermectin B_{1a} modulation	The macrocyclic lactone **avermectin B_{1a}** (AVM) isolated from *Streptomyces avermitilis* has potent **anti-helminthic** and **insecticidal** properties. AVM and its structural analogues **increase chloride ion permeability**† of vertebrate and invertebrate nerve and muscle membranes. When applied following GABA agonism†, AVM can act either as an antagonist† or as a synergist†. High-affinity AVM-binding sites exhibit a series of complex allosteric interactions with *other* (distinguishable) binding sites for benzodiazepines, barbiturates and picrotoxinin-TBPS *(briefly reviewed in ref.[5])*
Ethanol modulation	Ethanol modulation of the $GABA_AR$ (generally a potentiation† of GABA responses) may occur at some receptor subtypes but not at others. Ethanol modulation of $GABA_AR$ resembles some properties typical of barbiturate and benzodiazepine modulation (anticonvulsant†, anxiolytic† and sedative functions†). *For further details, see paragraphs 10-44-05 to 10-44-15 and Fig. 6*
Chlorinated hydrocarbon modulation	Certain chlorinated hydrocarbons such as **hexachlorocyclohexanes** are potent inhibitors of $I_{G\text{-}Actx}$ (typical $GABA_A$ current expressed from cerebral cortex mRNA in oocytes); e.g. with γ-hexachlorocyclohexane (**lindane** or **γ-HCH**) current suppression is detectable at concentrations as low as 50 nM[1]. By contrast, **α-HCH** and **δ-HCH** induce clear positive modulation of $I_{G\text{-}Actx}$ elicited by low (e.g. ~10 μM) concentrations of GABA. Atypical GABA receptors (bicuculline/ baclofen-insensitive $I_{G\text{-}Aret}$ expressed by retinal RNA in oocytes) resemble $GABA_A$ receptors in their sensitivity to γ-HCH but are largely insensitive to α-HCH and δ-HCH[1]
Unsaturated fatty acid modulation following liberation by **phospholipase A_2**	**Phospholipase A_2** (PLA_2) treatment of synaptosomal† membranes, which causes the release of unsaturated fatty acids *(see ILG K AA, entry 26)*, both inhibits chloride ion flux through $GABA_A$ channels in response to activation by agonists[132] and affects ligand binding to the receptor. **Arachidonic acid** and **oleic acid** has been shown to mimic the effect of PLA_2 treatment by enhancing **flunitrazepam** and **muscimol** binding but inhibiting binding of the 'neurosteroid' **TBPS** (*t*-butylbicyclophosphorothionate) in a dose-dependent manner[132]. *Note: External application* of a purified (neurotoxic) PLA_2 from the venom of the taipan snake to chick sensory neurones reversibly induces a picrotoxin-sensitive (DIDS-insensitive) anion current in a voltage-dependent† manner[133]

Table 5. *Continued*

Modulator class	
Clomethiazole modulation	The anxiolytic, anticonvulsant and sedative-hypnotic **clomethiazole**[134, 135] may exert part of its effect at the $GABA_AR$ complex. At low-doses, clomethiazole dose-dependently potentiates† GABA-activated currents while at high doses is capable of directly activating bicuculline-sensitive chloride flux. Clomethiazole may act at a site distinct from other sedative-hypnotic sites (barbiturate/benzodiazepine/GABA) *(for brief review, see ref.*[5]*)*
Chlordiaz-epoxide modulation	For details of the $GABA_A$-modulatory effect of the benzodiazepine **chlordiazepoxide** (where **one molecule** of bound GABA becomes sufficient to open the channel)[103], *see Dose–response, 10-36, and γ subunit data, this field*
Anion modulation	All $GABA_A$ modulatory sites (and their interactions) show dependence on (and modulation by) the anions Cl^-, Br^-, I^-, NO_3^-, SCN^- or ClO_4^- (briefly reviewed in ref.[5])
Excitatory amino acid receptor antagonist (indirect modulation)	Monosynaptically evoked *inhibitory* post-synaptic currents in hippocampal pyramidal slices diminish in the presence of **CNQX** (an AMPA-receptor-selective antagonist, *see ELG CAT GLU AMPA/KAIN, entry 07*) and **APV** (an NMDA-receptor-selective antagonist, *see ELG CAT GLU NMDA, entry 08*) following a train† of action potentials. However, responses to GABA applied by iontophoresis† do *not* change significantly[136]
Post-synaptic intracellular calcium (indirect modulation)	Localized physiological changes in **post-synaptic intracellular calcium** (post-synaptic spike† firing) has been shown to potently modulate synaptic $GABA_A$ inputs (i.e. reduce responses to GABA) within CA1 cells of the hippocampus[136]: Following a train of action potentials, spontaneous IPSPs† are transiently suppressed, which can not be accounted for by membrane conductance changes following the train or activation of a recurrent circuit. Amplitudes of spontaneous $GABA_A$ inhibitory post-synaptic currents (IPSCs) are also diminished following the action potential train. The Ca^{2+} channel agonist **BAY K 8644** enhances the suppression of IPSPs†, while buffering Ca^{2+} changes in with EGTA or BAPTA prevents suppression[136]
Other indirect modulation of $GABA_A$ responses	Other indirect modulators of $GABA_A$ function include **glutamate** (potentiating), intracellular Ca^{2+} (in porcine pituitary intermediate lobe) *(extracellular Ca^{2+}-influx - independent)*[137], external **Na^+ concentration**[138] and some **purines** and **peptides**

Note: *Not all of the sites listed in this table and elsewhere in this entry are necessarily present on any one $GABA_A$ complex. The large variety of component $GABA_A$ subunits (see Gene family, 10–05) and observed 'cell-type-specific' patterns of modulation indicate there is much scope for selective modulation of $GABA_A$ receptor–channels according to subunit composition at the single-cell level.*

channel itself (possibly the **picrotoxinin site**, *see Table 5*) but may include many other discrete sites. Features of a provisional **GABA$_A$ receptor subtype classification** (GABA$_{A1}$, GABA$_{A2}$, GABA$_{A3}$) *based on the functional properties* of their **allosteric modulatory centres**† has been summarized[4–6]. *Note:* Updates on **IUPHAR Nomenclature Subcommittee recommendations** on GABA$_A$ receptor–channels will appear via the 'home page' of the Cell-Signalling Network (CSN) available over the World Wide Web on the Internet (for description, *see Feedback & CSN access, entry 12).*

Potentiation of GABA$_A$ currents by ethanol

10-44-05: The '**positive-modulatory**' effects of **ethanol** on GABA$_A$ currents are reversible (in chick cerebral cortical neurones). In ~60% of neurones, ethanol causes a **potentiation**† of the membrane current elicited by GABA (threshold concentration 1 mM, maximal effect at 10 mM). At **higher concentrations** (40–50 mM), ethanol inhibits GABA-activated currents[139] but this effect may be subunit-specific: In other synaptoneurosomal preparation from rat cerebral cortex, ethanol stimulates Cl^- flux to ~250%, saturating at ~50 mM). Ethanol sensitivity of GABA$_A$ receptors in *Xenopus* oocytes requires eight amino acids contained in the **γ_{2L} subunit**[140]. **Ro154513** *(see Receptor inverse agonists, 10-52)* can antagonize certain effects of ethanol. A commentary on **ethanol modulation** of extracellular ligand-gated receptor–channels including the GABA$_A$R has appeared[141]. The comparative sensitivity of these receptor–channels (and that for voltage-gated Ca^{2+}-flux) is illustrated in Fig. 6.

Recombinant GABA$_A$ receptor responses to high concentrations of ethanol

10-44-06: Replacement of the γ_{2S} subunit by the alternatively spliced† **variant γ_{2L}** in recombinant† GABA$_A$ channels in *Xenopus* oocytes significantly stimulates the GABA response to *high concentrations* of ethanol (> 60 mM)[144]. The response to ethanol may also be dependent upon the precise subtypes specified at different stages of developmental expression (in adult/foetal dorsal root ganglion neurones).

Selective suppression of GABA$_A$R subunit transcripts by continuous exposure to ethanol

10-44-07: Chronic ethanol exposure has been reported to reduce the level of mRNAs encoding the α_1, α_2 and α_4 (but *not* α_3) GABA$_A$ subunits in brain. *Note:* The decreased responses to GABA following chronic ethanol treatment may play a role in the **neuronal hyper-excitability**† associated with **ethanol withdrawal** *(cited in ref.[141])*.

Neurosteroids increase GABA$_A$ single-channel open times

10-44-08: Neurosteroid† regulation of GABA$_A$ receptor single-channel kinetic† properties have been studied in cultured mouse spinal cord neurones[145]. Averaged† GABA$_A$ receptor currents are increased in the presence of the steroids (e.g. with 2 μM GABA plus **androsterone** (5α-androstan-3α-ol-17-one 10 nM–10 μM) or **pregnanolone** (5β-pregnan-3α-ol-20-one, 100 nM–10 μM). GABA$_A$ receptor current amplitudes† (main conductance level ~28 pS, ~96% of evoked current plus subconductance level of ~20 pS) are *unchanged* by the steroids, but average **channel open durations**† are increased (due to an increased relative proportion of the two longer open-duration time constants†)[145]. Average durations of **channel bursts**† (groups of openings separated by closures > 5 ms) are also increased by neurosteroid application *(see also next paragraph)*.

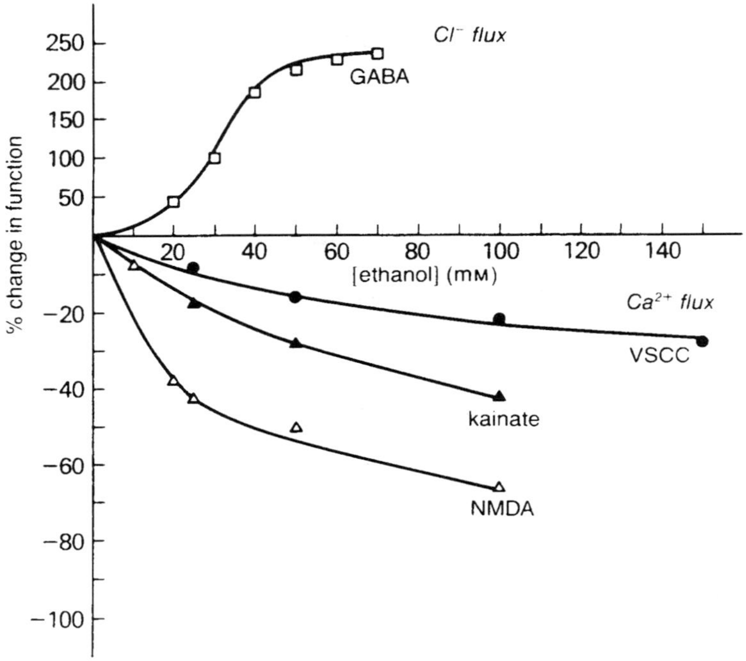

Figure 6. *Comparative modulatory effects of ethanol on ionic flux through GABA$_A$ receptor–channels (this entry), voltage-sensitive calcium channels (see VLG Ca, entry 41), kainate-selective glutamate receptor–channels (see ELG CAT GLU AMPA/KAIN, entry 07) and NMDA-selective glutamate receptor–channels (see ELG CAT GLU NMDA, entry 08). The open box symbols (□) describe ethanol stimulation of bicuculline-sensitive,* ***GABA$_A$-associated $^{36}Cl^-$ flux*** *from a rat synaptoneurosomal preparation from rat cerebral cortex measured in the absence of exogenous GABA agonists*[142]*. The open triangles (△) represent the effect of ethanol on* ***NMDA-stimulated $^{45}Ca^{2+}$-influx*** *in primary cultures of rat cerebellar granule cells*[143]*, while the closed triangles (▲) represent* ***kainate-stimulated $^{45}Ca^{2+}$-influx*** *in the same preparation*[143]*. The filled circles (●) show the effect of ethanol on* ***voltage-sensitive Ca^{2+} uptake*** *in a rat cortex synaptosomal preparation under depolarizing† conditions induced by 65 mM extracellular KCl (P.L. Hoffmann and B. Tabakoff, cited in ref.*[141]*). (Reproduced with permission from Gonzales (1991)* Trends Pharmacol Sci **12**: *1–3.) (From 10-44-05)*

Neurosteroid enhancement of $GABA_A$ channel subconductances
10-44-09: Androsterone and **pregnanolone** enhance averaged currents resulting from **subconductance levels** of $GABA_A$ channels. The mechanism for 'prolongation' of average open and burst† durations† is similar to that described for barbiturates, although significant changes in channel-opening frequency† has not been a prominent effect observed for barbiturates. These results suggest that **steroids** and **barbiturates** may regulate the GABA receptor–channel through at least one **common effector† mechanism**[145].

Subunit-selective modulatory responses – α subunit determinants
10-44-10: α subunit data: The 'exact' pharmacology of the **benzodiazepine response** (whether it conveys a benzodiazepine type I† or type II†-like pharmacology – *see ref.*[146]) varies according to the α **subunit type** expressed *(reviewed in ref.*[32]*; see also paragraphs in Domain functions, 10-29)*. The binding site for the partial inverse† benzodiazepine agonist **Ro154513** (ethyl-8-azido-5,6-dihydro-5-methyl-6-oxo-*4H*-imidazo[1,5*a*][1,4]benzodiazepine-3-carboxylate) is otherwise benzodiazepine-insensitive, and this binding site appears to require the α_6 **subunit**. Recombinant† receptors composed of α_6**,** β_2 **and** γ_2 **subunits** display a pharmacology similar to that of native† receptors immunoprecipitated by α_6 subunit antisera (i.e. high affinity† for the GABA structural analogue **muscimol** and **Ro154513**)[147]. Varying the α **subunit type** (with constant β and γ subunits) in complexes also affects **steroid modulation** of GABA responses, but does *not* apparently affect barbiturate, picrotoxinin or bicuculline sensitivity.

Subunit-selective modulatory responses – β subunit determinants
10-44-11: β subunit data: Replacement of a β_2 by a β_1 **subunit** in co-expressed combinations of $GABA_AR$ in the *Xenopus* oocyte expression system results in strongly increased **sensitivity to diazepam**[81]. The affinity of β_1 for GABA is usually greater than that for β_2. β subunits (β_1 or β_2) are *not* required for expression of GABA-gated ion currents displaying diazepam sensitivity[81]. The β_1, but *not* the β_2 subunit can form a picrotoxinin-sensitive, GABA-independent anion-selective channel, which can be suppressed by co-expression with any α subunit, but *not* by the γ_2 subunit *(for further information, see ref.*[81]*). Note:* β subunits are not *essential* for picrotoxinin block (as the complex $\alpha_3\alpha_5\gamma_2$ is picrotoxinin-sensitive).

Subunit-selective modulatory responses – δ subunit determinants
10-44-12: δ subunit data: Possibly associated with 'benzodiazepine-insensitive' receptors.

Subunit-selective modulatory responses – γ subunit determinants
10-44-13: γ subunit data: Native† γ **subunits** are necessary for **benzodiazepine sensitivity** of GABA receptor–channels (i.e binding and modulation – *see ref.*[3]). Heterologously† co-expressed $\alpha_1\beta_1$ **subunits lacking** γ respond to GABA but the response is only 'weakly and inconsistently' modified by benzodiazepines. Thus, although these *small* effects of benzodiazepines can be detected in the absence of γ_2 subunits in co-expression experiments[81], the presence of the γ_2 subunit is essential for *large* effects *(see also Dose–*

response, 10-36). For control of **ethanol sensitivity** by γ_{2L} **phosphorylation**, *see Protein phosphorylation, 10-32.*

Interactions of bretazenil and diazepam modulation

10-44-14: γ subunit data: *In vivo*, the modulatory efficacy of the anxiolytic† drug **bretazenil** is generally much lower than that of **diazepam**. In cultured HEK-293 (human embryonic kidney) cells transfected with γ_2 subunits and various isoforms† of α and β subunits, bretazenil efficacy is always lower than that of diazepam[148]. However, in cells transfected with γ_1 or γ_3 subunits and several isoforms† of α and β subunits, the efficacy of both diazepam and bretazenil is lower and always of similar magnitude. When bretazenil and diazepam are *applied together* to $GABA_A$ γ_2 subunit homomeric† receptor–channels, the action of diazepam is curtailed in a manner related to the dose of bretazenil[148].

Subunit-selective modulatory responses – ρ subunit determinants

10-44-15: ρ subunit data: ρ subunit expression is associated with **$GABA_C$** ('non-$GABA_A$/$GABA_B$) responses in **retinal bipolar cells** and is insensitive to bicuculline, barbiturates, and benzodiazepines and baclofen *(see Domain functions, 10-29).*

Equilibrium dissociation constant

Sites and subunits affecting affinity for GABA and its agonists

10-45-01: The **GABA-binding site** (GBS) on the $GABA_AR$ complex can be selectively labelled with radioligands† such as [^{3}H]-GABA or [^{3}H]-muscimol *(see Ligands, 10-47).* The GBS characteristically display both **'low-affinity'** labelling ($K_{d(GABA)}$ ~ μM range) and **'high-affinity'** labelling ($K_{d(GABA)}$ ~ nM range). The 'low-affinity' recognition site appears to be an **'antagonist-preferring site'** *(see ref.[149])* since it can also be selectively labelled by specific antagonists such as **(+)-bicuculline** or **SR95531** *(see Table 6 under Ligands, 10-47 and Table 8 under Receptor antagonists, 10-51).* The 'high-affinity' and 'low-affinity' GBS may represent **alternative forms** of the *same* receptor, since (i) *both* sites show similar drug specificities and (ii) 'positive modulators' such as **pentobarbital** (a benzodiazepine-site agonist† – *see Activation, 10-33)* increase the number of 'high-affinity' sites while *proportionally reducing* the numbers of 'low-affinity' sites *(for a review, see ref.[149]).* The 'high-affinity' GBS may represent a **desensitized† form** of the receptor.

10-45-02: β subunit data: In studies on cloned subunits expressed in *Xenopus* oocytes, replacement of the β_2 subunit by the β_1 subunit in a given subunit combination generally results in a decrease of $K_{a(GABA)}$. *Note:* For native $GABA_A$ receptors in cerebellar Purkinje cells, a **half-maximal response** is elicited by ~50 μM GABA.

Hill coefficient

10-46-01: Hill coefficients† for both vertebrate and insect $GABA_A$ receptors are generally greater than unity (e.g. 1.8 in native† cerebellar Purkinje cells or ~2 in cultured spinal cord neurone preparations), suggesting that at least two GABA molecules are required for activation. *(See also the modulatory effects of chlordiazepoxide under Dose–response, 10-36.)*

Ligands

10-47-01: Available radioligands† for the GABA$_A$ receptor are summarized in Table 6. **Ligand interactions** at the GABA$_A$ receptor complex have been reviewed[150]. Radioligands† for the **picrotoxinin/convulsant sites** and the **benzodiazepine sites** have been described, although there is a lack of radioligands† specific for the **steroid** or **barbiturate sites** (for indexing of data on these sites, *see Fig. 5*).

Table 6. *Available ligands for the GABA$_A$ receptor (From 10-47-01)*

Ligand type or binding site	Commercially available radioligand
GABA$_A$ competitive site	[^{3}H]-GABA (endogenous agonist) [^{3}H]-Muscimol, as used for photoaffinity† labelling of a GABA$_A$R 'β subunit' of ~56 kDa[a] [^{3}H]-SR95531 ([^{3}H]-2-(carboxy-3′-propyl)-3-amino-6-*p*-methoxyphenyl-pyridazinium bromide) [^{3}H]-Methyl-bicuculline methyl chloride (antagonist)
Benzodiazepine modulatory site (also known as ω-site)	[^{3}H]-Methyldiazepam (agonist) [^{3}H]-Methylflunitrazepam (~2 nM, agonist), as used for the original photoaffinity labelling of the GABA$_A$R in crude brain homogenates with a molecular weight of ~51 kDa[151] [^{3}H]-Flumazenil (0.9 nM); [^{3}H]-*N*-Methyl-PK11195 (antagonist, selective at '**peripheral**'† **BZ receptors**[b] [^{3}H]-7,9-Ro154513 (photoaffinity antagonist, selective at central† receptors). [^{3}H]-Methyl-Ro151788 (antagonist, selective at '**central**'† **BZ receptors**[b] [^{3}H]-Methyl-Ro54864 (agonist, selective at peripheral† receptors) [^{3}H]-Phenyl-2,6-Zolpidem (~7 nM, agonist, selective at central† BZ$_1$ receptors)
Other **non-competitive antagonist** sites	4-*tert*-butylbicycloortho[3′,4′-^{3}H-(2)]benzoate, 4′-cyano-4-*sec*-[3,4-^{3}H-(2)]butylbicycloorthobenzoate, and the 4′-ethynyl-4-n-[2,^{3}H-(2)3(2)]propylbicycloorthobenzoate[152]
Steroid modulatory site	[^{3}H]-Allopregnanolone ('neurosteroid' modulator)
GABA$_A$ chloride channels in open conformation	[^{35}S]-TBPS (4-*tert*-butylbicyclophosphoro-^{35}S-thionate, antagonist)

[a]Following photoaffinity-labelling and SDS–PAGE† analysis, several GABA$_A$-selective radioligands can discriminate 'microheterogeneous' protein bands from crude tissue preparations (according to solubilization and binding

continued on p344

continued from p343
conditions). This type of result is consistent with the existence of multiple ligand-binding subunits being encoded by a large related gene family.
[b]'Peripheral' benzodiazepine receptors are binding sites that are localized to the **outer mitochondrial membrane** of many tissues (including brain). 'Peripheral' benzodiazepine receptors are pharmacologically distinct from (and unrelated to) the $GABA_A$-associated **'central'** type benzodiazepine receptors, where the $GABA_A$ modulators (listed in *Channel modulation, 10-44, Receptor agonists, 10-50, and Receptor antagonists, 10-51*) exert their 'clinically relevant' effects. Notably, the 'peripheral type' BZ receptors have a similar distribution to the **VDAC (voltage-dependent anion channel**) isolated from mitochondria of various sources[90] (for further details, see *Protein interactions, 10-31, and MIT (mitochondria)*, entry 37. Some compounds listed in this table can distinguish between peripheral and central BZ receptor types in radioligand† binding assays (as indicated).

High-affinity ligands for benzodiazepine receptors
10-47-02: Imidazo(1,5-α)quinoxalin-4(5*H*)-one, 3-(5-cyclopropyl-1,2,4-oxadiazol-3-yl)-5-(1-methylethyl) or **U-78875** is one of a series of **imidazoquinoxaline-derivative ligands** which show high affinity for benzodiazepine receptors. A structure–activity study for U-78875 has been published[153].

Openers

10-48-01: *See Receptor agonists 10-50.* For effects of **volatile anaesthetics**† on elicitation of $GABA_A$ currents, *see Phenotypic expression (above).*

Receptor/transducer interactions

Distinction of $GABA_A$ and $GABA_B$ receptor subtypes
10-49-01: $GABA_A$ responses should not be confused with those from **$GABA_B$** receptors, which do *not* form integral ion channels but are coupled to Ca^{2+} or K^+ channels via G protein transducers *(see Resource A – G protein-linked receptors, entry 56 and refs.*[160–162]*)*. Activation of $GABA_B$ receptor proteins can inhibit the function of $GABA_A$ receptors, for example in the cerebellum[163]. Note also that '**$GABA_C$**' receptor-mediated responses (bicuculline- and baclofen-insensitive GABA receptors) have been characterized in retinal horizontal cells and linked to **GABA rho subunit** expression[84, 123, 164] *(see also Blockers, 10-43).*

Neuromodulator receptor potentiation of retinal $GABA_A$ responses
10-49-02: Vasoactive intestinal polypeptide (**VIP**) plays a neuromodulatory role in bipolar cells and ganglion cells of the rat retina by potentiating† $GABA_A$ responses[165]. Whereas VIP alone elicits no current response, VIP potentiation† of $GABA_A$ chloride currents controls the excitability† of inner retinal neurones and thus can modulate the efficacy of synaptic transmission in the retina[165] *(see also Abstract/general description, 10-01).*

Muscarinic receptor stimulation increases inhibitory post-synaptic potentials
10-49-03: The muscarinic receptor agonist **carbachol** (10–25 μM) can be used

experimentally to increase GABA-mediated IPSP frequency and simultaneously block 'potentially confounding' K^+ conductances during analysis of $GABA_A$ receptor-mediated processes[136].

Receptor agonists

Note: A diagrammatic summary of the important pharmacological modulatory 'sites' including those acting as ***receptor agonists*** is included under ***Channel modulation***, *10-44.*

Endogenous agonist activation of $GABA_AR$

10-50-01: By definition, $GABA_AR$–chloride channel gating is regulated by the binding of **gamma-aminobutyric acid** (GABA, and its analogues) to the extracellular GABA receptor portion of the intact complex. Since only **micromolar concentrations** of GABA are necessary to activate the integral chloride channel[154] it is likely that GABA exerts its physiological effects at the 'low-affinity' GABA binding site *(for description, see Equilibrium dissociation constant, 10-45).* It has been estimated[155] that one '**inhibitory quantum**' of GABA opens ~12–20 chloride channels and that GABA agonist molecules bind only once to post-synaptic receptors.

10-50-02: The principal synthetic and naturally occurring agonists† for the $GABA_A$ receptor are summarized in Table 7. $GABA_AR$–channel currents are generally enhanced by three classes of 'CNS depressant' drugs: the **benzodiazepines**, the **barbiturates** and the **anaesthetic steroids** (**'neurosteroids'**). The properties of the binding sites for these modulators are indexed in Fig. 5 under *Channel modulation, 10-44 and Table 7.*

Table 7. *The principal synthetic and natural agonists for the $GABA_A$ receptor (From 10-50-02)*

Site or receptor type	Agonist (notes)
GABA competitive site agonists[a]	**Muscimol** (a naturally occurring structural analogue of GABA from the mushroom *Amanita muscaria*; which induces ~2-fold increases in open times† of $GABA_AR$–channels). Isoguvacine and muscimol have been shown to preferentially activate **subconductance**† **states**[156]. **Isoguvacine** (a synthetic GABA agonist; which induces ~0.5-fold increase in open times† of $GABA_AR$–channels). **THIP** (4,5,6,7-tetrahydroisoxazolo-[4,5-c]pyridin-3-ol, a rigid bicyclic synthetic analogue of muscimol/GABA, which generally induces ~0.5-fold increase in open times† compared to GABA). Different agonists affect gating† properties and do not alter the single-channel conductance *see also* **THDOC** in Table 5. **Piperidine-4-sulphonate**, a synthetic structural analogue of GABA

Table 7. *Continued*

Site or receptor type	Agonist (notes)
Benzodiazepine modulatory site agonists[b]	**Flunitrazepam** -- *see Channel modulation, 10-44* **Zolpidem**, a benzodiazepine type I† -selective imidazo-pyridine hypnotic drug **Abecarnil** (a partial agonist)
Barbiturate site agonists	**Pentobarbital** -- *see Activation, 10-33* **Phenobarbital** -- *see Channel modulation, 10-44.*
'Anaesthetic steroid receptor site' agonists	**Alphaxalone** – *see Table 5.* The development of alphaxalone as a 'steroid anaesthetic' was based on the observed sedative† effects of the naturally occurring steroid hormone **progesterone**. Notably, **progesterone** and its structural analogues, as well as metabolites of progesterone, **testosterone** and **corticosterone** (like alphaxalone) can enhance GABA$_A$ receptor function. These metabolites do *not* activate intracellular transcription factor†-like receptors, and so this group of molecules may act as *endogenous physiological modulators* of GABA$_A$ receptor–channels[6]. This interpretation is consistent with several behavioural observations following steroid drug use, namely variations in CNS pharmacology related to the sex of subjects, oestrus† and diurnal† rhythms and drug-dependence phenomena
GABA-binding agonists	**Protopine-hydrochloride** is a GABA-binding alkaloid activator of GABA-gated channels. *See also Phenotypic expression, 10-14,* for effects of volatile anaesthetics on elicitation of GABA$_A$ currents
GABA$_A$ autoreceptor† agonists[c]	**di-*N*-Naphthyl-GABA** (pD_2 7.4) **δ-Amino-laevulinic acid** **Progabide** **Fengabine**

[a]Taurine, glycine and baclofen do *not* have detectable agonist activity at GABA$_A$ receptors; e.g. see characterizations in rat/cat dorsal root ganglion neurones[157] and bovine adrenal medulla chromaffin cells[158].
[b]Distinctions between the GABA$_A$-related **'central'-type** benzodiazepine receptors (as described in this entry) and the GABA$_A$-unrelated **'peripheral'-type** benzodiazepine receptors are outlined in footnote[b] to Table 6.
[c]GABA$_A$ autoreceptors appear *insensitive* to neurosteroid modulation and to the GABA$_A$ antagonist SR95531.

'Non-discrimination' of inhibitory agonists by mutant glycine receptor–channels

10-50-03: Gamma-aminobutyric acid and D-serine 'efficiently' activate certain **mutant glycine receptor–channels**, demonstrating a requirement for **aromatic hydroxyl groups** in ligand† discrimination† at inhibitory† amino acid receptors[159] *(cf. Domain functions under ELG Cl GLY, 11-29).*

Receptor antagonists

*Note: For a diagrammatic summary of the important pharmacological modulatory 'sites' including those acting as **receptor antagonists**, see **Channel modulation**, 10-44. For agents acting specifically or partially to block chloride flux, see **Blockers**, 10-43.*

10-51-01: The principal synthetic and naturally occurring antagonists† for the GABA$_A$ receptor are summarized in Table 8.

Table 8. *Principal synthetic and 'natural' antagonists for the GABA$_A$ receptor (From 10-51-01)*

Class or type of antagonist	Description of action or physiological effect
Competitive site antagonists	**Bicuculline**, a convulsant alkaloid† of plant origin, is a potent competitive† antagonist (pA_2 6.0) of vertebrate GABA$_A$ receptors; often applied as the methiodide. Other antagonists acting at the same site include the quaternary nitrogen analogues of bicuculline, **pitrazepin** and the amidine steroid **RU135**[a]. **SR95531** (2-(carboxy-3′-propyl)-3-amino-6-*p*-methoxyphenyl-pyridazinium bromide) is a competitive antagonist which is a pyridazinyl derivative of GABA and therefore shares close structural similarity with it *(see paragraph 10-51-03)*. **SR42641**. ***d*-β-Hydrastine** (isocoryne, an alkaloid† which is approximately twice as potent as bicuculline)
Benzodiazepine modulatory site antagonists[b]	**Flumazenil** (Ro151788). Behavioural studies *(reviewed in*[121]*)* suggest that flumazenil may be antagonistic to naturally occurring benzodiazepines like β-carboline-3-carboxylate or various peptide ligands of central† benzodiazepine receptors; flumazenil is anxiogenic† at high doses
Picrotoxinin-TBPS site antagonists	Receptor sites for plant-derived convulsant† **picrotoxinin** (sometimes called picrotoxin) are closely associated with the chloride ion channel of the GABA$_A$R. Picrotoxinin can act as a **channel blocker** by **'steric hindrance†'** of agonist-induced fluxes and under some conditions can act to *elicit* chloride flux. *For further details, see Domain functions, 10-29, Blockers, 10-43, and Channel modulation, 10-44* **TBPS** (*t*-butylbicyclophosphorothionate) is a **bicyclic 'cage'† convulsant†** which (like picrotoxinin) does not displace benzodiazepines from their high-affinity binding sites. High-affinity TPBS binding may be associated with the **'closed' conformation** of the integral chloride channel. *For further details, see Table 5 under Channel modulation, 10-44)*

Table 8. *Continued*

Class or type of antagonist	Description of action or physiological effect
Steroid antagonists	**DHEAS** (dehydroisoandrosterone-3-sulphate), has been characterized as an antagonist in astrocyte† (glial) neurones[166]
Antibacterial/ non-steroidal anti-inflammatory agent interactions	*In vitro*, the GABA$_A$-antagonistic effects of **enoxacin** (a quinolone antibacterial agent) are potentiated† ~80-fold in the presence of **felbinac**, a non-steroidal anti-inflammatory drug. This may provide a mechanism for known **convulsant reactions** following concomitant administration of these classes of drugs *in vivo*[167]
Other antagonists with convulsant activities (*some non-selective compounds also block voltage-sensitive Cl$^-$ channels)	**TETS** (**tetramethylenedisulphotetraamide** (a non-competitive† GABA-gated Cl$^-$ channel antagonist; anti-platelet agent) **Flucybene** (an organofluorate convulsant) **Cloflubicyne** (organofluorate convulsant) **Propybicyphat***, **Isobicyphat***, **Mebicyphat***, **Etbicyphat*** (organophosphorous non-competitive antagonist compounds which can induce epileptiform seizures) Several chemically distinct classes of **insecticide** are non-competitive antagonists of GABA$_A$ receptors and include the **trioxabicyclooctanes**, **dithianes**, and **cyclodienes** **Pentylenetetrazole** (PTZ) In addition to their principal target of Na$^+$ channels, *(see VLG Na, entry 55)* neuroactive insecticides such as the **pyrethroids** and **DDT** can contribute to hyperactivity by suppressing GABA$_A$ channel complexes[168]

[a]For pre-synaptic GABA$_A$ **autoreceptors**†, **bicuculline** is an order of magnitude more potent (pA$_2$ 8.0) as an antagonist than at the 'classical' post-synaptic receptors. A derivative of bicuculline, **WAY100359** acts as a more potent antagonist (pA$_2$ 9.14) in some preparations.
[b]Distinctions between the GABA$_A$-related **'central'-type** benzodiazepine receptors (as described in this entry) and the GABA$_A$-unrelated **'peripheral'-type** benzodiazepine receptors are outlined in footnote[b] to Table 6 under *Ligands, 10-47.*

Discrimination of GABA$_A$ from glycine inhibitory chloride receptor–channel

10-51-02: Strychnine can be used to distinguish the GABA$_A$ receptor from the glycine receptor Cl$^-$ channels *(see ELG Cl GLY, entry 11)*. pA$_2$ values against the agonists† **muscimol** and **glycine** are 5.3 and 6.0 (neonatal) or 8.0 (adult) respectively. *Note:* Some antagonistic action of strychnine has been reported for the GABA$_A$ receptor[169].

Mapping agonist/antagonist-binding sites to α subunits by site-directed mutagenesis

10-51-03: α subunit data: GABA$_A$ α **subunits** are generally associated with the **agonist/antagonist-binding sites** of intact receptor–channel complexes[79]. An **α1** Phe64 Leu substitution strongly decreases the apparent affinity for GABA-dependent channel gating (from 6 μM to 1260 μM) when the α subunit is co-expressed with **β2** and **γ2** subunits[79]. Homologous mutations in **α5** subunits show similar phenotypes, but homologous mutations in **β2** and **γ2** result in intermediate and small shifts in potency† (EC_{50}†) respectively. Apparent affinities of bicuculline methiodide and SR95531 *(see Table 8)* are decreased 60- to 200-fold by these mutations in α subunit cDNA† coding regions†. *Comparative note:* Agonist†/antagonist† affinities† are largely unaffected upon introduction of the homologous† mutations† in β2 and γ2 cDNAs†, or upon mutation of a 'neighbouring' α1 amino acid (Phe65 to Leu)[79] *(for clarification and supporting data, see Table 3 under Domain functions, 10-29).*

GABA$_A$ receptor antagonist binding kinetics and mechanism of block

10-51-04: The binding kinetics† and relative affinities† to bovine brain GABA$_A$ receptors have been determined for 25 GABA$_A$ antagonists, including four radiolabelled compounds[152] *(see Ligands, 10-47).* The low association rate constants† for all ligands ($\leqslant 3 \times 10^7$ M^{-1} min^{-1} at 25°C) is consistent with a **slow transition** to a blocked receptor conformation upon binding of these channel blockers. The 'association rate-controlled' affinities for the **trioxabicyclooctanes** and **dithianes**[152] suggest an **induced-fit model** in which binding of the ligand initiates a conformational change in the receptor complex to the blocked state[152].

Receptor inverse agonists

Note: For a diagrammatic summary of the important pharmacological modulatory 'sites' including those acting as ***receptor inverse agonists****, see* ***Channel modulation****, 10-44.*

β-Carboline compounds acting as GABA$_A$ inverse† agonists†

10-52-01: Several **β-carbolines** not only antagonize the effects of GABA$_A$ agonists by occupying the benzodiazepine receptor sites but also induce pharmacological effects that are *opposite* those of the 'classical' benzodiazepine agonists (i.e. they are **inverse† agonists**). For example, the β-carboline **DMCM** (methyl-6,7-dimethoxyl-4-ethyl-β-carboline-3-carboxylate) decreases GABA-induced whole-cell currents although it acts at the benzodiazepine modulatory site. The inverse agonist **Ro194603** and **Ro154513** (ethyl-8-azido-5,6-dihydro-5-methyl-6-oxo-*4H*-imidazo[1,5α][1,4]benzodiazepine-3-carboxylate, a partial inverse† benzodiazepine agonist) can also antagonize† certain effects of **ethanol** *(for details of ethanol modulation, see Channel modulation, 10-44). Note:* Inverse agonists have revealed striking differences between **neuronal** and **glial (astrocytic)** GABA receptor–channel responses *(for details, see Cell-type expression index, 10-08).*

A putative peptide precursor for an endogenous† ligand acting at $GABA_A$ benzodiazepine sites

10-52-02: DBI (**diazepam-binding inhibitor**) is a peptide isolated from brain tissue and a precursor of putative **natural ligands**† of benzodiazepine recognition sites[4, 170]. DBI has an action similar to DMCM *(see paragraph 10-52-01 and Fig. 7).*

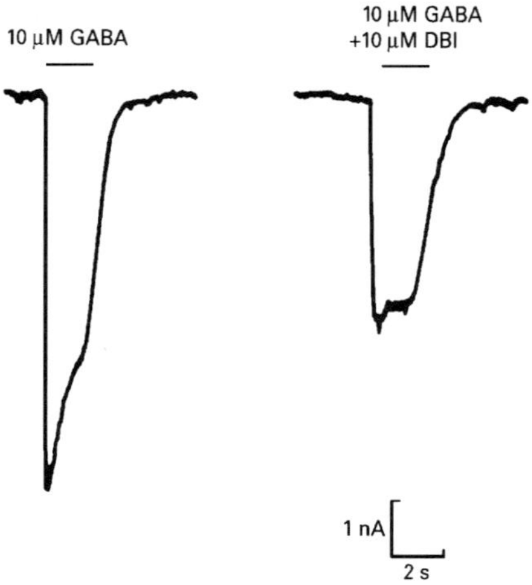

Figure 7. *Effect of brain-derived DBI (diazepam-binding inhibitor) peptide on native $GABA_A$ receptor–channel activity. (Reproduced with permission from Bormann (1985)* Regul Pept *(Suppl.)* ***4****: 33–8). (From 10-52-02)*

INFORMATION RETRIEVAL

Database listings/primary sequence discussion

10-53-01: *The relevant database is indicated by the lower case prefix (e.g. gb:); database accession numbers immediately follow the colon. Note that a comprehensive listing of all available accession numbers is superfluous for location of relevant sequences in GenBank® resources, which are now available with powerful in-built* ***neighbouring† analysis*** *routines (for description, see the Database listings field in the Introduction & layout of entries, entry 02). For example, sequences of cross-species variants or related gene family† members can be readily accessed by one or two rounds of neighbouring† analysis (which are based on pre-computed alignments performed using the BLAST† algorithm by the NCBI†). This feature is most useful for retrieval of sequence entries deposited in databases later than those listed below. Thus, representative members of known sequence homology groupings are listed to permit initial direct retrievals by accession number, author/reference or nomenclature. Following direct accession, however, neighbouring† analysis is strongly recommended to identify newly-reported and related sequences.*

GABA$_A$ α subunit database listings

Nomenclature	Species, DNA source	ORF[†] for original cDNA[†]	Accession	Sequence/ discussion
alpha – GABA(A) receptor alpha-subunit	Human	354 aa (partial)	gb: M22868 gb: X13584	Garrett, *Biochem Biophys Res Commun* (1988) **156**: 1039–45.
alpha-1 – GABA(A) receptor alpha-1 subunit; precursor	Chicken	455 aa	sp: 19150 em: X54244 pir: CHCHA1 prosite: PS00236	Bateson, *Mol Brain Res* (1991) **9**: 333–9.
alpha-1 – GABA(A) receptor alpha-1 subunit; precursor	Mouse	455 aa	gb: M86566 prosite: PS00236	Wang, *J Mol Neurosci* (1992) **3**: 177–84.
alpha-1 – GABA(A) receptor alpha-1 subunit; precursor	Bovine	456 aa	em: X05717 sp: P08219 pir: A27142 prosite: PS00236 *see also* pir: S06791	Schofield, *Nature* (1987) **328**: 221–7.
alpha-1 – GABA(A) receptor alpha-1 subunit; precursor	Mouse	455 aa	gb: X61430	Keir, *Genomics* (1991) **90**: 390–5.
alpha-1 – GABA(A) receptor alpha-1 subunit; precursor	Human	456 aa	sp: P14867 em: X13584 em: X14766 pir: A31588 pir: S03332 prosite: PS00236	Schofield, *FEBS Lett* (1989) **244**: 361–4. Garrett, K.M. (1988) *Biochem Biophys Res Commun* (1988)**156**: 1039–45.

$GABA_A$ α subunit database listings continued

Nomenclature	Species, DNA source	ORF† for original cDNA†	Accession	Sequence/ discussion
alpha-1 – GABA(A) receptor alpha-1 subunit; precursor	Rat	455 aa	sp: P18504 gb: M86566 gb: M63436 pir: A39062 pir: S03889 pir: JQ0158 prosite: PS00236	Lolait, *FEBS Lett* **246**: 145–8. Khrestchatisky, *Neuron* (1989) **3**: 745–53. Wang, *J Mol Neurosci* (1992) **3**: 177–84. Keir, *Genomics* (1991) **9**: 390–5.
alpha-1 – GABA(A) receptor alpha-1 subunit; precursor	Rat	455 aa	gb: L08490	Seeburg, *Cold Spring Harb Symp Quant Biol* (1990) **55**: 29–40. Draguhn, *Neuron* (1990) **5**: 781–8. Wisden, *Curr Opin Neurobiol* (1992) **2**: 263–9. Wieland, *J Biol Chem* (1992) **267**: 1426–9. Wisden, *J Neurosci* **12**: 1040–62. Laurie, *J Neurosci* (1992) **12**: 1063–76.
alpha-2 – GABA(A) receptor/benzo-diazepine receptor alpha-2 chain precursor	Rat	451 aa	sp: P23576 pir: JH0370 prosite: PS00236	Khrestchatisky, *J Neurochem* (1991) **56**: 1717–22.
alpha-2 – GABA(A) receptor alpha-2 subunit; precursor	Bovine	451 aa	em: X12361 pir: ACBOG2 prosite: P10063	Levitan, *Nature* (1988) **335**: 76–9.

$GABA_A$ α subunit database listings continued

Nomenclature	Species, DNA source	ORF[†] for original cDNA[†]	Accession	Sequence/ discussion
alpha-2 – GABA(A) receptor alpha-2 subunit; precursor	Bovine	451 aa	gb: X12361 prosite: PS00236	Schofield, *Nature* (1988) **335**: 76–9.
alpha-2 – GABA(A) receptor alpha-2 subunit; precursor	Mouse	451 aa	gb: M86567 sp: P26048 prosite: PS00236	Wang, *J Mol Neurosci* (1992) **3**: 177–84.
alpha-3 – GABA(A) receptor alpha-3 subunit; precursor	Bovine	492 aa	sp: P10064 em: X12362 pir: CHBOA3 prosite: PS00236	Levitan, *Nature* (1988) **335**: 76–9.
alpha-3 – GABA(A) receptor alpha-3 subunit; precursor	Rat	493 aa	gb: X51991 pir: A34130 sp: P20236 prosite: PS00236	Malherbe,*FEBS Lett,* (1990) **260**: 261–5.
alpha-3 – GABA(A) receptor alpha-3 subunit; precursor	Bovine	492 aa	gb: X12362 prosite: PS00236	Schofield, *Nature* (1988) **335**: 76–9.
alpha-3 – GABA(A) receptor alpha-3 subunit; precursor	Mouse	492 aa	sp: P26049 gb: M86568 prosite: PS00236	Wang, *J Mol Neurosci* (1992) **3**: 177–84.
alpha-4 – GABA(A) alpha-4 subunit precursor	Bovine	556 aa	sp: P20237 em: X61456 pir: S06838 prosite: PS00236 gb: Y07515	Ymer, *FEBS Lett* (1989) **258**: 119–22.

$GABA_A$ α subunit database listings continued

Nomenclature	Species, DNA source	ORF† for original cDNA†	Accession	Sequence/ discussion
alpha-4 – GABA(A) receptor alpha-4 subunit; precursor	Rat	552 aa	sp: P28471 pir: S17551 prosite: PS00236	Wisden, *FEBS Lett* (1991) **289**: 277–30.
alpha-5 – GABA(A) receptor/benzo-diazepine receptor alpha-5 chain precursor	Rat	464 aa	pir: B34130 sp: P19969 pir: JQ0159 prosite: PS00236	Khrestchatisky, *Neuron* (1989) **3**: 745–53.
alpha-5 – GABA(A) receptor alpha-5 subunit; subunit precursor	Rat	464 aa	gb: X51992 prosite: PS00236	Malherbe, *FEBS Lett* (1990) **260**: 261–5.
alpha-6 – GABA(A) receptor alpha-6 subunit; precursor	Mouse	443 aa	sp: P16305 em: X51986 pir: S08684 pir: S11396 prosite: PS00236	Kato, *J Mol Biol* (1990) **214**: 619–24.

Note: The listing has been made by reference to names used in sequence database entries; the designations **gabra1–gabra6** are also in use as names for genes encoding $GABA_A$ α subunits.

$GABA_A$ β subunit database listings

Nomenclature	Species, DNA source	ORF† for original cDNA†	Accession	Sequence/ discussion
beta – Invertebrate (Great pond snail) GABA(A) receptor beta-subunit	Mollusc (*Lymnaea*)	499 aa	em: X58638 sp: P26714 prosite: PS00236	Harvey, *EMBO J* (1991) **10**: 3239–45.
beta-1 – GABA(A) receptor beta-1 subunit; precursor	Human	474 aa	sp: P18505 gb: M59212 to M59216 incl. em: X14767 pir: S03333 mim: 137190 prosite: PS00236	Schofield, *FEBS Lett* (1989) **244**: 361–4. Kirkness, *Geomics* (1991) **10**: 985–95.
beta-1 – GABA(A) receptor beta-1 subunit; precursor	Bovine	474 aa	gb: X05718 pir: B27142 sp: P08220 prosite: PS00236	Schofield, *Nature* (1987) **328**: 221–7.
beta-1 – GABA(A) receptor beta-1 subunit; precursor	Rat	474 aa	em: X15466 sp: P15431 pir: S04464 prosite: PS00236	Ymer, *EMBO J* (1989) **8**: 1665–70.
beta-1 – human genomic DNA encoding the beta-1 subunit of the GABAa receptor (GABRB1)	Human	exons 1, 2 and 3 exon 4 exon 5 exons 6, 7 and 8 exon 9 (transl: 474 aa)	gb: M59212 gb: M59213 gb: M59214 gb: M59215 gb: M59216	Kirkness, *Genomics*, (1991) **10**: 985–95.
beta-2 – GABA(A) receptor beta-2 subunit mRNA, complete coding sequence	Human	474 aa	gb: S67368	Hadingham, *Mol Pharmacol* (1993) **44**: 1211–18.

$GABA_A$ β subunit database listings continued

Nomenclature	Species, DNA source	ORF† for original cDNA†	Accession	Sequence/ discussion
beta-2 – GABA(A) receptor beta-2 subunit; precursor	Rat	474 aa	em: X15467 sp: P15432 pir: S04465 prosite: PS00236	Ymer, *EMBO J* (1989) **8**: 1665–70.
beta-3 – GABA(A) receptor beta-3 subunit mRNA, complete coding sequence	Human	473 aa	gb: M82919 sp: 28472 MIM: 137192 prosite: PS00236	Wagstaff, *Genomics* (1991) **11**: 1071–8.
beta-3 – GABA(A) receptor beta-3 subunit precursor	Chicken	476 aa	em: X54243 gb: X54243 sp: P19019 pir: S11440 prosite: PS00236	Bateson, *Nucleic Acids Res* (1990) **18**: 5557.
beta-3 – GABA(A) receptor beta-3 subunit; precursor	Rat	473 aa	em: X15468 sp: P15433 pir: S03890 pir: S04466 prosite: PS00236	Ymer, *EMBO J* (1989) **8**: 1665–70.
beta-4 – GABA(A) alternatively spliced beta-4 and beta-4′ subunits	Chicken	488 aa	sp: P240454 gb: X56646 to gb: X56648 inclusive pir: JH0360 pir: JH0359 prosite: PS00236	Bateson, *J Neurochem* (1991) **56**: 1437–40.
beta subunit precursor – Invertebrate (*Drosophila*) GABA(A) receptor cyclodiene resistance protein	*Drosophila (see also Domain functions, 10-29)*	606 aa	gb: M69057 sp: P25123 flybase: 04244 prosite: PS00236	Ffrench-Constant, *Proc Natl Acad Sci USA* (1991) **88**: 7209–13.

Note: The listing has been made by reference to names used in sequence database entries; the designations **gabrb1–gabrb4** are also in use as names for genes encoding $GABA_A$ β subunits.

$GABA_A$ δ subunit database listings

Nomenclature	Species, DNA source	ORF[†] for original cDNA[†]	Accession	Sequence/ discussion
delta – GABA/ benzodiazepine receptor type A delta chain; also complete coding sequence	Rat	449 aa	pir: A34625 pir: JQ0076 sp: P18506 gb: M35162 prosite: PS00236	Shivers, *Neuron* (1989) **3**: 327–37. Zhao, *Biochem Biophys Res Commun* (1990) **167**: 174–82. Zhao, *Biochem Biophys Res Commun* (1990) **168**: 887.
delta – GABA(A) receptor delta-subunit gene, exons 1 to 9 respectively	Mouse	449 aa	gb: M60587 gb: M60588 gb: M60589 gb: M60591 gb: M60592 gb: M60593 gb: M60594 gb: M60595 gb: M60596 sp: P22933 pir: A36303 prosite: PS00236	Sommer, *DNA Cell Biol* (1990) **9**: 561–8.
delta subunit – GABA(A) receptor delta subunit mRNA, complete coding sequence	Mouse	449 aa	bbs: 111421 *(Medline identifier: 92370453)*	Wang, *Brain Res Bull* (1992) **29**: 119–23.

Note: The listing has been made by reference to names used in sequence database entries; the designation **gabrd** is also in use as the gene name for those encoding $GABA_A$ δ subunits.

$GABA_A$ γ subunit database listings

Nomenclature	Species, DNA source	ORF† for original cDNA†	Accession	Sequence/ discussion
gamma 1 – GABA(A) receptor gamma-1 subunit mRNA, complete coding sequence	Rat	465 aa	em: X57514 sp: P23574 pir: S12056 prosite: PS00236	Ymer, *EMBO J* (1990) **9**: 3261–7.
gamma 2 – GABA(A) receptor gamma-2 chain alternatively spliced precursor	Mouse	pir: 474 aa; gb: 466 aa (apparent conflict)	sp: P22723 gb: M86572 pir: JH0317 prosite: PS00236	Kofuji, *J Neurochem* (1991) **56**: 713–15.
gamma 2 – GABA(A) receptor gamma-2 subunit	Chicken	474 aa	em: X54944 sp: P21548 pir: S13086 prosite: PS00236	Glencorse, *Nucleic Acids Res* (1990) **18**: 7157.
gamma 2 – GABA(A) receptor gamma-2 subunit mRNA, complete coding sequence	Bovine	475 aa	gb: M55563 sp: P22300 pir: A39272 pir: B39272 prosite: PS00236	Whiting, *Proc Natl Acad Sci USA* (1990) **87**: 9966–70.
gamma 2 – GABA(A) receptor gamma-2 subunit precursor	Human	467 aa	sp: P18507 em: X15376 pir: S03905 MIM: 137164 prosite: PS00236	Pritchett, *Nature* (1989) **338**, 582–5.
gamma 2 – GABA(A) receptor gamma-2 subunit precursor	Rat	466 aa	gb: L08497 sp: P18508 pir: A37164 pir: JQ0077 prosite: PS00236	Shivers, *Neuron* (1989) **3**: 327–37. Malherbe, *J Neurosci* (1990) **10**: 2330–7.

$GABA_A$ γ subunit database listings continued

Nomenclature	Species, DNA source	ORF[†] for original cDNA[†]	Accession	Sequence/ discussion
				Seeburg, *Cold Spring Harb Symp Quant Biol* (1990) **55**: 29–40. Draguhn, *Neuron* (1990) **5**: 781–8. Wisden, *Curr Opin Neurobiol* (1992) **2**: 263–9. Wieland, *J Biol Chem* (1992) **267**: 1426–9. Wisden, *J Neurosci* **12**: 1040–62. Laurie, *J Neurosci* (1992) **12**: 1063–76.
gamma 2 – GABA(A) receptor gamma-2 subunit, complete coding sequence	Mouse	474 aa	gb: M62374	Wafford, *Neuron* (1991) **7**: 27–33. Sikela, listed as unpublished.
gamma 3 – GABA(A) receptor gamma-3 subunit	Rat	467 aa	gb: M81142 gb: X63324 pir: S19317 sp: P28473	Herb, *Proc Natl Acad Sci USA* (1992) **89**: 1433–7. Knoflach, *FEBS Lett* (1991) **293**: 191–4.
gamma 3 – GABA(A) receptor gamma-3 subunit; precursor	Mouse	467 aa	sp: P27681 em: X59300 pir: S15469 pir: S16915 prosite: PS00236	Wilson-Shaw, *FEBS Lett* (1991) **284**: 211–15.

Note: The listing has been made by reference to names used in sequence database entries; the designations **gabrg1–gabrg3** are also in use as names for genes encoding $GABA_A$ γ subunits.

$GABA_A$ ρ subunit database listings

Nomenclature	Species, DNA source	ORF† for original cDNA†	Accession	Sequence/ discussion
rho 1 – GABA(A) receptor rho-1 subunit, complete coding sequence	Human	473 aa	gb: M62400 sp: P24046 gb: M62323 pir: A38627	Cutting, *Proc Natl Acad Sci USA* (1991) **88**: 2673–7.
rho 2 – GABA(A) receptor rho-2 subunit, complete coding sequence	Human	465 aa	gb: M86868 sp: P28476 MIM: 137162 prosite: PS00236	Cutting, *Genomics* (1992) **12**: 801–6.

Note: The listing has been made by reference to names used in sequence database entries; conventionally, the designations **gabrr1–gabrr2** could be used as names for genes encoding $GABA_A$ ρ subunits.

Official nomenclatures of $GABA_A$ receptor–channel genes, subunits and native receptors

10-53-02: Updates on **IUPHAR Nomenclature Subcommittee recommendations** on $GABA_A$ receptor–channels will appear via the 'home page' of the *Cell-Signalling Network* (CSN) available over the World Wide Web on the Internet *(for description, see Feedback & CSN access, entry 12).*

Related sources & reviews (mainly >1988)

10-56-01: $GABA_A$ receptor properties and subtype reviews (major sources)[3,5,6]; molecular biology of $GABA_A$ receptors[172]; $GABA_A$ electrophysiology[173]; $GABA_A$ phosphorylation[92,174]; $GABA_A$ modulation by ethanol and endogenous benzodiazepines[141,175]; $GABA_A$ structure–function overview[176]; relations to psychiatric illness[177]; meeting reports/$GABA_A$ pharmacology[126,141,178–180]; definition of 'consensus' $GABA_A$ properties and a full dataset for co-expression of α_1, α_3, α_5, β_1, β_2 and γ_2 subunits[10,11,80,81]; properties of insect GABA receptors[181]; properties of 'high-affinity' and 'low-affinity' GABA-binding sites[149]; ligand interactions at the $GABA_A$ receptor complex[150]; benzodiazepine interactions[182–184]; comparisons of $GABA_A$ receptors with other ELG channels[64,65]; $GABA_A$ receptors as molecular targets of insecticides and other toxicants[168,185,186]; $GABA_AR$ in vertebrate glial cells[12]; subfamily nomenclature reviews[3,172]; subunit-sequence and function discussions[32,187]; $GABA_A$ research retrospective[188]; $GABA_A$ subunit mRNA quantitation methods[38,189,190]; steroid modulation[191,192]; relation to pharmacologically defined subtypes[5,6,187]; permeation pathways of neurotransmitter-gated ion channels[111]; GABA$_A$ overview and guide to earlier literature[193]. Reviews on aspects of specific GABA$_A$ subunits include: α subunits – function and pharmacology[32] and subunit-specific data[3]; β subunits – GABA$_A$ protein phosphorylation[92,194]; δ subunit-specific data[3]; ρ subunit-specific data[3].

Feedback

Error-corrections, enhancement and extensions

10-57-01: Please notify specific errors, omissions, updates and comments on this entry by contributing to its **e-mail feedback file** (*for details, see Resource J, Search Criteria & CSN Development*). For this entry, send e-mail messages To: **CSN-10@le.ac.uk,** indicating the appropriate paragraph by entering its **six-figure index number** (xx-yy-zz or other identifier) into the **Subject**: field of the message (e.g. Subject: 08-50-07). Please feedback on only **one specified paragraph or figure per message,** normally by sending a **corrected replacement** according to the guidelines in *Feedback & CSN Access* . Enhancements and extensions can also be suggested by this route (*ibid.*). Notified changes will be indexed via 'hotlinks' from the CSN 'Home' page (http://www.le.ac.uk/csn/) from mid-1996.

Entry support groups and e-mail newsletters

10-57-02: Authors who have expertise in one or more fields of this entry (and are willing to provide editorial or other support for developing its contents) can join its support group: In this case, send a message To: **CSN-10@le.ac.uk,** (entering the words "support group" in the Subject: field). In the message, please indicate principal interests (see *fieldname criteria in the Introduction for coverage*) together with any relevant **http://www site links** (established or proposed) and details of any other possible contributions. In due course, support group members will (optionally) receive **e-mail newsletters** intended to **co-ordinate and develop** the present (text-based) entry/fieldname frameworks into a 'library' of interlinked resources covering ion channel signalling. Other (more general) information of interest to entry contributors may also be sent to the above address for group distribution and feedback.

REFERENCES

1 Woodward, *Mol Pharmacol* (1992) **41**: 1107–15.
2 Levitan, *Nature* (1988) **335**: 76–9.
3 Burt, *FASEB J* (1991) **5**: 2916–23.
4 Barbaccia, *Neurochem Res* (1990) **15**: 161–8.
5 Sieghart, *Trends Pharmacol Sci* (1992) **13**: 446–50.
6 Macdonald, *Annu Rev Neurosci* (1994) **17**: 569–602.
7 Michelson, *Science* (1991) **253**: 1420–3.
8 Nicoll, *Physiol Rev* (1990) **70**: 513–65.

[9] Berger, *J Neurosci Res* (1992) **31**: 21–7.
[10] Mohler, *Neurochem Res* (1990) **15**: 199–207.
[11] Seeburg, *Cold Spring Harbor Symp Quant Biol* (1990) **55**: 29–40.
[12] Blankenfeld, *Mol Neurobiol* (1991) **5**: 31–43.
[13] Backus, *J Neurosci Res* (1989) **22**: 274–82.
[14] Blankenfeld, *Eur J Neurosci* (1991) **3**: 310–16.
[15] Bormann, *Proc Natl Acad Sci USA* (1988) **85**: 9336–40.
[16] Wisden, *Neurosci Lett* (1989) **106**: 7–12.
[17] Louiset, *Mol Brain Res* (1992) **12**: 1–6.
[18] Rorsman, *Nature* (1989) **341**: 233–6.
[19] Hales, *Mol Pharmacol* (1992) **42**: 197–202.
[20] Meier, *J Neurochem* (1984) **43**: 1737–44.
[21] Montpied, *Mol Pharmacol* (1991) **39**: 157–63.
[22] Montpied, *J Biol Chem* (1991) **266**: 6011–14.
[23] Perez-Velazquez, *Nature* (1993) **361**: 457–60.
[24] Laurie, *J Neurosci* (1992) **12**: 4151–73.
[25] Killisch, *Neuron* (1991) **7**: 927–36.
[26] Cole, *Nature* (1989) **340**: 474–76.
[27] Walton, *J Neurosci* (1993) **13**: 2068–84.
[28] Smart, *Br J Pharmacol* (1990) **99**: 643–54.
[29] Pritchett, *Nature* (1989) **338**: 582–85.
[30] Siegel, *Neuron* (1988) **1**: 579–84.
[31] Séquier, *Proc Natl Acad Sci USA* (1988) **85**: 7815–19.
[32] Lüddens, *Trends Neurosci* (1991) **12**: 49–51.
[33] Nicholson, *Neuroscience* (1992) **50**: 355–70.
[34] Shivers, *Neuron* (1989) **3**: 327–37.
[35] Limbird, *FASEB J* (1988) **2**: 2686–95.
[36] Cutting, *Proc Natl Acad Sci USA* (1991) **88**: 2673–7.
[37] Cutting, *Genomics* (1992) **12**: 801–6.
[38] Bovolin, *Proc Natl Acad Sci USA* (1992) **89**: 9344–8.
[39] Snodgrass, *FASEB J* (1990) **4**: 2775–88.
[40] Meldrum, *Epilepsia* (1986) **27**: S3–7.
[41] Becker, *Neuron* (1992) **8**: 283–9.
[42] Becker, *FASEB J* (1990) **4**: 2767–74.
[43] Wafford, *Science* (1990) **249**: 291–3.
[44] Varga, *Eur J Pharmacol* (1992) **214**: 223–32.
[45] Harris, *FASEB J* (1989) **3**: 1689–95.
[46] Jones, *J Physiol Lond* (1992) **449**: 279–93.
[47] Yang, *FASEB J* (1992) **6**: 914–18.
[48] Alkon, *Proc Natl Acad Sci USA* (1992) **89**: 11862–6.
[49] Nilsson, *Neuroscience* (1993) **54**: 605–14.
[50] Korpi, *Nature* (1993) **361**: 356–9.
[51] Wagstaff, *Am J Hum Genet* (1991) **49**: 330–37.
[52] Mcgowan, *Brain Res* (1993) **620**: 86–96.
[53] Zhang, *Science* (1993) **259**: 531–4.
[54] Baude, *Neuroscience* (1992) **51**: 739–48.
[55] Frosch, *Neurosci Lett* (1992) **138**: 59–62.
[56] Buckle, *Neuron* (1989) **3**: 647–54.
[57] Weiner, *Annu Rev Biochem* (1986) **55**: 631–61.

58 Stephenson, *FEBS lett* (1989) **243**: 358–62.
59 Sigel, *J Biol Chem* (1983) **258**: 6965–71.
60 Stephenson, *Eur J Biochem* (1982) **167**: 291–8.
61 Vandenberg, *Mol Pharmacol* (1993) **44**: 198–203.
62 Angelotti, *J Neurosci* (1993) **13**: 1418–28.
63 Angelotti, *J Neurosci* (1993) **13**: 1429–40.
64 Betz, *Neuron* (1990) **5**: 383–92.
65 Grenningloh, *Nature* (1987) **330**: 25–6.
66 Galzi, *Nature* (1992) **359**: 500–5.
67 Numa, *Harvey Lect* (1989) **83**: 121–65.
68 Noda, *Nature* (1983) **302**: 528–32.
69 Barnard, *Trends Biochem Sci* (1992) **17**: 368–74.
70 Hutton, *FEBS Lett* (1993) **326**: 112–16.
71 Schofield, *Nature* (1987) **328**: 221–7.
72 Ffrenchconstant, *Nature* (1993) **363**: 449–51.
73 Bormann, *J Physiol Lond* (1987) **385**: 243–86.
74 Pritchett, *Proc Natl Acad Sci USA* (1991) **88**: 1421–5.
75 Wieland, *J Biol Chem* (1992) **267**: 1426–9.
76 Smith, *Soc Neurosci Abstr* (1992) **18**: 264.
77 Olsen, *Neurochem Res* (1991) **16**: 317–25.
78 Smith, *J Biol Chem* (1994) **269**: 20380–7.
79 Sigel, *EMBO J* (1992) **11**: 2017–23.
80 Verdoorn, *Neuron* (1990) **4**: 919–28.
81 Sigel, *Neuron* (1990) **5**: 703–11.
82 Blankenfeld, *Neurosci Lett* (1990) **115**: 269–73.
83 Feigenspan, *Nature* (1993) **361**: 159–62.
84 Shimada, *Mol Pharmacol* (1992) **41**: 683–7.
85 Harvey, *EMBO J* (1991) **10**: 3239–45.
86 Squires, *Pharmacol Biochem Behav* (1979) **10**: 825–9.
87 Nielsen, *Nature* (1980) **286**: 606.
88 Birnir, *Proc R Soc Lond (Biol)* (1992) **250**: 307–12.
89 Staley, *J Neurophysiol* (1992) **68**: 197–212.
90 Bureau, *J Biol Chem* (1992) **267**: 8679–84.
91 Kimelberg, *J Neurosci* (1990) **10**: 1583–91.
92 Leidenheimer, *Trends Pharmacol Sci* (1991) **12**: 84–7.
93 Sigel, *FEBS Lett* (1991) **291**: 150–2.
94 Kellenberger, *J Biol Chem* (1992) **267**: 25660–3.
95 Leidenheimer, *Mol Pharmacol* (1992) **41**: 1116–23.
96 Moss, *J Biol Chem* (1992) **267**: 14470–6.
97 Whiting, *Proc Natl Acad Sci USA* (1990) **87**: 9966–70.
98 Swope, *FASEB J* (1992) **6**: 2514–23.
99 Wafford, *FEBS Lett* (1992) **313**: 113–17.
100 Macdonald, *J Physiol Lond* (1989) **410**: 479–99.
101 Twyman, *J Physiol Lond* (1990) **423**: 193–220.
102 Frosch, *J Neurosci* (1992) **12**: 3042–53.
103 Serfozo, *FEBS Lett* (1992) **310**: 55–9.
104 Belhage, *Neuroscience* (1993) **54**: 1019–34.
105 Oh, *Neuroscience* (1992) **49**: 571–6.
106 Porter, *Mol Pharmacol* (1992) **42**: 872–81.

107 Robello, *Neuroscience* (1993) **53**: 131–8.
108 Shirasaki, *J Physiol Lond* (1992) **449**: 551–72.
109 Osmanovic, *J Physiol Lond* (1990) **421**: 151–70.
110 Fatimashad, *Proc R Soc Lond (Biol)* (1992) **250**: 99–105.
111 Lester, *Annu Rev Biophys Biomol Struç* (1992) **21**: 267–92.
112 Newland, *J Physiol Lond* (1991) **432**: 203–33.
113 Khrestchatisky, *Neuron* (1989) **3**: 745–53.
114 Xie, *Nature* (1991) **349**: 521–4.
115 Westbrook, *Nature* (1987) **328**: 640–3.
116 Smart, *J Physiol Lond* (1992) **447**: 587–625.
117 Draguhn, *Neuron* (1990) **5**: 781–8.
118 Kilic, *Eur J Neurosci* (1993) **5**: 65–72.
119 Chow, *Br J Pharmacol* (1986) **88**: 541–7.
120 Twyman, *Biophys J* (1991) **59**: 256a.
121 Izquierdo, *Trends Pharmacol Sci* (1991) **12**: 260–5.
122 Lange, *Pflugers Arch* (1987) **410**: 648–51.
123 Drew, *Neurosci Lett* (1984) **52**: 317–21.
124 Study, *Proc Natl Acad Sci USA* (1981) **78**: 7180–4.
125 Edgar, *Mol Pharmacol* (1992) **41**: 1124–9.
126 Olsen, *Adv Exp Med Biol* (1988) **236**: 1–14.
127 Puia, *Neuron* (1990) **4**: 759–65.
128 Prince, *Biochem Pharmacol* (1992) **44**: 1297–1302.
129 Newland, *J Physiol Lond* (1992) **447**: 191–213.
130 Concas, *Brain Res* (1991) **542**: 225–32.
131 Hales, *Br J Pharmacol* (1991) **104**: 619–28.
132 Koenig, *Biochem Pharmacol* (1992) **44**: 11–15.
133 Possani, *Biochim Biophys Acta* (1992) **1134**: 210–16.
134 Hales, *Eur J Pharmacol* (1992) **210**: 239–46.
135 Moody, *Eur J Pharmacol* (1989) **164**: 153–8.
136 Pitler, *J Neurosci* (1992) **12**: 4122–32.
137 Mouginot, *J Physiol Lond* (1991) **437**: 109–32.
138 Akaike, *J Physiol Lond* (1987) **392**: 543–62.
139 Reynolds, *Brain Res* (1991) **564**: 138–42.
140 Wafford, *Neuron* (1991) **7**: 27–33.
141 Gonzales, *Trends Pharmacol Sci* (1991) **12**: 1–3.
142 Sudzak, *Proc Natl Acad Sci USA* (1986) **83**: 4071–5.
143 Hoffmann, *J Neurochem* (1989) **52**: 1937–40.
144 Sigel, *FEBS Lett* (1993) **324**: 140–2.
145 Twyman, *J Physiol Lond* (1992) **456**: 215–45.
146 Sieghart, *J Neurochem* (1983) **41**: 47–55.
147 Lüddens, *Nature* (1990) **346**: 648–51.
148 Puia, *Proc Natl Acad Sci USA* (1992) **89**: 3620–4.
149 Schumacher, *Mol Neurobiol* (1989) **3**: 275–304.
150 Knapp, *Neurochem Res* (1990) **15**: 105–12.
151 Möhler, *Proc Natl Acad Sci USA* (1980) **77**: 1666–70.
152 Hawkinson, *Mol Pharmacol* (1992) **42**: 1069–76.
153 Petke, *Mol Pharmacol* (1992) **42**: 294–301.
154 Segal, *J Neurophysiol* (1984) **51**: 500–15.
155 Ropert, *J Physiol Lond* (1990) **428**: 707–22.

[156] Mistry, *Pflugers Arch* (1990) **416**: 454–61.
[157] Robertson, *J Physiol Lond* (1989) **411**: 285–300.
[158] Peters, *Pflugers Arch* (1989) **415**: 95–103.
[159] Schmieden, *Science* (1993) **262**: 256–8.
[160] Bowery, *Trends Pharmacol Sci* (1989) **10**: 401–7.
[161] Gage, *Trends Neurosci* (1992) **15**: 46–51.
[162] Bowery, *Annu Rev Pharmacol Toxicol* (1993) **33**: 109–47.
[163] Hahner, *FASEB J* (1991) **5**: 2466–72.
[164] Qian, *Nature* (1993) **361**: 162–4.
[165] Veruki, *J Neurophysiol* (1992) **67**: 791–7.
[166] Chvatal, *Pflügers Arch* (1991) **419**: 263–6.
[167] Kawakami, *Biol Pharm Bull* (1993) **16**: 726–8.
[168] Narahashi, *Trends Pharmacol Sci* (1992) **13**: 236–41.
[169] Shirasaki, *Brain Res* (1991) **561**: 77–83.
[170] Mocchetti, *Neurochem Res* (1990) **15**: 125–30.
[171] Bormann, *Regul Pept (Suppl)* (1985) **4**: 33–8.
[172] Olsen, *FASEB J* (1990) **4**: 1469–80.
[173] Bormann, *Trends Neurosci* (1988) **11**: 112–16.
[174] Huganir, *Neuron* (1990) **5**: 555–67.
[175] Lister, *Neuropharmacology* (1991) **30**: 1435–40.
[176] DeLorey, *J Biol Chem* (1992) **267**: 16747–50.
[177] Zorumski, *Am J Psychiat* (1991) **148**: 162–73.
[178] Henley, *Trends Pharmacol Sci* (1991) **12**: 357–9.
[179] Olsen, *Annu Rev Pharmacol Toxicol* (1982) **22**: 245–77
[180] Nattel, *Drugs* (1991) **41**: 672–701.
[181] Rauh, *Trends Pharmacol Sci* (1990) **11**: 325–9.
[182] Haefely, *Neurochem Res* (1990) **15**: 169–74.
[183] Sieghart, *Trends Pharmacol Sci* (1989) **10**: 407–11.
[184] Farrant, *Neurochem Res* (1990) **15**: 175–91.
[185] Eldefrawi, *FASEB J* (1987) **1**: 262–71.
[186] Narahashi, *Adv Exp Med Biol* (1991) **287**: 61–73.
[187] Schofield, *Trends Pharmacol Sci* (1989) **10**: 476–8.
[188] Costa, *Neuropsychopharmacology* (1991) **4**: 225–35.
[189] Buck, *Biotechniques* (1991) **11**: 636–8.
[190] Grayson, *Methods Neurosci* (1993) **12**: 191–208.
[191] Kirkness, *Trends Pharmacol Sci* (1989) **10**: 6–7.
[192] Lambert, *Trends Pharmacol Sci* (1987) **8**: 224–7.
[193] Stephenson, *Biochem J* (1988) **249**: 21–32.
[194] Porter, *Neuron* (1990) **5**: 789–96.

ELG Cl GLY

Inhibitory receptor–channels gated by glycine

Edward C. Conley **Entry 11**

NOMENCLATURES

Abstract/general description

11-01-01: The amino acid **glycine** is an important **inhibitory**† **transmitter** in the **spinal cord** and **brainstem**, participating in the regulation of both **motor** and **sensory** functions. Upon release by an inhibitory† neurone, glycine binds and opens post-synaptic glycine receptor–channels (**GlyR**) causing neuronal **hyperpolarization**† through their permeability to **chloride ions**. Generally, glycine inhibits glutamate-evoked depolarization† and **depresses firing of neurones**.

11-01-02: Although gamma-aminobutyric acid (**GABA**) is generally considered to be the major inhibitory neurotransmitter in higher regions of the nervous system *(see ELG Cl GABA$_A$, entry 10)*, the wide distribution of GlyR subtypes (as detected by *in situ* hybridization, for example) also suggests important roles for glycinergic transmission in **higher brain functions**. GlyR are expressed in relaying centres for processing **pain** and **sensory information** and deficiencies in glycine receptors leads to **spasticity**† and loss of **motor control** in various neuronal disorders.

11-01-03: Post-synaptic inhibition of neuronal firing by GlyRs is selectively antagonized by the competitive antagonist **strychnine** and receptors specific for glycine were originally identified by strychnine radioligand†-binding studies[1,2]. Strychnine exerts **pro-convulsant effects** resulting from excitatory depolarization by Na^+-influx. Thus in **strychnine poisoning** normal glycinergic† inhibition is abolished, giving rise to the muscular convulsions by 'overexcitation' of the **motor system**.

11-01-04: Comparative studies suggest that both glycine and **GABA$_A$ agonist-activated channels** *(see ELG Cl GABA$_A$, entry 10)* act as multi-ion pathways and have similar permeation† characteristics (although several important pharmacological distinctions exist) *(see Table 7 under Selectivity, 11-40)*. The sequences encoding the chloride-selective glycine and **GABA$_A$ receptor–channel**s display multiple regions of homology†. A marked **outward rectification**† is observed when GlyR channels conduct in the steady state†. This rectification† is likely to reflect a voltage-dependence† of gating†, but not a voltage-dependence of ion permeation†.

11-01-05: Molecular cloning studies have revealed a heterogeneity of glycine receptor subtypes in the CNS. Native† GlyR are assembled from subunits encoded by five distinct but related genes (α_1, α_2, α_3, α_4 and β). Intact receptor complexes are likely to consist of a **pentameric assembly** of α and β subunits, and co-expression studies in *Xenopus* oocytes have established that the GlyR α/β hetero-oligomers† are present in an **invariant** (**3:2**) **stoichiometry**.

11-01-06: Purification of the GlyR to homogeneity† using affinity† ligands such as **2-aminostrychnine** typically resolves three polypeptides of **48 kDa** (α

subunit), **58 kDa** (**β subunit**) and **93 kDa** (**gephyrin**, *see Subcellular locations, 11-16*) following SDS–PAGE† analysis. Following detergent-solubilization†, the GlyR sediments as a **macromolecular complex** of approximately **250 kDa**.

11-01-07: Site-directed mutagenesis† of GlyR coding sequences have identified protein subunit domains† involved in **ligand discrimination**, **agonist efficacy**†, **antagonist binding** and **ion channel formation**: In common with other extracellular ligand-gated anion channels (e.g. the $GABA_AR$) **aromatic hydroxyl groups** appear crucial for **ligand discrimination** while the presence of net positive charges in GlyR–channel vestibules are associated with 'attraction' of anions into the hydrophilic† pore. The **α subunits** contribute the **strychnine-binding site** of intact pentameric GlyRs. The GlyR α subunit **M2 domain** is likely to 'line' the pore, which is similar in arrangement to other extracellular ligand-gated channels *(see Domain conservation, 11-28).*

11-01-08: Extracellular signals released from pre-synaptic neurones (such as **neurotransmitters** and **neuropeptides**) as well as certain **extracellular matrix proteins** may modulate GlyR function and receptor density *in vivo*. In common with other extracellular ligand-gated receptor–channels, the GlyR contains a large (putatively) intracellular loop that contains a number of consensus sites for **protein phosphorylation** *(see Protein phosphorylation, 11-32).*

11-01-09: Developmental variants of GlyR α subunit genes have been discovered via sequencing of cDNA clones *(see α_2 and α_2* under Gene family, 11-05).* Amino acid substitutions in the α_2* subunit can account for a 500-fold lower **strychnine sensitivity** observed in **neonatal** rats compared to adults. The α_2 subunit expression is predominant in foetal or newborn tissues and '**developmental switching**' from α_2 to α_1 occurs within 3 weeks following birth, accompanied by a strong post-natal increase in high-affinity **[^{3}H]-strychnine binding**. This developmental transition from 'neonatal' to 'adult' GlyR isoforms appears to be perturbed in the mutant mouse **spastic** which acquires a **severe motor disease** about 2 weeks after birth. *Note:* GlyR protein *coding regions* appear unchanged in ***spastic*** mice, suggesting that an aberrant regulation of ion channel gene switching underlies the disease phenotype†.

11-01-10: GlyR β subunits show wide distribution in brain and spinal cord with a common (non-variant) sequence. β subunits are *not* required for high-affinity agonist or antagonist binding and they show **constitutive expression** throughout development. Site-directed mutagenesis† studies have predicted the existence of '**assembly domains**' on the β subunit which may have evolved to prevent homo-oligomerization†. These domains may function in a '**sequential assembly**' pathway forming functional GlyRs (i.e. $\alpha/\beta + \alpha/\beta + \alpha \rightarrow \alpha_3\beta_2$). 'Pore-lining' residues within the M2 of the **β subunit** are major determinants of **picrotoxinin resistance**. The β subunit is expressed in areas of brain where *none* of the presently characterized subunits (or their mRNAs) can be detected, which suggests further ('uncloned') GlyR subunit genes exist.

Category (sortcode)

11-02-01: ELG Cl GLY, i.e. extracellular ligand-gated chloride channels activated by glycine. The suggested **electronic retrieval code** (unique embedded identifier or **UEI**) for 'tagging' of new articles of relevance to the contents of this entry is UEI: GLY-NAT (for reports or reviews on native† channel properties) and UEI: GLY-HET (for reports or reviews on channel properties applicable to heterologously† expressed recombinant† subunits encoded by cDNAs† or genes†). *For a discussion of the advantages of UEIs and guidelines on their implementation see the section on Resource J under Introduction and layout, entry 02, & for further details, see Resource J – Search criteria & CSN development, entry 65.*

Channel designation

11-03-01: The GlyR; the iGlyR; ionotropic† glycine receptors; GlyR-C (the glycine receptor–channel). The **neonatal (foetal) forms** of native GlyR channel complexes have been given the designation $GlyR_N$ to distinguish them from those predominantly expressed in the adult ($GlyR_A$) *(see Gene family, 11-05, and Developmental regulation, 11-11).*

Current designation

11-04-01: Of the form $I_{agonist}$, e.g. I_{Gly}, $I_{(Cl,\ Gly)}$.

Gene family

The GlyR gene family

11-05-01: The genes encoding glycine receptor–channels form part of the extracellular ligand-gated† (ELG) channel gene superfamily† *(described under ELG Key facts, entry 04).* Conventional cross-homology† screening has (thus far) identified five distinct but related genes encoding subunits which assemble into native GlyR complexes (α_1, α_2, α_3, α_4 and β – *see Table 1 for comparison of features*). Monoclonal† antibodies raised against the native† GlyR have been shown to immunoprecipitate both **α- and β-chains**, and limited proteolytic† cleavage of these proteins show similar patterns.

Variants derived from individual GlyR genes

11-05-02: Subtle **variants** of GlyR α subunit genes have been discovered via sequencing of cDNA clones. For example, the 'α_2* **neonatal**' subunit (sp: P22771, *see Table 1 and Database listings, 11-53)* is a novel cDNA variant identical to the human 'α_2 subunit', except for five amino acid substitutions at positions 18, 24, 37, 194 and 404. Substitution at position 194 (Gly194 to Glu) in the α_2* protein accounts for the 500-fold lower **strychnine sensitivity** observed in **neonatal** rats. Rat α_2* contains **Glu** at position 167, as compared to **Gly** at position 167 in α_2[3].

β subunit genes and proteins

11-05-03: GlyR β subunits show wide distribution in brain and spinal cord with a common (non-variant) sequence. Two β subunits are present in each five-subunit native GlyR complex *(for details of the stoichiometric† assembly patterns of GlyR, see Predicted protein topography, 11-30, and Table 1).* β subunits are *not* required for high-affinity agonist or antagonist

Table 1. *Genes encoding subunits present in native GlyR–channel complexes (From 11-05-03)*

Subunit	Description	Encoding[a]	Mol. wt (from SDS–PAGE†)
GlyR alpha-1 subunit = α_1	*Component of adult* '**nM strychnine-sensitive**' GlyR. Alpha-1 polypeptides are the principal ligand-binding subunits of **adult** GlyR complexes (**$GlyR_A$**). Binds glycine ligand and the convulsant strychnine. Different variants of α_1 have been isolated[5].	447–449 aa (human)	48 kDa (human)
GlyR alpha-2 subunit variant = α_2	*Variant of the neonatal (or 'foetal')*, '**nM strychnine-insensitive**' GlyR. Alpha-2 polypeptides are the principal ligand-binding subunit of **neonatal** GlyR complexes (**$GlyR_N$**)[3,6]. Contains a **glycine** residue at position 167. The α_2 gene generates two *apparently* isofunctional splice variants affecting residues in the N-terminal domain which are co-expressed throughout development *(see Gene organization, 11-20, and Developmental regulation, 11-11).*	452 aa (rat) Transcript size ~2.8 kb in rat spinal cord poly $(A)^+$ mRNA[7]	49 kDa (human/rat)
GlyR alpha-2* subunit variant = $\alpha_2{}^*$	*Variant of the neonatal (or 'foetal')*, '**nM strychnine-insensitive**' GlyR. Principal component in **neonatal** spinal cord GlyR complexes (**$GlyR_N$**). Contains a **glutamate** residue at position 167, which is sufficient to explain lower antagonist-binding properties than α_2[3]. $\alpha_2{}^*$ binds glycine ligand but has 500-fold lower sensitivity to strychnine when expressed in oocytes[3]. $\alpha_2{}^*$ is probably equivalent to the 'strychnine-insensitive' GlyRs in native neonatal rat spinal cord[3].	448 aa including signal peptide. 425 aa (mature) (rat)	48 kDa (human/rat)

GlyR alpha-3 subunit = α_3	Principal GlyR component in post-natal cerebellum[8]. Binds glycine ligand and the convulsant strychnine.	464 aa (rat)	48 kDa (rat)
GlyR alpha-4 subunit variant = α_4	The novel α_4 variant was identified during screening of mouse genomic clones[9]. The predicted α_4 polypeptide displays very high homology to the α_2 subunit. α_4 mRNA is expressed at very low levels in brain		48 kDa (rat) predicted
GlyR beta subunit variant = β	Presumptive 'structural' subunit without ligand-binding sites. The native GlyR is a pentamer. GlyR alpha/beta hetero-oligomers have an **invariant stoichiometry** (**3 alpha:2 beta**). The β subunit appears to be expressed in a **single form** throughout the brain at all developmental stages and is not required for ligand-binding[7]	Transcript size ~3.4 kb in rat spinal cord poly $(A)^+$ mRNA[3]	58 kDa (rat)
Peripheral membrane protein	Co-purifies with the GlyR on affinity columns[10]. Associated with the cytoplasmic domains of the receptor core – possibly a glycine receptor-to-tubulin linker protein (**gephyrin**)[11] *(for details, see Subcellular locations, 11-16)*		93 kDa (rat) (non-glycosylated)

[a]Shows the number of amino acid residues in the specified channel subunit as predicted from open reading frame† lengths of cDNA sequences.

binding and they show **constitutive expression** throughout development. *Note:* The β subunit is expressed in areas of brain where none of the presently characterized subunits (or their mRNAs) can be detected, which suggests further ('uncloned') GlyR subunit genes exist.

GlyR genes are highly related to those encoding $GABA_A$ receptors
11-05-04: On the basis of shared amino acid homology†, the chloride-selective glycine and **$GABA_A$ receptor–channel** genes are more similar to each other than either is to the nAChR or other cation-selective receptor–channels. Despite some key pharmacological differences, many similarities exist between glycine- and $GABA_A$-activated currents – e.g. see comparative study of these channel types in post-natal cultured hippocampal neurones[4]. *(Similarities in the selectivity characteristics of the GlyR and the $GABA_AR$ are given in Table 7 under Selectivity, 11-40).*

Trivial names

11-07-01: The inhibitory glycine receptor; the glycine receptor–channel; the strychnine-sensitive glycine receptor *(but note strychnine-insensitivity of foetal/neonatal GlyR forms – see Table 1)*; the ionotropic† glycine receptor.

EXPRESSION

Cell-type expression index

The glycine receptor–channel is abundant in the spinal cord and brainstem of vertebrates
11-08-01: The GlyRs are highly expressed in regions of the CNS associated with **spinal neurotransmission** and **motor control** *(see Developmental regulation, 11-11, mRNA distribution, 11-13, and Protein distribution, 11-15).* Extensive patch-clamp analysis has been performed on GlyRs of foetal and mature rodent spinal neurones *(e.g. see refs.[12–15] and most fields of this entry).* GlyRs have also been studied in isolated ganglion cells from the goldfish retina[16], isolated bipolar cells of the mouse retina[17], cultured hippocampal neurones[4], teleost Mauthner cells[18] and renal proximal cells[19].

Channel density

Clustering of GlyRs by specific interactions with cytoskeletal elements
11-09-01: Glycine receptors are known to be associated with **cytoskeletal elements** which co-purify using GlyR affinity ligands[11,20–23]. Such interactions may control channel mobility, subcellular distribution and **local expression density** *(see also gephyrin, under Protein interactions, 11-31).*

Comparable expression density of glycine- and GABA-activated channels in hippocampus

11-09-02: Responses to saturating concentrations of glycine and GABA agonists† in cultured hippocampal neurones suggests that they activate *comparable numbers* of '**anatomically-distinct**' channels with 'very similar' permeation properties[4] *(see also Selectivity, 11-40).*

Developmental regulation

Post-natal switching of GlyR α subunits – effects on antagonist binding and glycinergic† IPSC†

11-11-01: Glycine receptor subunits exist in 'foetal/newborn' (e.g. α_2 and $\alpha_2{}^*$) and 'adult' (α_1, α_3) isoforms *(see also Table 1 under Gene family, 11-05).* In rat spinal cord, α_2 subunit expression is predominant in foetal or newborn tissues; '**developmental switching**' from α_2 to α_1 occurs within 3 weeks following birth[3, 7, 24–26]. This 'switching' is accompanied by a strong post-natal increase in high-affinity **[^{3}H]-strychnine binding**[24]. This process of developmental switching of GlyR α subunits also **accelerates kinetics** of glycinergic inhibitory post-synaptic currents (IPSCs†, for example, in spinal neurones[27, 28]). *Note:* Functional maturation of the nicotinic acetylcholine receptor *(see ELG CAT nAChR, entry 09)* is executed by its gamma-to-epsilon subunit switching†.

GlyR pharmacological properties in neonates mediated by the GlyR $\alpha_2{}^*$ isoform

11-11-02: It has been known for over a hundred years that neonatal rats are relatively 'immune' (resistant) to **strychnine poisoning**[29]. Increases in sensitivity of spinal cord neurones to glycine and strychnine correlates with the disappearance of a low-strychnine-binding $\alpha_2{}^*$ subunit isoform (sp: P22771, *see Domain conservation, 11-28, and Database listings, 11-53)*[3]. The $\alpha_2{}^*$ variant shows a number of amino acid substitutions which have effects on intact GlyR function *(for structural variations, see Gene family, 11-05).* GlyR α_2 mRNA accumulates in **pre-natal development** and sharply decreases after birth, whereas transcripts for strychnine-sensitive α_1 and α_3 subunits appear only in **post-natal** brain structures[7, 8].

Constitutive expression of GlyR β subunit mRNA throughout development

11-11-03: Hybridization† signals of β subunit mRNA are observed in early embryonic stages and continuously increase to high levels in adult rats[7] *(see Table 1 and mRNA Distribution, 11-13). Note:* The *apparently* '**isofunctional**' α_2 splice variants *(see Gene organization, 11-20)* have been reported to be co-expressed throughout embryonic and post-natal development[6].

Regional accumulation of GlyR transcripts

11-11-04: Regional accumulation of mRNAs encoding GlyR α_1, α_2, α_3 during development has been observed (see ref.[7] for original *in situ* hybridization reporting the time course of regional expression in brain). Table 2 summarizes the relative signal intensities reflecting mean levels detected for the principal GlyR subunit transcripts† in developing spinal cord[7].

Table 2. *Relative signal intensities reflecting mean levels of GlyR subunit transcripts in developing spinal cord (From 11-11-04)*

	Developmental stage†					
Transcript	E14	E19	P0	P5	P15	P20
α_1	low	very low	moderate	moderate	high	high
α_2	high	high	high	low	very low	very low
α_3	undetectable	very low	very low	very low	very low	very low
β	low	low	high	high	high	high

Physiological consequences of 'sequential activation' of GlyR–channel gene expression

11-11-05: The developmental transition from the neonatal to the adult GlyR isoform appears to be perturbed in the mutant mouse ***spastic*** which acquires a **severe motor disease** about 2 weeks after birth[30] *(see Phenotypic expression, 11-14)*. For further details of the amino acid determinants for differential properties between GlyR α subunits switched during development, *see Domain functions, 11-29*. For further details of the electrophysiological changes observed during the transition from foetal-type to adult-type GlyR currents, *see Single-channel data, 11-41.*

mRNA distribution

GlyR α subunit mRNAs display differential spatial expression

11-13-01: As summarized in Table 3 several differences exist in **regional** and **developmental expression** of the different GlyR α subunit mRNAs. These 'highly-regionalized' patterns provide evidence for novel functions for GlyR proteins in the mammalian brain and spinal cord. The wide ('non-regionalized') distribution of GlyR β subunit transcripts† include regions lacking α subunit hybridization signals *(see Table 3)* and may therefore indicate the existence of *further* distinct GlyRs in mammalian brain (the genes for which have not yet been cloned).

Phenotypic expression

General phenotypic roles of glycine receptor–channels

11-14-01: The wide occurrence of the GlyR β and α isoforms throughout the CNS *(see mRNA distribution, 11-13)* also suggests several roles for **glycinergic transmission**† in higher brain functions[33]. In medullary dorsal horn and in the spinal dorsal horn (relaying centres for processing **pain** and **sensory information**), glycine inhibits glutamate-evoked depolarization† and **depresses firing of neurones**. Deficiency in glycine receptors leads to **spasticity**† and loss of **motor control** in various neuronal diseases[34–36].

Persistence of the neonatal isoform is associated with acquired motor disease in mice

11-14-02: The **developmental transition** from the neonatal to the adult GlyR isoform *(see Developmental regulation, 11-11)* is perturbed in the mutant

Table 3. *Distribution patterns of GlyR mRNAs (From 11-13-01)*

Transcript	General distribution patterns[a, b]
α_1	*In situ* hybridization of *adult* brain shows GlyR α_1 subunit mRNA to be abundant in spinal cord, superior and inferior colliculi[7]. α_1 subunit transcripts† are highly-expressed in adults *(see Developmental regulation, 11-11)*
α_2	Expressed in several forebrain regions including layer VI of the cerebral cortex, thalamus and hippocampus[7]. In rat spinal cord, α_2 subunit expression is predominant in foetal or newborn tissues; switching from α_2 to α_1 occurs within 3 weeks following birth[3, 7, 24–26] *(see Developmental regulation, 11-11)*
α_3	Expressed at low abundance in cerebellum, olfactory bulb and hippocampus[7]
β	Northern† analysis[31] and *in situ* hybridization[7] report high amounts of β subunit transcripts *throughout* the spinal cord and brain[7] (including cerebellum and cortex, where none of the presently characterized α polypeptides are expressed at comparable levels[c]

[a]See also original *in situ* hybridization† data and full breakdown of regionalized expression patterns[7].
[b]The spatial distribution of channel subunit-specific mRNAs presumably reflects a functional specialization in particular cell types. Mapping of expression patterns is a complex task and has to take many variables into account, such as *in situ* localization, developmental regulation, subunit stoichiometry, and factors regulating overlapping or co-expression. Notes on the integration of computer-based information resources able to cross-reference these diverse factors influencing expression can be found in Fig. 2 under *Feedback & CSN access, entry 12 and Appendix J – Search criteria & CSN development, entry 65.*
[b]The possibility of the β subunits participating in other ELG receptor–channel type complexes (e.g. $GABA_AR$, putative taurine/β-alanine receptors and NMDAR) has been discussed[32].

mouse ***spastic*** (***spa***) which acquires a **severe motor disease** about 2 weeks after birth[30]. Some key phenotypic features of this disorder are listed in Table 4 (limited to factors capable of disturbing the normal balance between inhibitory and excitatory ion channel function). The **neurophysiological** and **neuropharmacological** aspects of these disorders are further discussed in refs[30, 37].

Long-term potentiation† of inhibitory circuits and synapses in the CNS

11-14-03: Glycinergic† inhibition† in teleost Mauthner cells (produced by stimulation of the contralateral eighth nerve) exhibits long-term potentiation† following classical **tetanization†** of the pathway[18]. Potentiation† (enhancement) occurs at the synapses between **primary afferents†** on to second-order **interneurones†** and the connections between

Table 4. *Some key phenotypic features of the acquired motor disease observed in the mutant mouse spastic (From 11-14-02)*

Phenotypic feature	Observation/description/inference	Refs
Time of onset and gross physiological characteristics of *spa* mice	About 2 weeks following birth, homozygotic† *spa* mutants develop **severe motor defects** including **tremor** and **episodic spasms**, walking on extended toes and in short steps. Homozygous† *spa/spa* mice exhibit spontaneous **myoclonic**† **jerking** movements, arched back posture and rapid tremor of trunk and limbs	***38***
Stereotyped† movements of contralateral† limbs in *spa* mice	Electrical stimulation of limbs in *spa* mice can excite a 'burst response' with stereotyped† timing and latency in the contralateral† limb, indicating involvement of *spinal* motor circuits† *Note:* spinal neurotransmission is dominated by glycinergic pathways *(see Abstract/general description, 11-01)*	***39***
Selective deficit of the adult GlyR isoform in *spa* mice	Levels of GlyR (as measured by [^{3}H]-strychnine binding and electrophysiological procedures) are drastically reduced in homozygous† *spa/spa* mice. The *spastic* mutation selectively interferes with the **post-natal accumulation** of mRNA encoding the adult isoform $GlyR_A$. Perinatal† expression of the neonatal isoform $GlyR_N$ is *apparently* unaffected. Although mRNA 'levels' encoding GlyR 'adult' α_1 subunits are normal, the selective effect of the *spa* mutation *may* act at both the transcriptional† level (i.e. affecting receptor stability, turnover or interactions with other proteins) and/or the post-transcriptional† levels	***30*** and citations therein
Mimicking of *spa* phenotype by GlyR antagonists	Symptoms of **strychnine poisoning** at subconvulsive doses in normal mice resemble the *spastic* phenotype. Administration of **picrotoxinin** *(see Blockers under ELG Cl $GABA_A$, 10-43) fails* to induce these symptoms	***39***
Apparent *lack* of GlyR gene structural changes in *spa* mice	Changes in function or structure of the GlyR proteins appear *unaffected* in *spastic* mice, factors which have been taken as further evidence for a *regulatory* rather than a structural effect of the spastic mutation. Ligand-binding properties, subunit composition and synaptic localization of the GlyR are unchanged in *spa* mice compared to wild-type†	***38, 40***

Table 4. *Continued*

Phenotypic feature	Observation/description/inference	Refs
GlyR-mediated currents in motoneurones of *spa* mice	Strychnine-sensitive chloride conductances (which are spontaneously active in normal mice) are rare in *spa* mice. Current amplitudes† are small, *resembling* GlyR currents inhibited by strychnine	*41*
'Compensatory expression' of $GABA_AR$ for deficits in GlyR expression in *spa* mice	A significant increase in **$GABA_A$ receptor** density in the lower CNS may serve as a **compensatory function**, partially 'counteracting' losses of glycinergic† function. Pharmacological facilitation† of $GABA_A$ responses, for example by administration of **benzodiazepine agonists** or **aminooxyacetic acid** (which reduces degradation of endogenous GABA by inhibiting **GABA transaminase**) also alleviates symptoms of affected mice *(see Phenotypic expression under ELG Cl $GABA_A$, 10-14)*. **Benzodiazepines** (e.g. **diazepam**, **flunitrazepam** and **Ro11-6896** *(see ELG Cl $GABA_A$, entry 10)* alleviate symptoms, and lead to a long-lasting relaxation of muscle rigidity	*30* and citations therein; *42, 43*
Correlation of *spa* mouse symptoms with motor disorders in humans	Human **hyperexplexia**† and some types of **spastic paraplegia**† show symptoms *resembling* those of the *spa/spa* mouse	*44–46*
Correlation of *spa* mouse symptoms with similar disorders in cattle	**Inherited myoclonus**† of cattle is characterized by **hyperaesthesia**† and **myoclonic jerks** of the skeletal musculature that occur spontaneously and in response to sensory stimuli. **Inherited myoclonus** is also associated with a deficiency of GlyR in brainstem and spinal cord, although **altered motor symptoms** are also apparent in *pre-natal* calves	*37*

these inhibitory cells and the Mauthner neurone. Increases in gain still occur following pharmacological blockade of potentiation† at the excitatory synapse with glutamate antagonists. These findings imply that *in vivo*, learning can alter the 'balance' between excitation and inhibition within a network by modifying one or both of them[18] *(see also Developmental regulation, 11-11, and Receptor antagonists, 11-51). For further descriptions of long-term potentiation† processes, see Fig. 1 under ELG CAT GLU NMDA, entry 08.*

'Cytoprotection' by glycine and strychnine in renal proximal tubule cells

11-14-04: At certain concentrations both glycine and strychnine *(see Receptor antagonists, 11-51)* have **cytoprotective functions** in **renal proximal tubule cells** treated with 1 μM **antimycin A**[19] as estimated by lowered release of **lactate dehydrogenase**† (a marker of cell death/lysis). A critical role for **chloride-influx pathways** in promoting **cell death** has been discussed[19].

Protein distribution

[^{3}H]-Strychnine binding in CNS sections displays a pronounced rostro-caudal gradient

11-15-01: Extensive autoradiographic studies employing *in situ* radiolabelled† antagonist† binding have consistently shown a *gradient of binding levels* along the **rostro-caudal axis**†, with highest binding in the brainstem and spinal cord regions.

Protein and mRNA localization techniques report 'overlapping subsets' of total GlyR proteins

11-15-02: Binding patterns of strychnine radioligands† and immunocytochemical studies with specific monoclonal† antibodies demonstrate that GlyR subunits are concentrated in **spinal cord**, **brainstem** and other areas of the **lower neuraxis**†. In keeping with the detection of GlyR mRNA expression in some *higher* brain regions *(see mRNA distribution, 11-13)* immunocytochemical probes for GlyR proteins also detect low signals in **olfactory bulb**, **midbrain**, **cerebellum**[47] and **cortex**. In general, mRNA distribution studies[7] have detected a considerably wider distribution pattern of GlyR subunit expression than that revealed by immunochemical/autoradiographic methods in previous studies. This apparent discrepancy may be explained by (i) not all GlyRs binding the antagonist† radioligands† with high affinity and (ii) not all GlyRs possessing epitopes† for the antibodies in use.

Subcellular locations

The GlyR channel is an immobilized† post-synaptic peripheral membrane protein

11-16-01: The 93 kDa polypeptide (**gephyrin**) which is co-purified using affinity† ligands† of the GlyR[10] *(see Protein molecular weight (purified), 11-22)* has been localized as a **peripheral membrane protein** by immunoelectron† microscopy. The 93 kDa subunit has a cytoplasmic location in post-synaptic membranes[48]. Gephyrin is thought to form a bridge or **linker protein** between cytoplasmic domains of the GlyR and **tubulin**, thereby immobilizing the GlyR receptor–channel complex. The 93 kDa component has been cloned and sequenced[11] *(see Database listings, 11-53).*

Transcript size

11-17-01: *See Table 1 under Gene family, 11-05.* Selective **mRNA hybrid-arrest**† techniques were originally used to define *three* distinct mRNA species encoding GlyRs in developing rat CNS[49].

SEQUENCE ANALYSES

Note: The symbol [PDTM] denotes an illustrated feature on the channel protein domain topography model (Fig. 1).

Chromosomal location

The gene defective in mouse spastic mutants has not been co-localized to GlyR structural genes

11-18-01: The *spastic* mutation affecting the regulation of glycinergic† inhibition† *(see Phenotypic expression, 11-14)* is an **autosomal† recessive† trait†** carried by a single Mendelian† gene with **full penetrance†** of the phenotype in homozygotes†[38]. The *spastic* gene locus *spa* has been located on mouse chromosome 3[50]. An allele of the *spa* gene displaying a very similar (but more severe) phenotype in mouse has been defined as spa^{Alb}[51]. The mouse GlyR α_2 and α_4 genes have been mapped to the X chromosome[9].

Gene organization

The genomic organization of GlyR α subunit genes is conserved across subtypes

11-20-01: Analysis of genomic† clones covering the coding regions† of the murine GlyR α_1, α_2 and α_4 subunit genes has shown that all genes contain **eight intronic† regions** with *precisely conserved* boundaries[9].

Alternative splicing of a_1 transcripts

11-20-02: A variant rat α_1 cDNA has been identified originating from the selection of an alternative splice†-acceptor† site at an exon† encoding the cytoplasmic domain adjacent to transmembrane domain M3. Splice variant α_{2ins} has an extra eight amino acid segment containing a novel consensus phosphorylation site *(for further details, see [PDTM], Fig. 1 and Protein phosphorylation, 11-32).*

Apparently 'isofunctional' N-terminal substitutions in α_2 introduced by alternative splicing

11-20-03: In the rat, alternative use of two closely spaced versions of **exon 3** in the α_2 gene can produce two α_2 subunit isoforms, which differ by only two **isofunctional† substitutions** in the N-terminal extracellular region[6]. The α_2 splice variants† are co-expressed throughout embryonic and post-natal development[6].

Protein molecular weight (purified)

Affinity purification of GlyR subunits

11-22-01: Purification of the GlyR to homogeneity† has been achieved using matrices† derivatized† with **2-aminostrychnine** as an affinity ligand†. SDS–PAGE† analysis resolves three polypeptides of **48 kDa** (**α subunit**), **58 kDa** (**β subunit**) and **93 kDa** (**gephyrin**, *see Subcellular locations, 11-16*). Following detergent-solubilization†, the GlyR sediments as a **macromolecular complex** of approximately **250 kDa**[21,52]. *(See also chemical cross-linking studies under Predicted protein topography, 11-30.)*

Molecular weight determinations following heterologous expression
11-22-02: Human homomeric† α_1 GlyR channels expressed in the baculovirus† expression system can be gated† by glycine, but not in the presence of strychnine. An immunoreactive 48 kDa protein is apparent following SDS–PAGE† of whole-cell lysates† with maximal expression 3 days post-infection[53].

Protein molecular weight (calc.)

11-23-01: For predicted (non-glycosylated†) subunit molecular masses determined from cDNA sequences, see the respective subunit gene name in Table 1 *under Gene family, 11-05.*

Sequence motifs

The disulphide loop motif is conserved across a large number of ELG channels
11-24-01: An *invariant feature* of all members of the extracellular ligand-gated channel-receptor superfamily is the presence of an extracellular **disulphide loop motif**† *(e.g. see ELG Cl GABA$_A$, entry 10, ELG CAT 5-HT$_3$, entry 05, ELG CAT nAChR, entry 09).* As may be expected from its occurrence in several ligand-specific receptors, this small domain does not form part of an agonist-binding site as receptors with mutations *within* the loop (e.g. Lys143 → Ala143) display few functional changes[54] *(see [PDTM], Fig. 1 and Table 6 under Domain functions, 11-29).*

Consensus sites for signal peptide cleavage, disulphide bonds and glycine-binding sites
11-24-02: In common with other ELG channels, GlyR primary† amino acid sequences show a number of consensus† sites for the formation of disulphide bonds, *N*-glycosylation and protein phosphorylation *(see Table 1 under ELG Key facts, entry 04, and Protein phosphorylation, 11-32).* A number of consensus† sites are listed in Table 5 *(for an illustration of the approximate positions of these motifs on native pentameric GlyRs, see [PDTM], Fig. 1).*

Table 5. *Example motif positions as reported in GenBank® entries (From 11-24-02)*

Subunit sequence (rat)	Signal† peptide cleavage motifs	Disulphide bond formation motifs	*N*-glycosylation motifs
GlyR α_2*	1–27	172-bond-186 232-bond-243	72, 103
GlyR α_3	1–33	171-bond-185 231-bond-242	71
GlyR β	1–22	183-bond-197	54, 242

Note: For sequence motifs† and species not shown, refer to entries retrievable by the accession numbers in the *Database listings, 11-53.*

STRUCTURE & FUNCTIONS

Note: The symbol [PDTM] denotes an illustrated feature on the channel protein domain topography model (Fig. 1).

Amino acid composition

11-26-01: Amino acid hydropathicity† analyses of all GlyR α and β subunits show a pattern of hydrophobic† domains (M1–M4) typical of the extracellular ligand-gated ion channel superfamily *(see [PDTM], Fig. 1)*.

Domain arrangement

Architecture of the glycine-binding domain

11-27-01: A **multisite model** of the GlyR ligand-binding region (based on analysis of agonist responses in site-directed† mutants) which includes both high- and low-affinity agonist subsites within an extended binding domain has been proposed[28]. In this model, '**ligand recognition**' implies that several side-chains fold together in the assembled GlyR complex, including those in the second half of the extracellular N-terminal domains of α subunits *(see [PDTM], Fig. 1)*. This pattern appears to be conserved amongst other ligand-gated ion channels such as the nAChR and $GABA_A$ receptor–channels *(see ELG CAT nAChR, entry 09 and ELG Cl $GABA_A$, entry 10, respectively)*.

Net positive charges are associated with GlyR vestibules

11-27-02: The existence of **channel vestibules**† containing **net positive charges** are consistent with permeability† properties of glycine receptor–channels (i.e. anion-selective) and may function to **attract anions**[12] (compare with the 'rings' of negatively charged residues within the vestibules† of the *cation-selective* nicotinic acetylcholine receptor – *see Domain arrangement under ELG CAT nAChR, 09-27*). The GlyR α subunit **domain M2** is likely to 'line' the hydrophilic† pore, which is similar in arrangement to other extracellular ligand-gated channels *(see Domain conservation, 11-28)*.

Domain conservation

Conserved features of the M2 (pore-lining) domain

11-28-01: In general, there is a high degree of sequence conservation of transmembrane segments M1–M3 between the GlyR and the $GABA_AR$, indicating their potential importance in chloride channel formation[55,56]. The **M2 domain** shows the *highest* conservation *(see Domain functions, 11-29)*. By analogy to other ELG channels, the M2 domain can be covalently† labelled by crosslinked† non-competitive† channel blockers and determines the single-channel conductance† properties *(a comparison of permeation† properties of these two channel types appears under Selectivity, 11-40)*.

Conserved residues and motifs between GlyR and $GABA_A$ channels
11-28-02: Both glycine and $GABA_A$ receptor–channels display (i) an invariant proline residue at mid-position in the M1 domain; (ii) a common hydroxy-rich sequence Thr–Thr–Val–Leu–Thr–Met–Thr (Ser) and a total of eight Ser or Thr in each M2 domain; (iii) a proline residue at the fourth position preceded by a phenylalanine residue in the M4 domain; (iv) relatively **high positive charge density** within eight residues of the ends of the transmembrane domains on the extracellular sides (by contrast, *cation-selective* ELG channels display negatively charged residues in addition to positive charges bordering M2[57,58]); (v) potential to form a ***β*-loop**† between Cys139 and Cys153, with 8 of 15 positions being identical or highly conserved in all subunits[59].

Amino acid conservation among GlyR subunits
11-28-03: The α_3 ligand-binding GlyR subunit[8] exhibits ~82–83% **amino acid identity** to the previously characterized rat and human α_1 and α_2 subunit sequences. In *Xenopus* oocytes, α_3 homomultimers† form functional glycine-gated channels which show low glycine affinity and small responses to taurine[8].

Domain functions (predicted)

Properties of heterologously† expressed GlyR subunits resemble native† receptor–channels
11-29-01: Glycine can elicit large chloride currents following transient† or stable† heterologous expression of α_1, α_2 or α_3 GlyR subunit cDNA in *Xenopus* oocytes or mammalian cells in culture[31,60]. Generally, electrophysiological and pharmacological characteristics of heterologously† expressed channels are similar to those observed in cultured spinal neuronal cells *(see, e.g., ref.[12])*.

Delineation of amino acid residues involved in antagonist binding in GlyR α subunits
11-29-02: α subunits contribute the **strychnine-binding site** of intact pentameric GlyRs. Studies involving proteolytic† digestion of [^{3}H]-strychnine-labelled glycine receptors[61,63] have shown that covalent incorporation of the antagonist† occurs between amino acid positions 170 and 220 of the α_1 subunit, close to the first transmembrane region *(see [PDTM], Fig. 1)*. Photoaffinity-labelling has further localized this site to between residues 177 and 220 *(see ref.[63]) (see also below)*. Based on *structural* considerations, a stretch of charged residues around two cysteines preceding the first transmembrane segment has been implicated in ligand binding to the GlyR[62]. *Functional analysis* of transiently expressed†, mutated GlyR cDNAs in mammalian (HEK-293) cells has also shown that residues from two *separate* domains form the α_1 subunit strychnine-binding site. The first domain includes the amino acid residues Gly160 and Tyr161, and the second domain includes the residues Lys200 and Tyr202[64] *(see below)*.

Distinct amino acid residues critical for GlyR agonist and antagonist binding

11-29-03: *Independent reports* have concluded that binding of the GlyR antagonist† **strychnine** requires an interaction with residues Lys200 and Tyr202 of the GlyR[64] with an **agonist-binding site** of the GlyR being located at the residue Thr204[65]. These results demonstrate that agonist- and antagonist†-binding sites are composed of *distinct* amino acid residues (in contrast to other studies of ligand-gated receptor–channels). Chemical modification of lysine and histidine residues of the GlyR abolishes agonist (glycine) but not antagonist (strychnine) binding[63, 66]. *Note:* Interactions of the GlyR with agonists are likely to be mediated by **hydrogen bonding**† and not by ionic interactions.

Residues important for taurine activation – distinct agonist subsites on α_1 and α_2 GlyRs

11-29-04: Whereas α_1 receptors are efficiently gated† by β-alanine and taurine, α_2 GlyRs show only a low relative response to these agonists, which also display a reduced sensitivity to inhibition by the glycinergic† antagonist† and convulsant† **strychnine**. Construction of an α_2/α_1 subunit chimaera† and site-directed mutagenesis† of the extracellular region of the α_1 sequence has identified amino acid positions 111 and 212 as important determinants of taurine activation[28] *(see [PDTM], Fig. 1)*. These results suggest that the **ligand-binding pocket** is formed from discontinuous domains of their extracellular region and indicate the existence of distinct subsites for agonists on α_1 and α_2 GlyRs[28]. *(A diagrammatic representation of predicted agonist and antagonist interactions with the GlyR is included as an inset to the [PDTM], Fig. 1.*

Critical residues for agonist efficacy†

11-29-05: Mutational exchange† of the amino acid residues Phe159 and Tyr161 in the α_1 (ligand-binding) subunit of the rat GlyR increases efficacy† of amino acid agonists[67]. Doubly-mutated (Phe159 to Tyr, Tyr161 to Phe) α_1 homomeric† GlyRs require ~0.7 mM β-alanine for a half-maximal response when expressed in *Xenopus* ocytes (an affinity ~110-fold greater than that of the wild type†). *Note:* β-Alanine displayed greater potency† than glycine in these experiments[67].

Anion selectivity of GlyR–channels depends on terminal positive charges in the M2 segment

11-29-06: The hydrophilic† lining of the GlyR chloride channel includes eight conserved, **hydroxylated side -chains** and a series of **positively charged amino acids** bordering M2. Synthetic peptides which correspond to segment M2 of the GlyR α_1 chain of rat (including its N- and C-terminal-adjoining arginine residues) form randomly gated channels following the incorporation into lipid bilayers[68]. Ion selectivity of these 'synthetic channels' can be modified following 'inversion' of the terminal charges of the peptide, suggesting anion binding to these residues at the mouth of the channel (compare the 'rings of negatively charged residues' bordering the M2 domain of the nicotinic receptor which determine channel conductivity – *see ELG CAT nAChR, entry 09, and ref.*[69]).

Conversion of cation- to anion-selective ELG channels by site-directed mutagenesis

11-29-07: Introduction of three amino acids 'from' the glycine receptor M2 segment (or the $GABA_A$ receptor) 'into' the M2 segment of an α_7 nicotinic receptor is sufficient to convert the **cation-selective** nAChR into an **anion-selective channel** gated by acetylcholine. The critical mutation in these studies was the insertion of an *uncharged* residue at the N-terminal end of M2, indicating the importance of protein geometrical constraints on ion selectivity[70] *(see Selectivity under ELG CAT nAChR, 09-40).*

A conserved Asp148 is important for GlyR assembly and influences antagonist binding

11-29-08: Site-directed mutagenesis† of the GlyR α subunit coding sequence† predicts that an invariant residue (Asp148) forms part of a **receptor assembly/ antagonist-binding site** *(see Table 6).* This structural motif† may play a role in antagonist binding to other related receptors[54].

Table 6. *Effects of selected mutations on GlyR function (From 11-29-08)*

Mutation	Functional changes/ observations	Conclusion/inference
Lys143Ala	Essentially unaltered GlyRs with small decreases in strychnine affinity, glycine displacement of strychnine binding, and glycine activation of chloride currents[54]	Lys143 does not play a major role in either agonist or antagonist binding or agonist activation of the GlyR
Asp148Ala *or* Asp148Asn	No binding sites or ion channels were expressed on the cell surface[54]	The mutation disrupts the expression and/or assembly of the GlyR. Note the conservation of the Asp148 residue *(see above)*
Asp148Glu (Conservative substitution)	Reduced efficiency of expression; a single order of magnitude decrease in strychnine affinity; no change in glycine displacement of strychnine binding or glycine activation of chloride currents[54]	Asp148 plays an important role in GlyR assembly and in antagonist binding, but no significant role in agonist binding

Residues in M2 of α subunits determine the main conductance state of GlyR channels

11-29-09: The **main conductance state**† of GlyR α subunit homo-oligomers† expressed in HEK-293 cells depends on residue 221 which is located within transmembrane segment M2. The mutant α_1 G221A gives rise to a main state† of 107 pS, recorded at symmetrical Cl^- concentrations of 145 mM. (cf. 86 pS for wild-type† homo-oligomers†)[71] *(see Single-channel data, 11-41).*

Residues in M2 of β subunits determine picrotoxinin resistance of GlyR

11-29-10: Site-directed mutagenesis† has identified **pore-lining residues** within the second predicted transmembrane segment (M2) of the ***β* subunit** as major determinants of **picrotoxinin resistance**[72, 73] *(see Receptor antagonists, 11-51).* When divergent residues of the β M2 region (which differs markedly from that of the GlyR α_1 or $GABA_A$ M2 segment) are replaced with side-chains from the α_1 subunit, heteromeric† α/β channels generated with the modified β subunit are inhibited by **picrotoxinin**. In addition to reduced single-channel conductance properties upon co-assembly of the β subunit[71] *(see Single-channel data, 11-41),* these results indicate that the M2 segment is critical for both binding of blockers and ion flux through GlyR chloride channels.

Predicted protein topography

Basic features of GlyR protein assemblies

11-30-01: Amino acid composition† analysis of intact, chemically cross-linked GlyR subunits has shown the GlyR (i) to consist of a **pentameric** assembly of membrane-bound subunits; (ii) to possess three copies of the α subunit per complex and two copies of the β subunit per complex *(see [PDTM], Fig 1, and paragraph 11-30-02).* This quaternary† stucture resembles that of the nicotinic acetylcholine receptor *(see ELG CAT nAChR, entry 09)* and other extracellular ligand-gated channels. *Note:* An additional 93 kDa non-glycosylated **peripheral membrane protein** appears to influence channel topography† *in situ (see Subcellular locations, 11-16, and Protein interactions, 11-31).*

Functional GlyR α/β hetero-oligomers have an invariant (3:2) stoichiometry

11-30-02: Co-expression studies in *Xenopus* oocytes have established that GlyR α/β hetero-oligomers have an **invariant (3:2) stoichiometry**[22]. By contrast, *variable* subunit ratios can be distinguished when low-affinity mutants of the α_2 subunit are co-expressed with wild-type α_1 and α_2 subunits[74].

Amino acid sequence motifs governing GlyR subunit stoichiometry† and assembly

11-30-03: Construction of **chimaeric receptors** between α and β subunits has revealed that differences in **subunit assembly ratios** are determined by the N-terminal (putatively extracellular) regions of the subunits[75]. Substitution of residues diverging between the α and β subunits has identified a subdomain of the β subunit N-terminal region (residues 1–153) that is essential for **stoichiometric subunit assembly** *(see [PDTM], Fig. 1). Comparative note:* The *transmembrane* domains of the α and β subunits appear *unimportant* for

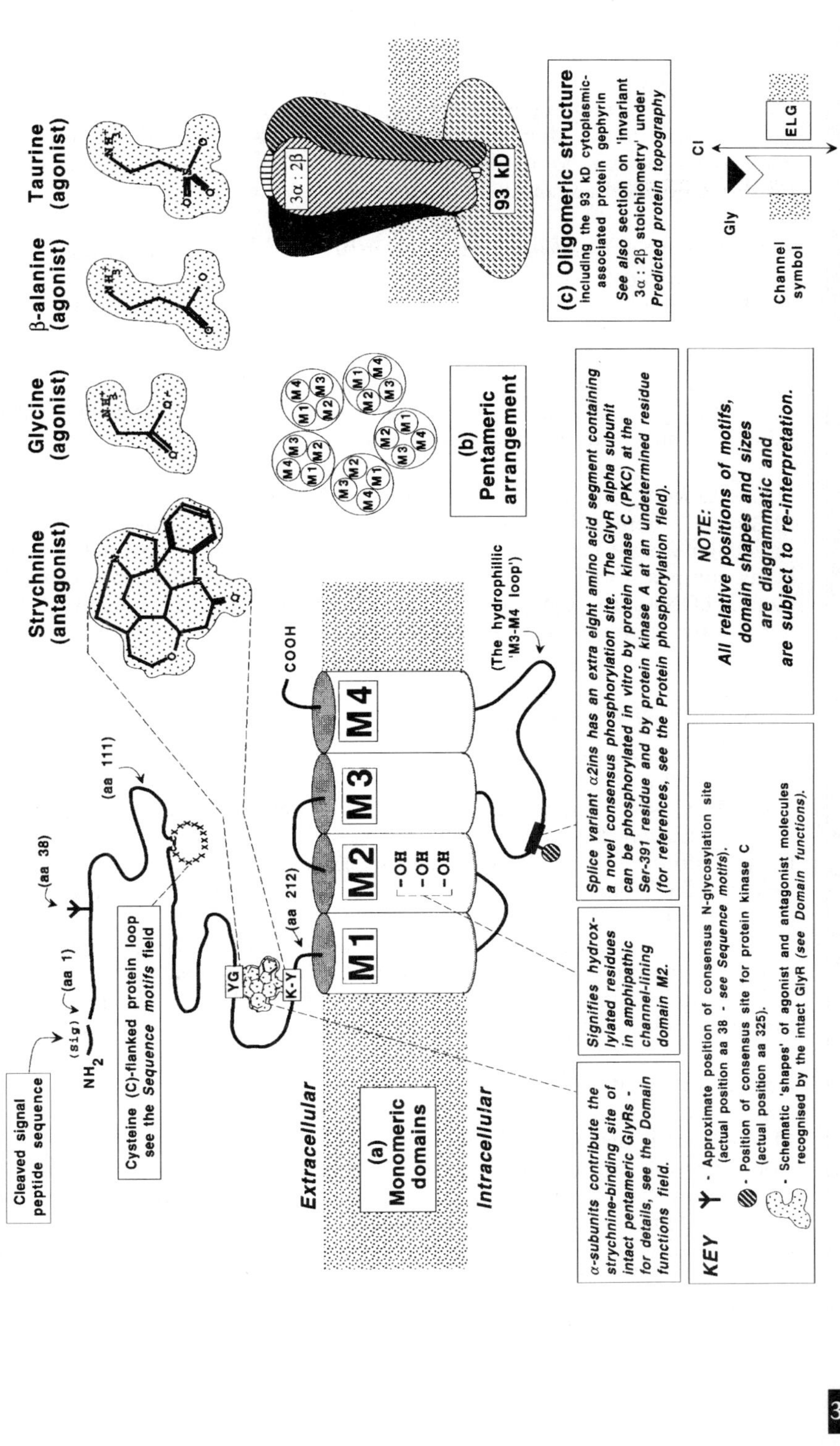

Figure 1. *Monomeric protein domain topography model [PDTM] for the glycine receptor–chanel (GlyR) subunit α_1. (From 11-30-01)*

correct subunit stoichiometry†. On the basis of agonist† dose–response-changes with combinations of GlyR carrying mutations in various '**assembly boxes**', it was deduced that the properties of stoichiometric† assembly and homo-oligomer† formation were 'mutually exclusive'. These findings may imply that the '**assembly domains**' of the β subunit have evolved to *prevent* homo-oligomerization, and that this may be essential for the '**sequential assembly**' of functional GlyR (i.e. $\alpha/\beta + \alpha/\beta + \alpha \rightarrow \alpha_3\beta_2$)[75].

Arg219 is essential for correct assembly and biogenesis of α_1 homomeric GlyRs

11-30-04: Site-directed mutagenesis† of Arg219 at the cytoplasmic terminus of domain M2 in the GlyR α_1 subunit generated homomeric proteins which were only **core-glycosylated**†. These mutant regions were retained within **intracellular compartments**, and aggregated to high molecular weight complexes[76]. *Note:* Arg219 corresponds to an arginine/lysine conserved in other ELG superfamily receptor–channels.

Cross-linked pentameric assemblies of GlyR

11-30-05: Chemically cross-linked† glycine receptor subunits have a molecular weight consistent with the assembled receptor–channel complex being composed of five subunits[22]. The principal molecular features of the cloned cDNAs for the GlyR[31,77] are illustrated on the channel protein domain topography model *[PDTM] (Fig. 1).*

Protein interactions

Receptor-associated proteins can 'optimize' functional properties of GlyR–channels

11-31-01: Co-expression of the GlyR with the tubulin-linker protein **gephyrin** *(see Subcellular locations, 11-16)* changes the agonist†- and antagonist†-binding affinities of GlyRs generated by α_2 subunit expression in HEK-293 kidney cells[78]. This has been interpreted as contributing to an 'optimization' of the post-synaptic neurotransmitter response[78]. The 93 kDa component of the native GlyR purified by affinity chromatography with **2-aminostrychnine** probably represents the GlyR linked to tubulin via gephyrin[11] *(see Subcellular locations, 11-16).* The GlyR purified *without* gephyrin is functional when reconstituted in lipid vesicles[23].

Co-assembly of GlyR α and β subunits – dependence on the type of expression system

11-31-02: Although assembly of GlyR α and β subunits is inferred from their **co-purification** during affinity† chromatography and immunoprecipitation†, *transient* co-expression of α with β subunits in *Xenopus* oocytes produces *no significant alterations* in agonist-induced gating†, but channels have relatively low agonist affinity[31]. *Stable* co-expression of α with β subunits in HEK-293 cells results in large whole-cell currents with altered (lowered) picrotoxinin sensitivity when compared to α subunit homomultimers† (see *Domain functions, 11-29, and Receptor antagonists, 11-51).*

Putative interactions of β subunits with other ELG channel subunits
11-31-03: The **ubiquitous expression** of β subunits in brain *(see mRNA distribution, 11-13)* has suggested the possible existence of **native chimaeric† receptors** containing β subunit components (e.g. as part of $GABA_AR$, putative taurine/β-alanine receptors and NMDAR). These proposals have been discussed[32,33].

Protein phosphorylation

Alternative splicing of α_1 transcripts can include or exclude a novel phosphorylation site
11-32-01: Sequence analysis of alternative splice† variants from the GlyR α_1 gene show variation in the presence or absence of a novel **consensus† phosphorylation site**[79] (the α_{1ins} variant containing an eight amino acid insert in the M3–M4 intracellular loop, close to M3 *(see [PDTM], Fig. 1)*. This feature raises the possibility that alternative splicing† can contribute to functional diversity of the GlyR through regulating susceptibility to **protein kinases**. *Comparative note:* An equivalent eight amino acid sequence insertion (LLRMFSFK) has been identified for the γ_2 subunit of the $GABA_A$ receptor which has been shown to confer an *in vitro* substrate for protein kinase C[80] *(see Protein phosphorylation under ELG Cl $GABA_A$, 10-32).*

'Clustering' of consensus phosphorylation sites in the putative M3–M4 loop
11-32-02: Potential phosphorylation-modulatory sites exist in the **hydrophilic loop** separating transmembrane segments M3 and M4 in GlyR monomers *(see [PDTM], Fig. 1). Note:* The M3–M4 loop regions would be *expected* to provide most of the cytoplasmic mass of the intact GlyR complex.

In vitro *phosphorylation of GlyR subunits*
11-32-03: The GlyR α subunit can be phosphorylated *in vitro* by **protein kinase C** (PKC) at the Ser391 residue close to the fourth transmembrane domain[81]. The same subunit is a substrate for phosphorylation by protein kinase A (PKA) at an undetermined residue[82]. Incubation of intact rat spinal cord neurones with specific PKC or PKA activators leads to increased phosphorylation of the GlyR α subunits, strongly suggesting a physiological role for this modification. The treatment of oocytes expressing GlyRs from pooled rat brain $poly(A)^+$ mRNA with phorbol esters† *decreases* glycine-evoked currents whereas treatment with dibutyryl cAMP *enhances* glycine-evoked currents[82].

Protein kinase A phosphorylation increases GlyR currents in medullary dorsal horn neurones
11-32-04: Protein kinase A phosphorylation via a cholera toxin-sensitive G protein (G_s) dramatically increases glycine-induced Cl^- currents through the GlyR of spinal trigeminal neurones by increasing the channel P_{open}[83].

ELECTROPHYSIOLOGY

Current–voltage relation

Rectification properties of whole-cell GlyR channel currents

11-35-01: A pronounced **outward rectification**† is observed when GlyR channels conduct in the steady state†. This rectification† is likely to reflect a voltage-dependence† of gating†, not a voltage-dependence of ion permeation†. The open probability† (P_{open}) depends upon the **charge** on the **cytoplasmic face** of the membrane (when it is negatively charged, P_{open} will be low). Upon depolarization of the neuronal membrane, the cytoplasmic face will become more positive and P_{open} of the GlyR will increase (opposing the depolarizing stimulus). *(See also Voltage sensitivity, 11-42).*

Selectivity

GlyR–channels share many ionic selectivity characteristics with $GABA_A$R–channels

11-40-01: Many comparative studies suggest that both glycine and **$GABA_A$ agonist-activated channels** *(see ELG Cl $GABA_A$, entry 10)* act as multi-ion pathways and have similar permeation† characteristics[84]. A systematic comparison of the selectivity† mechanisms of GlyR and $GABA_A$R in mouse spinal cord neurones has appeared[12] – some of the data applicable to GlyR ion permeation properties are summarized in Table 7.

Estimated open-pore diameters of GlyR

11-40-02: A pentameric arrangement of α-helices has been predicted to form a **central bore** of ~0.58 nm in diameter, a value which closely matches estimated **effective pore diameters** using bi-ionic reversal potential† measurements (compare 0.52 nm and 56 nm for the open GlyR and $GABA_A$R channels respectively)[12]. The structure–function relationships† for **permeation pathways**† in a range of neurotransmitter-gated ion channels has been reviewed[85]. For critical amino acid residues in controlling ionic selectivity, *see Domain functions, 11-29.*

Single-channel data

Characteristic 'bursting' and subconductance behaviour of native GlyRs

11-41-01: Glycine elicits **'bursts'**† of single-channel openings displaying **multiple subconductance**† **states** in cultured cells[12, 13, 15, 86], native spinal neurones[14] and in cells expressing recombinant† GlyR *(see Fig. 2)*. In the falling phase of glycinergic† inhibitory synaptic currents, single-channel currents can be resolved as **discrete steps**. In cell-attached† patches, the single-channel slope conductances (close to 0 mV membrane potential) are 29, 18 and 10 pS[12]. In outside-out† patches (with equal extracellular and intracellular concentrations of 145 mM Cl^-), conductance states of 46, 30, 20 and 12 pS are observed, with the most frequently occurring substate† being 46 pS[12]. *(See also Selectivity, 11-40.)*

Table 7. *Similarity of selectivity characteristics for inhibitory $GABA_A$ and glycine receptor–channels (From 11-40-01)*

Ion species	Selectivity characteristics
GlyR and $GABA_AR$, chloride ion currents	The E_{rev} of whole-cell currents shifts 56 mV per tenfold change in internal Cl^- activity, indicating activation of Cl^--selective channels
GlyR and $GABA_AR$, selectivity over K^+ ions	The permeability ratio of K^+ to Cl^- (P_K/P_{Cl}) is <0.05
GlyR and $GABA_AR$, small anion relative permeability series	$SCN^- > I^- > Br^- > Cl^- > F^-$
GlyR and $GABA_AR$, small anion relative single-channel conductance series	Single-channel conductances† measured in equal 140 mM concentrations of small anions on both membrane faces reveal a **conductance sequence**† of $Cl^- > Br^- > I^- > SCN^- > F^-$ for both anion channels. *Comparative note:* A near-inverse **permeability sequence**† *(see above)* indicates the presence of binding sites for these ions in the channels
GlyR, polyatomic anion permeability	Phosphate and propionate ions are *not* measurably permeant in GlyR–channels although they support small currents through $GABA_AR$–channels. The permeability sequence for large polyatomic anions through GlyR is formate > bicarbonate > acetate

Shifts in single GlyR–channel behaviour with increasing agonist concentration

11-41-02: GlyRs in cultured mouse spinal cord neurones display increases in total current, single-channel opening frequency and longer open times† with increasing agonist† concentration (0.5–2.0 μM). These changes are due primarily to shifts in relative occurrence of **opening frequencies** from the shortest open states† to two **long open states**[15].

Conductance values for homo-oligomeric and hetero-oligomeric GlyR–channels in HEK cells

11-41-03: When recorded at symmetrical Cl^- concentrations of 145 mM in HEK-293 cells, α_1, α_2 and α_3 homo-oligomers† display **main-state**† single-channel conductances† of 86, 111 and 105 pS, respectively[71]. Main-state† conductances of hetero-oligomers† show reductions to 44 pS (α_1/β), 54 pS (α_2/β) and 48 pS (α_3/β). *Note:* Since the lower values are similar to those found in spinal neurones, this has been taken as evidence for native GlyRs existing as **α/β hetero-oligomers**. Co-expression of native α_1 with mutant β subunits shows that residues within (and close to) segment M2 of the β

subunit determine the conductance differences between homo- and hetero-oligomers[71] *(see Domain functions, 11-29).*

Developmental changes in open-time kinetics of single-channel GlyR currents

11-41-04: In rat spinal cord, α_2 **subunit expression** is predominant in foetal or newborn tissues, and **developmental switching** from α_2 to α_1 occurs within 3 weeks following birth[3,7,24–26] *(see Developmental regulation, 11-11).* In direct comparisons, mean channel open times† of GlyR α_1 and mature native† glycine receptors (rat spinal cord) are equally short, whereas both the recombinant α_2 and (native) foetal receptors show significantly longer open times†[27]. GlyRs in native spinal cord neurones show a striking *continuous fall* in mean open time† following birth (typically from 40–50 ms at birth to 2–10 ms at age 20 days)[27]. Consistent with these results, the decay time of the glycinergic† inhibitory post-synaptic currents† (IPSCs) in spinal neurones becomes shorter during post-natal development.

Comparison of single-channel conductance levels for typical pre- and post-natal GlyR components

11-41-05: In outside-out† patches from *Xenopus* oocytes expressing GlyR α_1 or α_2 subunit cRNAs†, application of glycine induces single-channel currents exhibiting **multiple amplitudes**[27] *(see Fig. 2a and c).* By plotting **single-channel amplitude histograms**† for currents supported by homomeric† α_1 receptors and homomeric† α_2 receptors, **distinct single-channel (peak) conductances**† can be resolved at 75, 59, 43 and 25 pS for α_1 receptors *(Fig. 2b)* and 88, 72, 42 and 24 pS for α_2 receptors *(Fig. 2d).* With the exception of the 88 pS level (observed only for the 'foetal' α_2 receptor) the *conductance* levels for channels supported by these α_1 or α_2 homo-multimers† are broadly similar. Analysis of glycine-gated channel currents of **native**† **GlyRs** in rat spinal neurones at foetal stage (E20) and post-natal stage (P16) also show broadly similar conductance levels except for a **94 pS level** which is observed only in the foetal neurones[27]. *Methodological note:* By fitting for Gaussian curves† using the least-squares method†, single-channel conductances† can be estimated from the peak amplitude† of the Gaussian curve†.

Voltage sensitivity

Gating processes of GlyR–channels are voltage-sensitive

11-42-01: Current–voltage (I–V) relations† of transmitter-activated currents obtained from whole-cell measurements in mouse spinal cord neurones, show **outward rectification** (with 145 mM Cl^- intracellularly and extra-cellularly)[12]. In voltage-jump† experiments, the instantaneous† I–V relations are linear, and the steady-state† I–V relations are outwardly rectifying†, indicating that the gating of GlyR channels is **voltage-sensitive** *(see also Current–voltage relation, 11-35).*

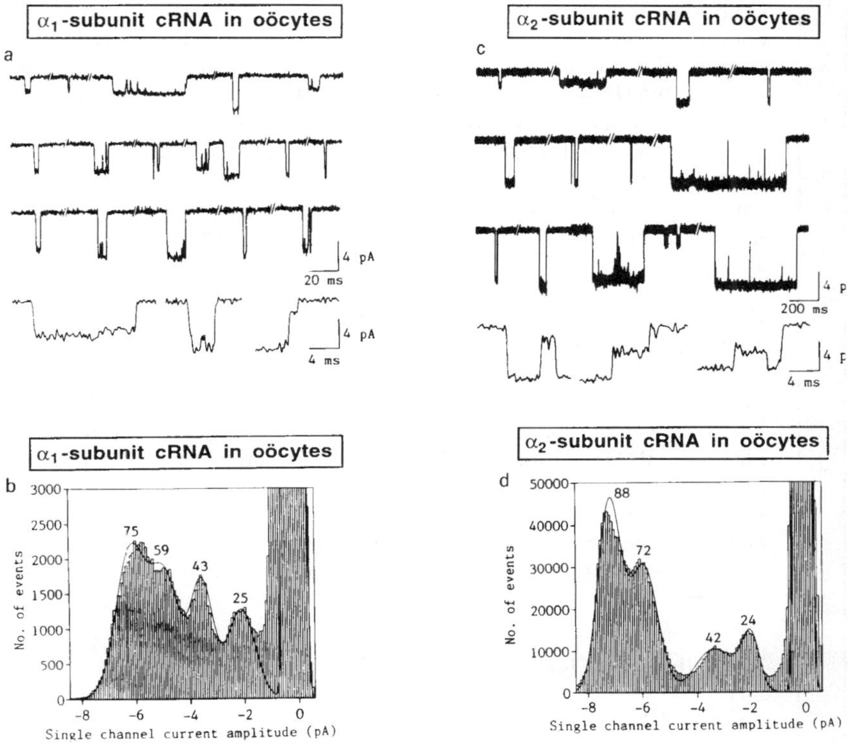

Figure 2. *Comparison of single-channel conductance levels for typical pre- and post-natal GlyR components. (Reproduced with permission from Takahashi et al. (1992)* Neuron **9**: 1155–61). (From 11-41-05)

PHARMACOLOGY

Blockers

Picrotoxinin sensitivity is greatly reduced in hetero-oligomeric GlyRs

11-43-01: Inhibition of α homo-oligomeric† GlyRs by **picrotoxinin** (a non-competitive blocker of ion flow, and sometimes described as 'selective' for $GABA_A$ chloride channels) is reduced 50- to 200-fold for α/β hetero-oligomeric† receptors generated by co-transfection[72] . *Note:* For molecular determinants of **picrotoxinin resistance** on β subunits, *see Domain functions, 11-29.*

Channel modulation

Importance of phosphomodulation for GlyR–channel function

11-44-01: In common with other extracellular ligand-gated receptor–

channels, the GlyR contains a large (putatively) intracellular loop that contains a number of consensus recognition sites for protein kinases *(see Protein phosphorylation, 11-32)*. **Extracellular signals** released from the presynaptic neurone, such as **neurotransmitters** and **neuropeptides** as well as an **extracellular matrix protein** *(see note below)*, are known to regulate several functional characteristics of ligand-gated ion channels by phosphorylation *(see review[87] and ELG Key facts, entry 04)*. *Note:* Co-expression† of the GlyR/tubulin-linker protein **gephyrin** with the GlyR influences agonist- and antagonist-binding affinities *(see Protein interactions, 11-31)*.

Penicillin G-induced potentiation of I_{Gly}

11-44-02: Penicillin G (PenG) has been observed to potentiate† GlyR-mediated chloride current in rat ventro-medial hypothalamic neurones[88]. When PenG is applied simultaneously with glycine (Gly), PenG depresses I_{Gly} like a Cl^- channel blocker (from ~30 units/10 ml to a maximum blockade at 600 units/ml). When test solutions containing PenG plus Gly are rapidly substituted with one containing glycine alone, a new 'rebound-like transient current' (I_T) is observed which passes through the Cl^- channel. The peak amplitude† of I_T induced by PenG (>100 units/10 ml) is greater than that induced by glycine alone (i.e. PenG-induced potentiation of I_{Gly})[88].

Equilibrium dissociation constant

Antagonist affinities

11-45-01: The 'high-affinity' binding of **[^{3}H]-strychnine antagonists**† to membranes of rat spinal cord has a typical K_d in the range of 2–12 nM.[1,2,89] This binding is displaced by the agonists† **glycine**, **β-alanine**, **taurine** and **β-aminoisobutyric acid** *(see Domain functions, 11-29)*.

Differential agonist affinities of cloned subunits

11-45-02: Different GlyR α subunit isoforms (e.g. α_1, α_2, α_2*, α_3) form receptors with **distinct agonist affinities**, allowing their discrimination by dose–response criteria[3,6,8,26,28].

Hill coefficient

Co-operativity in agonist-induced gating

11-46-01: α_1 subunit RNA or poly(A)$^+$ RNA from brain or spinal cord injected into oocytes supports the expression of GlyR which display Hill coefficients† of $n = 2.5–3.3$ *(e.g. ref.[28])*, suggesting that a *minimum* of three glycine molecules are required to open the channel. High **co-operativity** may indicate a mechanism for gating† within a narrow extracellular agonist† concentration range.

Dependence of Hill coefficient on receptor subunit stoichiometry

11-46-02: Dose–response curves† of whole-cell† currents in HEK-293 cells expressing recombinant α_1 subunits display an average Hill coefficient† of

4.2. Co-expression of GlyR α_1 and β subunit cDNAs in HEK-293 cells markedly increases glycine-gated whole-cell currents[71] and alters the mean Hill coefficient to 2.5 *(see Single-channel data, 11-41).*

Ligands

11-47-01: The principal radioligand† used for GlyR binding assays is **[^{3}H]-strychnine**, which is antagonized by (in order of relative potency) **glycine** > **β-alanine** > **taurine** ≫ **L-alanine**, **L-serine** > **proline**. Cross-linking† of [^{3}H]-strychnine to native and heterologously† expressed GlyRs have determined the **ligand-binding site** to be located on α subunits but not β subunits *(for details, see Domain functions, 11-29).*

Mechanism of ligand discrimination

11-47-02: Gamma-aminobutyric acid (**GABA**) and **D-serine**, amino acids that do *not* activate wild-type α_1 homomeric glycine receptors, efficiently gate† a mutant† glycine receptor–channel (a F159Y, Y161F amino acid exchange of the GlyR α_1 subunit demonstrating that **aromatic hydroxyl groups** are crucial for **ligand discrimination** at inhibitory amino acid receptors[67] *(see also Domain functions, 11-29).*

Receptor agonists (selective)

The GlyR is activated by a range of α and β amino acids

11-50-01: In general, potency† of GlyR activation by agonists† decreases in the order **glycine** > **β-alanine** > **taurine** ≫ **L-alanine**, **L-serine** > **proline**[5,91]. In rat brain slice substantia nigra neurones, **taurine** acts at the *same* recognition site as glycine on the GlyR–channel[92].

Comparative note – glycine is also a co-agonist for NMDA receptor–channels

11-50-02: Glycine *or a glycine-like molecule* is an important modulator for the NMDA receptor–channel *(see ELG CAT GLU NMDA, entry 08).* Desensitization† of NMDA receptor-mediated currents elicited by glutamate in newly excised† outside-out patches is reduced in the presence of *saturating concentrations* of glycine, probably due to the **allosteric coupling** of the glutamate- and glycine-binding sites[93] *(see Receptor agonists under ELG CAT GLU NMDA, 08-50).*

Functional note – glycine is a structurally simple molecule

11-50-03: The agonists for inhibitory receptor–channels (glycine and GABA) are of relatively simple structure *(see [PDTM], Fig. 1)* and in consequence, the number of specific interactions they can have with their respective receptors is likely to be limited[65]. This factor is of significance in determination of **structure–function relationships**† of the GlyR, as removal of any one of these specific interactions can be expected to result in dramatic reductions in agonist affinity[65].

Receptor antagonists (selective)

The plant alkaloid strychnine is a well-characterized, high-affinity antagonist for adult GlyRs

11-51-01: The GlyR can be identified by the convulsant **strychnine**[1], which binds at a site *distinct* from glycine[65] *(see Domain functions, 11-29)*. In **strychnine poisoning**, normal glycinergic† inhibition is abolished, giving rise to **muscular convulsions** by 'overexcitation' of the **motor system**. Strychnine can be used to distinguish the $GABA_A$ receptor from glycine receptor Cl^- channels (compare pA_2 values against muscimol and glycine of 5.3 and 6.0 (neonatal) or 8.0 (adult) respectively). *Note:* Some antagonistic action of strychnine has been reported at the **$GABA_A$ receptor**[90].

INFORMATION RETRIEVAL

Database listings/primary sequence discussion

11-53-01: *The relevant database is indicated by the lower case prefix (e.g. gb:), which should not be typed (see Introduction & layout of entries, entry 02). Database locus names and accession numbers immediately follow the colon. Note that a comprehensive listing of all available accession numbers is superfluous for location of relevant sequences in GenBank® resources, which are now available with powerful in-built neighbouring† analysis routines (for description, see the Database listings field in the Introduction & layout of entries, entry 02). For example, sequences of cross-species variants or related gene family† members can be readily accessed by one or two rounds of neighbouring† analysis (which are based on pre-computed alignments performed using the BLAST† algorithm by the NCBI†). This feature is most useful for retrieval of sequence entries deposited in databases later than those listed below. Thus, representative members of known sequence homology groupings are listed to permit initial direct retrievals by accession number, author/reference or nomenclature. Following direct accession, however, neighbouring analysis is† strongly recommended to identify newly-reported and related sequences.*

Nomenclature	Species, cDNA source	Original isolate	Accession	Sequence/ discussion
Glycine receptor **alpha-1** chain precursor	Human (principal adult GlyR in spinal cord); Glycine ligand-binding subunit	421 aa (48 kDa protein)	gb: X52009 eb: Y00276 gb: M63915 pir: A27141 pir: JN0014 prosite: PS00236 sp: P07727	Grenningloh, *Nature* (1987) **328**: 215–20. Sontheimer, *Neuron* (1989) **2**: 1491–7.

Nomenclature	Species, cDNA source	Original isolate	Accession	Sequence/ discussion
				Ruiz-Gómez, *Biochemistry* (1990) **29**: 7033–40. Malosio, *J Biol Chem* (1991) **266**: 2048–53. Langosch, *Eur J Biochem* (1990) **194**: 1–8.
Glycine receptor **alpha-1** chain precursor	Human	449 aa	gb: X52008 sp: P23415 pir: S12382 mim: 138491 prosite: PS00236	Grenningloh, *EMBO J* (1990) **9**: 771–6.
Glycine receptor **alpha-1** chain precursor	Rat	449 aa	gb: D00833	Akagi, *Neurosci Res* (1991) **11**: 28–40.
Glycine receptor **alpha-1** chain precursor	Rat	457 aa	gb: 55246	Kuhse, Malosio, Betz, Unpublished (1990).
Glycine receptor **alpha-2** chain precursor	Human *'foetal, nM strychnine-sensitive GlyR'*	452 aa **Gly** at aa 167 (cf. Glu167 in α_2*)	gb: X52008 sp: P23416 MIM: 305990 prosite: PS00236	Grenningloh, *EMBO J* (1990) **9**: 771–6.
Glycine receptor **alpha-2A**	Rat	452 aa	gb:X61159	Kuhse, *FEBS Lett* (1991) **283**: 73–7.

Nomenclature	Species, cDNA source	Original isolate	Accession	Sequence/ discussion
Glycine receptor **alpha-2*** chain precursor	Rat, Wistar, 10 day old (principal neonatal GlyR in spinal cord); Glycine ligand-binding subunit; 'foetal, nM strychnine-insensitive GlyR'	452 aa (48 kDa protein) **Glu** at aa 167 (cf. Gly167 in α_2)	gb: X57281 em: X57281 PIR: S14816 sp: P22771 prosite: PS00236	Akagi, *FEBS Lett* (1991) **281**: 160–6. Kuhse, *Neuron* (1990) **5**: 867–73. Kuhse, *FEBS Lett* (1991) **283**: 73–7.
Glycine receptor **alpha-3** chain precursor	Rat (principal adult cerebellar isoform)	464 aa (48 kDa protein)	gb: M55250 gb: M38385 pir: A23682 sp: P24524	Kuhse, (1990) *J Biol Chem* **265:** 22317–20
Glycine receptor **alpha-4** gene	Mouse genomic	not found	not found	Matzenbach, *J Biol Chem* (1994) **269**: 2607–12.
Glycine receptor **beta** chain precursor	Rat	496 aa (58 kDa protein)	pir: JH0165 PROSITE: PS00236 sp: P20781	Grenningloh, (1990) *Neuron* **4:** 963–70
Glycine receptor **peripheral membrane component**	Co-purifies with the GlyR on affinity columns[10]; gephyrin	(93 kDa protein, non-glycosylated, cytoplasmic location)	not found	Prior, *Neuron* (1992) **8**: 1161–70.

Related sources & reviews

11-56-01: Major quoted sources – GlyR structure–function[34, 94–96]; general reviews[5]; neurotransmitter action and opening of extracellular ligand-gated ion channels[97]; phosphorylation of ELG channels, including the GlyR[87]; permeation pathways of neurotransmitter-gated ion channels[85].

Feedback

Error-corrections, enhancement and extensions

11-57-01: Please notify specific errors, omissions, updates and comments on this entry by contributing to its **e-mail feedback file** (*for details, see Resource J, Search Criteria & CSN Development*). For this entry, send e-mail messages To: **CSN-11@le.ac.uk,** indicating the appropriate paragraph by entering its **six-figure index number** (xx-yy-zz or other identifier) into the **Subject**: field of the message (e.g. Subject: 08-50-07). Please feedback on only **one specified paragraph or figure per message,** normally by sending a **corrected replacement** according to the guidelines in *Feedback & CSN Access* . Enhancements and extensions can also be suggested by this route (*ibid.*). Notified changes will be indexed via 'hotlinks' from the CSN 'Home' page (http://www.le.ac.uk/csn/) from mid-1996.

Entry support groups and e-mail newsletters
11-57-02: Authors who have expertise in one or more fields of this entry (and are willing to provide editorial or other support for developing its contents) can join its support group: In this case, send a message To: **CSN-11@le.ac.uk,** (entering the words "support group" in the Subject: field). In the message, please indicate principal interests (see *fieldname criteria in the Introduction for coverage*) together with any relevant **http://www site links** (established or proposed) and details of any other possible contributions. In due course, support group members will (optionally) receive **e-mail newsletters** intended to **co-ordinate and develop** the present (text-based) entry/fieldname frameworks into a 'library' of interlinked resources covering ion channel signalling. Other (more general) information of interest to entry contributors may also be sent to the above address for group distribution and feedback.

REFERENCES

[1] Young, *Proc Natl Acad Sci USA* (1973) **70**: 2832–6.
[2] Young, *Mol Pharmacol* (1974) **10**: 790–809.
[3] Kuhse, *Neuron* (1990) **5**: 867–73.
[4] Fatimashad, *Proc R Soc Lond [Biol]* (1992) **250**: 99–105.
[5] Langosch, *Eur J Biochem* (1990) **194**: 1–8.
[6] Kuhse, *FEBS Lett* (1991) **283**: 73–7.
[7] Malosio, *EMBO J* (1991) **10**: 2401–9.
[8] Kuhse, *J Biol Chem* (1990) **265**: 22317–20.
[9] Matzenbach, *J Biol Chem* (1994) **269**: 2607–12.
[10] Schmitt, *Biochemistry* (1987) **26**: 805–11.
[11] Prior, *Neuron* (1992) **8**: 1161–70.
[12] Bormann, *J Physiol Lond* (1987) **385**: 243–86.
[13] Hamill, *Nature* (1983) **305**: 805–8.
[14] Takahashi, *Neuron* (1991) **7**: 965–9.
[15] Twyman, *J Physiol Lond* (1991) **435**: 303–31.
[16] Cohen, *J Physiol Lond* (1989) **418**: 53–82.
[17] Suzuki, *J Physiol Lond* (1990) **421**: 645–62.
[18] Korn, *Proc Natl Acad Sci USA* (1992) **89**: 440–3.
[19] Miller, *Life Sci* (1993) **53**: 1211–15.
[20] Kirsch, *J Biol Chem* (1991) **266**: 22242–5.
[21] Pfeiffer, *J Biol Chem* (1982) **257**: 9389–93.

[22] Langosch, *Proc Natl Acad Sci USA* (1988) **85**: 7394–8.
[23] Garcia-Calvo, *Biochemistry* (1989) **28**: 6405–9.
[24] Becker, *EMBO J* (1988) **7**: 3717–26.
[25] Akagi, *Science* (1988) **242**: 270–3.
[26] Akagi, *FEBS Lett* (1991) **281**: 160–6.
[27] Takahashi, *Neuron* (1992) **9**: 1155–61.
[28] Schmieden, *EMBO J* (1992) **11**: 2025–32.
[29] Falck, *Arch Ges Physiol* (1884) **34**: 375–81.
[30] Becker, *Neuron* (1992) **8**: 283–9.
[31] Grenningloh, *Neuron* (1990) **4**: 963–70.
[32] Malosio, *EMBO J* (1991) **10**: 2401–9.
[33] Betz, *Trends Neurosci* (1991) **14**: 458–61.
[34] White, *Nature* (1982) **298**: 655–7.
[35] Hayashi, *Ann Neurol* (1981) **9**: 292–4.
[36] Hall, *Neurology* (1979) **29**: 262–7.
[37] Gundlach, *FASEB J* (1990) **4**: 2761–6.
[38] Becker, *FASEB J* (1990) **4**: 2767–74.
[39] Heller, *Brain Res* (1982) **234**: 299–308.
[40] White, *Nature* (1985) **298**: 655-657.
[41] Biscoe, *J Physiol* (1986) **379**: 275-292.
[42] Biscoe, *Br J Pharmacol* (1982) **75**: 23–5.
[43] Biscoe, *J Physiol* (1980) **379**: 275–92.
[44] Saenz-Lope, *Ann Neurol* (1984) **15**: 36–41.
[45] Heller, *Brain Res* (1982) **234**: 299–308.
[46] Harding, *J Neurol Neurosurg Psych* (1981) **44**: 871–83.
[47] Pfeiffer, *Proc Natl Acad Sci USA* (1984) **81**: 7224–7.
[48] Triller, *J Cell Biol* (1985) **101**: 683–8.
[49] Akagi, *Proc Natl Acad Sci USA* (1989) **86**: 8103–7.
[50] Eicher, *J Hered* (1980) **71**: 315–18.
[51] White, *J Neurogenet* (1987) **4**: 253–8.
[52] Graham, *Biochemistry* (1985) **24**: 990–4.
[53] Cascio, *J Biol Chem* (1993) **268**: 22135–42.
[54] Vandenberg, *Mol Pharmacol* (1993) **44**: 198–203.
[55] Betz, *Neuron* (1990) **5**: 383–92.
[56] Grenningloh, *Nature* (1987) **330**: 25–6.
[57] Numa, *Harvey Lect* (1989) **83**: 121–65.
[58] Noda, *Nature* (1983) **302**: 528–32.
[59] Barnard, *Trends Biochem Sci* (1992) **17**: 368–74.
[60] Sontheimer, *Neuron* (1989) **2**: 1491–7.
[61] Graham, *Eur J Biochem* (1983) **131**: 519–25.
[62] Grenningloh, *Nature* (1987) **328**: 215–20.
[63] Ruiz-Gómez, *Biochemistry* (1990) **29**: 7033–40.
[64] Vandenberg, *Proc Natl Acad Sci USA* (1992) **89**: 1765–9.
[65] Vandenberg, *Neuron* (1992) **9**: 491–6.
[66] Ruiz-Gómez, *Biochem Biophys Res Commun* (1989) **160**: 374–81.
[67] Schmieden, *Science* (1993) **262**: 256–8.
[68] Langosch, *Biochim Biophys Acta* (1991) **1063**: 36–44.
[69] Imoto, *Nature* (1988) **335**: 645–8.
[70] Galzi, *Nature* (1992) **359**: 500–5.

71 Bormann, *EMBO J* (1993) **12**: 3729–37.
72 Pribilla, *EMBO J* (1992) **11**: 4305–11.
73 Pribilla, *EMBO J* (1994) **13**: 1493.
74 Kuhse, *Neuron* (1993) **11**: 1049–56.
75 Berendes, *Science* (1993) **262**: 427–30.
76 Langosch, *FEBS Lett* (1993) **336**: 540–4.
77 Grenningloh, *Nature* (1987) **328**: 215–20.
78 Takagi, *FEBS Lett* (1992) **303**: 178–80.
79 Malosio, *J Biol Chem* (1991) **266**: 2048–53.
80 Whiting, *Proc Natl Acad Sci USA* (1990) **87**: 9966–70.
81 Ruiz-Gómez, *J Biol Chem* (1991) **266**: 559–66.
82 Vaello, *J Biol Chem* (1994) **269**: 2002–8.
83 Song, *Nature* (1990) **348**: 242–5.
84 Fatimashad, *Proc R Soc Lond [Biol]* (1993) **253**: 69–75.
85 Lester, *Annu Rev Biophys Biomol Struc* (1992) **21**: 267–92.
86 Smith, *J Membr Biol* (1989) **108**: 45–52.
87 Swope, *FASEB J* (1992) **6**: 2514–23.
88 Tokutomi, *Br J Pharmacol* (1992) **106**: 73–8.
89 Becker, *Neurochem Int* (1988) **13**: 137–46.
90 Shirasaki, *Brain Res* (1991) **561**: 77–83.
91 Henzi, *Mol Pharmacol* (1992) **41**: 793–801.
92 Hausser, *Brain Res* (1992) **571**: 103–8.
93 Lester, *J Neurosci* (1993) **13**: 1088–96.
94 Betz, *Q Rev Biophys* (1992) **25**: 381–94.
95 Betz, *Biochemistry* (1990) **29**: 3591–9.
96 Betz, *Adv Exp Med Biol* (1991) **287**: 421–9.
97 Unwin, *Cell* (1993) **72**: 31–41.

FEEDBACK & CSN ACCESS

Feedback and access to the Cell-Signalling Network (CSN)

Entry 12

FEEDBACK

Consolidation of contents

12-01-01: This edition of *The Ion Channels FactsBook* has been compiled by 'scanning' the literature up to the end-of-year 1993 (for vol. 1), to near the end-of-year 1994 (for vol 2) and to mid-1995 for vols 3 and 4. Indexed information supplementary to the entries (i.e. appearing after the publication deadlines) will be accessible over the Internet† using a World Wide Web† browser from mid-1996 (*see below*). This arrangement facilitates a mechanism for both **error correction** and **consolidation** of the information set. To begin with, incremental changes to entries will be logged on a host computer within the CMHT (Leicester, UK – *see computer address below*) but other sites will contribute to the project in due course. Each entry or resource is supported by a unique e-mail feedback file. The feedback format is intended to provide an 'open' and informal mechanism for entry validation by specialists on each subtopic. The method also provides an efficient means to communicate **changes** and **extensions** to entries, and offer a forum for debate. Most of all, 'communities' linked by the core interest in the entries can contribute to the evolution of the dataset as a comprehensive 'informatic tool' by structured development of the field contents *(see Resource J – Search & criteria & CSN development, entry 65).*

The need for feedback

12-01-02: An objective of *The FactsBook* is that the scope and arrangement of the information should be *constantly refined to contain what is most useful and authoritative* – feedback from individual users is an essential part of this process. Inevitably, with so much diverse and detailed information, specialists will discover errors, misinterpretations and significant omissions from the entries as presented. However, if communicated to the appropriate e-mail feedback file (*see below*), these points will be attended to directly.

> *To feedback an error correction or notify a significant omission:* For each individual comment or point, send a separate e-mail message to the feedback file supporting the entry in question as indicated in Feedback *(field 57)* (e.g. **To: CSN-19@le.ac.uk**). Indicate the appropriate paragraph under discussion by entering its **six-figure index number** (xx-yy-zz) into the Subject: field of the message (e.g. **Subject: 19-29-02**). For optimal correction, updating or for signalling a significant omission, include a piece of 'corrected' text or send a re-drawn figure designed to **replace** the existing text or figure without loss of information.

GUIDELINES ON THE TYPES OF FEEDBACK REQUIRED

1 If a *field, or 'significant fact' is not covered* for a particular channel, yet the relevant information is available from a published source, please advise on the best accessible reference (but see note 2).
2 'Updates' for the purposes of 'feedback' are published 'facts' which actually *invalidate a statement as presented within the entries.* New papers in the published literature following the *FactsBook* deadlines do *not* require notification as they will normally be incorporated as part of the literature 'scanning' and updating procedure.
3 If a 'fact' is *incorrectly referenced* please indicate accordingly. *Full* referencing for each separate 'fact' has proven impractical for a printed format but further specific citations could be incorporated if considered useful or necessary.
4 The 'running order' of entries within *The FactsBook* is *not* absolute, and subsequent descriptions of novel channel types (or the cloning of the genes that encode them) are likely to extend and modify the present 'working arrangement'. There will be a shift towards classification by gene structure when more channel genes are cloned and a true consensus on classification and nomenclatures is reached – see the IUPHAR entries under the CSN *(below)* for developments in this area.
5 Electronic mail (e-mail) is strongly preferred for receiving feedback, however formatted text on disk and graphical materials illustrating or enhancing the entry text can be sent by surface mail to: Dr Edward C. Conley, c/o Ion Channel/Gene Expression, University of Leicester/ Medical Research Council, Centre for Mechanisms of Human Toxicity, PO Box 138, Hodgkin Building, Lancaster Road, Leicester LE1 9HN, UK.

12-01-03: Consolidation of the existing database in this way is vital to achieve the original aim of systematic coverage for ion channel molecular properties. Following the establishment of this information framework, an increasing proportion of 'incremental' entries (containing new data) will be maintained via an international network of specialist authors on each type of channel or cell-signalling molecule *(see below).*

THE CELL-SIGNALLING NETWORK (from–Mid-1996)

Posting of updates on the Cell-Signalling Network (CSN)

12-01-04: Following receipt of update suggestions by electronic mail, new or modified entry paragraphs will be posted in computer files as part of the CMHT's **Cell-Signalling Network (CSN)**. Anyone with access to Internet† 'browsing†' software such as Netscape™ can access, search and down-load these corrected/upgraded paragraphs over the WWW (World Wide Web†) network. 'Hot-links†' between different parts of the database (activated by 'double-clicking' objects or words highlighted in boldface underlined type) can be configured between files compiled in multiple centres around the world – expansion by 'double-clicking' will automatically trasverse the 'Web' to the appropriate server† updated by specialist contributors or centres. 'Navigation' through the Web is straightforward – 'double-clicking' items of interest will open new 'linked windows' as you go.

Accessing the Cell-Signalling Network (from–mid-1996)

12-01-05: In addition to updates, the CSN 'home (index) page' (URL: **http://www.le.ac.uk/csn/index.html**) *(Fig. 1)*, provides access to several appendices and reference indexes supporting the contents of *The FactsBook (for a listing of these on-line support appendices, reference indexes and glossary items, see the descriptions under Figs 2 and 3)*. In time, the CSN will aim to provide a convenient index for all information resources that are of potential relevance to cell signalling and that are presently available from Internet† sites. In common with other WWW resources, the CSN can be accessed from any type of computer. Users unfamiliar with Internet browsing† should ask their local network administrator about WWW/Netscape®. Once launched using the above URL† (the **uniform resource locator**†, an 'address' or 'pointer†' to another document on the network), the 'Welcome to the CSN' file on the 'home page' will provide all of the other information needed to use the service *(see facsimile, Fig. 1)*.

Exploring the CSN by hypertext† links

12-01-06: An important criterion for the CSN is that many different groups of specialists, expert in each type of cell-signalling molecule or topic will contribute to the development of an *integrated* dataset *(see the sections entitled 'The CSN as a co-operative project' and 'How to contribute to the CSN', below)*. Thus, each contributor or group of contributors will aim to compile an authoritative 'repository' of information on a molecule or topic of interest 'hyperlinked†' to other information already on the network. Compilation of such a comprehensive 'consensus archive' will inevitably take some time to implement properly, but the process has begun. The availability of free software tools designed to assist high-quality document preparation and exchange over the Internet will make the whole process relatively straightforward. For instance, the graphical user interface† (GUI) provided by the Mosaic or Netscape browser† programs, integrates the process of finding, cross-referencing and down-loading information of interest. Other 'browsers', 'gophers†' and document-exchange software (both commercial and public domain) will become available in due course. Following publication of *The FactsBooks*, the **latest information** on the development of the CSN using these and other tools will be made available directly from the its 'home page' *(Fig. 1)*.

The CSN as a co-operative project

12-01-07: Consolidation of knowledge about all of the different classes of molecules affecting signal transduction in cells remains a significant organizational challenge. Researchers are constantly faced with the problem of **integrating new information** into what is already known (perhaps about a molecule or phenomenon which is not within their immediate field of expertise). However, genuine 'breakthroughs' in understanding often come from relating unexpected developments in parallel fields. Furthermore, in many areas it has become difficult (even for experts) to 'keep pace' with emerging information, which can leave significant comparative data undiscovered. The **relational linking** inherent in the CSN at least offers scope for penetration of unfamiliar fields, together with facilities for integration and communication of the latest developments in familiar ones.

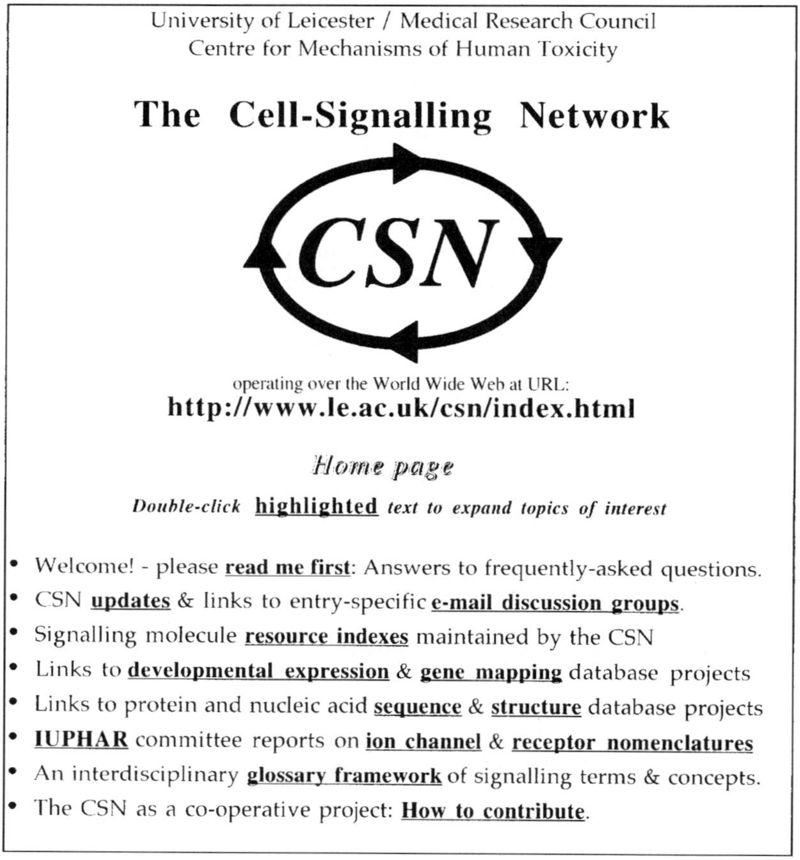
University of Leicester / Medical Research Council
Centre for Mechanisms of Human Toxicity

The Cell-Signalling Network

CSN

operating over the World Wide Web at URL:
http://www.le.ac.uk/csn/index.html

Home page

Double-click **highlighted** *text to expand topics of interest*

- Welcome! - please **read me first**: Answers to frequently-asked questions.
- CSN **updates** & links to entry-specific **e-mail discussion groups**.
- Signalling molecule **resource indexes** maintained by the CSN
- Links to **developmental expression** & **gene mapping** database projects
- Links to protein and nucleic acid **sequence** & **structure** database projects
- **IUPHAR** committee reports on **ion channel** & **receptor nomenclatures**
- An interdisciplinary **glossary framework** of signalling terms & concepts.
- The CSN as a co-operative project: **How to contribute**.

Figure 1. *The CSN 'home page', used to index and inter-relate other information resources on cell-signalling and provide on-line support for* The Ion Channel FactsBook. Note: the 'home page' shown above represents a *prototype* compiled at the time of going-to-press; the details and contents will probably differ in its final form, depending on available support and interest.

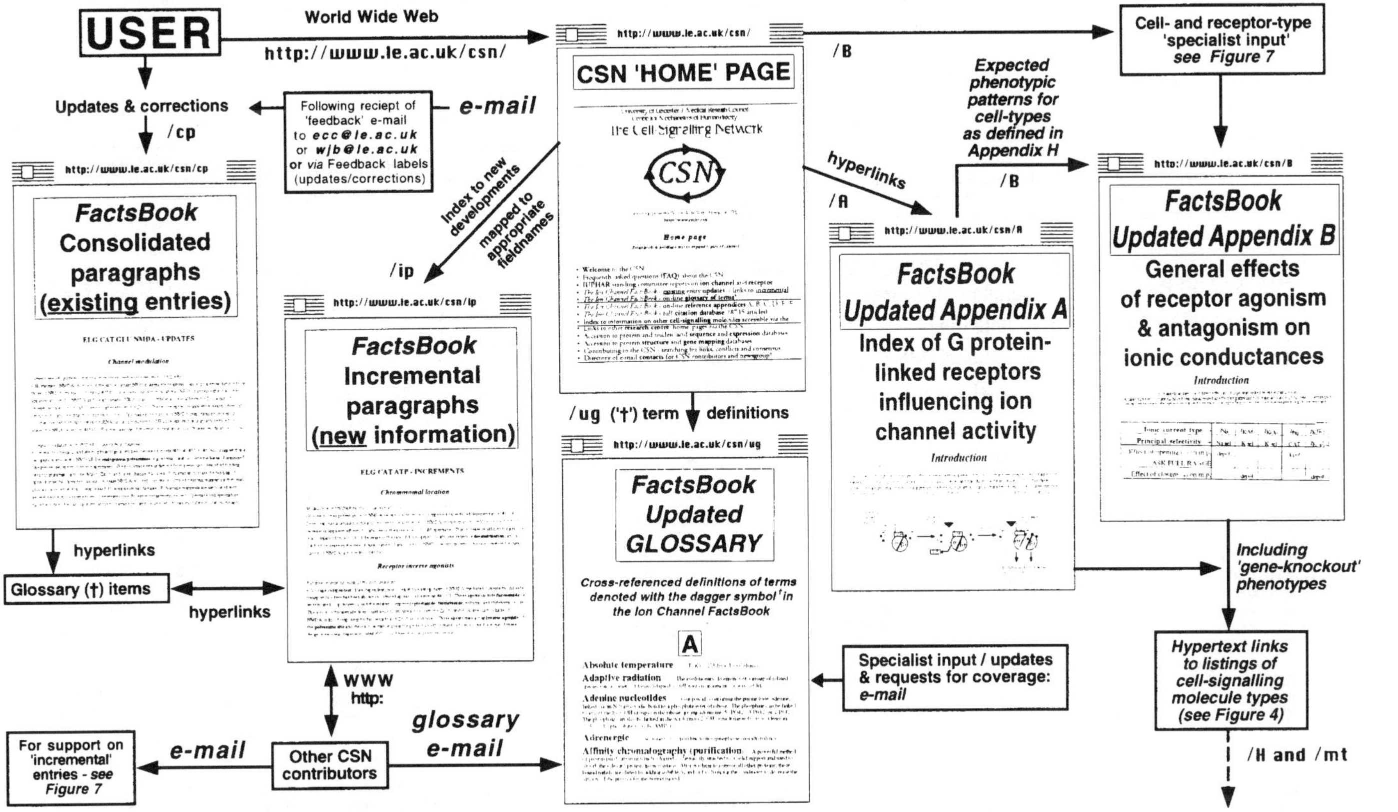

Figure 2. *Presently planned (prototype) indices for consolidation of the Ion Channels FactsBook entries (accessible via the CSN 'home page' from mid-1996). The lines and e-mail addresses are illustrative, and may not reflect the actual organization. In particular, users are requested to direct feedback to the address given in field 57 of each main entry.*

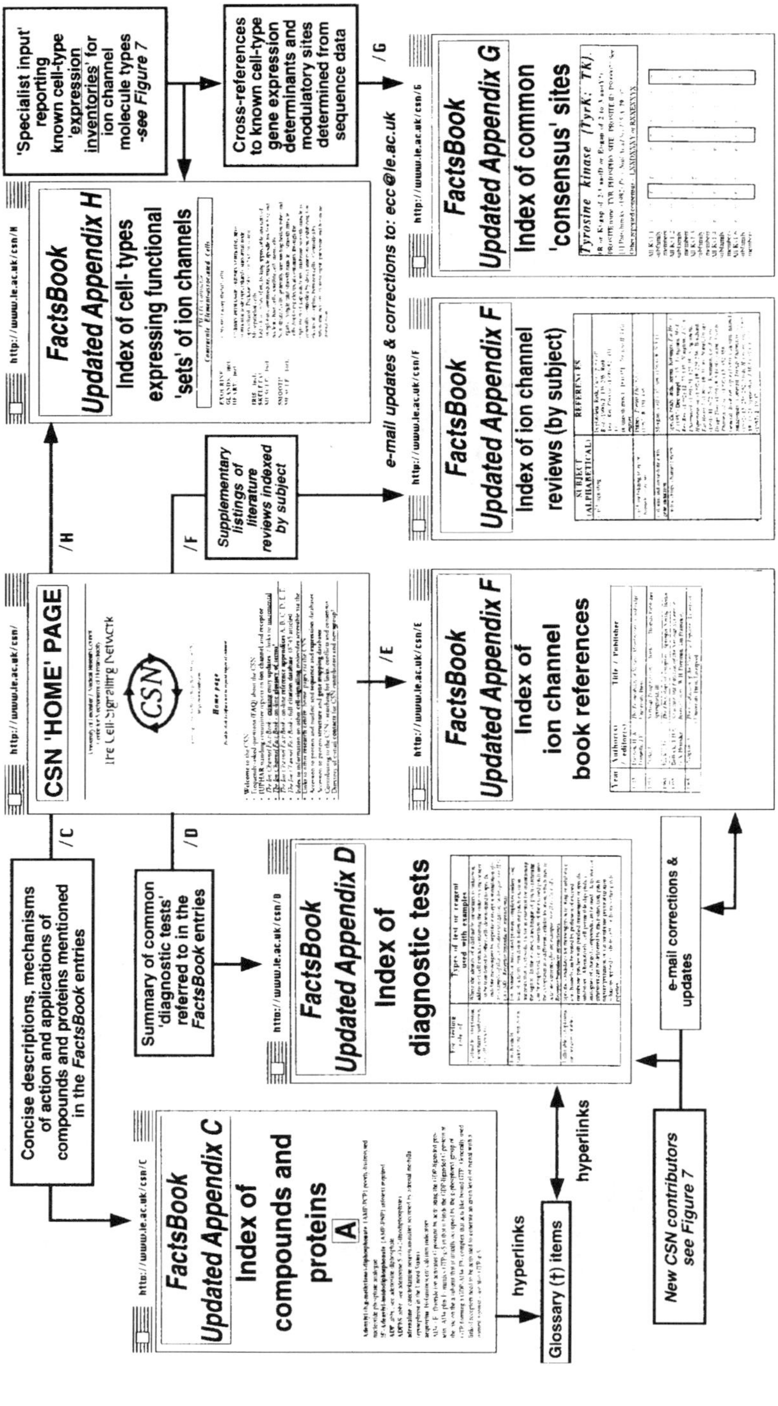

Figure 3. *Prototype FactsBook support appendices (continued). See comments in legend to Fig. 2.*

Realization of such a relational dataset depends on several factors, not the least of which are the perceived value of the exercise, a single 'platform' for information exchange and the contribution of many individuals, each an expert in an area of the literature.

A common, rational framework for dissemination of research results in cell signalling

12-01-08: As already described, The Cell-Signalling Network (using the hypertext facilities of the World Wide Web†) could offer part of the solution to the above problems. However the utility of the **collective resource** held across different sites will depend on some form of common 'architecture' to prevent the data accumulating in a haphazard way. To provide a framework for developers, some of the **key applications** of the CSN and their requirements are outlined below. *For detailed discussion of these issues, see Resource J – Search criteria & CSN development, entry 65.* Conceivably, each of these resources should be addressable from within any document already on the network by means of a URL† embedded within the source document. The versatility of embedded '**pointers**†' to other documents is illustrated in Fig. 4, which describes the integration of applications relating different aspects of cell-signalling molecule **gene expression**. In this example, complex datasets such as those reporting patterns of *in situ* hybridization, immunolocalization and developmental lineage can be related to basic information about gene/protein molecular type and physiological function. The same hierarchy allows **cross-referencing** of the molecular type to substantial existing information resources covering the genetic determinants of **cellular pathology** (e.g. the on-line version of McKusick's *Mendelian Inheritance in Man*) and/or to databases supported by the HMGW (**Human Gene Mapping Workshop**) project.

12-01-09: The choice of cell-signalling *molecules* as the organizing principle also allows direct integration of information resources relating aspects of protein **structure and function**, as illustrated in Fig. 5. In these examples, pointers† to documents containing primary sequence and crystallographic co-ordinates (available from established databases such as the **Brookhaven Protein Databank**, PDB) can be incorporated into source documents. In order to keep pace with the large number of results emergent from mutational/functional analysis of specified proteins, some form of interactive protein domain topology modelling and protein structure/function mapping could be developed as shown (*Fig. 5*).

A single reference source for official nomenclatures of receptors, ion channels and other cell-signalling molecules

12-01-10: An important application of the CSN is in standardization and dissemination of **official nomenclatures** for cell-signalling molecules and in **classifications of drugs** that act at them (*as exemplified in Fig. 6*). The great diversity of cell-signalling molecules and the 'avalanche' of new

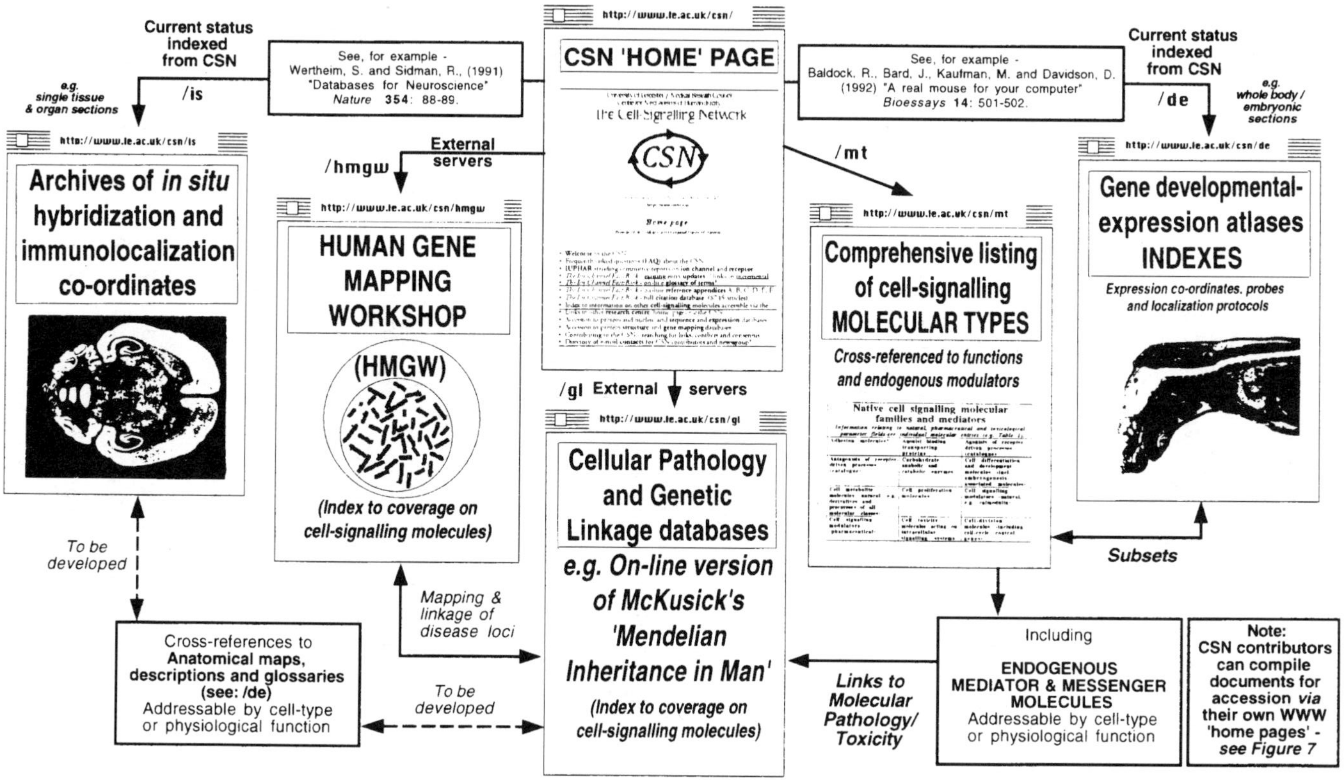

Figure 4. *Proposed (prototypic) integration of external information resources relating aspects of cell-signalling molecule gene expression. For further discussion of the requirements of more information resources and their interlinkage, see entry 65.*

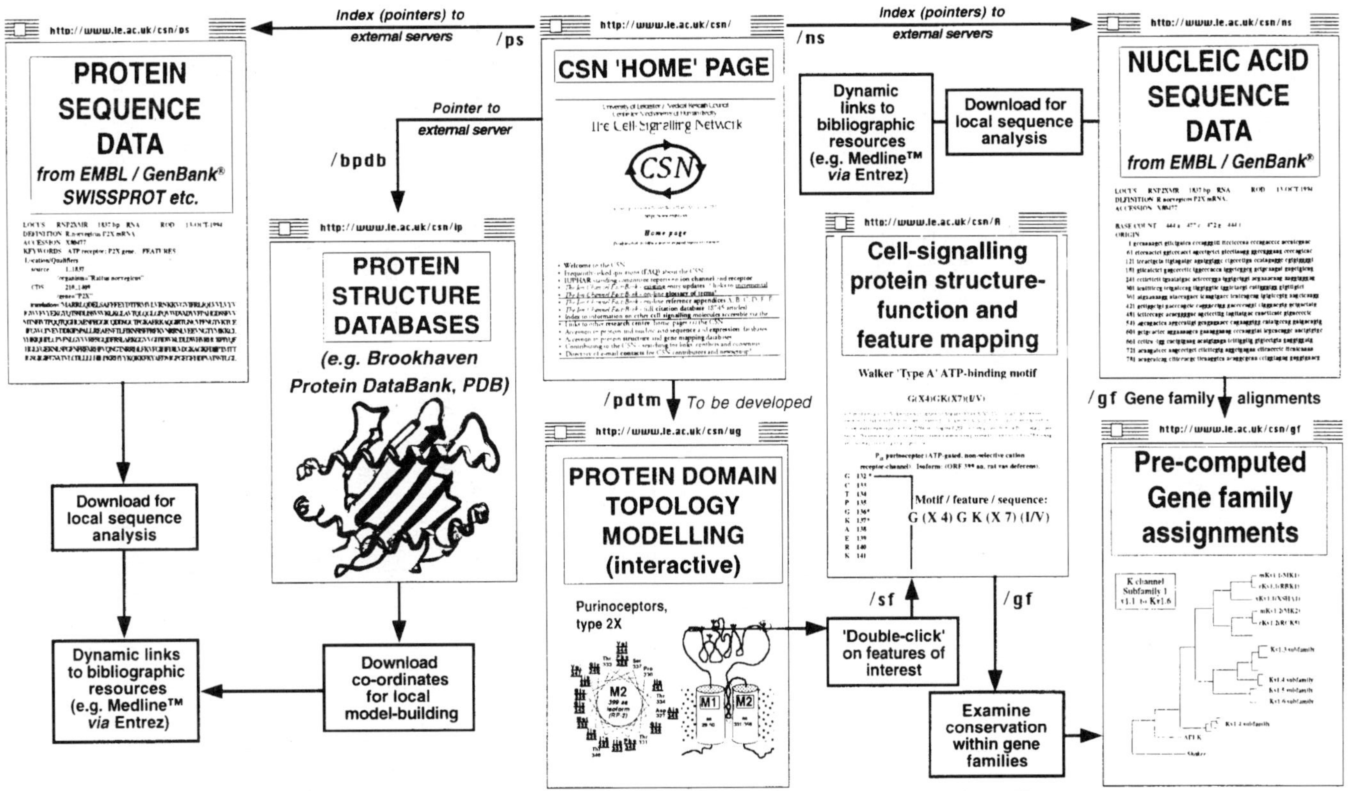

Figure 5. *Proposed integration of information resources relating aspects of cell-signalling molecule structure and function. Diverse data types could be managed by specialized electronic media as part of 'data-type extensions' to the existing entry/fieldname index numbers.*

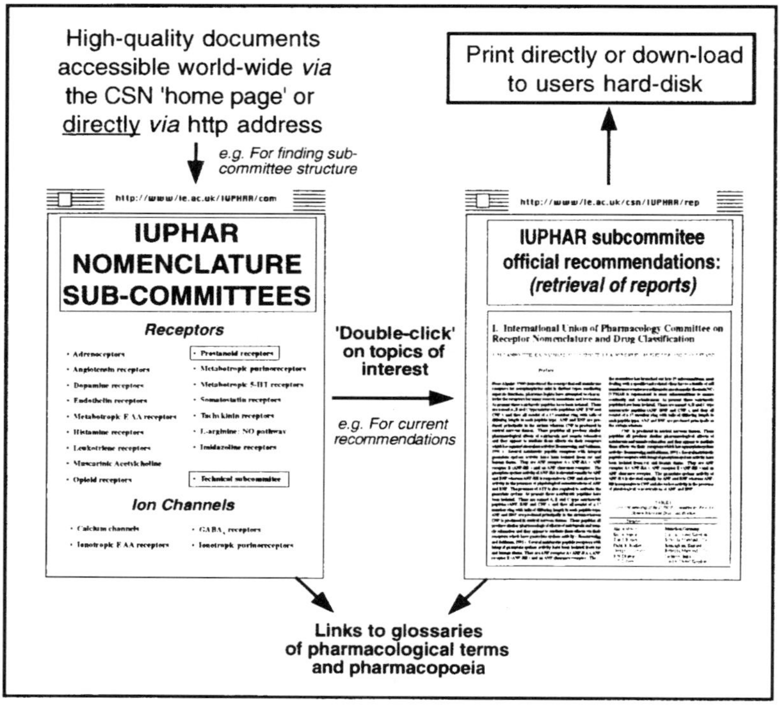

Figure 6. *Hypothetical application of CSN 'pointers' to standardization and dissemination of cell-signalling molecule nomenclature.*

information forthcoming from refined methods for their identification and characterization has led to significant problems for those concerned with proposing current and accurate nomenclatures (see, for example, Vanhoutte *et al.* (1994) *Pharmacol Rev* **46**: 111–16). In particular, the rapid cloning of genes encoding multiple isoforms† of receptors and channels has outpaced the ability of the research community to characterize their function in native† tissues. Furthermore, since the proposal of an official nomenclature essentially involves multiple criteria (e.g. structural, functional and understanding of signal transduction mechanisms) it necessitates the input of expertise from a wide range of disciplines. Finally, once a nomenclature committee has produced a consensus, there still remains the problem of efficient dissemination of the proposals, so that researchers around the world can follow them.

12-01-11: The **relational linking** structure of the CSN can help alleviate many of these problems, if only by providing a convenient and automatic means to distribute 'high-quality' update documentation electronically to interested parties. In common with other proposed applications of the CSN *(Figs 1–5)*, relational links to **on-line glossaries** (e.g. pharmacological terms and pharmacopeia) could provide important supplementary information for the application of official nomenclatures. Overall, the implementation of the CSN approach would help expert subcommittees (e.g. those of the IUPHAR) involved in **proposing drug, receptor** and **channel nomenclatures** to create, exchange and then disseminate official recommendations directly to researchers, as well as providing an important forum for **feedback** and **debate** on current proposals.

Integration with on-line bibliographic resources and conventional 'hard-copy' publishing

12-01-12: By analogy to the use of **accession numbers** for retrieval of primary sequence data from databases, the growth of **on-line bibliographic resources** in recent years makes it likely that retrieval of bibliographic data by accession number will become commonplace (minimally as citation data and abstracts, but in a few instances as 'electronic journals', and 'printable' facsimiles of articles). For the time-being, retrospective, searchable compilations of the literature offer opportunities for creating **relational links** with cell-signalling molecule data. The integration of sequence database entries (in GenBank™ for instance) 'pointing to' related papers in Medline™ via eight-figure accession numbers has already been implemented in the *Entrez* project produced by the NCBI, and both sequences and articles can be used for 'neighbouring analysis' *(for further description, see Introduction & Layout section under Field number 53: Database listings/primary sequence discussion)*. In principle, such **relational indexing** could be incorporated in compilations of documents within the CSN, thus providing an unambiguous link from the molecular database to electronic retrieval of abstracted or full papers. Alternatively, supporting documents that are too unwieldly for inclusion in conventional papers (such as gene-family alignments, large-scale *in situ* localizations or molecular structure/function models) could be referenced and retrieved by means of a URL†. **Portable document software** which makes the free exchange of high-quality electronic documents straightforward irrespective of computer platform (e.g. Adobe Acrobat™, Microsoft Internet Assistant™/Word Viewer™ and several others in development) are therefore likely to be of great utility in the CSN project.

How to contribute to the CSN

12-10-01: Some suggestions on how individuals or groups can contribute **new information resources** as part of the network of available information are outlined in Fig. 7. As well as individual contributions, **collaborative (multi-author) documents** can be created on personal computers or networks (**contributor networks**) for editing and consolidation. Following achievement of '**consensus**', one can publish the document as an option accessible from your own 'home' or 'index' page on a computer connected to the WWW/Internet (computer network administrators will tell you how to do this). Contributions on **cell-signalling molecules** not already covered are particularly valuable, but some effort should be made to refine the

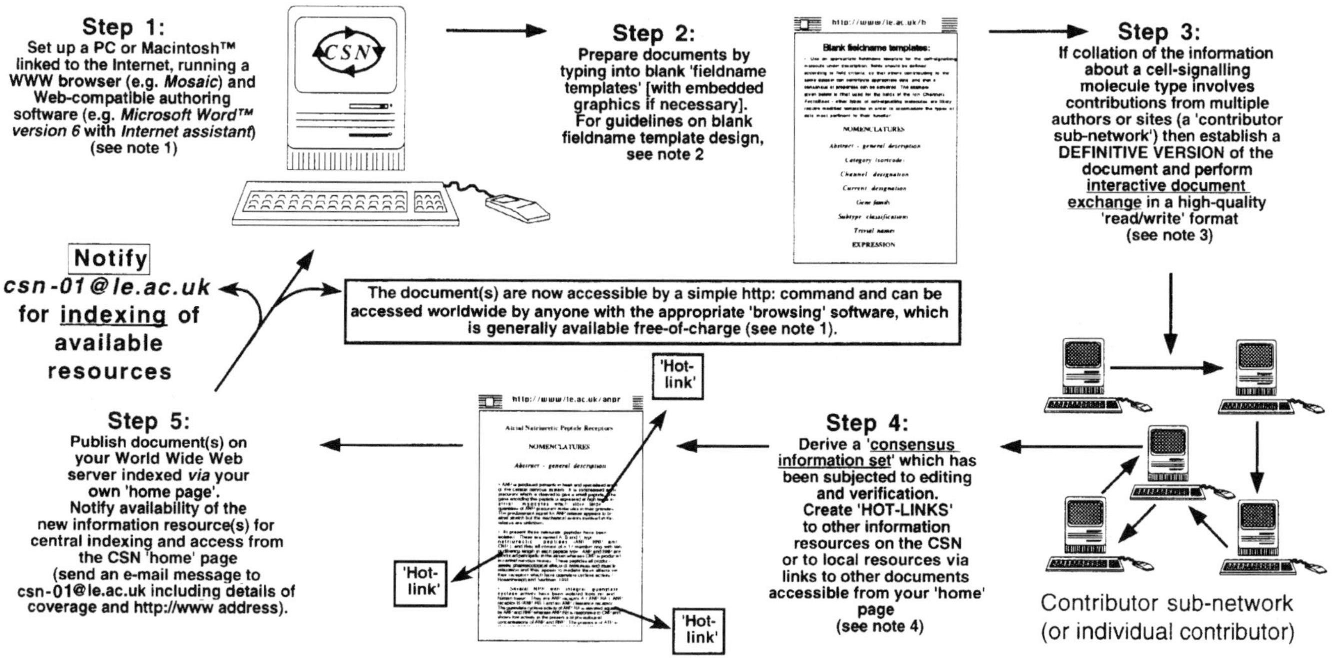
Step 1:
Set up a PC or Macintosh™ linked to the Internet, running a WWW browser (e.g. *Mosaic*) and Web-compatible authoring software (e.g. *Microsoft Word™ version 6* with *Internet assistant*) (see note 1)
CSN
Step 2:
Prepare documents by typing into blank 'fieldname templates' [with embedded graphics if necessary]. For guidelines on blank fieldname template design, see note 2
http://www/le.ac.uk/b
Blank fieldname templates:
NOMENCLATURES
EXPRESSION
Step 3:
If collation of the information about a cell-signalling molecule type involves contributions from multiple authors or sites (a 'contributor sub-network') then establish a DEFINITIVE VERSION of the document and perform interactive document exchange in a high-quality 'read/write' format (see note 3)
Notify
csn-01@le.ac.uk
for indexing of available resources
The document(s) are now accessible by a simple http: command and can be accessed worldwide by anyone with the appropriate 'browsing' software, which is generally available free-of-charge (see note 1).
'Hot-link'
http://www/le.ac.uk/anpr
NOMENCLATURES
'Hot-link'
'Hot-link'
Step 5:
Publish document(s) on your World Wide Web server indexed *via* your own 'home page'. Notify availability of the new information resource(s) for central indexing and access from the CSN 'home' page (send an e-mail message to csn-01@le.ac.uk including details of coverage and http://www address).
Step 4:
Derive a 'consensus information set' which has been subjected to editing and verification. Create 'HOT-LINKS' to other information resources on the CSN or to local resources via links to other documents accessible from your 'home' page (see note 4)
Contributor sub-network
(or individual contributor)

information through **inter-laboratory links** and by co-operation with individuals working in the same field (so that a truly **comprehensive and authoritative** 'consensus' of properties is reached). Authors should take care **not to duplicate unnecessarily** information already available (coverage to date will be posted on the CSN 'home page'). Errors and misinterpretations should be communicated back to contributors by e-mail, as this minimizes paperwork. The CSN may also be used to **raise awareness** of particular topics within the broad field of cell signalling and, via their own home page, can publicize the activities of research centres, laboratories or other institutions. *When new resources of information become available or if existing resources are not indexed already, send details of scope, authors and the host institution's URL/http: address in an e-mail message to the* ***CSN-01@le.ac.uk*** *feedback file and they will be made accessible via an appropriate* ***resource index*** *on the CSN 'home page'*. Many further details on the scope and development of the CSN can be found in *Resource J – Search criteria & CSN development, entry 65.*

Figure 7. *How to contribute to the CSN on the World Wide Web. Notes: 1. Several Internet/WWW 'authoring' software packages are becoming available, although the most convenient options will be those that can automatically convert word-processed documents (with graphics) directly to the HTML format (as used to send documents over the WWW). At the time of going to press, Microsoft announced an option for Word 6 (both Mac and PC versions) which can simplify the preparation and publication of pages destined for the WWW. The Microsoft Internet Assistant program is available from the Microsoft WWW 'home page', as is the WordViewer program, which enables documents to be read by other users (even if MS Word is not installed). 2. Selection of 'fieldnames' should follow a set of criteria for placement of material (as exemplified by the FactsBook fields in Introduction and layout, entry 02) so that other contributors can consolidate the dataset according to the same rules. The actual field categories will largely depend on the type of cell-signalling molecule being described, and they may change in time. A good 'rule of thumb' is to base all fieldnames on molecular characteristics or properties emergent from them. Large datasets may be accommodated by 'splitting' the task across a 'contributor sub-network' (see Step 3), which can also be used to discuss the most appropriate fieldnames in the first instance. 3. Interactive document exchange can be performed with workgroup software, with the advantage that formatting will remain intact from file creation to publication on the Web. Consolidated versions of documents following further 'circulation' around the contributor sub-network may be published as 'incremental files' or as 'updates' (replacing an earlier version). 4. For help on the structuring of HTML file directories and the creation of 'hot-links' to other resources, see your network administrator. Specialist support for initiating or maintaining ion channel entries would be gratefully received (e-mail to the appropriate support group, see field 57 of each entry).*

Entry Number Rubric

Entry 13

Entry (channel type) numbers form the **first two numbers xx** of the **six-figure index number (xx-yy-zz)** according to the rubric below. **yy** is the field i.d. number (see the accompanying field number rubric) and **zz** is the **datatype** i.d. number within a field (in the current entries zz simply indicates 'paragraph running order' but this number will eventually define and index **specific** types of data 'falling under' each fieldname). 'Coverage' under each page header/sortcode as listed below is indicated under *Cumulative contents. entry 02.*

Entry	page header / sortcode *See also accompanying PAGE HEADERS RUBRIC*	Entry e-mail feedback file
Entry 01-yy-zz	**Cumulative tables of contents**	CSN-01@le.ac.uk
Entry 02-yy-zz	**Introduction & layout of entries**	CSN-02@le.ac.uk
Entry 03-yy-zz	**Abbreviations**	CSN-03@le.ac.uk
Entry 04-yy-zz	**ELG [key facts]**	CSN-04@le.ac.uk
Entry 05-yy-zz	**ELG CAT 5-HT_3**	CSN-05@le.ac.uk
Entry 06-yy-zz	**ELG CAT ATP**	CSN-06@le.ac.uk
Entry 07-yy-zz	**ELG CAT GLU AMPA/KAIN**	CSN-07@le.ac.uk
Entry 08-yy-zz	**ELG CAT GLU NMDA**	CSN-08@le.ac.uk
Entry 09-yy-zz	**ELG CAT nAChR**	CSN-09@le.ac.uk
Entry 10-yy-zz	**ELG Cl $GABA_A$**	CSN-10@le.ac.uk
Entry 11-yy-zz	**ELG Cl GLY**	CSN-11@le.ac.uk
Entry 12-yy-zz	**Feedback & CNS access** – *See also Entry 65.*	CSN-12@le.ac.uk
Entry 13-yy-zz	**Rubrics**	CSN-13@le.ac.uk
Entry 14-yy-zz	**ILG [key facts]**	CSN-14@le.ac.uk
Entry 15-yy-zz	**ILG Ca AA-LT[C]$_4$ [native]**	CSN-15@le.ac.uk
Entry 16-yy-zz	**ILG Ca Ca $InsP_4S$ [native]**	CSN-16@le.ac.uk
Entry 17-yy-zz	**ILG Ca Ca RyR-Caf**	CSN-17@le.ac.uk
Entry 18-yy-zz	**ILG Ca CSRC [native]**	CSN-18@le.ac.uk
Entry 19-yy-zz	**ILG Ca $InsP_3$**	CSN-19@le.ac.uk
Entry 20-yy-zz	**ILG CAT Ca [native]**	CSN-20@le.ac.uk
Entry 21-yy-zz	**ILG CAT cAMP**	CSN-21@le.ac.uk
Entry 22-yy-zz	**ILG CAT cGMP**	CSN-22@le.ac.uk
Entry 23-yy-zz	**ILG Cl ABC-CF**	CSN-23@le.ac.uk
Entry 24-yy-zz	**ILG Cl ABC-MDR/PG**	CSN-24@le.ac.uk
Entry 25-yy-zz	**ILG Cl Ca [native]**	CSN-25@le.ac.uk
Entry 26-yy-zz	**ILG K AA [native]**	CSN-26@le.ac.uk
Entry 27-yy-zz	**ILG K Ca**	CSN-27@le.ac.uk
Entry 28-yy-zz	**ILG K Na [native]**	CSN-28@le.ac.uk
Entry 29-yy-zz	**INR K [key facts]**	CSN-29@le.ac.uk
Entry 30-yy-zz	**INR K ATP-i [native]**	CSN-30@le.ac.uk
Entry 31-yy-zz	**INR K G/ACh [native]**	CSN-31@le.ac.uk
Entry 32-yy-zz	**INR K [native]**	CSN-32@le.ac.uk
Entry 33-yy-zz	**INR K [subunits]**	CSN-33@le.ac.uk
Entry 34-yy-zz	**INR K/Na I_{fhq} [native]**	CSN-34@le.ac.uk
Entry 35-yy-zz	**JUN [connexins]**	CSN-35@le.ac.uk
Entry 36-yy-zz	**MEC [mechanosensitive]**	CSN-36@le.ac.uk

Entry 37-yy-zz	**MIT [mitochondrial, native]**	CSN-37@le.ac.uk
Entry 38-yy-zz	**NUC [nuclear, native]**	CSN-38@le.ac.uk
Entry 39-yy-zz	**OSM [aquaporins]**	CSN-39@le.ac.uk
Entry 40-yy-zz	**SYN [vesicular]**	CSN-40@le.ac.uk
Entry 41-yy-zz	**VLG key facts**	CSN-41@le.ac.uk
Entry 42-yy-zz	**VLG Ca**	CSN-42@le.ac.uk
Entry 43-yy-zz	**VLG Cl**	CSN-43@le.ac.uk
Entry 44-yy-zz	**VLG K A-T**	CSN-44@le.ac.uk
Entry 45-yy-zz	**VLG K DR**	CSN-45@le.ac.uk
Entry 46-yy-zz	**VLG K eag**	CSN-46@le.ac.uk
Entry 47-yy-zz	**VLG K Kv-beta**	CSN-47@le.ac.uk
Entry 48-yy-zz	**VLG K Kv1-Shak**	CSN-48@le.ac.uk
Entry 49-yy-zz	**VLG K Kv2-Shab**	CSN-49@le.ac.uk
Entry 50-yy-zz	**VLG K Kv3-Shaw**	CSN-50@le.ac.uk
Entry 51-yy-zz	**VLG K Kv4-Shal**	CSN-51@le.ac.uk
Entry 52-yy-zz	**VLG K Kvx (Kv5.1/Kv6.1)**	CSN-52@le.ac.uk
Entry 53-yy-zz	**VLG K M-i [native]**	CSN-53@le.ac.uk
Entry 54-yy-zz	**VLG K minK**	CSN-54@le.ac.uk
Entry 55-yy-zz	**VLG Na**	CSN-55@le.ac.uk
Entry 56-yy-zz	**Appendix A – G Protein-Linked Receptors**	CSN-56@le.ac.uk
Entry 57-yy-zz	**Appendix B – Electrical Effectors**	CSN-57@le.ac.uk
Entry 58-yy-zz	**Appendix C – Compounds & Proteins**	CSN-58@le.ac.uk
Entry 59-yy-zz	**Appendix D – Diagnostic Tests**	CSN-59@le.ac.uk
Entry 60-yy-zz	**Appendix E – Book References**	CSN-60@le.ac.uk
Entry 61-yy-zz	**Appendix F – Subject Reviews**	CSN-61@le.ac.uk
Entry 62-yy-zz	**Appendix G – Consensus Sites & Motifs**	CSN-62@le.ac.uk
Entry 63-yy-zz	**Appendix H – Cell-Types**	CSN-63@le.ac.uk
Entry 64-yy-zz	**Appendix I – Cell-Signalling Molecular Types**	CSN-64@le.ac.uk
Entry 65-yy-zz	**Appendix J – Search Criteria & CSN Development**	CSN-65@le.ac.uk
Entry 66-yy-zz	**Multidisciplinary Glossary Framework**	CSN-66@le.ac.uk
Entry 67-yy-zz	**Cumulative Subject Index**	CSN-67@le.ac.uk

Note: Entry 'running order' is only of significance in book-form publications; computer-compiled updates will use the xx-yy-zz numbers as **hyper-relational pointers**.

Field Number Rubric

Entry 13

Field numbers form the **third and fourth numbers** **yy** of the **six-figure index number (xx-yy-zz)**according to the rubric below. **zz** is the **datatype** i.d. number within a field (in the current entries zz simply indicates 'paragraph running order' but this number will eventually define and index **specific** types of data 'falling under' each fieldname). Omission of a field number within the main entries implies information was 'not applicable' or was 'not found' during compilation..

NOMENCLATURES SECTION
xx-**01**-zz **Abstract – general description**
xx-**02**-zz **Category (sortcode)**
xx-**03**-zz **Channel designation**
xx-**04**-zz **Current designation**
xx-**05**-zz **Gene family**
xx-**06**-zz **Subtype classifications**
xx-**07**-zz **Trivial names**

EXPRESSION SECTION
xx-**08**-zz **Cell-type expression index**
xx-**09**-zz **Channel density**
xx-**10**-zz **Cloning resource**
xx-**11**-zz **Developmental regulation**
xx-**12**-zz **Isolation probe**
xx-**13**-zz **mRNA distribution**
xx-**14**-zz **Phenotypic expression**
xx-**15**-zz **Protein distribution**
xx-**16**-zz **Subcellular locations**
xx-**17**-zz **Transcript size**

SEQUENCE ANALYSES SECTION
xx-**18**-zz **Chromosomal location**
xx-**19**-zz **Encoding**
xx-**20**-zz **Gene organization**
xx-**21**-zz **Homologous isoforms**
xx-**22**-zz **Protein MW (purified)**
xx-**23**-zz **Protein MW (calc)**
xx-**24**-zz **Sequence motifs**
xx-**25**-zz **Southerns**

STRUCTURE & FUNCTIONS SECTION
xx-**26**-zz **Amino acid composition**
xx-**27**-zz **Domain arrangement**
xx-**28**-zz **Domain conservation**
xx-**29**-zz **Domain functions (predicted)**
xx-**30**-zz **Predicted protein topography**
xx-**31**-zz **Protein interactions**
xx-**32**-zz **Protein phosphorylation**

ELECTROPHYSIOLOGY SECTION
xx-**33**-zz **Activation**
xx-**34**-zz **Current type**
xx-**35**-zz **Current-voltage relation**
xx-**36**-zz **Dose-response**
xx-**37**-zz **Inactivation**
xx-**38**-zz **Kinetic model**
xx-**39**-zz **Rundown**
xx-**40**-zz **Selectivity**
xx-**41**-zz **Single-channel data**
xx-**42**-zz **Voltage sensitivity**

PHARMACOLOGY SECTION
xx-**43**-zz **Blockers**
xx-**44**-zz **Channel modulation**
xx-**45**-zz **Equilib. dissoc. constant**
xx-**46**-zz **Hill coefficient (n)**
xx-**47**-zz **Ligands**
xx-**48**-zz **Openers**
xx-**49**-zz **Receptor/transducer ints**
xx-**50**-zz **Receptor agonists**
xx-**51**-zz **Receptor antagonists**
xx-**52**-zz **Receptor inverse agonists**

INFORMATION RETRIEVAL SECTION
xx-**53**-zz **Database listing**
xx-**54**-zz **Gene map. locus desig.**
xx-**55**-zz **Miscellaneous information**
xx-**56**-zz **Related sources & reviews**
xx-**57**-zz **Feedback**
In-press updates

REFERENCES SECTION

Index

The following is a conventional index, note however that a cross referenced subject index for topics appearing within fields covering all entries, appears under the section cumulative page index, as a cumulative topic index. For rapid location of information see entry and field number rubrics, under entry 13. Also note italic page numbers represent entries in figures and tables.